P9-APT-594

LIVING with the EARTH

Second Edition

Concepts in Environmental Health Science

LIVING with the EARTH

Second Edition

Concepts in Environmental Health Science

GARY S. MOORE

LEWIS PUBLISHERS

A CRC Press Company

Boca Raton London New York Washington, D.C.

Library of Congress Cataloging-in-Publication Data

Moore, Gary S.
Living with the earth : concepts in environmental health science / Gary S. Moore.—2nd ed.
p. cm.
Includes bibliographical references and index.
ISBN 1-56670-585-1
1. Environmental health—Computer assisted instruction. 2. Industrial medicine—Computer-assisted instruction. 3. Industrial hygiene—Computer-assisted instruction. 4. Industrial safety—Computer-assisted instruction. I. Title.

RA565 .M665 2002
615.9′02—dc21 2001050316

Direct all inquiries to CRC Press LLC, 2000 N.W. Corporate Blvd., Boca Raton, Florida 33431.

Lewis Publishers is an imprint of CRC Press LLC

International Standard Book Number 1-56670-585-1
Library of Congress Card Number 2001050316
Printed in the United States of America 1 2 3 4 5 6 7 8 9 0
Printed on acid-free paper

Author

Gary Moore, M.S., Ph.D., is a faculty member in the Department of Environmental Health Sciences in the School of Public Health and Health Sciences at the University of Massachusetts. He is also the academic distance learning coordinator (DLC) for the School of Public Health based on his extensive experience in computer graphics, Web-based training, and membership on national committees dealing with distance learning. He is an advisory board member to the national Center for Internet Technology Education (CITE) located in Denver, CO, and an advisory board member to the distance education steering committee of the National Association of Schools of Public Health in Washington, D.C. Dr. Moore teaches several graduate-level environmental courses, including courses on Air Quality Assessment, Environmental Regulations and Compliance, and Environmental Health Practices. He has over 75 publications in the area of health effects from environmental insults. Many of his recent book publications have Web-enhanced components, including:

- *You CAN Teach Online — Building Creative Learning Environments: A Web-Enhanced Resource,* McGraw-Hill, February 2001.
- *Environmental Compliance — A Web-Enhanced Resource*, Lewis Publishers, Boca Raton, FL, January 2001.
- *Living with the Earth — Concepts in Environmental Health Science: A Web-Enhanced Book,* Lewis Publishers, Boca Raton, FL, 1999. (This book is one of *Choice Magazine's* Outstanding Academic Books of 1999, judged on distinguished scholarship. *Choice* is a publication of the Association of College and Research Libraries.)

Dr. Moore has received several awards for his efforts in providing training and certification programs to the Boards of Health in Massachusetts during 1998 and 1999, including recognition from the Massachusetts Senate, the Massachusetts House of Representatives, The Massachusetts Association of Health Boards (MAHB), and the Robert C. Huestis/Eric W. Mood award from the New England Public Health Association (NEPHA). The certification and training programs for boards of health have been very successful. Since 1995, Dr. Moore has assisted MAHB in establishing the groundwork for this first-in-the-nation program. This project is a strong foundation for further collaboration between the university and MAHB, providing practical educational opportunities to people serving every community in the Commonwealth.

Dr. Moore is committed to providing the maximum learning experience and highest possible profile for his graduate students. Some of those graduate students have contributed significantly to the preparation of the first edition of this book and are listed below. Others have contributed to this second edition, and their contributions and names are also included.

Student Contributors

Kathleen Bell, M.P.H., earned her Bachelor of Arts degree in biology from Cornell University. Her work experience includes environmental claim management. She earned her Masters of Public Health in Environmental Health Sciences from the University of Massachusetts School of Public Health.

Christopher Landry, M.P.H., earned his B.S. in biology from the University of Alabama in 1993. As an undergraduate, Landry was invited to serve as a project supervisor for a Mexican mission. He then went to El Salvador, where he served as a water sanitation technician for the Peace Corps. He earned his Masters degree in Public Health from the University of Massachusetts at Amherst. Currently Landry is pursuing a career as a public health advisor to Latin American countries.

Amy Tomasello, M.P.H., graduated from the University of Rhode Island, Kingston, where she received her Bachelor of Science in pharmacy. She worked as a consultant pharmacist for 2 years before returning to school. In 1998, she received her Masters of Public Health in Environmental Health Sciences from the University of Massachusetts at Amherst.

SECOND EDITION

The following students are credited with contributing to updating chapters in this second edition:

Sean Jobin and Glenna Vitello updated Chapter 1, Ecosystems and Energy Flow; Danielle Bachand and April McNally updated Chapter 2, Human Population; Xiaohan Zou updated Chapter 3, Environmental Degradation and Food Security; Kenneth Huber updated Chapter 4, Environmental Disease; Pradeep Sukumar updated Chapter 5, Toxicity and Toxins; Fred Hooven updated Chapter 6, The Trouble with Pests; Vipion Aggarwal and Erin Grenier updated Chapter 7, Emerging Dseases; Heather Burelle updated Chapter 8, Foodborne Illness; Nick Constantino and Ian Cambridge updated Chapter 9, Water and Wastewater; Keith Rego and Tim Lockhart updated Chapter 10, Air Quality and Radiation; Mary Rothermich updated Chapter 11, Solid and Hazardous Waste; Mary Wicks and Jason Jenkins updated Chapter 12, Risk Assessment and Management; and Amy Crosby and Pam Hogan updated Chapter 13, Environmental Regulations.

Preface

This book is a college-level textbook for introductory courses in environmental health and environmental sciences. It is suitable for undergraduate- and graduate-level courses for students of environmental health, environmental sciences, community health education, epidemiology, civil engineering, community medicine, general medicine, and public health. The book is also highly recommended for members of boards of health, health officers and health inspectors, and citizens who are members of proactive environmental groups such as Sierra Club, Clean Water Action, Environmental Defense Fund, and public interest research groups.

This book contains 13 carefully illustrated and narratively exciting chapters on the subject of environmental health sciences. The contents are derived from a core course Dr. Moore instructs at the University of Massachusetts titled "Environmental Health Practices," and this book incorporates the traditional concepts associated with environmental health along with new, emerging, and controversial issues associated with environmental threats to human health. Emphasis is placed on biological, chemical, and physical sources of pollution and the methods of controlling or limiting those exposures. Careful attention is paid to offering a balanced view representing opposing scientific perspectives on major issues ranging from the greenhouse effect to reproductive problems associated with endocrine disruptors.

Key features of the book include:

- A wide offering of instructor support materials, including Microsoft Powerpoint® presentation slides in digital format, a Web-enhanced course site, and an examination bank
- Numerous detailed technical illustrations and photographs throughout the text and the Web-enhanced course site
- A new chapter about emerging diseases, including the most significant reasons for their emergence and the major etiological agents associated with emerging diseases
- Measures of population dynamics, including the contrasts of various cultural philosophies regarding overpopulation and the new problems of underpopulation in developed nations
- New developments in the emerging problems of asthma and associated air pollutants; the genetic basis of cancer and the roles of behavior and pollution in cancer risk; and the biology of genetics, mutation, birth defects, and much more

Using the Web Site

The Web site has been constructed to provide maximum usefulness for the student and the instructor. The site has been actively tested by students and modified to meet student and instructor expectations. This is an ongoing and dynamic effort, and you are encouraged to suggest improvements to the Web site that would make it more useful and interesting. The site has been designed to work with most popular browsers and is fast-loading, with most images linked to text and selected for viewing by the user. This design is intended to speed up the use of the site and to let users reach their destination quickly, without being overwhelmed by graphics that slow down the process. Future versions are planned that feature streaming audio and some video clips.

STUDENTS

The Web site address is **<http://www-unix.oit.umass.edu/-envhl565/>**. With this book, you will receive a user ID and a password from your instructor. These will be changed annually. Connect to the site by entering the correct ID and password, and the possible use of the site will be immediately obvious. The most useful section to you will be the site location featuring *A Book Study Guide*. This section features chapter-by-chapter detailed notes and associated Web links with each chapter. Just scroll down to the chapter of interest and click on the highlighted chapter number. This will bring the detailed notes section for that chapter to your screen. You may print these out directly. The associated figures for the chapter are in full color and may be displayed by clicking on the highlighted figure number. These may also be printed for your use. These are copyrighted figures and are not to be used for electronic or printed distribution in any form or manner without the publisher's permission.

Associated with each chapter are links to useful sites maintained on that subject by the U.S. government, international and national organizations, academic institutions, and private groups. You may take a trip through the Amazon rain forest with still and video images by entering the National Aeronautics and Space Administration (NASA) Web site *Live from the Rainforest* under the Chapter 1 Web links. If you want to know what industries near your home are producing and/or emitting hazardous chemicals and the amount and type of chemicals, enter the site called *Scorecard Home.* These and other sites offer a wealth of information that the student is encouraged to explore. There are also a number of searchable databases where you can find nearly any environmental topic by keyword.

INSTRUCTOR MATERIALS

The instructor for this course will have acess to the same features that are available to the student and will also have access to chapter-by-chapter Powerpoint files that may be downloaded directly from the Web site. There are nearly 1100 slides with more than 250 images in full color taken directly from the book. The slides are based on the detailed course notes condensed from each chapter. The Powerpoint files may be shown using an LCD projection system in combination with a computer, printed on transparencies, or sent to a service bureau for digital imaging to 35 mm slides. The digital files may be modified by the instructor in Powerpoint to meet classroom needs. A test bank of exam questions on a chapter-by-chapter basis is also included in the instructor materials section. The questions include multiple choice, true and false, and short essays. The answers are supplied, along with the page numbers in the text where the answers may be found. The test bank materials are in rich text format (RTF) that may be opened with any common popular word processor. The instructor will also benefit from the searchable databases provided as Web links to obtain the latest information on environmental health topics. Additionally, I provide a hot topics section where new materials on important environmental issues are placed, making the book and its associated Web site current. The instructors are encouraged to contact me with information they would like to see included in the hot topics section, as well as recommendations to improve the usefulness of the Web site.

This Web site is password protected. Instructors may acquire a user name and password by submitting a request on official university or departmental stationary, signing it, and sending the request to the attention of CRC Press LLC, Textbook Sales Department, 2000 N.W. Corporate Blvd., Boca Raton, FL, 33431-9868 (1-800-272-7737, Ext. 2202). Please include the ISBN number, 1-5667-05851. For technical support, please call CRC Press Technical Support Department, 1-800-272-7737, Ext. 6066, or E-mail techsupport@crcpress.com.

Acknowledgments

I thank my wife, Lucille, for her unwavering support during my long periods of absence from her side. She accepted my dedication to this project and kept me fed and clothed in the interim. I extend sincere thanks to Chris Landry, a graduate student in our department, for his excellent preparation of the text in Chapter 8 entitled Food-Borne Illness; to Kathleen Bell, another graduate student in our department, who wrote most of the text for Chapter 9 on Water Quality and Chapter 12 on Risk Assessment; and to Amy Tomasello, a third graduate student who contributed to Chapter 5 on Toxicity and Toxins and provided detailed proofreading for most of the book. These students all benefited enormously from the experience and grew in confidence and capabilities. I provided instruction in the areas of writing and organization, page layout, and graphic design. Each student took something of importance with them. They are remembered in this book with pictures and text about who they are.

Dr. John Edman, an entomology professor at the University of Massachusetts, provided slides for most of the insects used in Chapter 6. The pictures featuring African children and scenes were provided by Marie Krause Cote, who spent several months in Africa helping to teach tribeswomen to produce cloth goods for market. Many of the pictures on wildlife, wetlands, and biomes came from U.S. government departments, including the U.S. Fish and Wildlife Service (USFWS), the National Aeronautics and Space Administration (NASA), and the Environmental Protection Agency (EPA).

Students enrolled in my course on Environmental Health Practices were instructed to submit articles about any of the subjects covered in this book. They were required to review and provide a synopsis about the topic. Many of these articles became the basis for information in this second edition. They also made recommendations and provided sources of pictures for the information covered in the book and on the Web site.

Contents

CHAPTER 1

Ecosystems and Energy Flow

INTRODUCTION

We are immersed in life. We breathe it in, we walk on it, we touch it. Each footstep on a fertile lawn or forest mat will send tremors to trillions of bacteria, millions of algae, fungi, and protozoa, and hundreds of insects and worms. The skin on our bodies, when viewed microscopically, is a teaming matrix of tiny caverns filled with bacteria, viruses, and mites. So dense are the unseen life forms on our

OBJECTIVES FOR THIS CHAPTER

A student reading this chapter will be able to:

1. **Discuss and define the concepts of biosphere and climate**
2. **List and explain the factors influencing climate**
3. **Define the term biome; list the major global biomes, and discuss their primary features**
4. **Describe the flow of energy through ecosystems**
5. **Describe and explain the various trophic levels**
6. **List and explain the various nutrient cycles including the carbon, nitrogen, and phosphorous cycles**
7. **Define the term *succession*, explain the mechanisms of succession, and discuss the types of human intervention that interfere with succession**

bodies that they form an almost complete shell about each of us. Every breath draws in untold numbers of fungal spores, bacteria, viruses and other microbes. Life abounds most everywhere inhabited by humans. Life thrives on the nutrients in soil and water, the oxygen and carbon dioxide in the air, and on the sunlight that ultimately powers most life. In those areas of earth where nutrients are depleted, oxygen is rare, sunlight is extinguished, or moisture is diminished, life becomes reduced or absent. Conditions for most life are found in a layer about the globe that extends from approximately five miles in the atmosphere (where some microbial spores and insects may be found) to five miles below the ocean's surface, where some unusual life forms adapted to darkness and high pressure survive. This theoretical "layer of life" is called a biosphere because life is thought not to exist outside this area. Most life occurs in a much narrower layer extending from about a 600-foot depth in the ocean where sunlight is able to penetrate, to the summer snow line of high mountain peaks where a thin layer of soil supports plant life such as lichens and mosses. Within this biosphere, the forms and quantities of life vary dramatically. Surface- or land-based life may be categorized into major regions known as biomes. Biomes are based on the dominant types of vegetation that are strongly correlated with regional climate patterns. We will explore the various types of biomes in short order, but first let us look at the driving force of these regional differences — the climate.

CLIMATE

What Is It?

Climate varies in latitudinal bands around the globe. Relatively constant warm and humid weather affects most equatorial countries; as you travel north or south, the numbers of warm months diminish, and freezing months escalate. Climate can be viewed as average weather within a geographical area viewed over years or even centuries. Like weather, climate includes temperature, precipitation, humidity, wind velocity and direction, cloud cover, and associated solar radiation. Climate does change, but over centuries or millennia. Some of these climatic changes are dramatic. There have been many ice ages with the formation of glaciers resulting in major climatic changes globally. These climatic changes have occurred from different phenomena that may include: (1) changes in ocean temperatures; (2) changes in the earth's orbital geometry; (3) volcanic activity with increased atmospheric dust and reduced sunlight penetration; (4) variations in solar radiation; or (5) increases in atmospheric gases that absorb heat energy.[1]

Volcanic eruptions have been thought a major factor in studies that examine changes in the environment. These volcanic eruptions can cause sudden and severe environmental changes. Eruptions that emit large volumes of sulfur, rather than particulate matter, can change climate and depress temperatures around the world. In one study conducted by Mass and Portman in 1989, they analyzed the impact of nine volcanic eruptions between 1883 and 1982. The study discovered two main points. First, volcanic eruptions are associated with surface temperature anomalies that usually fall within the normal seasonal statistical variation. Second, there is a lot of data suggesting that volcanic eruptions are related to temperature and circulation anomalies, but there are questions still unanswered as to the scale and patterns of the cooling.[5] Many of these gases are produced by human activities such as the burning of fossil fuels, which release large amounts of heat-absorbing carbon dioxide.

Today, a huge amount of heat-trapping greenhouse gases are being released into the atmosphere at a relatively fast rate. Scientists from the U.S. and Australia report that during the last 100 million years there may have been a sudden release of greenhouse gases from the ocean floor. It is believed that gradual warming caused a change in ocean currents, which caused warm water from the surface to plunge deeper into the ocean. The solid methane that was trapped in the ocean sediments became gaseous. The methane then exploded from the sediments that resulted in mudslides. These mudslides allowed the methane to escape into water and then into the atmosphere. The methane reacted with the oxygen and formed CO_2, causing heat to be trapped in the atmosphere. The end result was an increase in ocean temperature by 5–9°F. This warming was so severe that it wiped out many species of marine life. Even though these conditions happened without humans around, it seems that we are creating the same conditions that caused this climate change. The main greenhouse gas that caused this phenomena was CO_2. Since the Industrial Revolution, CO_2 has been released, causing the atmospheric content to rise by 30% since the 18th century.[6] These climate-changing gases will be discussed in more detail in Chapter 10.

How Is Climate Affected?

Climate is most affected by temperature. By the end of the next century, scientists predict that atmospheric CO_2 levels will be double that of the preindustrial era. They also suggest that it could be much greater if not for a continuous evolution of energy sources away from carbon producers such as wood and coal to more efficient sources like natural gas and burgeoning hydrogen. Regardless of our decreasing rate of CO_2 to energy ratio, population explosions and economy increases have started a continuous rise in CO_2 production. As a result, we can expect extreme changes in the world's climate, weather and lifestyle.[7]

The power that modifies, influences, and controls climate originates with the sun. The amount of sunlight striking a particular area of the globe determines the level of warmth and ultimately the movement of air and the amount of precipitation. Because the Earth is a globe that rotates about an axis that is tilted, the amount of sunlight striking the Earth varies by region and time. The intensity or energy from the sun diminishes with the distance from the sun. More specifically, the intensity declines inversely with the square of the distance from the sun. Consequently, our distance of 93 million miles from the sun means we receive a very small fraction of the sun's intensity (about one two-billionths of the sun's energy output pointed in this direction).[2]

Earth's distance from the sun has little to do with the seasons. The seasons are caused by the tilt of the Earth on its axis as it revolves around the sun. The Earth is tilted at a 23.5° angle from a vertical axis drawn perpendicular to the plane of the earth's orbit around the sun. This tilt causes some parts of the Earth to get slanting rays of sunlight some of the year and vertical rays of sunlight at other times. When a hemisphere of the Earth is tilted toward the sun, it is summer in that hemisphere. Moreover, because the Earth is tilted, seasonal differences will occur as the Earth rotates about the sun. As seen in Figure 1.1, the equator receives vertical rays of the sun so that the largest amount of solar energy is absorbed in a band circumventing the globe at the equator. Further, the angle at the equator doesn't vary much as the Earth rotates about the sun so that there are no significant seasonal changes, and temperatures remain fairly warm and constant throughout. As an example, New England is tilted more towards the sun in the summer months and receives vertical rays, but receives slanted rays in the winter because of the earth's tilted axis (Figure 1.1). Conversely, Australia, which is in the southern hemisphere, experiences just the opposite seasons. The sun impacts the Earth in bands of decreasing energy extending north and south from the equator. Warmer air can hold more moisture and therefore may deposit that moisture as precipitation. Therefore, as you move further from the equator, the seasons develop and months of colder and drier air increase, producing major climatic regions following approximate latitudinal bands around

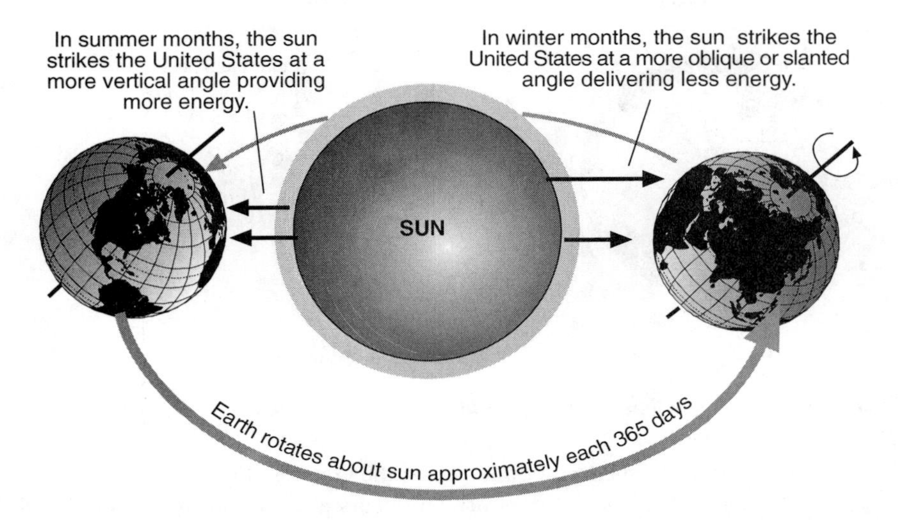

Figure 1.1 Variations in the angle of the sunlight striking the Earth influences season.

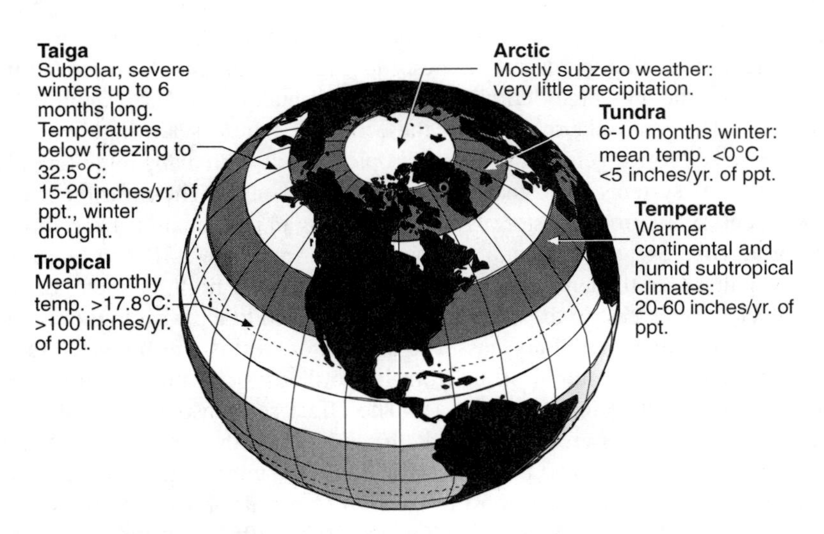

Figure 1.2 Variations in global regional temperatures and precipitation.

the Earth (Figure 1.2). Sir George Hadley provided the first written explanation of global air circulation in 1735.[1] Although modified in recent years, the basic concept still holds. The sun warms air at the equator. As the molecules of air absorb heat energy, they spread apart, the gases expand and become less dense. This expanded air rises, traveling away from the equator. The warm air loses its heat by radioactive and convective cooling and subsides back to the Earth where it flows towards the equator and completes the Hadley cell. (Figure 1.3A). More recent models show that there are multiple Hadley cells known as the three-zone model.[1] In this model,

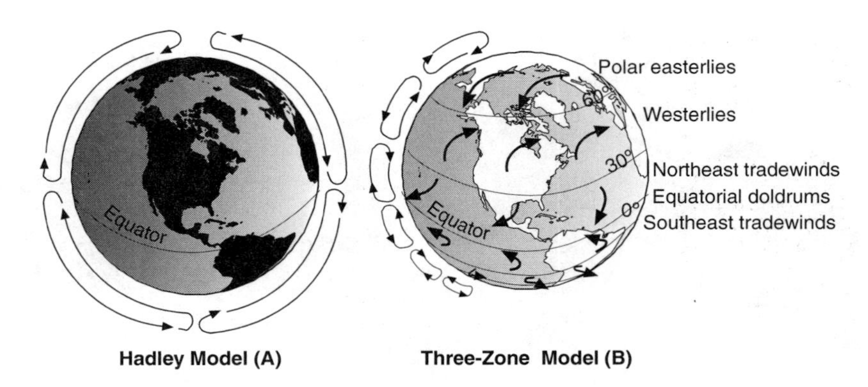

Figure 1.3 Hadley and three-zone models of global circulation.

Hadley-type cells are formed on both sides of the equator (Figure 1.3B). Thermal energy is lost, and subsidence of cool dry air occurs at about 30° North and South latitudes. The falling dry air compresses and warms. The warmer dry air absorbs moisture, creating some of the driest land areas on Earth at these latitudes. The Sahara Desert in North Africa and the Chilean coastal deserts are representative of these areas. There are other reasons for desert formation, discussed later in this chapter under the section on desert biomes. Moving north or south of these latitudes brings you into the temperate latitude including the bulk of the U.S., Central Asia, and Europe. The atmospheric flows in these middle latitudes are highly variable because there is a convergence of polar and tropical fronts with many migrating high- and low-pressure systems that cause frequent weather changes. The collision of warm and cold fronts often produces storm systems resulting in precipitation and ultimately highly productive agricultural land. As can be seen in Figure 1.3B, the large-scale wind flow patterns in the temperate climates are not directly north or south, but tend to be deflected to the east in the northern hemisphere and to the west in the southern hemisphere. The deflection of air masses to the east or west is a result of the Earth's rotation, causing the deflection of air from its northerly or southerly path. This is known as the Coriolis effect (Figure 1.4).[1] The effect seen in the temperate North American zone produces prevailing winds that flow from the west to the east, and these are known as westerlies. Because of this effect, there is a tendency for storm systems to travel from the West Coast to the East Coast of the United States.

We saw hurricanes like Floyd throughout the 1940s, 1950s and 1960s. Scientists are now questioning why they seem to be returning. Many meteorologists agree that the climate of the Earth shifts naturally and that this trend is normal. Others, however, believe that there is another factor involved — a man-made factor, global warming. As we release more and more greenhouse gases into the atmosphere, both the air and ocean temperatures rise significantly. Hurricanes gain their strength from thermal energy found in air and water, and the effect of these increasing temperatures creates the perfect conditions for what scientists refer to as *hypercanes*. Storms of this size and power would make Hurricane Floyd seem like nothing. Storms like these have been linked to the possible destruction of the dinosaurs 65 million years ago. Strong

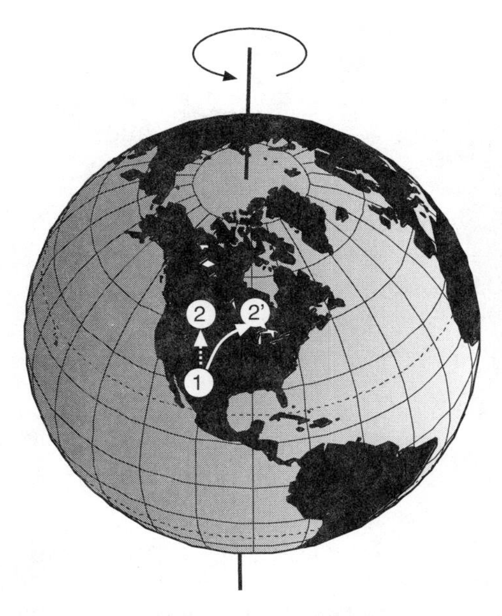
2
2'
1
Air is deflected from its direct northerly direction to the northeast in the northern subtropical Hadley cell. This deflection is caused by the earth's rotation along with friction.

Figure 1.4 The Coriolis effect produces an easterly or westerly deflection in the expected trajectory of the wind. (Adapted from Godish, T., *Air Quality,* 3rd ed., Lewis Publishers, Boca Raton, 1997.)

winds that approach 150 to 160 mph stir up cold waters from deep below the ocean's surface. This causes a cooling effect that will eventually destroy the storm. For this reason, we almost never see a Category 5 hurricane. We might never see a hypercane, but it is true that temperatures are rising, ice caps are melting, and most of the U.S. population resides in costal communities.[8] Ultimately, climates are determined by the position of the Earth with respect to the sun. This influences how much solar energy is absorbed, what wind patterns are established, and how much moisture is deposited. Temperature and moisture combine to become very important factors in establishing climates, which determine the major regional grouping of plants and animals known as biomes.

ECOSYSTEMS AND BIOMES

Ecosystems

Ecosystems are often a component of a biome. The relationship of biosphere, biomes, ecosystems, and populations is shown in Figure 1.5. A system may be viewed as a collection of components that are interdependent or "working together." For example, a collection of gears and springs as you might find in a wall clock or nondigital watch may be viewed as a system that functions in a very precise way to monitor time. Changes or alterations to any component gear may alter or halt this function. The concept of a system can be made very inclusive, such as a society with a system of government, or an individual with a system of organs or tissues

WEB PICK **Go to http://www-unix.oit.umass.edu/~envhl565/index.html** **Click on "Chapter Web Links" to find** **Climate Diagnostics Center under Chapter 1**	
Chapter 1	Ecosystems and Energy Flow
Whose Site?	NOAA-CIRES Climate Diagnostics Center
URL	http://www.cdc.noaa.gov
What's Here?	This is a site maintained by the National Oceanic and Atmospheric Administration (NOAA) of the Centers for Disease Control (CDC). Climate diagnostics involve analyzing the air and water temperature and looking at measurements of atmospheric pressure to identify naturally recurring atmospheric and oceanic features such as El Niño. This leads to better predictions of the future climate. Without a good understanding of climate phenomena based on past observations, there would not be good predictions of climate change. Climate change may contribute to flooding, emerging diseases, crop loss, and natural disasters. This site provides climate data, conceptual and mathematical models used for prediction, information on El Niño/La Niña, long-range forecasts, and excellent interactive graphics/maps on precipitation and other climate phenomena.

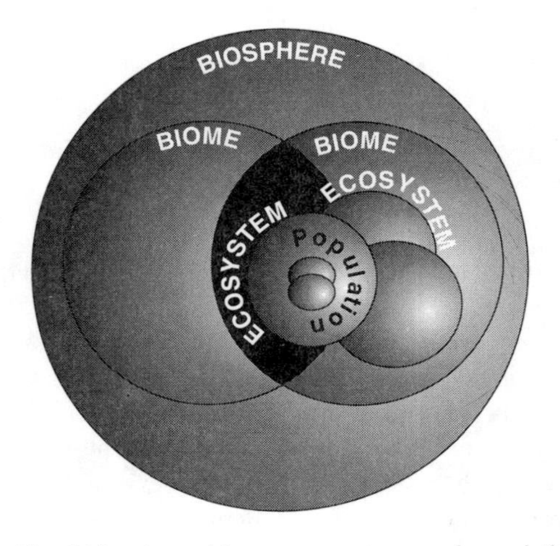

Figure 1.5 Relationship of biosphere, biome, ecosystem, and populations.

such as the immune system. The level at which you wish to focus depends on what you wish to study. Ecosystems refer to identifiable areas within nature where the organisms interact among themselves and their physical environment and exchange nutrients largely within that system. An ant farm inside a glass aquarium can be an ecosystem. Maggots and microbes causing decay of garbage in a collection container can be an ecosystem. A pond, tidal pool, a river valley, or the planet itself can be considered an ecosystem. All of these ecosystems consist of biotic and abiotic components. The biotic components include living organisms and the products of these organisms, including waste products and decay such as urine, feces, decaying leaves and twigs, bones, and flesh. The physical or abiotic components of the ecosystem include such things as water, air, sunlight, minerals, and their interaction to produce climate, salinity, turbulence, and other conditions that influence the physical conditions under which organisms survive. The interaction of abiotic and biotic components of the environment is so closely intertwined that they cannot be easily separated for study. A pound of soil, for instance, consists of fragments of stone and pulverized rock, water, decaying plants and animals, and numerous living organisms such as bacteria, fungi, protozoa, insects, and worms. Without biotic components, soil could not support the growth of plants and would be lifeless dust.

Biomes

Biomes may be seen as groupings of plants and animals on a regional scale whose distribution patterns depend heavily on patterns of climate. The biome is identified by the climax vegetation or community. A climax community is identified as one that forms in an undisturbed environment and continues to grow and perpetuate itself in the absence of further disturbance. For instance, a mature climax forest of hardwoods (oak, beech, elm, maple) may take a century or more to develop from an abandoned northeastern farm field as a series of different plants succeed each other to this final stage. The biome is a community of interacting vegetation, soil types, and animal populations that are adapted to the physical environment of the region. The major biomes of the Earth include tundra, taiga, temperate forest and grassland, deserts, and tropical biomes (Tables 1.1a; 1.1b).[2–4] These major biomes are described briefly in Chapter 3 under the topic of soils, with greater attention paid to conditions of soil depth and fertility. In this section we will discuss the biomes in more detail with regard to climate conditions, typical climax vegetation, and predominant mammal populations.

Tundra

The tundra is limited to the upper latitudes of the northern hemisphere and forms a belt around the Arctic Ocean (Figure 1.6). The continents do not extend southward sufficiently to create a similar biome in the southern latitudes. The tundra appears as a barren and treeless land consisting mostly of mosses, lichens, sedges, dwarf shrubs, and perennial forbs. The scattered shrubs and woody bushes lie close to the ground where they can preserve some warmth. The growing season is very short, lasting 6 to 10 weeks. The winters are long, dark, and cold with mean monthly

Table 1.1 Major Earth Biomes

BIOME and LOCATION	RAINFALL	TEMPERATURE	CLIMAX VEGETATION	ANIMALS
Tundra High latitudes of N. hemisphere in a belt around Arctic Ocean.	Less than 5 inches per year, strong drying winds.	6 to 10 weeks growing season. Winters are dark and long (6 to 10 months). Mean monthly temp. below freezing.	Small shrubs, mosses, sedges, lichens, some grasses.	Muskox, arctic hare, arctic fox, migratory waterfowl, and caribou.
Taiga Boreal forest in N. hemisphere forming a continuous belt of coniferous trees across N. America through Eurasia. Overlying formerly glaciated area	15 to 20 inches annually. Low evaporation rates make area humid in summer. Winter drought.	Subpolar. Severe winter up to 6 months. Extreme temperatures of below freezing to +32°C.	Needleaf, coniferous tress (spruce, fir, pine).	Fur-bearing predators such as lynx, wolverine, fisher, and sable. Mammalian herbivores such as snowshoe hares, lemmings, and voles. Large herbivores such as elk, moose, and beaver.
Temperate Forest Western and central Europe, eastern Asia, eastern North America.	20 to 60 inches per year.	Warmer, continental and humid subtropical climates. Drought during cold winters.	Temperate broadleaf deciduous forest. Oak, maple, beech, hickory, elm. Autumn leaf fall provides for abundant, rich humus.	Herbivores such as white- tail deer, gray squirrels, and chipmunks. Omnivores such as raccoons, skunks, and black bears. Carnivores largely eliminated such as timber wolves, mountain lions, and bobcats.
Temperate Grasslands N. America in central lowlands and high plains. Steppes of Ukraine, eastward through Russia and Mongolia. Pampas of S. America. Veldt of Republic of S. Africa.	10 to 20 inches per year. Much falls as snow.	Semiarid, warm to hot summer.	Rich mix of perennial grass underlain by fertile soils. Members of sunflower and pea families. Scattered woodlands in wetter areas.	Low in diversity. Bison and pronghorn in N. America. Rodent herbivores including ground squirrels, and gophers. Carnivores/omnivores include coyotes, badgers, and black-footed ferrets.
Tropical Savannas E. Africa, Venezuela, Columbia, Brazil, Central America.	Dry and wet seasons. 30 to 50 inches with dry season less than 4 inches a month.	Mean monthly temperature above 17.8°C.	Perennial grasses, dry zones, with sporadic woods in wetter areas.	Elephants, zebras, and wildebeests. Hoofed mammals such as antelopes, gazelles, and giraffes. Carnivores such as lions, leopards, jackals, and hyenas.

Table 1.1 (continued) Major Earth Biomes

BIOME and LOCATION	RAINFALL	TEMPERATURE	CLIMAX VEGETATION	ANIMALS
Tropical Forest Broadleaf forest or rainforest includes areas in Central and S. America, Central Africa, South and Southeast Asia.	In excess of 100 inches a year.	Mean monthly temp. above 17.8°C.	Widely spaced emergent trees of 100 to 120 feet,. A closed canopy of 60 to 80 foot trees. Air plants, woody vines, climbers, and	Most complex biome with numerous birds, insects, and primates.
Deserts West coast of continents between 20° and 30° latitudes including Baja, Ca., Atacama in S. America, and Namib in Southwest Africa. Interior positions of continents such as Gobe of Mongolia, Australian desert, and the Great Basin desert in the U.S. Rainshadow areas of high mountain ranges such as Death Valley, CA and Peruvian desert of S. America.	Less than 10 inches precipitation per year, arid climate in which evaporation exceeds precipitation annually.	May exceed 37.8°C in summer afternoons but drop to below freezing at night. Winters may be cool to cold.	Evergreen, or small-leaved deciduous plants, typically with thorns or spines and shallow, extensive root systems. Some plants with long tap roots (20 to 30 feet) to reach groundwater.	Animals with small body sizes and long appendages. Bright plumage to reflect sunlight. Reptiles have waterproof skin, produce uric acid instead of urine, and hard-shelled eggs.

temperatures below freezing (<0°C) and precipitation less than 5 inches per year. Dryness is further encouraged by strong, drying winds. Beneath the thin layer of topsoil lies a permanent layer of ice or frozen soil known as permafrost. The permafrost, along with the cold weather, prevents the growth of larger plants and trees, prevents drainage, and leads to the formation of bogs and marshes.

The soils are shallow and poor in fertility because of the limited growing season combined with the slow decomposition of organic matter in extreme cold temperatures. Consequently, the soils and the associated plant life tend to be very fragile, and damage done to the tundra is usually long-lasting.

In 1998, a team of researchers at Vostok, Antarctica, drilled 3263 meters into the ice and brought up ancient ice samples. From these samples it was possible to gather information such as age, temperature at the time the ice formed, and composition of the atmosphere. The composition of ancient atmospheres is shown through the analysis of tiny air bubbles trapped in the ice. Glaciologist Claude Lorius found a strong positive relation between CO_2 and methane. Both of these chemicals are

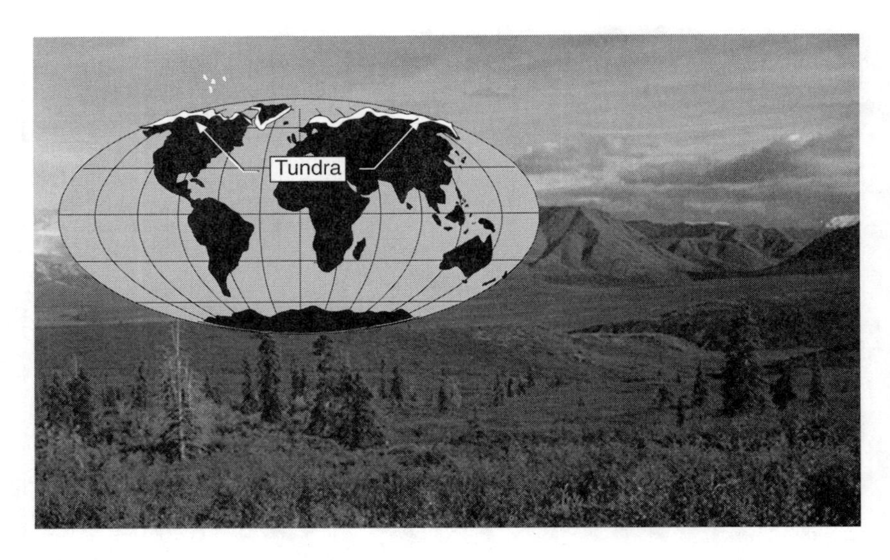

Figure 1.6 Typical tundra biome and location(s).

known as greenhouse gases. They have shown the rapid temperature changes of the last ice age and also relate to the current increase in global temperature. This could be due to the increased levels of greenhouse gases in the atmosphere. Computers have been very useful when making paleoclimate models and finding climate mechanisms.[9]

Scientists were able to date the ice core with an ice flow model using electrical conductivity measurements, ice accumulation changes, and correlation with other paleoclimate records. The data on greenhouse gases led Lorius to conclude that the greenhouse gases and the ice sheets played a significant part in the glacial/interglacial climate changes. There is also the problem of which came first — the greenhouse gases or the climate changes. At this time it is thought that the climate change is a result of the increased greenhouse gas concentration.[9]

The Arctic supports great amounts of wildlife, which appears as large numbers of few species. These species have adapted to the extremely harsh conditions by morphological, physiological, and behavioral adaptations. Many animals have developed large, heavy bodies with a thick layer of fur (muskox) or feathers. Some species accumulate fat during a shortened growing season (bear) while others adapt by experiencing periodic population crashes and explosions as in the lemming population. Some species such as caribou and waterfowl migrate southward from the tundra during the harsh winter conditions. Some of the better known Arctic inhabitants include muskox, caribou, grizzly bears, polar bears, wolves, foxes, snowshoe hares, weasels, and mink. A few species of birds (about six or seven) such as the snowy owl and ptarmigan also survive here. A number of tundra species can also be found at higher elevations in the mountainous regions of the northern hemisphere where thin soils and cold temperatures prevent the growth of most middle latitude trees

Figure 1.7 Typical Taiga biome and location(s).

and allow tundra species to grow above the tree line. Here the alpine tundra and alpine grasses flourish with sedges, perennial grasses, and dwarf shrubs.

Taiga

The taiga, or boreal forest, is an almost continuous band of coniferous (cone-bearing) trees extending in a giant arc from Alaska and Canada through Europe and Siberia.[3] It exists over areas that were once glaciated (Figure 1.7). Such areas tend to have long, severe winters lasting up to six months, with temperatures below freezing and growing seasons between 50 to 100 days. The rainfall tends to be 15 to 20 inches annually, but poor evaporation and poor drainage result in the formation of bogs often overgrown with sphagnum moss. The cone-bearing trees include fir, spruce, and pine. These needle-leaf trees are adapted to the taiga through their conical shape that sheds snow, and the waxy, thin needles that reduce water loss and protect against drying winds. The dark green needles absorb heat and are not shed so that photosynthesis can start immediately with the beginning of the shortened growing season.[4]

Fur-bearing predators including wolverines, sable, fisher, mink, and lynx seek mammalian herbivores such as snowshoe hares, red squirrels, lemmings, and voles. Many of these species develop white coats during the winter to blend with the snow. A number of larger herbivores may also be found in or near the taiga, including moose, elk, and red deer. Year-round birds of the taiga include finches, sparrows, and ravens. Some birds such as the grosbeak and crossbill become migratory during years when pinecones are reduced in numbers.

Figure 1.8 Typical broadleaf temperate forest biome and location(s).

Temperate Areas

Temperate Broadleaf Deciduous Forest

The temperate broadleaf deciduous forests (Figure 1.8) are located in Western and Central Europe, Eastern Asia and Eastern North America. Climax deciduous forests include oak, maple, hickory, chestnut, elm, linden, and walnut trees. The temperate forests receive 20 to 60 inches of precipitation distributed evenly throughout the year. Snow and ice conditions during winter months make water unavailable and create droughts during the cold season. The broadleaf trees lose their leaves (deciduous) during the winter to conserve moisture and resist the cold. As the cold fall weather settles in, the leaves begin the deciduous process (leaf fall). The leaves turn to brilliant reds, oranges, and yellows as chlorophyll formation is halted; and the leaves finally die and fall to the ground, forming a litter that decays further to a rich organic mat called humus.

Mammals in this biome include acorn and nut feeders such as squirrels and chipmunks; omnivores such as raccoons, skunks, black bear, and opossum; and rare carnivores such as mountain lions, bobcats, and wolves. Except for the recent emigration of coyotes from the grasslands, the carnivores have been mostly eliminated by habitat destruction and hunting. Birds include insect eaters such as hummingbirds, thrushes, warblers, and wrens. Omnivores such as ravens and cavity nesters such as woodpeckers and chickadees also abound.[3,4] Most of the bird species are migratory and fly south for the winter while a number of mammalian species enter into hibernation.

Temperate Evergreen Forests

In temperate areas where soil is poor and droughts and fires are frequent, the predominant species tend to be evergreens including coniferous, needle-leafed trees with some broad-leafed evergreens. In such areas, spruce, fir, and ponderosa pine proliferate. These are known as temperate evergreen forests. In cool coastal climates where there is considerable rainfall or frequent heavy fogs, one is likely to see forests of giant trees such as redwoods, which characteristically reach heights of 198 feet (60 meters) to 297 feet (90 meters). These are temperate rainforests, and such areas exist along the Pacific Coast of North America stretching from Northern California to Sitka, Alaska. Chile and New Zealand also have these temperate forests.

Chaparrals

Areas of moderately dry climate and limited summer precipitation are further characterized by small (3 to 15 feet) shrubs with leathery leaves that contain aromatic and flammable substances. Consequently, these chaparrals often catch on fire, allowing shrubs to regrow from parts of the plant that survive near the ground. Unfortunately, when such growth is allowed to proliferate near houses that are rapidly expanding into these areas, they become likely fuel for wildfires.

Temperate Grasslands

Temperate grasslands are known by a number of names that include North American prairie, Asian steppes, South African veldt, and the South American pampas (Figure 1.9).[3,4] These semiarid regions tend to be located inland along the middle latitudes and experience only 10 to 20 inches of precipitation a year, much of which falls as snow. The predominant plant forms are perennial grasses, forbs,

Figure 1.9 Typical temperate grassland biome and location(s).

and members of the sunflower and pea families.[4] The perennial grasses grow from buds just below the ground surface and are adapted to fire, drought, and cold, while the well-developed root systems are able to trap moisture and nutrients. The more humid grasslands are populated by sod-forming grasses that have rhizomes or stems under the ground from which new plants are formed. Areas that are drier tend to favor bunch grasses that reproduce by seed. Biodiversity in temperate grasslands is limited. Most commonly seen in temperate grasslands are the herbivore rodents including ground squirrels, prairie dogs, and pocket gophers. Large animals such as bison, pronghorn, grizzly bears, and wolves have been largely confined to isolated preserves or the mountains as humans destroy more habitats.[3]

The Tropical Rainforest

The tropical rainforest is complex in terms of species diversity and structure. The combination of constant warmth, with average monthly temperatures above 17.8ºC and precipitation greater than 100 inches per year, encourages rapid plant growth throughout the year. The high humidity and continued warmth promote rapid growth and great diversity. There are relatively few numbers of any particular species, but myriad species exist in the tropical rainforest. These species account for nearly three quarters of the total number of species on Earth. Further, more than 40% of the world's plants and animals grow in the tropical rainforests, which account for less than 7% of Earth's surface.[3,4] There are no seasons in the rainforest, so there is no annual rhythm. Each species flowers or fruits at different times of the year on their own cycle. The majority of plants are tall trees with thin trunks that produce branches and leaves near the top. Many of these leaves are elongated at the tips to encourage dripping or shedding of water ("drip-tip" leaves), which reduces mold and insect attack.

The life of the forest occurs in the canopy. The canopy is an almost unbroken sea of green produced by leaves of trees at 70 to 100 feet above the forest floor (Figure 1.10a). Most of the sunlight is absorbed in the canopy, and it is where the bulk of photosynthesis occurs, resulting in more than 10 tons of new growth per acre per year. Much of the animal life also lives in the canopy. The leathery green leaves sprouting from branches at the top of trees help to shield the forest floor from the direct impact of intense showers. Occasionally, a taller tree penetrates the lower canopy to form an umbrella-like canopy over the trees below. These trees reach heights of 130 feet and are called *emergent trees*. Temperatures in the canopy often reach 35.5°C (96°F) with a humidity of 50 to 60%, while the temperature on the floor of the forest are normally lower at 27.7°C (82°F) with a humidity of 90%.

Researchers in Costa Rica have discovered evidence that increasing temperatures have slowed down the growth of tropical forests. It has been suggested that the tropical rainforests have actually added billions of tons of greenhouse gases into the atmosphere during the 1990s. Although trees take in CO_2 and then release O_2 during photosynthesis, they also release some CO_2 as a byproduct of respiration. When growing rapidly, plants take up more than they can produce. But if growth is slow, then the balance will shift. In 1984, researchers Deborah Clark and David Clark began researching the growth rates of adult tropical rainforest trees. The samples

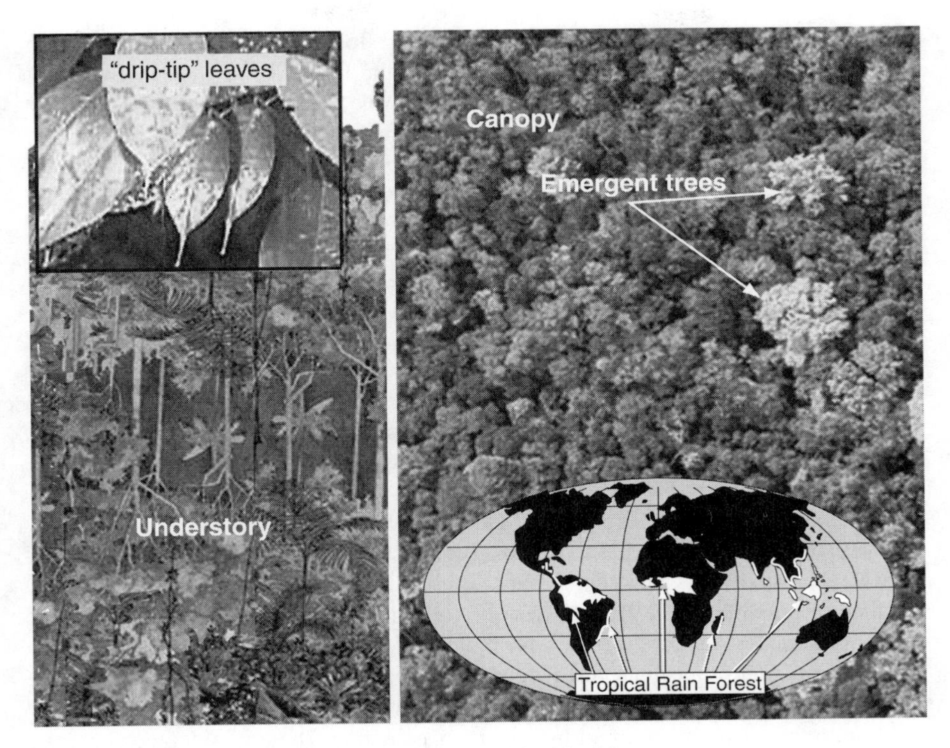

Figure 1.10 Typical tropical rainforest biome and location(s).

included six different tree species with both fast- and slow-growing tree types. They found that the growth of the trees changed each year. In years with high levels of CO_2, the trees grew slowly; in years when the level was small, the trees grew rapidly.

Less than 10% of the sunlight reaches the forest floor, and there is intense competition for available energy at this level. There is sparse vegetation in the inner sections of the undisturbed rainforest that must also compete in poor soil conditions. This part of the rainforest is known as the *understory* (Figure 1.10b). The soil is often soggy and quite shallow because the tremendous biological activity quickly decomposes organic matter and converts it into new life. The soils may be only a few inches deep covering red clay. This has caused many trees to adapt by forming shallow root systems and bases that spread at the bottom of the tree much like a buttress. This growth provides support in thin soil and may enhance the ability of the tree to secure nutrients over a broader area.

The reduced light beneath the canopy has encouraged a number of plant strategies to reach the top of the rainforest. Air plants or epiphytes grow on branches high in the trees. Supported by the tree limbs, epiphytes extract moisture from the air and trap leaf fall and wind-blown dust for nutrients. Bromeliads (pineapple family), orchids, cactuses, and ferns are all represented in the canopy as epiphytes. Many plants such as members of the Figure family begin life as epiphytes in the canopy and send their root systems downward to the forest floor. Such plants are then called stranglers. Many woody vines grow quickly up the tree trunks when a temporary gap opens in the canopy. Here they flower and fruit in the treetops. These woody vines are called lianas and are often deciduous. Few animals exist on the forest floor. Beetles, butterflies, and other insects abound in the canopy along with a rich array of birds, monkeys, and frogs. Tropical rainforests have been linked to possible cures for diseases and viruses. Typical rainforests circumvent the globe in a band from 20° South to 30° North and are found in Central and South America, Eastern and Central Africa, Malaysia, Indonesia, the Philippines, Papua New Guinea, and the Solomon Islands.

Oceans

Oceans are also a rich resource for pharmaceuticals. Animals from the ocean such as horseshoe crabs, starfish, bacteria and ocean molds have produced a few potential cures for cancers, Alzheimer's disease and other viruses and diseases. The ocean covers almost 70% of the Earth's surface. The organisms living in the ocean have primitive life systems and are built for survival. They are examples of the early stages of life with simple mechanisms to fight diseases. Biologists William Fenical and Bob Jacobs from the University of California at Santa Barbara have been looking to the ocean as a source of possible cures. They have been going to reefs in the Caribbean. They are studying animals that are "loaded with biological active molecules" that they think have natural defense mechanisms. Fenical has already discovered a substance from a sea fan that reduces inflammation from sunburn or chemical burns. This substance is much more effective than current hydrocortisone creams. These pseudopterosins may offer the opportunity to cure arthritis and psoriasis and may even limit damage from the sun. Fenical has also tested a substance found in the sea off of Australia, eleutherobin, which could possibly cure a cancer by disrupting malignant cell division.

Other marine biologists have found an anti-cancer chemical, halichondrin B. It comes from a yellow-slimy sponge found off of the New Zealand coast. Another, Bugula, an organism that is found on the bottom of boats, is under investigation as a possible treatment for leukemia and kidney cancer.

Deserts

Deserts are defined by their arid climates, which average less than 10 inches of precipitation a year. Evaporation tends to exceed this precipitation, causing a negative annual water budget. When rain falls, it tends to be very localized and unpredictable. Further, some years may witness much more rainfall than others. Deserts can easily reach temperatures higher than 37.8°C (100°F) on summer days while plummeting to –6.7°C (20°F) at night. Some deserts may not experience frost, while others may have extended periods of subzero temperatures and even snow.[3,4]

Deserts are created by specific geographic and climate conditions.

1. Within the latitudes of 20° South to 30° North and along the western coasts of continents, the winds tend to be easterly (coming from the east) (see Figure 1.3B) and consequently keep moist air that rises off the oceans from reaching the coast. This causes some of the world's driest deserts such as the Sahara in Africa, Baja

Figure 1.11 Typical desert biome and location(s).

California in North America, and the Atacama in South America (Figures 1.11 and 1.11a).

2. Near the 30° latitude, subtropical air descends in association with the Hadley cell (Figure 1.3B). As it descends it compresses, causing the formation of heat and dry, warm air. This air then absorbs moisture, resulting in little precipitation and creating such areas as the Sahara and Australian deserts.
3. Temperate deserts are generally located in areas known as rainshadows. Rainshadows occur on the lee side of high coastal mountain ranges (Figure 1.12). As moisture-laden air approaches the windward side of the mountains, the air rises,

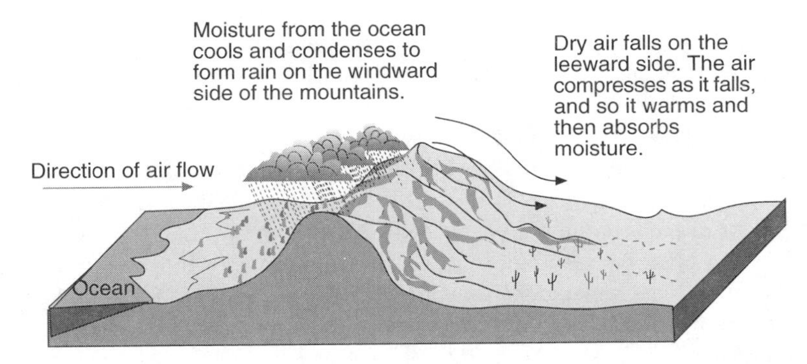

Figure 1.12 Rainshadow deserts are created on the lee side of coastal mountain ranges. (Adapted from Turk, J., and Turk, A., *Environmental Science,* Saunders College Publishing, Philadelphia, 1988.)

cools, and creates a low-pressure condition that causes precipitation. The windward side receives large amounts of rain and snow. The air falls on the leeward side, compresses and warms, and absorbs moisture, producing an arid environment and creating such deserts as Death Valley in California and the Peruvian Desert.

4. The Gobi Desert of Mongolia and the Great Basin Desert of the United States are located in the interiors of continents where they are distant from sources of moist air and where the rainshadow effect may also be important.[3,4]

The deserts are not completely lacking in water, and many life forms have adapted to the arid conditions. The major plant life in deserts consists of evergreen shrubs that have small drought-resistant leaves, often have spines or thorns, and emit sweet-smelling oils. The root systems usually extend outward by great distances to capture whatever moisture falls. Other plants have taproots that extend to 30 feet below the ground to reach underground water. Still other plants known as succulents store water in their tissues to be used during dry spells. Depending on the species, water may be stored in different parts of the plant including stem, leaf, fruit, or roots. For example, the familiar American cactus is a stem succulent while the aloe plant is an African leaf succulent.

Desert fauna have also adapted to aridity. Body sizes are small with long appendages that maximize heat radiation. Animals may have skin without sweat glands. They may concentrate urine into uric acid, have deposits of fat in tails or humps, or glands that secrete salt without moisture loss. They may be nocturnal or may even be dormant during summer months.

ENERGY FLOW

Energy Source

All life requires energy, and ultimately the energy comes from the sun. Sunlight travels across the galaxy at 186,000 miles per second to reach us, and in a period of less than 10 minutes it loses all but the smallest fraction of its radiation intensity. Close to 99.9% of the sun's energy reaching the Earth is reflected into space, absorbed as heat, or evaporated as water. Plants absorb a portion (about 0.1%) of this incoming light energy, particularly in the blue and red wavelengths of the visible spectrum. This energy is used by the plants for photosynthesis to create simple sugars from carbon dioxide and water with the release of oxygen (Figure 1.13). The simple sugars are used to store chemical energy, and the amount of energy stored by the plant is normally expressed as kilocalories of chemical energy stored per square meter of plant tissue per year (kcal/m2/yr). Some 20% of this stored energy is used by the plant for respiration, and it therefore escapes in the form of heat to the environment. The remainder (from 50% to 80%) is converted for growth, thereby increasing its mass. The flow of energy and the conservation of matter must obey certain fundamental physical laws (Table 1.2). The accumulation of organic material in an ecosystem is referred to as biomass. Some ecosystems, such as tropical

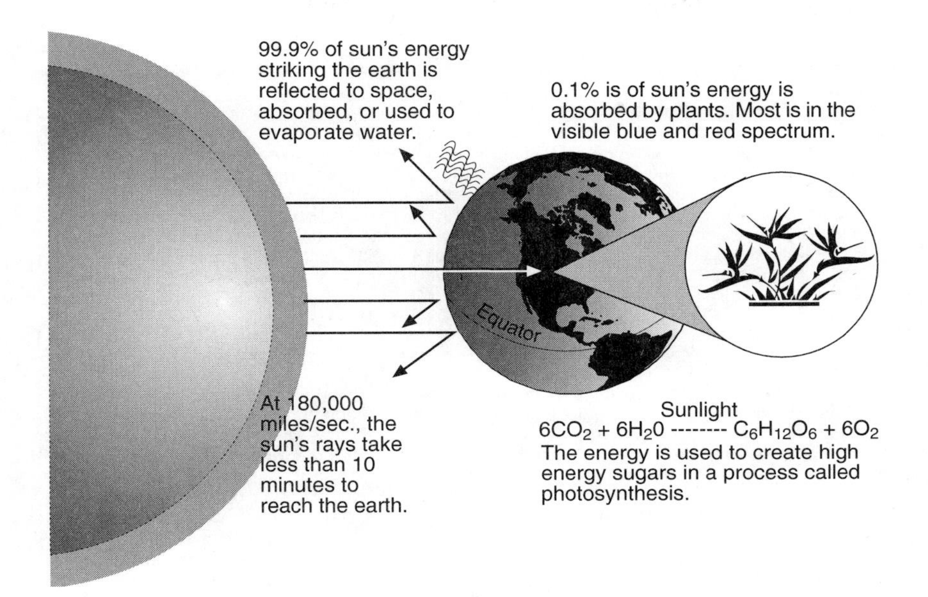

99.9% of sun's energy striking the earth is reflected to space, absorbed, or used to evaporate water.
0.1% is of sun's energy is absorbed by plants. Most is in the visible blue and red spectrum.
Equator
At 180,000 miles/sec., the sun's rays take less than 10 minutes to reach the earth.
Sunlight
6CO2 + 6H20 -------- C6H12O6 + 6O2
The energy is used to create high energy sugars in a process called photosynthesis.

Figure 1.13 Distribution of sun's energy striking the earth.

rainforests or marshes, are highly efficient in producing new plant growth because of the availability of nutrients, sunlight, warmth, and moisture. Consequently, the biomass in these areas is much greater than in a desert.

Within the biosphere, energy flows through the interconnecting web of life from organism to organism. The plant produced may be available to a variety of insects and animals, including humans, who must consume the plant to obtain energy-rich molecules since animals cannot photosynthesize. These consumers, or heterotrophs, convert about 10% of the consumed kilocalories into animal flesh or organic matter (Figure 1.14). Ninety percent of the remaining energy is consumed by heterotrophs for respiration necessary for the energy of motion, the maintenance of body functions, or as excreta.

As energy is transferred through the food chain, about 90% of that available energy is lost with each transfer while the consumer uses 10%. Consequently, an animal such as a wolf, which consumes deer or rabbits that eat grass, would be a secondary consumer and would receive ($1/10 \times 1/10 = 1/100$) of the available energy in the plant (Figure 1.14). In an example that demonstrates the efficiency of eating at a lower trophic level, 3000 pounds of corn and grain would support the energy requirements of one 1500-pound steer, which in turn would support the energy requirements of one 150-pound human (Figure 1.15). If the grain were eaten directly by humans, it could likely support up to 20 people or more in the same amount of time. Only wealthy countries can support a human population where meat is one of the main foods consumed. A primarily vegetarian society can feed many times the number of humans with the same amount of grain that is used to feed cattle. Countries

Table 1.2

SOME NATURAL PHYSICAL LAWS

Law of Conservation of Matter: While undergoing normal changes in physical or chemical reactions, matter cannot be created or destroyed. Practically, this means the earth has finite amounts of chemicals and nutrients that must be recycled if life is to continue. Oxygen can exist as iron oxides, carbon dioxide, nitrogen oxides, molecular oxygen, water, alcohols, within organic compounds and numerous other forms–but the total amount of oxygen available on the planet remains the same. The Law of Conservation of Matter means that everything is accounted for and that there is a mass balance. As we breath in oxygen and catabolize plant sugars in the process of respiration, we give off carbon dioxide and water. The plants are able to use the carbon dioxide and water in the process of photosynthesis which results in the production of sugars and oxygen.

$$\text{Photosynthesis: } 6CO_2 + 6H_20 \text{ -------- } C_6H_{12}O_6 + 6O_2$$

$$\text{Respiration: } C_6H_{12}O_6 + 6O_2 \text{ -------- } 6CO_2 + 6H_20 + \text{energy}$$

Without this exhange or recycling of oxygen, the oxygen would soon be used up and life as we know it would be gone. As another example, if we burn 1000 tons per day of solid waste in an incinerator, we will get 1000 tons per day out of the incinerator. The materials discharged will be ash, some unburnable solids, particulates, and gases–but the total matter will weigh the same as that which entered.

The First Law of Thermodynamics: Energy is defined as the capacity or potential to perform work. There are many forms of energy including kinetic, gravitational, chemical and thermal energy. Energy can be transformed from one form to another but cannot be created or destroyed. In practical terms, the energy of the sun is converted by plants to chemical energy in the form of sugars and plant tissues that can be used by herbivores as a source of energy. Plants use about 0.1% of the sun's energy reaching the earth. The rest is radiated back to space, absorbed as heat, or used to evaporate water. Animals that eat the plants convert about 10% for their use, with about 1% of this converted to animal tissue and the remainder used in respiration to maintain organ systems, provide motion, and produce heat. The energy used in respiration escapes as thermal or heat energy. Thus, energy was converted from light energy to chemical energy to thermal energy–so the total amount of energy remains the same. However, the energy so produced is not useful to plants or animals for growth in its final form. This relates to the Second Law of Thermodynamics.

The Second Law of Thermodynamics: This law is based on the concept of **entropy**, which means that matter or energy will disperse toward maximum disorder in an undisturbed environment. Just as dye particles will evenly distribute in a cup of water, so will energy move toward a greater state of dispersion that is less useful or unavailable. As light energy is converted to chemical energy and then to heat energy, the heat is radiated ultimately to space where the heat is distributed uniformly throughout. Consequently, there must be a constant supply of energy from the sun to drive the life forces on Earth. Although the total energy has not declined, it moves unidirectionally through the world to become transformed and dispersed into unuseful forms.

such as India, China, and most of the lesser developed countries (LDCs) as in Africa are predominantly vegetarian for this reason.

Consumption Types and Trophic Levels

Animals that eat only plants are *herbivores*. Other animals called *carnivores* eat primarily animal flesh and may include lions, tigers, hyenas, frogs, or even ladybugs (they feed on aphids). Animals that eat plants and animals are termed *omnivores* and include rats, bears, humans, hogs, and foxes.

All energy comes from the sun and flows in one direction from plants, which are *producers* and belong to the *first trophic level*. *Primary consumers* (herbivores) belong to the *second trophic level*. Caterpillars, grasshoppers, cattle, and elephants

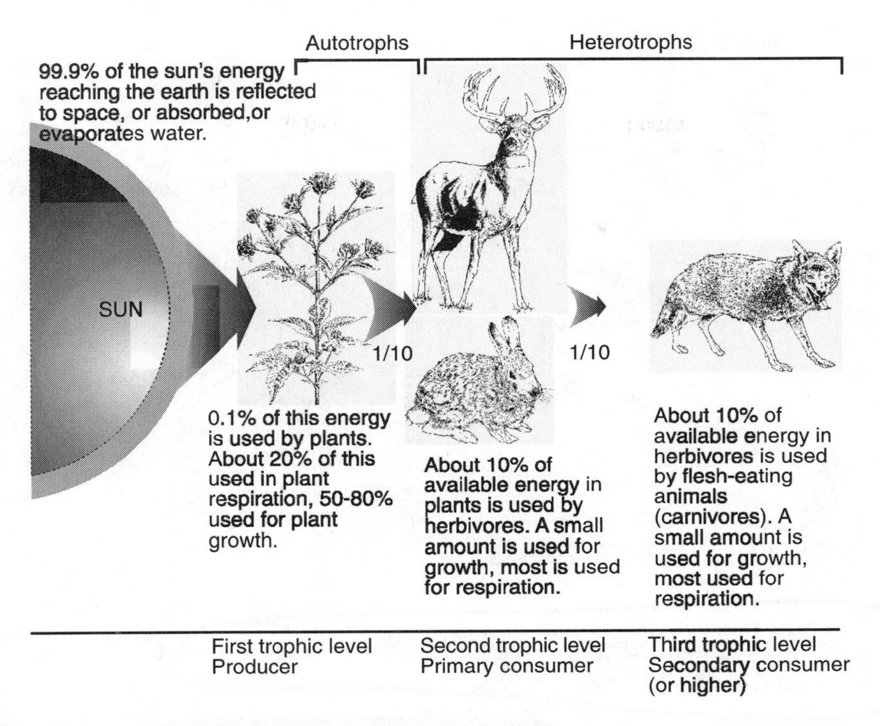

Figure 1.14 Relationships of energy flow in an ecosystem.

Figure 1.15 Comparison in efficiencies of feeding at various trophic levels.

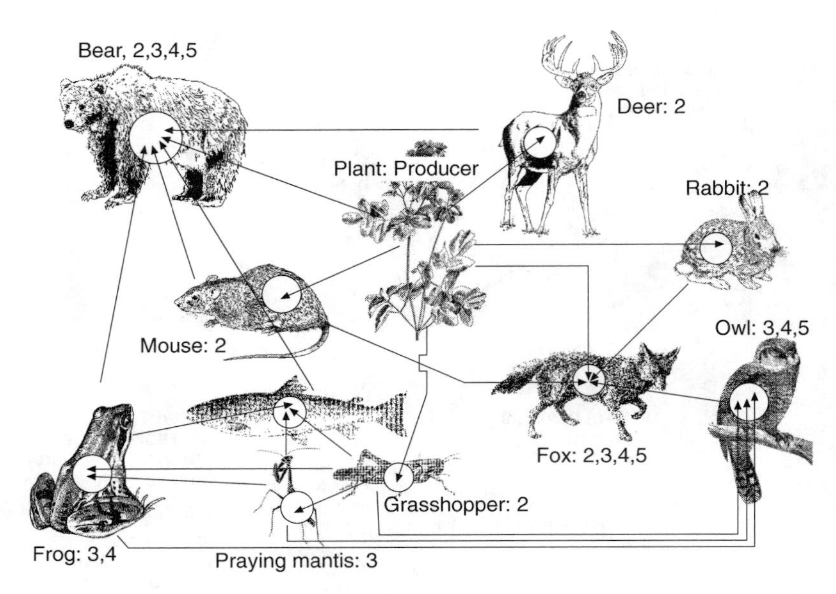

Figure 1.16 Simplified food web showing some of the possible trophic levels (represented by numbers) for various species. The arrows show direction of consumption. (Adapted from Turk, J., and Turk, A., *Environmental Science,* Saunders College Publishing, Philadelphia, 1988.)

are primary consumers because they are herbivores that feed primarily on plants. *Secondary consumers* (carnivores) belong to the *third trophic level* (or higher). Each stage through which this energy travels is called a trophic level. Humans, along with many other animal species, are omnivores and may consume plants or animals. This means that many species may belong to several trophic levels. Bears, for instance, may belong to trophic levels two, three, four and five. Bears will consume berries (herbivore, trophic level two), mice (trophic level three), frogs (trophic level four), or shrews (trophic level five). An example of this complex interrelationship is given in Figure 1.16. The term *food web* has been used to refer to these complex patterns of consumption among many trophic levels. This is important from a survival point of view, since those animals capable of feeding at many trophic levels are able to survive if portions of their food supply are decreased or eliminated. Because the amount of available energy (kcal) is reduced by a factor of 10 at each trophic level, there is a practical limit to the number of trophic levels that can be sustained. This number seems to be limited to five. As you move from primary consumers to higher trophic levels, the numbers and biomass (dry weight per unit area) of land organisms decrease with each successive level. The plants or producers far outnumber the next trophic level, or herbivores, which in turn far outnumber and have greater biomass than carnivores. This is because at each higher trophic level, there is less useful energy available. Consider that there are fewer grizzly bears than caribou, or fewer foxes than rabbits. As previously mentioned, more animals (including humans) can survive as herbivores than as carnivores.

Nutrients

Recycling

Although there is a constant supply of sunlight, the nutrients tied up in these trophic levels need to be recycled. Some animals may eat and even prefer dead or decaying animal matter. Many of us have seen crows or ravens settle down on a road to consume the remains of a small, killed animal. We know that vultures are also primarily scavengers because they prefer to feed upon the dead remains of animals. Some animals that are predominantly omnivores or carnivores may also become opportunistic scavengers. Hyenas and foxes will feed upon the remains of animals that have died or have been killed. The author has witnessed hogs consuming dead rats, which also blurs the separation of scavenger from omnivores or carnivores. Unlike energy, which is rapidly dissipated through trophic levels, nutrients consist of 40 of the 92 known chemical elements that are essential to most forms of life. When nutrients are recycled, the process is called *biogeochemical cycling*. Nearly all life consists of large percentages of sulfur, phosphorous, carbon, oxygen, hydrogen, and nitrogen, known as *macronutrients*. There are also many elements required in tiny amounts such as zinc, manganese, chlorine, iron, and copper that are termed *trace elements*.[2] The cycling of these nutrients is essential because they would soon be depleted by accumulation in dead matter that prohibits the development of new life. Remember from the Law of Conservation of Matter (Table 1.2) that matter cannot be created or destroyed. Another group of organisms essential to the process of biogeochemical cycling are the *decomposers*. Decomposers are insects, bacteria, fungi, and protozoans that sequentially break down complex organic materials into low energy mineral nutrients that once again may be reabsorbed and used by plants. This process is normally sequential in those specific types of saprophytic organisms, which feed primarily on dead organic matter, first attacking dead plants or animals. As these organisms process components of the dead tissues, they leave behind undigested materials that are in turn processed by the next group of decomposers until finally there is nothing left but energy-poor mineral nutrients. This sequence is so predictable that the science of forensic entomology has developed; practitioners can judge the time of death of a human from the types of decomposers evident at the time of discovery of the body.

Nutrient Cycles

The recycling of nutrients from inorganic to organic and back again in a living system was previously defined as biogeochemical cycling. This cycling may take two forms. The nutrients may be *sedimentary* and originate from soil or rocks, or *gaseous* and originate from air or water. The gaseous cycle is very efficient compared to the sedimentary cycle, which may be why the major constituents of most life consist of carbon, nitrogen, hydrogen, and oxygen. Representative substances from the gaseous and sedimentary cycles are discussed below.

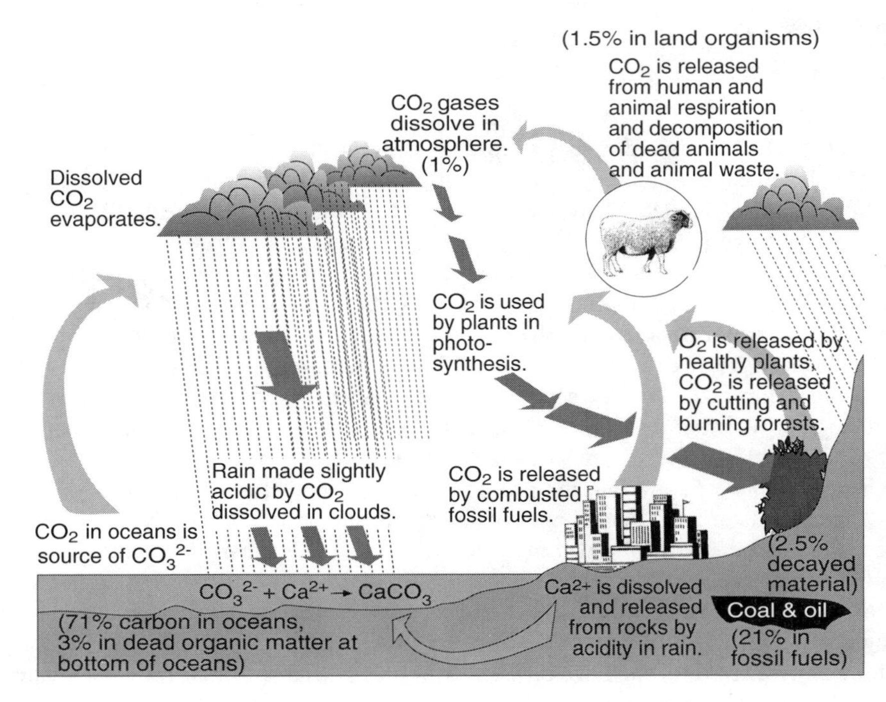

Figure 1.17 A simplified carbon cycle. (Percentages refer to total global carbon.)

Carbon Cycle

Humans, along with most animals and plants, are carbon-based life forms. It is not surprising then that carbon is available in such a wide spectrum of forms including the physical states of gas, liquid, or solid, and the chemical forms including organic and inorganic (Figure 1.17). Inorganic sources of carbon include gaseous carbon dioxide, which appears in the atmosphere at about 0.03% (300 ppm). Carbon dioxide is released to the atmosphere from (1) the respiratory process of animals and plants that consume oxygen and release carbon dioxide and water; (2) the combustion of fossil or organic fuels such as coal, wood, oil, and gas; and (3) the decomposition of organic matter (leaves, carcasses, animal waste) by insects and microbes. Carbon dioxide is also dissolved in water, where it forms a weakly acidic solution. Small amounts of carbonate and bicarbonate ions are produced, while more than 99% of the carbon dioxide remains dissolved in the water without conversion. Therefore, the water and precipitation are only slightly acidified by carbon dioxide. However, the amount of carbon dioxide in the water is enormous; estimates place the level of carbon dissolved in the oceans as 70 to 100 times greater than present in the atmosphere.[2,3] The carbonate ions formed in the water can react with calcium to form calcium carbonate, which mineralizes to form limestone. Limestone can also be formed from shellfish that create an insoluble calcium carbonate as part of their shells. The shells accumulate as sediment and eventually form sedimentary rock.

Plants convert inorganic carbon and water to useful energy-rich sugars through the process of photosynthesis. The plants then become food for many forms of animal life, which create new and more complex organic carbon compounds. The carbon is released back to the atmosphere through respiration and the decomposition of plant or animal matter or animal waste products. Although much of the decay of plant and organic matter may occur in weeks or months, decay may sometimes be incomplete because the materials are buried, lying in peat bogs or moorlands, at the bottom of oceans, or beneath the ground where geological processes convert this material into the fossil fuels of coal, oil or natural gas. These latter materials may be recovered and burned to release carbon dioxide back into the atmosphere. The combination of burning fossil fuels and forests is releasing greater and greater quantities of carbon dioxide into the atmosphere. Many scientists and world leaders think the increase in atmospheric carbon dioxide at the global level contributes to global warming. This is a topic to be covered in more detail in Chapter 10 on air pollution.

Nitrogen Cycle

Although nitrogen is the major component of the Earth's atmosphere comprising 78% of the air we breathe, it cannot be used directly by animals or most plants. Useable forms of nitrogen are in the forms of nitrates (NO3-), nitrites (NO2-), and ammonia ions (NH3+). Therefore, natural mechanisms must exist that convert atmospheric nitrogen to useable forms. There are two known major natural mechanisms (Figure 1.18). One method involves the conversion of atmospheric nitrogen to nitrates (fixation) by (1) bacteria such as *Rhizobium*, which live symbiotically in the root nodules of leguminous plants, or (2) some species of organobacteria that can convert gaseous atmospheric nitrogen to useable nitrate forms. Another natural method occurs when lightning causes atmospheric nitrogen to react with oxygen, converting it through a number of steps to nitrates. Nitrates are also produced in the atmosphere from nitrogen oxides created within the high-temperature confines of boilers or internal combustion engines. Nitrates are also released from the erosion of nitrate-rich rocks into soil and water. Nitrogenous compounds are continually recycled from animal wastes and the decomposition of organisms, which are acted upon by soil microbes to produce ammonia and then converted to nitrates by nitrifying bacteria, which is a form usable by plants. Nitrates and ammonia are also synthesized from nitrogen and hydrogen by heat and pressure to create agricultural fertilizers that can be added to soils to increase fertility. This process is energy consuming and consequently expensive. Once nitrogen is formed into complex organic molecules, it may be recycled back to atmospheric nitrogen by denitrifying bacteria, which live in the mud and sediment of lakes, ponds, streams, and estuaries and carry out these activities anaerobically (in the absence of oxygen).

Inland waterways, estuaries and coastal ecosystems around the world have been subjected to a long-term increase in organic matter that has been washed off the land. A major consequence of this is an oxygen-depleted zone off the coast of Louisiana. Many experts say that overfertilization from organic runoff has surpassed pollutants released from industries as a threat to marine ecosystems.[11]

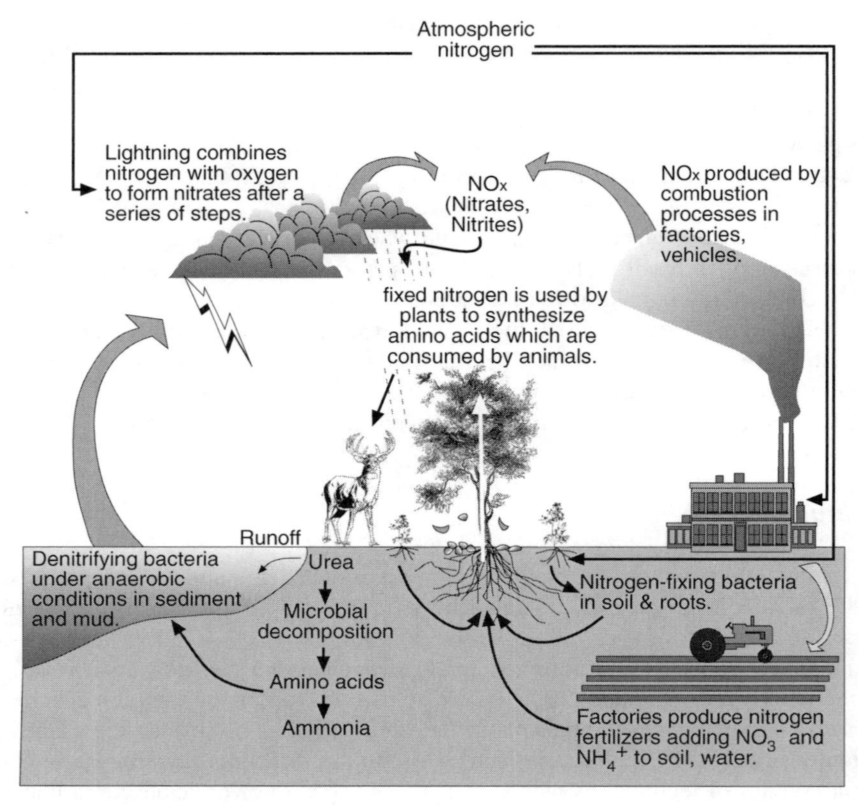

Figure 1.18 The basic nitrogen cycle. (Adapted from Turk, J., and Turk, A., *Environmental Science,* Saunders College Publishing, Philadelphia, 1988.)

Phosphorous Cycle

Calcium, iron, and phosphorous are examples of nutrients that flow through living systems in a pattern of recycling that is measured in eons of time or on a geological scale. These substances are gradually leached from sedimentary rock by the actions of rain or erosion from water in a process referred to as the *sedimentary cycle*. The phosphorous cycle is representative of this type of cycle. Phosphorous is the main element in compounds such as adenosine triphosphate (ATP). ATP is important in the transfer of chemical energy in animals. Phosphates are also a major component of desoxyribonucleic acid (DNA) and ribonucleic acid (RNA) and may be found in large amounts in shells, bones, and teeth. Phosphates do not occur as gas but are bound in sedimentary rocks where they are released in small amounts from weathering. Plants take up the released phosphates in the roots, which may then be consumed by animals. Animal wastes and decomposing animals release phosphorous back to the soil for reuse by plants in a fairly efficient manner when not disturbed. However, the processes of mining and agriculture can erode soil and carry phosphorous into streams, estuaries, and oceans where it settles in a mostly insoluble form to the bottom and becomes generally unavailable. Consequently, phosphate fertilizers, which are produced by mining phosphate rock, must enrich most agricultural soils. Phosphate rocks are a nonrenewable resource created millions of years ago. Therefore, sources of phosphates are being depleted rapidly. We are facing the potential of infertile soils, and it may take thousands of years for phosphorous to be replaced through geological processes of weathering sedimentary rock.

THE PROCESS OF SUCCESSION AND RETROGRESSION

Succession

In Chapter 3 you will read about the degrading of land by overgrazing or overuse leading to retrogression. *Retrogression* is a negative process that reduces the diversity of an ecosystem, leading to fewer and fewer organisms, weakening the food web, and placing greater numbers of species at risk. The normal undisturbed process of nature is one of succession. *Succession* refers to the predictable and gradual progressive changes of biotic communities toward the establishment of a climax community. A *climax community* is one that perpetuates itself with no further succession within an undisturbed ecosystem. Succession usually begins when a disturbance has occurred that kills many or all of the life forms within a region. Complete destruction of areas can occur in association with volcanic eruptions, retreat of glaciers or impacts from large meteorites. In these events, all life, including that in the topsoil, is destroyed; primary succession must take place by first creating soil on the barren lava or exposed rock surfaces. This is accomplished initially through the weathering action of rain and dissolved carbon dioxide, which create fragments and dissolved minerals. Dust is captured in cracks and crevices along with microscopic organisms and seeds carried by the wind or deposited by small animals and birds. Mixtures of fungi and algae grow together and are known as *lichens*. These are highly resilient life forms capable of surviving under sparse nutrient conditions. They gain a foothold and excrete organic acids that further deteriorate the rock surface to release more minerals. The organic material provided by decaying lichens and the inorganic minerals released from the rocks serve as nutrients for mosses that gradually replace the lichens, creating a denser growth that supports insects and other invertebrates while adding more organic matter to the mass. Eventually, enough organic matter is present to grow the seeds of rooted plants. After many years, perhaps hundreds, greater sizes and varieties of plants and animals move into the once-barren area; and eventually a climax community may develop. The reason for the succession of plants is that each stage of plants fails to compete successfully with the next stage. The tall grasses have more extensive root systems that draw moisture and other nutrients away from the wildflowers, while the grasses grow taller and reduce the available sunlight to these same flowers. Ultimately, the grasses are in turn overshadowed by the new growth tress, and so on.[2,3]

When soil conditions are disrupted but topsoil and some limited vegetation remain, succession can take place much more quickly. This process has been termed *secondary succession*. A common example of this is the return of an abandoned farm field to a climax forest. Since the fertile soil is already present, seeds and rhizomes in the soil grow quickly. Plants rapidly move in, sprouting from the present seeds or rhizomes, or carried to the site by wind, birds, and grazing animals. These early plants are also known as *pioneer plants* and may include wildflowers followed by tall grasses and compact woody bushes. These grasses will eventually crowd out the wildflowers, but new growth trees such as dogwood and sumac eventually replace themselves. After a time, perhaps a century or two, the climax forest trees will move into a typical temperate forest and include such trees as oak, maple, hickory, and elm. These trees will replace the new forest trees.[2,3]

Retrogression

Not all disrupted ecosystems undergo a progressive return to stability with climax vegetation. Human intervention, particularly, has threatened the stability of ecosystems throughout the world. Activities that cause changes in natural cycles or population, such as destroying species or adding new ones, can be a threat to environmental stability. Stable ecosystems are ones in which materials are constantly recycled within the system through growth, consumption and decomposition. These processes tend to balance each other so that there is little net loss over long periods of time in a process called *dynamic equilibrium*. Although minor disruptions occur to most ecosystems, and a true steady state is seldom achieved, there is a long-term balance within stable ecosystems whereby nutrients revolve within that system with little overall loss. However, when the climate changes, soils are poisoned or eroded, or species are eliminated or introduced, the internal compensating mechanisms of the ecosystem may not be sufficient to return the ecosystem to its previous stable state; and a new combination of plant and animal life may be established.

In September of 1996, two hurricanes released almost three feet of rain on eastern North Carolina. This rain turned the region into polluted floodwaters caused by decomposing chickens and hogs, rotting farm fields and ruined neighborhoods. This was one of the most disastrous floods in this region's history. Scientists now feel that the flood also started an ecological situation that could change the biological rich waters of the Outer Banks from the mainland of Carolina. These bays and inlets are the second-largest estuaries in the country. The main problem is that the flood picked up huge amounts of organic matter including decomposing vegetation, topsoil, fertilizers, raw sewage, and waste created from hogs and farms. This went directly into the estuary, which turned the water into a coffee color. This flood may have already killed many marine creatures such as fish, worms, crustaceans and crabs. This situation may take years to normalize.[11] As plots of land are overgrazed or intensively farmed, tropical forests burned and then farmed, water resources depleted, or other poor land management techniques employed, complex ecosystems are often replaced. This new ecosystem state may result in fewer overall species through retrogression. Further, the species remaining may be less desirable from a human point of view. Simple systems in which fewer species exist or diversity reduction takes place are at greater risk of failure. The lack of cross-links in an ecosystem food chain means fewer alternatives for feeding; and when a species is reduced in number or dies off, its predators also die off if other alternatives do not exist.

The Earth's climate has been changing rapidly and the temperature has been increasing. As the climate changes, due to human impact, there is an increased danger to human health through direct and indirect effects. Direct effects include heat stroke, hypothermia and deaths from hurricanes, floods and tidal waves. Indirect effects are the result of changing environments and ecosystems such as crop failure, algae blooms in the world's seas, and mental health disorders. Diseases such as malaria, dengue fever and cholera are moving into northern countries due to the warming of climates and ecosystems. These threats are primarily due to human actions.

The human population is exerting enormous pressure upon ecosystems throughout the world as it continues to multiply in logarithmic proportions and develop energy-intensive technologies resulting in the discharge of dramatic levels of toxic substances into the air, water, and land. Most biotic communities are proving unable to respond to the unrelenting pressures of disruption, causing major losses in species, soil degradation, desertification, contaminated water, possible climate changes, and other changes in global ecosystems that are not in the best interests for human survival or quality of life.

REFERENCES

1. Godish, T., *Air Quality,* 3rd ed., Lewis Publishers, Boca Raton, 1997, chap. 4.
2. Nadakavukaren, A., *Our Global Environment,* 4th ed., Waveland Press, Inc., Prospect Heights, IL, 1995, chap. 1.
3. Turk, J. and Turk, A., *Environmental Science,* Saunders College Publishing, Philadelphia, 1988, chap. 5.
4. Woodward, S., Introduction to biomes, *Geography,* 235, June 1998. <http://www.runet.edu/~swoodwar/classes/geog235/biomes>
5. Sadler, J.P. and Grattan, J.P., Volcanoes as agents of past environmental change, *Global and Planetary Change,* 21, 181, 1999.
6. Stevens, W., Lessons from ancient heat surge, *The New York Times,* Nov. 23, 1999.
7. Stevens, William, K., Carbon fuel use down globally, *Sunday Republican,* A-24, Oct. 31, 1999.
8. Nash, Madeline, Wait till next time, *Time Magazine,* 39, Sept. 27, 1999.
9. Schoen, D., Learning from polar ice core research, *Environmental Science and Technology News,* April 1, 1999.
10. Sargent, William, *The Year of the Crab, Marine Animals in Modern Medicine,* Norton & Company, NY, 1998.
11. Glanz, James, After the storm, an ecological bomb, *The New York Times,* November 30, 1999.
12. Kingsnorth, Paul, Human health on the line, *The Ecologists,* vol. 29, March/April, 1998.

CHAPTER 2

Human Populations

INTRODUCTION

This is a text on environmental health, so why is the subject of human population of any concern here? The answer is brief and startlingly cold. Human numbers have risen with such ferocity that we have overwhelmed the carrying capacity of sections of the Earth and threatened the sociopolitical structures of civilization in some parts of the world. We must come to understand "the environment" and realize that environmental destruction is the national security issue of the early 21st century. The world's billions of impoverished people are surging in an enormous tidal wave into the urban centers around the globe as they desperately seek relief and opportunity from the crumbling ecosystems of rural life. These billions are fleeing from deforestation, soil erosion, water depletion, natural disasters, hunger, emerging diseases, and hopelessness to find themselves entrenched in uncontrolled settlements of never-ending shanty towns built with corrugated metal shacks or cardboard. Armies of undernourished and pot-bellied children run through puddled streets with

OBJECTIVES FOR THIS CHAPTER

A student reading this chapter will be able to:

1. **Define the attributes of populations including birth and death rates, growth rate, density, and mobility (immigration and emigration)**
2. **Calculate rate of natural increase from birth and death rates, and mathematically demonstrate the effects of age–sex composition on a population**
3. **Define biotic potential and maximum growth rate, and list the various limits to growth**
4. **Identify, list, and explain the population growth forms**
5. **Recognize and explain the concept population explosion with respect to complete and incomplete demographic transition**
6. **Define population implosion and discuss the conditions that lead to this phenomenon**
7. **Explain the role of urbanization in influencing sustainability of populations**
8. **Explain global population projections and differentiate between developed and lesser developed countries with respect to these projections**
9. **List and discuss the various options for fertility control methods while contrasting the effectiveness, risks, and benefits of each type**

floating garbage where malarial-infested mosquitoes swarm in great black clouds. Electricity, sewage systems, and running water are rare. It is little wonder that in these environs criminal anarchy is the endemic illness. Overpopulation, infectious disease, unprovoked crime, few resources, and the influx of more refugees increases the erosion of nation–states, leading to the empowerment of private armies, security firms, and international drug cartels. This is the vision in many parts of the lesser developed countries (LDCs), and it threatens to expand with the continued growth of human populations.[1]

Since you are reading this text, you are probably sharing in the educational resources of the well-fed and technology-rich western society, where recent studies indicate that overpopulation may no longer be a problem. In much of Western Europe and Japan, declining populations are causing major concerns about the economy resulting from a diminishing work force, falling land and housing prices, and a dwindling economy. Where will the industrial workers come from? Who will care for the aged? In the U.S., demographic shifts are reshaping the racial landscape. There is an increasing divergence of generations, creating a gap that pits the "grays" against the "skateboarders." By 2025, the number of elderly white Americans will increase by 67% to 47 million, while less than half the children will be white. During the end of the same period, 38% of the working American population will be nonwhite (Asian, Black, and Hispanic). This mostly white elderly population will expect retirement and medical benefits to be paid by a racially diverse group who must yield a greater proportion of their limited income to pay for sharply increasing costs in caring for the elderly.

The world is becoming increasingly bifurcated with the "haves" and "have-nots," and there is growing realization that surging populations, environmental degradation and ethnic conflict are strongly intertwined.[1] Therefore, understanding the dynamics of human populations is a first order of business in beginning the study of environmental health.

THE CHARACTERISTICS OF POPULATIONS

It is usually easy to recognize a population. They have similar physical characteristics that identify them as a species and tend to inhabit specific areas. A species is normally considered to be a group of organisms that can breed together with the production of viable and fertile offspring. However, simply because members of species can breed together does not mean they do. Normally, organisms breed in much smaller collections of like organisms, which are called populations. A population is considered to be the breeding group for an organism. Different species not only have differing physical attributes, but they also differ in the population characteristics. Most species of rats give birth to several litters a year and have a relatively short life span of 1 to 2 years. Elephants have long gestation periods lasting nearly 2 years, may live more than 70 years, and rarely give birth to more than one at a time.

Each population has characteristics that help to identify it. Some of these characteristics are birth rate, death rate, age distribution, and sex ratio. *Birth rate* refers to the number of individuals added to a population through reproduction (live births) and is normally expressed as the number of live births per 1000 population (counting the population at the midpoint of the year). The example provided in Figure 2.1 shows a birth rate of 16/1000. *Death rate* is also similarly calculated using total deaths divided by the mid-year total population. The rate of natural increase is determined by subtracting the death rate from the birth rate. The *rate of natural increase* reflects the growth rate in which migration is not considered. The growth of a population in the absence of migration must depend on the birth rate being higher than the death rate. The calculation of the rate of natural increase depends on other factors beyond this simple calculation. A more accurate estimation of population growth may be made if the sex ratio and age distribution of the population

Birth rate equals the number of live children born in a year per 1000 total population

$$\text{Birth rate in year Y} = \frac{\text{Number of live children born in year Y}}{\text{Midyear population in year Y}}$$

$$\text{Birth rate in year 1998} = \frac{\text{4,345,600 (children born in 1998)}}{\text{271,600,000 (population in mid-1998)}} = 16/1000$$

$$\text{Death rate in year Y} = \frac{\text{Number of deaths in year Y}}{\text{Midyear population in year Y}}$$

$$\text{Death rate in year 1998} = \frac{\text{2,172,800 (deaths in 1998)}}{\text{271,600,000 (population in mid-1998)}} = 8/1000$$

Rate of natural increase in year 1998 = (Birth rate - Death rate) = 8/1000 or 0.8 %*

*These are approximate numbers for the U.S. used only for example.

Figure 2.1 Calculations of birth rate, death rate, and rate of natural increase.

is known. The age–sex composition of the population has a profound effect on the birth and death rates of a country because the probability of dying or giving birth within any given year depends upon the age and sex of the population members. Clearly, an aging population is going to have a higher death rate and lower birth rate than a population with many young women of reproductive age. The sex ratio refers to the number of males relative to the number of females in the population, while the age distribution describes the percent of the population in various age categories from birth to death. In most societies, it is useful to separate the age groups into prereproductive juveniles, reproductive adults, and postreproductive adults. Some approximate differences in age–sex composition for different countries are shown in Figure 2.2. These graphs are created by plotting the number of males and females for age groups in 5- or 10-year increments within a population. The lesser developed countries of the world such as Mexico and Africa, and much of Asia, more closely approach the broad-based pyramid-shaped graph in Figure 2.2. The distribution for these countries has a very broad base, indicating that birth rates have been very high in recent years. Further, there are many young children in the population who will likely reach reproductive age in the next few years and rapidly expand the population of these countries in the next generation. It has been estimated that the group of young women entering childbearing years will be the largest ever. Additionally, there will be over one billion young people between the ages of 15 and 24 and 3 billion under age 25.[2] Those countries with more stable populations such as the U.S. or France have approximately an equal number of people in the age groups from 0 to 35. As the younger groups mature, there will be approximately the same number entering the reproductive years so that the population will remain fairly constant. This is complicated in the U.S. by the more than one million people immigrating into the U.S. each year. It is the only industrialized country where large

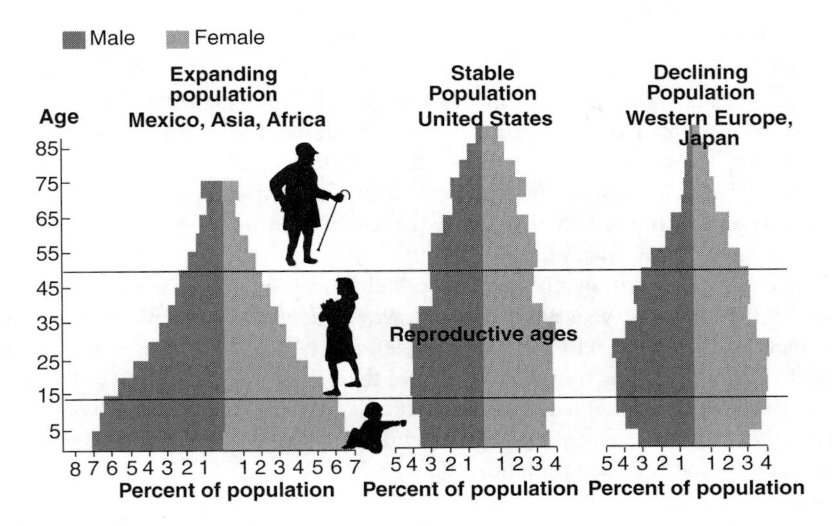

Figure 2.2 Age distribution types in human populations.

population increases are projected.[3] Many Western European countries and Japan are experiencing a decline in growth rates. This is characterized by an age distribution where there are fewer infants and children than young adults. In the years to follow, there will be fewer people in the reproductive age groups, and a population decline will occur in these countries if current trends continue.

Even with this information, predicting what will happen to future populations is a difficult task. Imagine that a large population of youngsters enters the pyramid at the base and travels upward in a group of diminishing size (because of deaths) into reproductive age and then even smaller numbers into postreproductive age. If this were a constant behavior of the population, predictions would be easy. However, wars, disease, famine, and reproductive intervention may cause the next generation to be reduced in number. This creates a constrictive formation or pinching at the base as seen in Western Europe (Figure 2.2). Since it is not easy to predict war, disease, or even human behavior with regard to reproduction, estimating future populations is a matter of some difficulty. Many countries are laboring with issues of unexpected population explosion or implosion right now.

Although it is apparent from this discussion that accurate population predictions may be difficult, there are other indicators useful in estimating future population changes. These include emigration, immigration, infant mortality, and total fertility rate.

In nature, when the density of organisms becomes too great, the intense competition for food, water and other resources damages the entire population. Some species have the ability to disperse or migrate out of the area and, in doing so, temporarily relieve the overcrowding. This process is called *emigration*. The pressure for resources may be the consequence of rapid population growth or the disappearance of essential resources brought on by natural phenomena such as drought. When species emigrate from an area, they must *immigrate* or enter into another area. Driven by natural disasters, war, disease, and disappearing resources, the numbers of refugees worldwide may exceed 15 million, with about 880,000 to 1.4 million immigrants entering the U.S. each year, including more than 200,000 who enter illegally. The number of immigrants is likely to remain high for many years, and this is contrasted with an American fertility rate that has been below replacement level for 20 years. Immigration has, therefore, had a disproportionate impact on population growth in the U.S. The consequences of immigration to the U.S. will be discussed later in this chapter.

As discussed earlier, death rates are calculated by dividing the total number of deaths by the total mid-year population. However, deaths of persons in the pre- and reproductive age groups will have a greater effect on future population growth than deaths in the postreproductive ages. Deaths of the young will mean a loss of potential births. Because of this impact, infant mortality is often a measured population characteristic that is used in conjunction with death rates. Infant mortality can be determined by the following relationship:

$$\text{Infant mortality (\%)} = (\text{No. infant deaths/No. live births}) \times 100\%$$

The birth rate is useful for determining how many births are occurring in the total population but does not tell what the current fertility patterns are. A useful predictor of reproductive behavior in a population is total fertility rate. Total fertility rates (TFR) represent the number of children a woman in a given population is likely to bear during her reproductive lifetime providing that birth rates remain constant for at least a generation. Recent figures show that the world TFR is 2.79, with the more developed nations having a TFR of 1.59 and the less developed countries (excluding China) having a TFR of 3.08.[4,5] The TFR is useful in determining the replacement level, which is defined as a TFR that corresponds to a population exactly replacing itself. The replacement TFR level for most countries is accepted as 2.1 — although some LDCs have such high infant mortality due to poor nutrition, disease, and medical services that the TFR may be as high as 2.7. Almost all European countries are experiencing below-replacement fertility rates, with levels as low as 0.9 in the Czech Republic. While many of the LDCs have reduced TFRs, most are still well above the replacement level of 2.1 or even 2.7.[6]

If it is known how many females of reproductive age are in a population based on the age–sex distribution and the average number of births per female from the TFR, then it is easy to predict the next generation population, providing the birth rates do not change. Of course, birth rates often do change as a consequence of changing social behavior, wars, famine, disease, or migration. The inability to predict these phenomena or their effects on the TFR contributes to inaccurate demographic predictions.

POPULATION DYNAMICS

There are periodic upsurges in many populations that lead to overwhelming numbers. Whether these population explosions occur in rabbits, lemmings, soldier ants, or locusts, there are always some natural pressures that bring the population back into balance with their natural surroundings. Bacteria reproducing by binary fission can grow to millions of organisms within a few hours given the right conditions, while many fungi grow cubically and could easily grow to fill an entire room overnight if pressures didn't exist to limit their growth. Fish and amphibians produce thousands of eggs, plants release millions of seeds, and many animal species produce multiple litters of offspring in a year. Every species has an inherent reproductive capacity to produce offspring that is generally many times greater than needed to replace the parents that die.

Clearly, if these organisms and offspring all survived, we would soon be overwhelmed by the sheer numbers of these species. In fact, even the human population growing at a rate of 1.6% for the next few hundred years would completely cover every square inch of the Earth and its oceans. The unrestricted growth of populations resulting in the maximum growth rate for a particular population is called its *biotic potential*. The biotic potential of species differs markedly and is influenced by (1) the frequency of reproduction; (2) the total number of times the organism reproduces; (3) the number of offspring from each reproductive cycle; and (4) the age at which reproduction starts. Some bacteria reproduce every 20 minutes by splitting in two (binary fission), while polar bears give birth only once in every 3 of 4 years, which gives them a much lower biotic potential.

This biotic potential is never actually reached for any species because the growth is limited by one or more factors that are collectively referred to as *environmental resistance*. Environmental resistance refers to those pressures that limit population and may include such factors as disease, wars, predatory behavior, toxic waste accumulation, or species competition. The actual rate of increase is then described by the biotic potential minus the environmental resistance as shown in Figure 2.3.[7] The sum of these limiting factors serves to control the numbers of a species that can survive in a defined area over time, and this represents the carrying capacity of that area for a particular species. The carrying capacity of a defined system can be demonstrated in the laboratory by cultivating a bacterial organism in a nutrient broth at 37°C and removing organisms hourly in order to count them. This is accomplished

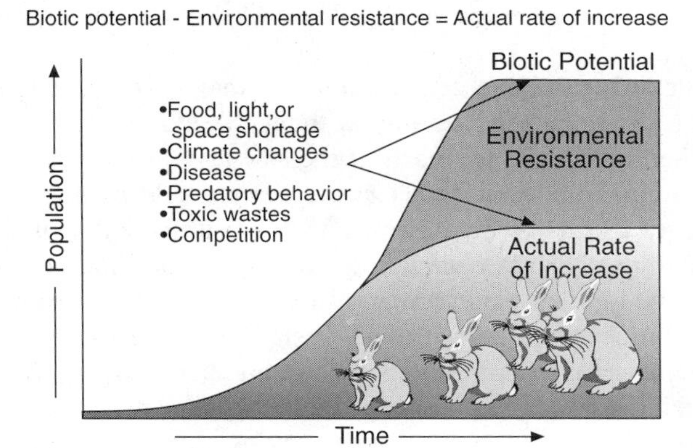

Figure 2.3 Environmental resistance represents the difference between biotic potential and the actual rate of increase.

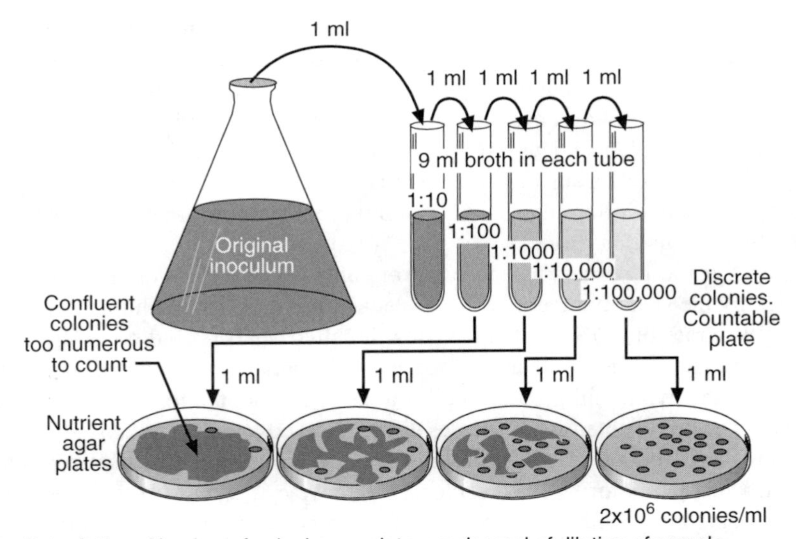

Figure 2.4 The plate counting technique is one counting process that may be used to count bacteria for generating a bacterial growth curve.

by making serial dilutions of the original broth culture and plating 1.0 ml of each dilution onto separate nutrient agar plates. Incubation of these plates results in the growth of discrete colonies that can be counted to determine the population at hourly intervals (Figure 2.4). By plotting these numbers, a typical sigmoid or "S" growth curve can be demonstrated (Figure 2.5). The growth curve initially shows a lag phase in which the organisms show no increase in growth rate but are undergoing changes that ready them for rapid proliferation. This is followed by a rapid increase in growth where the population doubles every few hours. The population increases at a rate proportional to its size, or exponentially. This is known as the *exponential growth phase*.

To demonstrate logarithmic growth, consider that a single bacteria splits every 30 minutes. If a single bacteria starting at time "0" divides in 30 minutes to form two bacteria, then in 30 minutes, each of the two divides to form four bacteria. If this were to continue unabated for 24 hours, there would be 48 generations for a population of over 16 million organisms. Can you determine the number that would be produced if allowed to grow uninterrupted another four hours?

Eventually, the limiting environment of the culture tube restricts the growth of the bacteria by an accumulation of toxic wastes and lowered pH. The number of organisms dying eventually comes to equal the number of living organisms so that a stationary level is reached. This also represents the carrying capacity of this culture tube environment. If left to grow without removing wastes or adding nutrients, the bacterial population will die off rapidly in what is known as the *death phase*. This "S" type growth curve was described in 1932 by the Russian ecologist G.F. Gausse

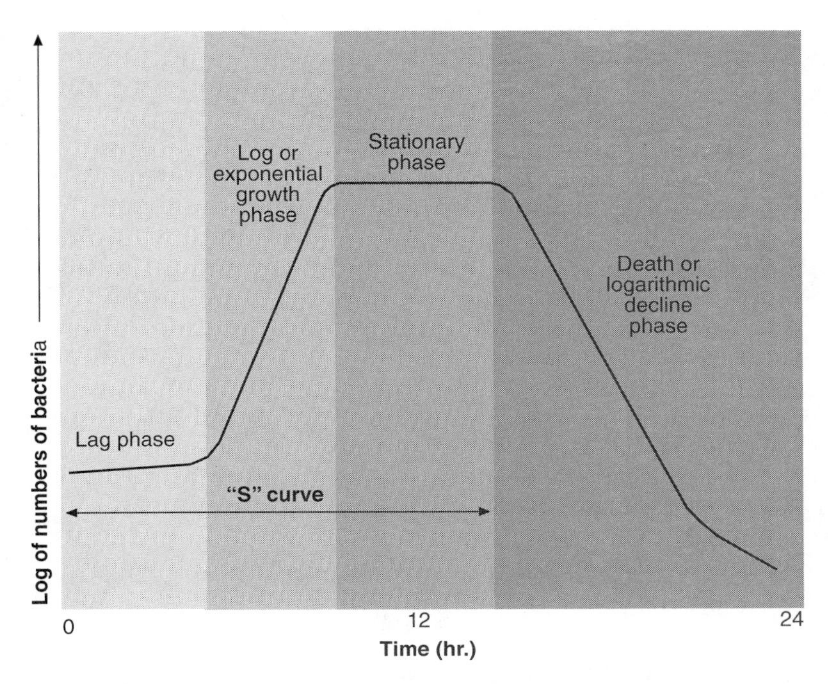

Figure 2.5 Bacterial growth curve showing various phases of growth including lag, exponential, stationary, and death phases.

when he studied the growth of experimental populations of Paramecium caudatum, a freshwater protist.

If an organism grows too rapidly and the population escalates beyond the carrying capacity of the environment in which it is located, a "J" type growth curve may develop (Figure 2.6). This curve is somewhat misleading, because it is truncated before the outcome of the event is revealed and may more accurately be referred to as geometric growth. There are many examples in nature where insects, mammals or other life forms have an explosive growth in population that far exceeds the carrying capacity of their habitat. The consequence of this is a precipitous decline in the population once food supplies are gone, disease intervenes, or the habitat is changed in some other way to create hostile conditions. This behavior sometimes oscillates every few years as in the case of lemmings that inhabit the arctic tundra north of the Canadian forest. Every 3 to 4 years the population explodes, then crashes the following year, followed by a 2-year cycle of slow recovery (Figure 2.7).[8]

There are other patterns of growth as well. Predator-prey interactions such as lynx and rabbit are likely to produce an oscillating pattern for both species, with the predator showing increases and declines following such changes in the prey. As the rabbit population increases, the lynx will have more prey to feed upon and its population will grow. Eventually, there will be sufficient predators to cause a reduction in the rabbit population, and its decline will cause a decline in the predator lynx population soon after as seen in Figure 2.8A. Another type of growth curve is seen in Figure 2.8B in which a population becomes extinct. This type of curve can be drawn for more and

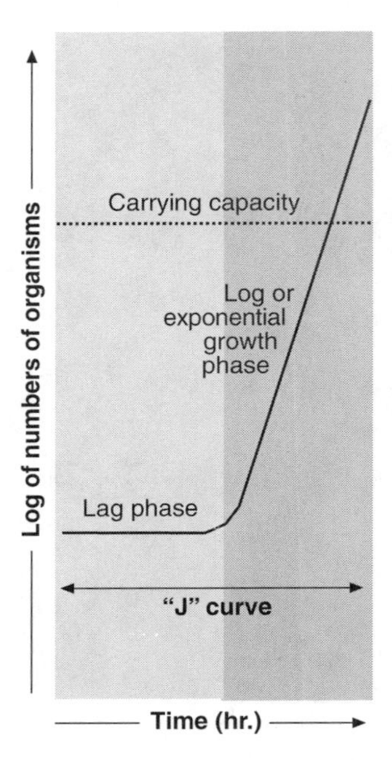

Figure 2.6 "J" type growth curve showing growth extending beyond carrying capacity of environment.

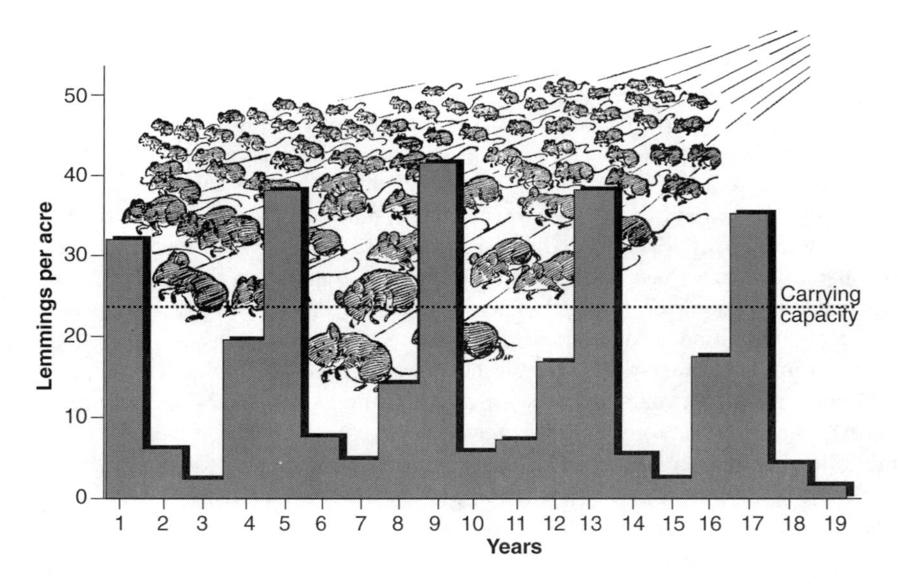

Figure 2.7 Typical population crashes seen in lemming populations every 3 to 4 years.

more species every day as human intervention eliminates species, habitats, spreads pesticides, and hunts down larger species for human uses. The disappearance of wildlife and species extinction will be discussed in more detail in Chapter 3.

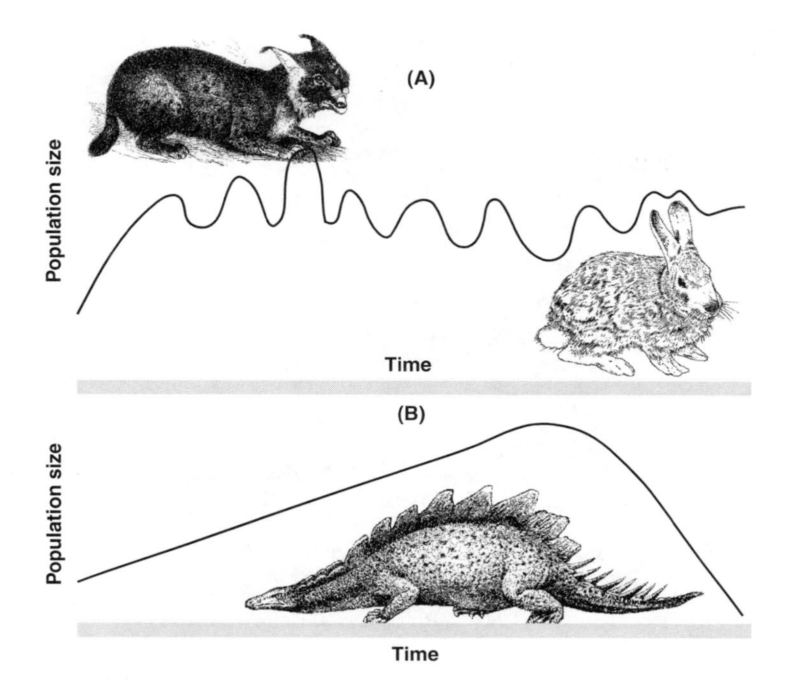

Figure 2.8 (A) Predator-prey and (B) extinction type growth curves.

The science of demography has categorized two different reproductive strategies for living organisms. One strategy is known as a *k-strategy* in which large organisms with relatively long life spans have only a few offspring, but they devote their energies to protecting and nurturing the offspring to enhance their individual survival until they can reproduce. Examples falling into this group are elephants, deer, large cats, swans, and humans. These populations normally stabilize at the carrying capacity of their environments where density-dependent factors limit growth. Density-dependent factors include such items as food supply, which becomes more limiting as the size of the population grows. The pattern for mortality of these species can be demonstrated by plotting a survivorship curve in which the fate of young individuals is followed to the point of death in order to describe mortality at different ages.[9] This curve can be used to predict the probability that a newborn will reach a specific age that is useful in determining the life expectancy within a population. The *type I* curve typical for the k-strategy survival population shows that the majority of infants reach sexual maturity, and many survive to maturity, eventually dying of diseases from old age (Figure 2.9). This is in sharp contrast to *r-strategy* populations that are typically small, short-lived organisms, which produce large numbers of offspring and receive little or no parental care. This is characteristic of species of invertebrates, fish, mollusks, plants, fungi, gypsy moths, locusts, grasshoppers, and some rodents. Large numbers are produced, but few offspring survive. There is high mortality among the offspring as depicted by the Type III survivor curve (Figure 2.9). These organisms

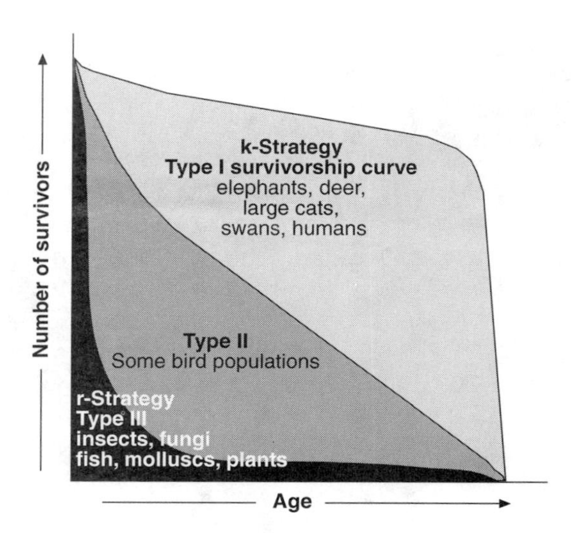

Figure 2.9 Characteristic survivor curves showing Type I (prosperous human population), Type II (impoverished human population and certain bird populations), and Type III (fish, plants, fungi, etc.). (Adapted form Turk, J., and Turk, A., *Environmental Science,* Saunders College Publishing, Philadelphia, 1988.)

are limited by density-independent factors such as a drought that dries up a pond, or sudden climactic changes such as El Niño, which alters the temperature of the water and making it uninhabitable for certain species. Such populations often demonstrate wildly fluctuating numbers with population explosions and crashes.

A *Type II* survival curve that lies between I and III is characterized by high mortality in the infant or beginning stages of life from disease, genetic defects, or predation. When this initial die-off is passed, the survivors die at a smaller but constant rate over the lifespan of the population.[9] This is characteristic for some bird species and for human populations under the constant stress of poor sanitation, infectious diseases, and inadequate food supplies. In such populations, there is a tendency to produce more offspring because of the higher mortality in infants or offspring.

POPULATION TRENDS IN THE WORLD

Demographers use the information on population size, fertility rates, migration, birth and death rates, growth rates, infant mortality, density, age–sex composition and other factors to statistically characterize human populations. Their purpose is to predict what will happen to that population over time. You may question why this is so important. What is the value of this information? The statistical knowledge of human populations certainly dates back thousands of years to the beginnings of civilization, when support systems for maintaining large groups of interdependent people became necessary. Government structures developed and there was a need to raise taxes, maintain armies, harvest and store foodstuffs, provide education or training, build roads, and promote commerce. These activities provide the infrastructure on which society depends to this day. Even in the smallest towns, the local officials must have census data to know how to plan for possible expansions in schools, sewer and septic systems, road construction, power lines, building lots, and land conservation. Humans require significant infrastructure to maintain a healthy, educated, and physically comfortable lifestyle. This takes significant planning and knowledge about how many people for which to account. There is little that goes on in society today that does not depend on a knowledge of human numbers. Marketing products, predicting enrollments, estimating tax incomes, analyzing insurance risks, and determining future costs for elderly care are just a bare mention of the uses for population numbers by age, race, sex, interests, disease, and so forth. Understanding the growth patterns of human populations is vital to a nation's interest and perhaps to the survival of the human race.

Historical Trends

Core samples taken from arctic ice and deep-sea beds reveal a chaotic history of extremes in the Earth's surface temperature over several hundred thousand years. Then, as if almost a fluke, the Earth's temperature warmed and stabilized about 10,000 years ago. Humans began to develop agriculture and domesticate animals. The ability to feed larger groups of people from planted crops and domesticated herds allowed civilizations to develop. Nomadic existence was no longer necessary, and soon the world's population began to increase as population groups could sustain more children. Although accurate predictions of human populations are questionable before circa 1700, estimates of primitive civilizations around 8000 BCE (before Christian era) place the number of humans at about 5 million (Figure 2.10). The advent of agriculture is thought to have promoted the growth of human populations to about 500 million by 1650. The world's population then doubled in the next 200 years to about one billion people. During this time the industrial revolution began,

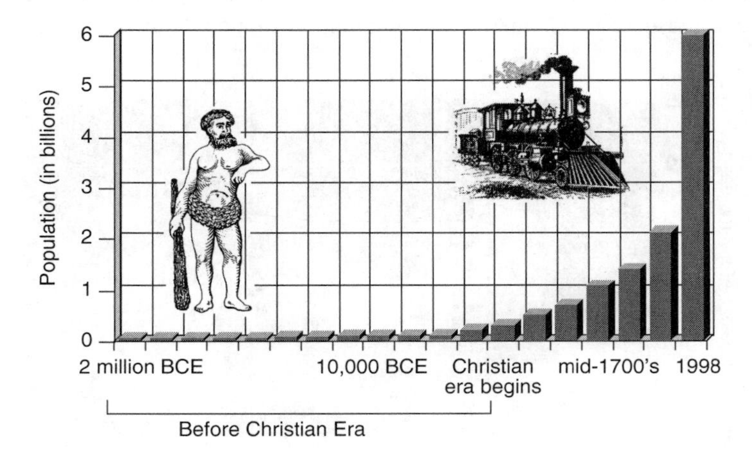

Figure 2.10 History of human population from early man, *Homo habilis,* to the present showing logarithmic growth beginning around the Industrial Revolution. (Adapted from Turk, J., and Turk, A., *Environmental Science,* Saunders College Publishing, Philadelphia, 1988.)

introducing mechanization into agriculture with increased production and delivery of foodstuffs, the production of fabrics and hardware, and the construction of domiciles. The ability to house and feed more people caused an eruption in the birth rates and numbers of people. By 1930 there were nearly 2 billion on the planet, taking less than 80 years to double the world's population. In the late 1800s the beginning concepts of sanitation emerged with the chlorination of public water supplies. The introduction of public sewers, pasteurization of milk, and food protection programs soon followed. Certainly by the early 1900s, most major cities in the U.S. and Europe understood these concepts and employed the principles of sanitation. Consequently, the spread of communicable diseases through milk, water, and food was dramatically reduced. Death rates plummeted and growth rates exploded into the 20th century. New discoveries in antibiotic therapy, vaccines, and medical treatment saved many more lives and increased life spans. The consequence was alarming in its effect on the growth rate. In 1975, the world's population had increased to 4 billion people, taking less than 45 years for the population to double from 1930. In the fastest population spurt in history — fewer than 40 years for the population to double — the world reached 6 billion people on October 12, 1999.

The rate of growth of the world population from the mid-1700s until the present is shown in Figure 2.11. The growth rate peaked in 1970 at 2.06% and has declined since then to a level of 1.4%.[5] This decline in growth rate is directly related to declines in births prompted by international efforts beginning during the 1960s to stem the tide of the growing global population. This concern escalated from private groups and then spread to national governments and international agencies. The majority of the population increases were seen in the LDCs, and it became clear that the population increases were so rapid and so large that they were passing the carrying capacity of these countries and threatening their stability. The majority of these nations are racing to reverse this trend and stabilize the populations before major calamities occur. However, the efforts to achieve population stability should have occurred decades sooner. The combination of population momentum and a

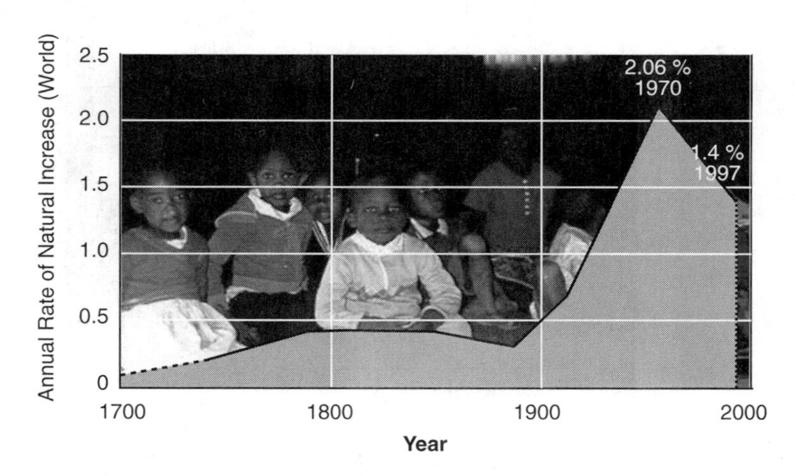

Figure 2.11 The world population growth rate from 1700 to 1997.

large population base in pre- and reproductive ages makes this effort very difficult. In order to understand how we reached this point, it is helpful to look at the relationships of growth rate, doubling time, demographic transition, and incomplete demographic transition.

Growth Rate

A rate is normally expressed as one quantity in relation to another. Runners like to keep track of the time it takes to run a 10-kilometer course in minutes per mile. This is a rate measurement, and runners can study their progress by decreasing the minutes per mile that it takes to run the course. In a similar way, growth rate can be used to assess additions or deletions to the population. People enter a population by births or immigration and leave a population by deaths or emigration. The rate of births is the ratio of births to the population, and death rates represent the ratio of deaths to the population. Growth rate is then determined by the birth rate minus the death rate. If the growth rate is positive, there will be an increase in population. Even if the rate of growth is slowed (declines), the population will accumulate. Similarly, even if the runner's pace is slowed, he or she will still accumulate miles. Consequently, we are still contributing large numbers of people to the population despite a declining growth rate. This addition becomes exaggerated when the base of people is larger. In 1975, the growth rate was close to 1.75% with a global population of 4 billion people, leading to 4 billion × 0.0175 = 70 million new infants each year. On October 12, 1999, the population reached 6 billion people with a growth rate of 1.4%, leading to 6 billion × 0.014 = 84 million new infants per year. Despite the successes in reducing the worldwide growth rates, the population has grown so much that even the smaller growth rates lead to additions of larger numbers of people to the global population (Figure 2.12). This is referred to as *population momentum*. For example, in Botswana, a country experiencing a drop in life expectancy, the population is expected to double by 2050 due to population momentum.[7]

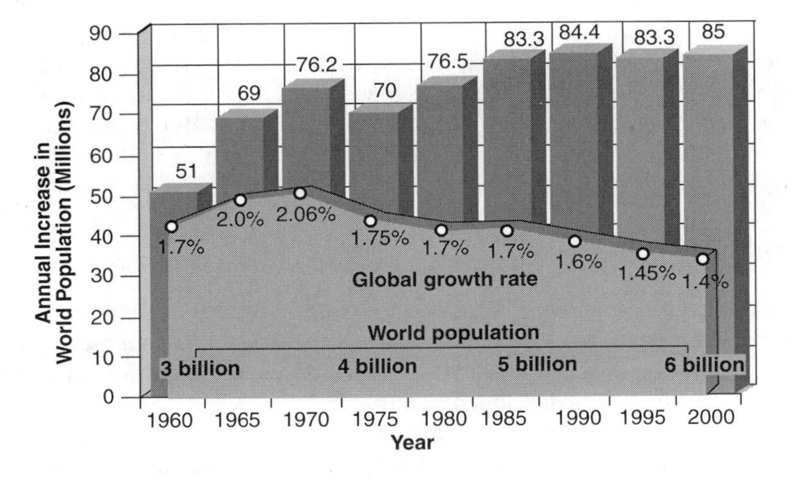

Figure 2.12 Annual increase in world population versus global growth rate.

Added to the problem of population momentum are the differences in age–sex distribution in LDCs vs. the developed countries. Figure 2.2 compares the age–sex distribution for the lesser and more developed countries. The very broad base for LDCs indicates that birth rates have been high and that young children represent the largest percent of the population. As these children reach reproductive age and bear their own children, the population will rapidly expand. Huge populations that have a large percent of young pre-reproductive people have a population momentum that will foster significant population increases through the next generation.

Doubling Time

Another useful way to demonstrate growth rate is to present it as *doubling time*, or the number of years for a human population to double its size. The doubling time can be calculated according to the following relationship:

$$\text{doubling time} = 0.70/\text{growth rate}$$

This relationship is derived from the formula for calculating compound interest at the bank and is an approximation. It is useful to remember that a growth rate in any population compounds itself since the individuals entering the population will ultimately reproduce and add more people to that population. As the growth rate increases, the doubling time decreases. Populations from 8000 BCE to 1650 doubled about every 1500 years as evidenced by the relatively slow rate of population increase (Figure 2.9). The population next doubled in 200 years, then 80 years, then 45 years, and 36 years when the world reached 4 billion people in 1975. Viewing population growth in this way is useful because almost everybody can understand what life would be like if the number of people in our particular location were doubled within a generation. A graphical presentation of doubling time vs. the rate of natural increase is shown in Figure 2.13. A rate of 4% annual increase will double the population of a country in 18 years. The World Bank estimated Kenya to have such a rate in 1979, dropping to 3.7% in 1988 with a population of 23.3 million.[10] The population in Kenya in 1998 is 28.8 million with an annual rate of increase of 3.3%, and its doubling rate has been increased to 21 years.[11] Although this may be viewed as improvement, such an increase in so short a time will likely surpass the carrying capacity of that country. It will be incredibly challenged to build roads, houses, schools, and sanitation facilities to accommodate this many people in so short a time.

Demographic Transition

Understanding population trends in the world requires an explanation of the differences in growth rates occurring between the developed countries of Western Europe, North America, and Japan vs. the lesser developed countries in Latin America, Asia, and Africa. The developed countries have exhibited slowly declining death rates over the last century with the gradual introduction of sanitary principles (chlorinated water supplies, pasteurized milk), increased agricultural production, and

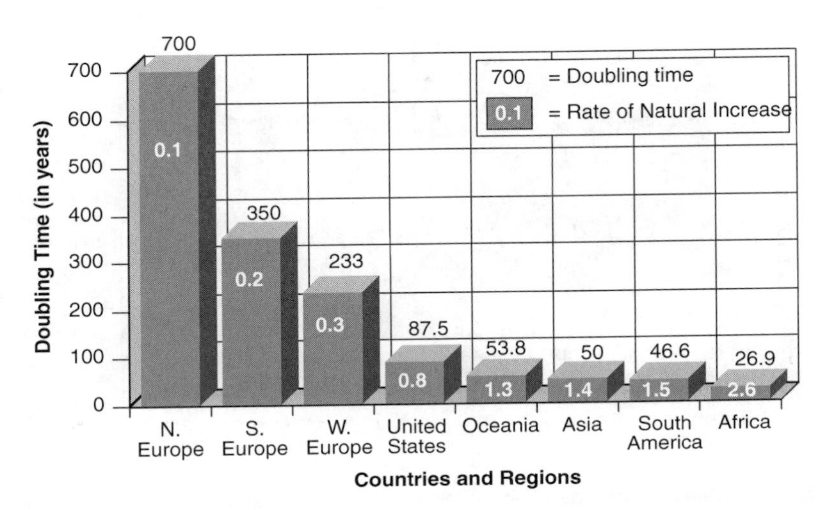

Figure 2.13 Population doubling times and rates of natural increase comparing certain developed and lesser developed regions and countries.

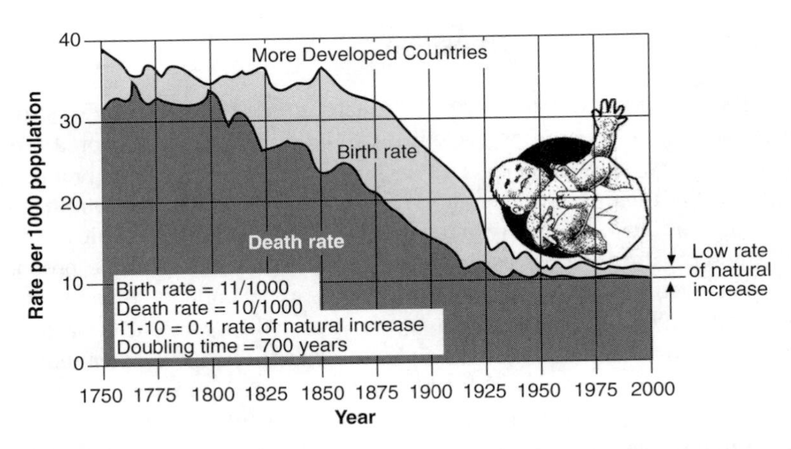

Figure 2.14 Demographic transition for developed countries showing declining death rates and birth rates resulting in a low rate of natural increase. (Adapted from United Nations Population Fund, *The State of the World Population 1999: 6 Billion, A Time for Choices,* 1999.)

improved medical care (vaccines, antibiotics). During this time, the industrialization of society and mechanization of agriculture reduced the economic benefit of more children, while those children who were born had a better chance at survival. Therefore, as the death rate declined, so did the birth rate. This has resulted in a diminishing difference between birth rates and death rates and a very low rate of natural increase, resulting in a stable population with very long doubling times (Figure 2.14).

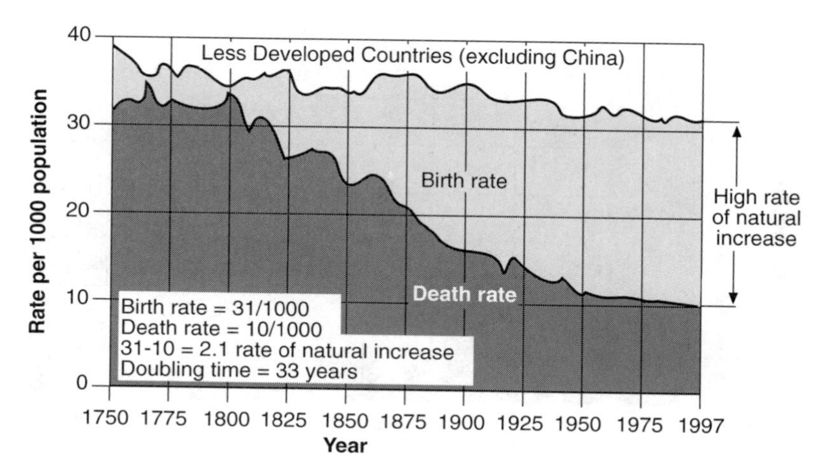

Figure 2.15 Demographic transition for lesser developed countries showing declining death rates with birth rates that remain high resulting in a high rate of natural increase. (Adapted from United Nations Population Fund, *The State of the World Population 1999: 6 Billion, A Time for Choices,* 1999.)

Incomplete Demographic Transition

The lesser developed countries had a much different experience. Most of these countries are and were agrarian societies without government-sponsored programs for social security. These are subsistence economies where rural inhabitants extract a living directly from plants and animals requiring human labor to accomplish simple tasks without the aid of electricity or tap water. Children in the household are required to cultivate crops, care for livestock, cook food, and produce simple products for market. Such activities may require up to 6 hours a day collecting wood, water and fodder. Children in India as young as 6 are expected to perform these duties. As resources diminish, there is a need for even more hands to collect water and firewood that are further from the home.[12] Such children, especially males, are also viewed as a source of security in old age because the governments in LDCs do not have the resources to institute social security, or they have unstable policies that fail to capture the trust of its citizens.[13] The populations in such countries had remained stable for thousands of years with high birth rates and high death rates. It was not uncommon for half of the children in a family to die, thereby maintaining a relatively low growth rate.[12] Then suddenly the developed nations introduced into these developing countries better nutrition, modern medicine, insecticides, improved water quality, and methods to isolate human wastes. The result has been a startling decrease in deaths — but without time for adjustment in birth rates. The number of births has remained high for many of the reasons discussed above. The combination of high birth rates and lowered death rates has resulted in very high rates of natural increase (Figure 2.15).

WEB PICK Go to http://www-unix.oit.umass.edu/~envhl565/index.html. Click on "Chapter Web Links" to find PopNet under Chapter 2.	
Chapter 2	Human Population
Whose Site?	Population Reference Bureau. This is a site maintained by the Population Reference Bureau with funding assistance from the U.S. Agency for International Development (USAID).
URL	http://www.popnet.org
What's Here?	PopNet is a resource for worldwide population information. Search this site by topic, keyword, organization, or world region map for population-related Web sites. You can find information here on demographic statistics, economics, education, environment, gender, policy, and reproductive health. The resources include government and international organizations, university centers, associations, and organizations.

Current Population Trends

The world's population is growing at a rate of 1.8% annually and reached 6 billion people in the middle of 1999.[3] This growth rate is a remarkable decline from the peak 2.06% growth rate reported in 1970, and we entered the year 2000 with a declining rate. According to the United Nations, rates are expected to continue to fall gradually for the next 20 years, and more rapidly thereafter.[3] However, we may have already set in motion the potential for destructive impacts to global ecosystems and some sociopolitical structures as the population momentum sweeps around the world. These problems are not equally balanced, nor are the population increases equally distributed. It is expected that the human population will reach 10.4 billion by the year 2100.[14]

The U.S. media has announced in bold headlines that the "population explosion is over," and that the real crisis ahead is a "population implosion" as many countries now face fertility rates that are below the replacement level of 2.01.[15] This may be true in the developed nations, where the United Nations predicts that during the next century industrialized nations will have fewer people than they have today.[2] From a different view, over 95% of the annual increase in the world's population is occurring in the LDCs, and more than 74 of the LDCs are expected to double their population in less than 30 years, adding more than 80 million people to Earth annually. A clear division has developed between the developed and less developed countries that will likely have enormous geopolitical consequences. We will examine the current population trends based on these differences.

Population Decreases in the Developed Countries

Declines in Fertility

In 1970 there were 19 countries reporting declining fertility rates, while in 1997 over 61 countries reported below-replacement fertility rates.[3] Nearly all of the European countries have below-replacement fertility rates, as do the Asian countries of China (1.8), Japan (1.4), Thailand (1.6), South Korea (1.7), and Hong Kong (1.3).[14] In North America, the countries of Canada (1.61) and America (1.96) are also at below-replacement levels. Those countries with the lowest fertility rates include the Czech Republic (0.9), Spain (1.15), Latvia (1.16), and Bulgaria (1.24). The European birth rates of the 1980s were already remarkably low, declined precipitously by another 20% to 1.4 in the early 1990s, and plummeted to 0.0 on the eve of 2000. These rates cannot sustain a population. Even if the European rate were to return to the replacement rate of 2.1 by the year 2060, Europe will have lost almost 25% of its population. Nearly 50 years ago, about 32% of the world's population lived in modern nations of Europe, North America, and Japan. Today, 20% of the world's population lives in these countries; and by 2020 it is projected to be 16%. In 2050 that figure will have dropped to 12% of the world's population.[15] The population in these regions fell so rapidly and so far that it was unprecedented in world history. The potential for major economic, geopolitical, and environmental changes remains to be discovered but is likely to be dramatic. These low fertility countries will have to look to active, older people and immigrants to supply needed services and contribute to the economy.[3] There was such concern that, for the first time in history, the U.N. convened experts and leaders from around the world in December 1997 to consider aspects of the low and rapidly declining fertility rates.[15]

The global fertility rate in the years 1950 to 1955 was 5.0 and on course to explode the world's population beyond its carrying capacity. The rate remained steady for another 15 years until the period 1975 to 1980, when the global TFR had fallen to four children. In 1995 the rate had fallen to slightly less than three, and fell below 2.79 in 1997. Demographers had earlier made projections of human populations reaching nearly 10 billion people by the year 2050. Now their projections show a population that is 650 million short of the original figure because there are people that will never be born. Now the United Nations predicts that the population will not reach 10 billion until the year 2100, following a 21st century medium fertility scenario.[16]

Concerns About Decline

The industrialized countries are deeply concerned about the static or declining work force. By the year 2025 Japan will lose 12.2 million people in the age group

15 to 64, while Germany will lose 4.5 million and Italy will see a loss of 4 million. These developed countries are experiencing a rapidly aging population and a significant decline in the ratio of working-age people. In Western Europe there is an average of less than 4.3 workers in the age of 15 to 64 years for each elderly person. The ratio of workers to elderly is projected to decline further by the year 2025 with Britain at 2.9 to 1, followed by France (2.7), Germany (2.5), and Japan (2.3), where almost a third of the population will be elderly.[15] If the present trend of a tumbling population in Japan continues, its population in 2100 will be 55 million people, down nearly 50% from the 125 million people today.[17] In developed countries, populations are projected to slowly rise until 2025, and then decline so that the total number in 2050 will be about the same as it is today.[14] There is a concern throughout Europe and Japan that the declining population will result in decreasing housing and land prices as the demand declines along with the population. A smaller work force and declining economy are also viewed as creating a less important country. Further, a high proportion of elderly non-wage earners in countries where governments support the elderly place an ever-increasing burden on the diminishing ratio of wage earners. Though these concerns have not been validated, there is a growing effort among some governments to reverse this trend.

Vastogirardi, a small village in southern Italy, has experienced a decline in fertility rate for many reasons, including many women choosing to have a career and men enjoying their bachelorhood.[18] The people of Vastogirardi fear that they will lose their culture, traditions, and schools. Although birth rates are low all over Europe, Italy and Spain have the lowest of all: an average of 1.2 bpw, slightly lower than Greece and considerably lower than the U.S.' rate of 2.1 bpw.[18] One of the main reasons Italy's birth rate is so low is because state aid to families with children is lower in Italy than in other countries such as Germany or France. Since Italian women who work are often still responsible for domestic chores, it is not uncommon to avoid having a second child in order to fulfill their obligations to work and the home. This village has also witnessed a variety of business closures due to people moving to larger towns. There was only one child born in Vastogirardi last year.[18] The major has even gone as far as to promote women to become single mothers, but few women have.

Officials in Kyokushi, a small town in the southern island of Kyushu, Japan, are offering a gift of $5000 to parents who have a fourth or subsequent child. Other Japanese towns are offering cash rewards and are considering reduced land prices and cash incentives for newlyweds.[17] It seems that very few couples are impressed, feeling that the amounts offered are too small. Many feel that the children are costly to raise and interfere with their free time. Others feel pressured to raise perfect children and do not want the burden of having to be perfect parents. Many children who feel this burden of being raised with such stringent expectations often rebel. The declining fertility rate may also be attributed to the growing number of women who marry later in life or not at all. The total fertility rate dropped from 1.57 in 1989 to 1.39 in 1997.[19]

In Japan the average woman marries at 27, and nearly 17% of women in their early 30s are unmarried.[20] Makoto Atoh, director of the General Institute of Population Problems in Tokyo, points to the advancement of women as the major factor in

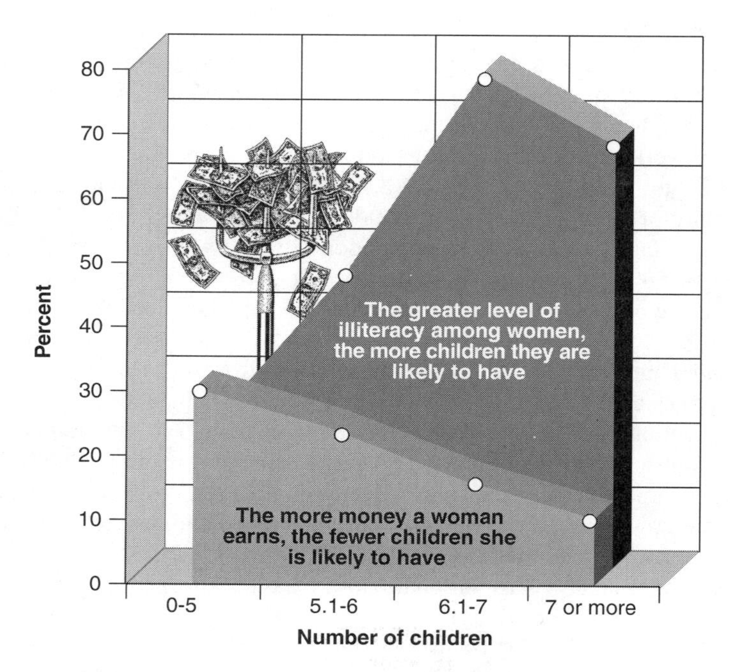

Figure 2.16 The number of children a woman is likely to have based on her level of literacy, or the earned income she contributes to the household. (Adapted from Kristof, N.D., *New York Times International,* Oct. 6, 1, 1996.)

fertility decline. In Japan, the failure of husbands to participate in child care and housework is also considered a significant factor in influencing women to have fewer children.[17] Higher education for women, with new aspirations and higher incomes, is considered to be a factor for declining fertility rates in many countries.[15] In fact, as illiteracy among women decreases in a country, the average number of children born to those women declines. Conversely, the more income a woman brings into the family, the fewer children she is likely to bear (Figure 2.16).[12] Also, pay is often reduced during a woman's maternity leave, making it even less likely that she will have many children, if any.[19] Another factor that may decrease fertility rates is increased urbanization. As more people move to cities, children become more expensive to maintain and are unnecessary for farm work. Improved methods of contraception, later marriages, more divorces, and substantially reduced infant mortality (giving parents confidence their children will survive) may all play a role in decreasing fertility rates.[15]

Is the concern about declining birth rates realistic? A large population is not a necessary condition for global power and influence as illustrated by the most rapidly growing nations in Africa, Asia, and Latin America. These nations are growing the most rapidly but are also the most poverty stricken. The population increases are expected for three reasons: declines in mortality, population momentum, and a fertility rate that is still above the replacement level (despite a dropped rate from 6 bpw to 3 bpw).[14] Certainly if a population were the answer to global power, India and China would be the world's two superpowers, with Israel and Great Britain essentially without power or influence.[11] Dr. Nafis Sadik illustrates that Brazil and Japan had similar growth in economy between 1960 and 1984. Brazil's per capita GNP (gross national product) doubled in that time, while Japan escalated by 7.4 times. The high fertility rate consumed much of the expanding economy to provide for the larger population, while Japan's economic and political clout rose in the pressure of a declining fertility rate.[20] The concerns about declining populations may not be warranted, but the concerns are apparent. As countries act on those concerns, the future may see upward shifts in fertility for some of those countries as governments look for successful ways to influence fertility behavior.

Increasing Death Rate

Some countries in Eastern Europe and the former Soviet Union have experienced mortality increases. The trends influencing the rise in adult mortality have been cited as increased poverty, inadequate nutrition, increased stress and overcrowding, and a greater incidence of infectious diseases.[3] These factors have been most likely brought about due to the current political changes and instability in the region; if continued, they could be a source for concern.

Fertility Rates in the U.S.

The U.S. has had a demographic history of rapidly changing patterns of fertility with stable mortality (Figure 2.17). Subtle changes in social attitudes appear to produce rather significant changes in fertility rates and subsequently cause marked changes in the rate of natural increase. It is not surprising that the Great Depression resulted in a major decline in fertility, reaching a low of 2.2 in 1936 and then rapidly

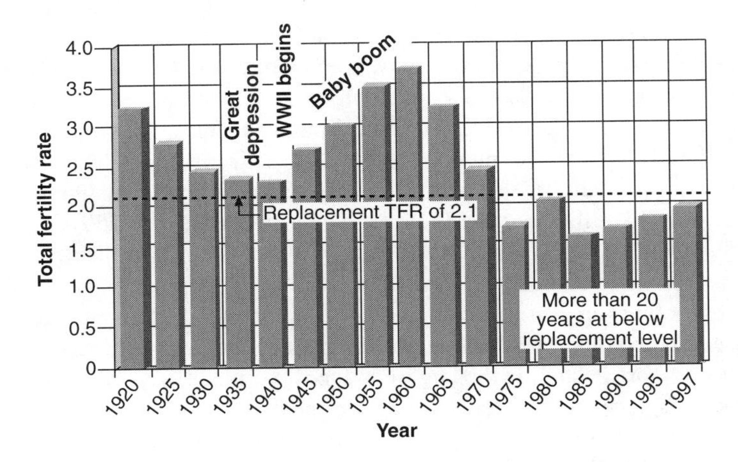

Figure 2.17 Total fertility rate (TFR) for U.S. from 1920 to 1997. (U.S. Bureau of the Census.)

increasing during and after World War II. The post-war demographers however, predicted a fertility decrease, but were surprised by the continued and rapid increase in population peaking in the early 1960s. The population then quickly declined to a below-replacement level in 1971. The demographers had predicted that the birth rates would remain high after the 1960s but once again were unable to predict the decrease. There are explanations given retrospectively that attribute the decline to a more educated female population seeking greater career opportunities and challenges. The ideal American family size grew smaller as couples realized that large families were an economic disadvantage. The availability of inexpensive and effective birth control methods made it easier for couples to reach this goal. Unpredictable changes in social attitudes had produced dramatic changes in fertility and the rate of natural increase. The TFR continued to drop from 1971 through 1976. It rose again in the 1975 to 1980 period to 2.1 for unexplained reasons and then fell again in 1986 to a low of 1.7. Since then the TFR has continued to rise to the present level of 1.96 while remaining below the 2.1 replacement level for the past 22 years.

Immigration and the Changing Racial Landscape in the U.S.

Although the TFR has remained below replacement levels, the U.S. has one of the most liberal immigration policies in the world. Immigration adds at least another 850,000 to 1.2 million people each year to the country's population.[11] If federal immigration policies do not change, these numbers will likely go unchecked for the next 50 years. Since TFR has been at or below replacement level for 27 years, immigration is considered to be among the most important demographic factors in the growth rate of the country. About 82 million of the projected 390 million people living in the U.S. in 2050 will either have immigrated since 1991 or be children of those immigrants. This group will then represent 21% of the nation's population, and total growth due to natural increase (0.8%) and net migration (0.2%) is expected to double the nation's population in less than 70 years.

The majority of immigration to the U.S. for the last 30 years has originated from Latin America and Asia. Hispanic women have been averaging more than three children each, black women more than two, and white women less than two. Within the next 25 years, the under-65 population of Hispanic and Asian Americans is expected to double as the number of white Americans under 65 will be fewer than in 1995 (Figure 2.18).[15] The number of black Americans under 65 is expected to increase by one third during this period, and the white elderly population will balloon by 67% to over 47 million people. In the next 25 years, more than 75% of the elderly, but less than half of the children in America, will be white. This divergence in race and age is leading to a situation in which an expanding population of elderly white will be expecting support from a working population of tremendous diversity. Another and coinciding event is that the aging Baby Boomers (those born between 1946 and 1964) are becoming the largest and most politically potent group of elderly in history. This generation of retirees will depend on a proportionally smaller group of racially diverse workers to support them through mandated government programs such as Social Security and Medicare.[15]

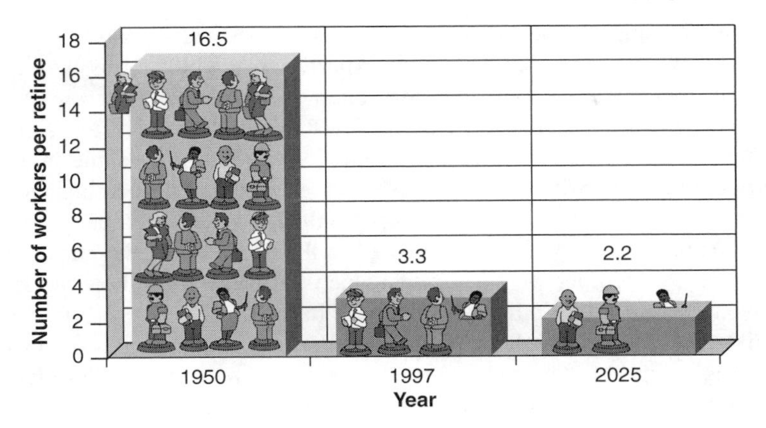

Figure 2.18 Number of workers per retiree in the U.S. from 1950 to 2025.

The Social Security system was established on a pay-as-you-go basis such that the money a person pays each year is not saved for their pension but is used to pay the pension of an older person alive at the time. The system was established assuming the ratio of workers to retirees would remain high, but this is not happening. There were 16 workers per retiree in 1950, and by the year 2020 there will be only 2.2 workers per retiree (Figure 2.18). Consequently, the percentage of worker income going to the Social Security system has increased and will likely increase more drastically, with a probability that one of the supporting workers will be white and the other nonwhite. Such a trend may create a gap between racial generations on issues of school funding and social security. The powerful, elderly, white electorate is demonstrating its unwillingness to fund an increasingly minority school-age population across America while consuming greater amounts of limited resources for expensive health care.[21] In the state of Florida, Medicaid nursing home costs multiplied more than five-fold from $200 million to $1.2 billion in the eleven years from 1985 to 1996. It is expected to reach a budget-destroying $8 billion by the year 2000.[21] The education and support of our increasingly racially diverse children is critical if we are to maintain a position of economic competitiveness in the world and if we are to expect ungrudging support of the elderly white.

Current Population Trends in Less Developed Countries

In the previous 12 years, fertility has fallen in a number of countries. Kenya dropped from a TFR of 7.5 to 4.4, and Cote d'Ivoire dropped from 7.4 to 6.15. The continent of Africa now averages a fertility rate of 5.3 and is likely to double its population within the next 25 years. Bangladesh almost cut its fertility rate in half from 6.2 in 1985 to 3.57 children per woman in 1997. The TFR in Turkey has declined from 4.1 to 2.5, in Syria from 7.4 to 5.9, and in Pakistan from 6.5 to 5.2. Western Asia now has a TFR of 3.82 and South Central Asia is at 3.42, with population doubling times of 31.8 and 38.8 years respectively.[4,6]

Despite a continuing decline in fertility among many of the world's LDCs, the populations of 74 of these countries are expected to double in less than 30 years because of the forces of population momentum and incomplete demographic transition (see page 50). More than 80% of the world lives in lesser developed countries, and about 98% of the annual increase in population of 85 million people has occurred in these countries.[6] Nearly 3 billion people will enter their child-bearing ages within the next 20 years, while 2.8 billion people will die — leaving nearly 1.2 billion people (600 million couples) to produce 1.8 billion children if the current fertility rate of three children per woman is maintained. Of these 1.8 billion children born in the next 20 years, 1.76 billion will be born in the LDCs.

India represents an excellent example of the role of population momentum in maintaining a high growth rate in spite of declining fertility. The fertility rate in India is now 3.2, having dropped from 4.3 children per woman in less than 12 years.[6,22] However, its enormous population size and large percentage of women in their reproductive age resulted in a population increase of 25 million newborns in 1997. In general, India's population grows about 16 million a year, which means approximately 72,000 Indians are born every day.[24] India's fertility rate remains well above replacement level; and when combined with its current population momentum, the country is expected to overtake China within the next 30 years as having the world's largest population. The Indian population, which continues to grow 2% annually, will reach at least 1.5 billion before it stabilizes.[23] India has four times as many people as the U.S. but is only 40% as big.[24] Despite these remarkable numbers, the most critical regions for population growth continue to be Sub-Saharan Africa, along with Western and South Central Asia as seen in Table 2.1.[4] These countries are witnessing environmental degradation, deterioration of urban centers, economic stagnation, increased morbidity and mortality from infectious diseases, malnutrition, and violence.[1,6] The contention by the U.S. news media that the "population explosion is over" may be premature. There remain a number of reasons to be concerned about the world's expanding population.[6]

- The world's population is expected to increase by 85 to 100 million people per year for many years because of population momentum and continued high fertility rates among the LDCs.
- Although fertility rates have declined globally, 33 countries still have fertility rates in excess of six children per woman — accounting for almost half a billion people — and continue to grow geometrically.
- In the last 27 years, more than 2 billion people were added to the global population despite the declining fertility rates in many countries. Ninety-eight percent of these annual increases in populations were in LDCs.
- Nearly a quarter of the world's people are so impoverished that they must survive on less than a dollar a day. Nearly one quarter lack drinking water or sanitation, while 20% lack adequate food. More than half a billion are homeless or without adequate shelter, and 86 countries are presently classified by the U.N. Food and Agriculture Organization as low-income and food-deficient because they lack the resources to purchase sufficient food.[6]
- There are 4.8 billion people in developing countries and three fifths of them lack basic sanitation.[3]

Table 2.1 Demographic Indicators of World and Regional Population Development

	TOTAL POP. 1997	PROJECT. POP. 2025	AVE. RATE OF NATURAL INCREASE 1995-2000	TOTAL FERTILITY RATE	DOUBLING TIME IN YEARS
WORLD TOTAL	**5,848**	**8,039**	**1.4**	**2.79**	**50.0**
Developed Regions[1]	1,178	1,220	0.3	1.59	233.0
Less Developed Regions[2]	4,670	6,819	1.7	3.08	21.2
AFRICA	**759**	**1454**	**2.6**	**5.31**	**26.9**
Eastern	234	480	2.9	6.05	24.1
Middle	88	189	2.7	6.01	25.9
Northern	165	257	2.0	3.67	35.0
Southern	50	83	2.2	3.92	31.8
Western	222	447	2.8	5.95	25.0
ASIA	**3538**	**4785**	**1.4**	**2.65**	**50.0**
Eastern	1447	1696	0.9	1.78	77.7
South Eastern	498	692	1.6	2.86	43.7
South Central	1418	2100	1.8	3.42	38.8
Western	175	297	2.2	3.82	31.8
EUROPE	**729**	**701**	**0.0**	**1.45**	**-**
Eastern	309	284	-0.3	1.41	-
Northern	94	96	0.1	1.73	700.0
Southern	144	137	0.2	1.34	350.0
Western	182	184	0.3	1.46	233.0
LATIN AMERICA & CARIBBEAN	**492**	**689**	**1.5**	**2.65**	**46.6**
Caribbean	37	48	1.1	2.59	63.6
Central America	128	189	1.9	3.04	36.8
South America	327	452	1.5	3.04	46.6
NORTH AMERICA	**302**	**439**	**0.8**	**1.93**	**87.5**
Canada	30	36	09	1.61	77.7
United States	272	333	0.8	1.96	87.5
OCEANIA	**29**	**41**	**1.3**	**2.46**	**53.8**
Australia/New Zealand	22	29	1.1	1.91	63.6

[1]More developed regions comprise North America, Japan, Europe, Australia and New Zealand

[2]Less developed regions comprise all regions of Africa, Latin America and the Caribbean, Asia (excluding Japan), and Melanesia, Micronesia, and Polynesia.

Source: United Nations Population Fund, *The State of the World Population 1999: 6 Billion, A Time for Choices,* 1999.)

- Death rates are increasing because of aging populations in developed countries. In developing countries, a return of infectious and parasitic diseases, AIDs, war and other violence is increasing the death rate. Ninety-seven percent of the deaths from emerging and reemerging infectious diseases (malaria, dengue fever, tuberculosis, cholera) have occurred in the LDCs, causing more than 25% of the 68 million deaths in these countries since 1995.
- Ninety-five percent of HIV-infected persons live in LDCs.[25]
- AIDs has killed more than 4.5 million people in Sub-Saharan Africa alone since 1985. HIV infects more than 10% of the adult population in these countries and has left more than 10 million children orphaned. AIDS is now the leading cause of death in Africa.[6,25,26]
- As the world's billions scramble for resources to feed, clothe, and house themselves, they are devastating natural resources, causing pollution of the oceans, lakes, streams and air; deforestation; desertification; the extinction of numerous plant and animal species; the elimination of wetlands; the decline in arable land resulting in diminished food supplies; and mass migration to cities, creating the phenomena of urbanization.
- Many areas with high fertility rates are beginning to see rising death rates due to the spread of HIV, depletion of aquifers, and a shrinking amount of cropland

available to support each person. For example, in Botswana, life expectancy has fallen from 61 years in 1990 to 44 years in 1999. It is expected to drop to 39 years by 2010 due to HIV.[27]

Many of these concerns are covered in the following chapter; but since populations are likely to increase the problems in these areas, this seems the right place to discuss future population trends.

Predicted Future Trends in Populations

We have already seen how difficult it is to predict population growth. There is, however, general agreement that the world's population will continue to grow substantially over the next 50 years with most of the growth occurring in the lesser developed regions of Latin America, Africa, and parts of Asia. The world's population is predicted to grow from its current 5.84 billion to over 8 billion in the next 25 years.[5] The developed regions are expected grow from 1.17 billion to 1.22 billion or 3.5%, while the LDCs will increase by more than 47.8% from 4.6 to 6.8 billion people. Current projections show the world's population leveling off at slightly more than 11.5 billion approaching the year 2150 (Figure 2.19). The population of the developed nations will be stable at nearly 1.5 billion, leaving nearly 98% of the population originating in the LDCs. The population of Africa will almost double in the next 25 years to 1.45 billion (Table 2.1). Asia will increase its population by 1.2 billion to 4.78 billion people, and Latin America will grow by 40% to 689 million. Contrast this with Europe, which will show a decline from 729 to 701 million. North America will show a rate of natural increase in population during the next 25 years from 272 to 333 million for a 22% increase in population excluding the effects of migration.[5,6] These predictions are risky and depend on accurately forecasting trends

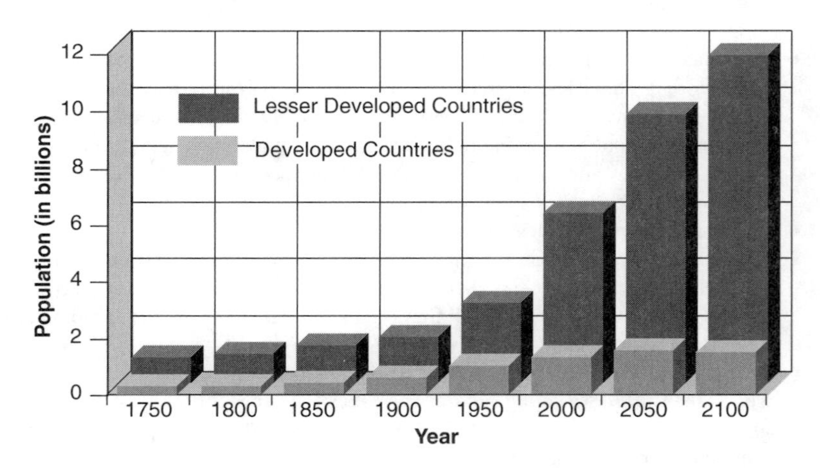

Figure 2.19 Population comparisons between developed and lesser developed countries from 1750 to 2100. (Adapted from Population Reference Bureau, 1997 World Population Data Sheet, 1997, and the United Nations Population Fund, the right to choose reproductive rights and reproductive health, *The State of the World Population, 1997,* New York, 1997.

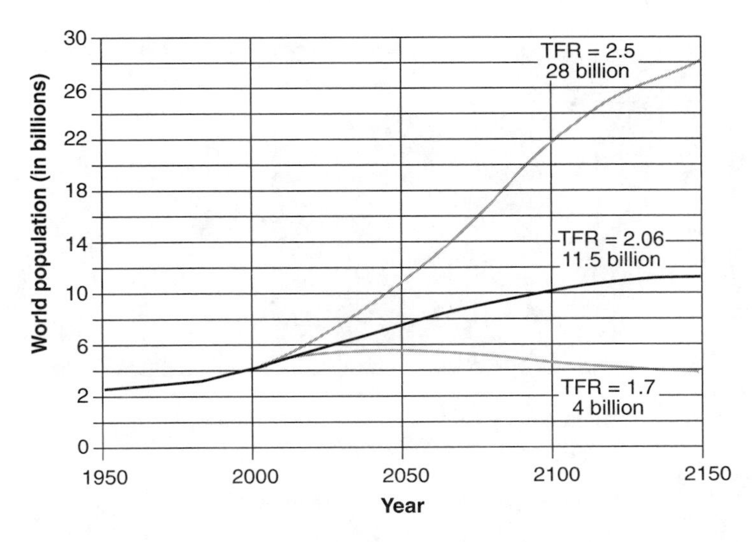

Figure 2.20 Projected world populations to 2150 based on various total fertility rate possibilites. (Adapted from Doyle, R., Global fertility and population, *Sci. Am.*, March 26, 1997; and Motavalli, J., Contents under pressure, *E Mag.*, Nov./Dec. 28, 1996.)

in mortality rates and the child-bearing behavior of millions of couples. The demographers have, therefore, projected several possible outcomes for global population allowing for deviations in fertility rates. The median or "best estimate" by the United Nations is that the world population will stabilize at 11 billion people around the year 2150 if the world fertility rate drops to 2.06 and life expectancy is 85 years (Figure 2.20). If fertility rates stabilize at 1.7 children, the world population would fall to 4 billion, down from the nearly 6 billion people today. Conversely, if the TFR falls only to 2.5 globally, the world population will escalate to 28 billion people and continue to climb.[28,29] The social, political, and ecological consequences would be greatly exaggerated by such a population tidal wave and would far outstrip the Earth's carrying capacity, resulting in unprecedented human calamities. Even if the median prediction of 11.5 billion people is approached in the next 150 years, the major effects will be felt in the developing countries. Since 98% of the expected increase in the world population will occur in the poorest countries, world leaders and demographic experts argue that the population growth rates should be stabilized much before they reach these theoretical levels projected by the United Nations.[5]

Urbanization

What is Urbanization?

There is another dynamic to the world's increasing population. The mass migration of people to the cities is known as *urbanization*. In 1950, when black-and-white television was beginning to become commonplace in American homes, 30% or 750 million of the world's 2.2 billion people lived in cities. There was only one city in the developing world that had more than 5 million people. Forty years later, in

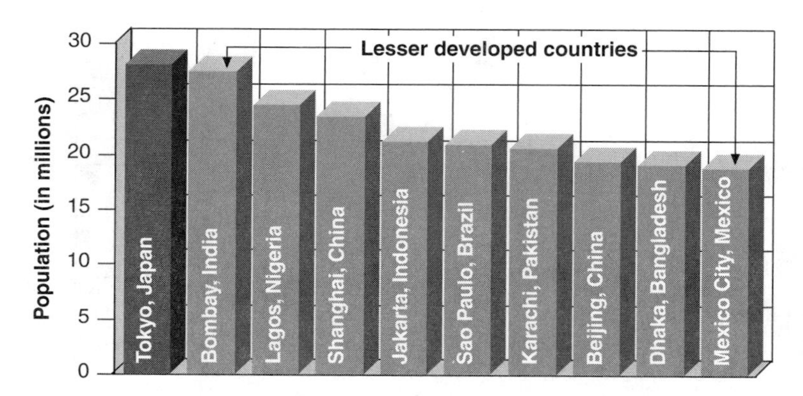

Figure 2.21 World's 10 largest cities projected for 2015. (Adapted from The environment, a new positive image, *The Weekly Review,* Kenya, Nov. 10, 24, 1995.)

1990, there were nearly 2.3 billion living in cities — nearly 43% of the global population. This number will reach 50% by the year 2000 and is expected to escalate to 61% or 5.12 billion people in the year 2020. By that time, 95% of the world's growth will be in the cities and 80% of the world's urban residents will live in the LDCs by the year 2025.[6] Megacities, defined as having a population of more than 10 million, will be commonplace by the year 2015, with nine of the ten largest cities in the developing countries. (Figure 2.21).[30] To put this population trend in perspective, megacities such as Lagos, Nigeria and Dhaka, Bangladesh are growing at a rate 10 times faster than New York, or Los Angeles. In 1975, there were five megacities; in 1995, 14 megacities; and the number of megacities is expected to reach 26 by the year 2015.[31]

The movement of masses of people into urban centers is prompted by the rapid destruction of rural ecosystems as increasing hordes of people attempt to eke out an agricultural life on land of diminishing quality. The depletion of resources and the lack of jobs and opportunities drive countless millions of desperately impoverished people to the cities.

In many of these countries, the rate of urbanization exceeds the ability of local and national governments to provide necessary employment and services such as sewage disposal and clean water supplies. Uncontrolled settlements, erupting as unwelcome growths around nearly all of the major developing cities, continue to grow faster than the core urban centers. These shanty towns often are dense clusters of corrugated metal shacks or cardboard walls covered with plastic wrap.[1] According to a World Resources report, an estimated 20 to 50% of urban inhabitants in developing countries live in impoverished slums or squatter settlements without access to adequate water, sanitation, and refuse collection.[31]

Few residents have electricity, proper sewage disposal, or clean water supplies. People urinate and defecate in streams that are filled with garbage and which incubate clouds of swarming mosquitoes.[1] Whether viewed by satellite or from windows of a car, internal African borders are disintegrating into huge megalopoli as evidenced by the countries of Ghana, Togo, Benin, Nigeria, and the Cote d'Ivoire (Figure 2.21). Borders no longer clearly exist as shanty towns merge into each other, creating an

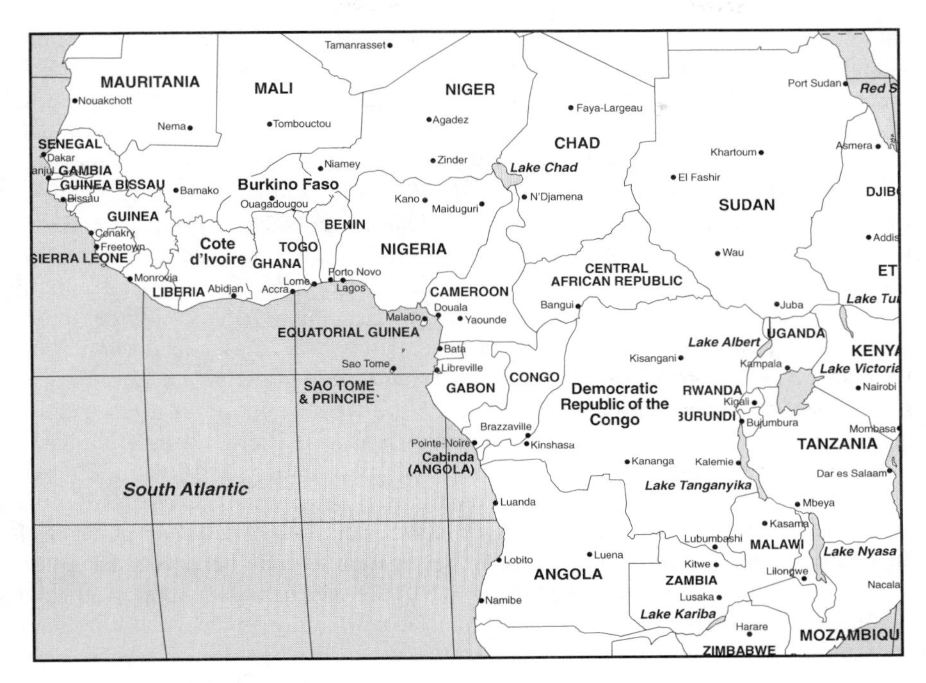

Figure 2.22 Areas of western Africa where traditional borders have broken down by merging megacities.

endless obscene vista of abandoned cars, wire mesh, metal shacks, and stores built of shipping containers facing into puddled streets of floating garbage.[1] These urban growths are breeding grounds for disease, violence and environmental degradation.

The Role of Urbanization in the Spread of Disease

Three quarters or 8 million of the 12 million HIV-positive people worldwide are located in Africa, where the disease is accelerated by war, refugee movements, and uncontrolled settlements linked by improved road systems. Tuberculosis, cholera and malaria have returned with a vengeance. Substandard housing, inadequate sanitation, and unsafe water in densely populated cities account for 10 million deaths globally each year, with 4 million infants and children dying from waterborne diseases alone.[30] Such disease outbreaks produce dramatic economic losses. As an example, a 1994 outbreak of the plague in the Indian city of Surat killed 54 people; affected 5000 others; caused economic damages approaching $1.5 billion in U.S. dollars; and caused nearly half a billion people to leave the country.[30]

Violence in Developing Urban Centers

Violence in densely populated urban centers of developing countries is not a certainty. However, there are a number of countries where violence appears to be on the increase, consistent with significant population increases. In an African tribal community, it is customary and natural to eat at any table with any family. An entire village or community is considered to be the family. However, as exodus to urban areas occurs, traditions and customs often erode and communal existence evaporates. Young men seek survival through joining gangs and descending into a criminal, violent process, making cities such as West Africa some of the unsafest places in the world. Gunmen roamed the streets with AK-47s in Somalia in 1992 and cruised the streets of Mogadishu in their armed "technical" rusting pick-up trucks, terrorizing the country. Somalia had become a country with no working economy, no police force, and no government.[32] Internal clan disputes warring over limited natural resources have prevented stability despite the intervention of American forces under Operation Restore Hope. This effort erupted into a disastrous armed conflict on October 3, 1993, when U.S. rangers raided a site near the Olympic Hotel where followers of the warlord Mohammed Farrah Aidid were meeting.[33] In 1994, gunfire and mortar attacks triggered terror and stampedes at the Zaire–Rwanda border in one of the world's most vicious war zones. The million refugees of Rwanda's civil war stretched for miles as they accumulated in death camps and succumbed by the thousands to cholera, dysentery, measles, and other diseases. The Prime Minister of Rwanda, Faustin Twageramunga, tried in desperation to coax his people back into the capital city, where violence had disrupted all services including phones; stopped the flow of water and electricity; and turned it into an armed camp with military checkpoints.[34]

Urban poverty is a socially destabilizing force where hordes of the impoverished people fleeing decaying rural ecosystems to urban centers are threatened by the loss of native customs in expanding cities. Here, their values are undermined, physical necessities are absent, and the spread of diseases is imminent.[1] Many countries in Africa, and others outside, are witnessing the rise of criminal anarchy, unprovoked crime, and the erosion of nation–states and international borders as disease, over-population and disappearing resources become prominent.

Environmental Degradation

In countries such as Cote d'Ivoire, Sierra Leone, Guinea, Ghana, and Brazil, the destruction of primary rainforests is proceeding at blinding speed. Less than 6% of the primary rainforest of Sierra Leone remains compared to the 60% that was forested 37 years ago. Deforestation has led to soil erosion, more flooding, and a proliferation of mosquitoes. As the tentacles of the urban population expand outward, it calls for greater and greater resources. Soil degradation is most pronounced in those areas of the world with the densest populations, such as the West African coast, the Middle East, the Indian subcontinent, China, and Central America. The degradation of soils is readily apparent in China, where land has been destroyed by deforestation, loss of topsoil, and salinization. Water supplies have also been heavily contaminated, forcing large numbers of people to the cities and to coastal China.[35] The results have been crime surges, growing regional disparities, and internal conflict with the looming possibility of government decentralization.[1]

The environment is very likely the national security issue of the early 21st century. Expanding populations, reemerging diseases, soil erosion, depletion of water, air pollution, hazardous chemicals, and radiation will likely force mass movements of people and incite group conflicts. Environmental disputes may very well merge with ethnic and historical disputes, resulting in a core foreign policy that drives the political issues. There is concern by some that the doubling and tripling of populations in the developing nations over the next 25 years will escalate the erosion of natural resources, prompting greater likelihood of inter- and intranational conflicts.[1]

THE CONTROL OF POPULATION

Empowerment or Force?

The International Conference on Population Development (ICPD) was held in Cairo beginning September 5, 1994. The major idea or topic behind the conference was women's rights. The draft program of action proposed that a full range of reproductive and health care services, including contraception and sex education, should be made available to all.[36] One of the most crucial points of the conference was the need for participation of women in the global work force at different levels. Education and employment were discussed as a tool to make women more powerful. The stabilizing of human population is no longer a mystery; it is best accomplished by improving women's education, providing them with paid employment, and making effective and inexpensive contraception available.[12,36] It has been estimated that if women had the power to make decisions about sexual activity and its consequences, 80 million unwanted pregnancies, 20 million unsafe abortions, 500,000 maternal deaths, and 333 million sexually transmitted infections could be avoided.[25] Unfortunately, this is not an easy solution in many countries where poverty, cultural traditions, and lack of educational opportunities contribute to illiteracy of nearly one billion people worldwide. Two thirds of the illiterate population are women, due to the practice of denying education to young girls. In most of the Asian, and in Muslim and Catholic countries, a woman's body may not be considered her own, nor the child in her womb. Collective decisions are often made about a woman by parents and in-laws that are accepted because the extended family is the institution that works. The state is often not trusted.[36] Countries attempting to bring population growth under control without first empowering women and providing effective birth control have often resorted to oppressive population control policies.

Population Policies in Some Countries

India

India was the first country to introduce family planning in 1951. The programs focused on the use of the unreliable "rhythm" methods, followed 10 years later by the use of intrauterine devices (IUDs) as the TFR continued to climb. The use of IUDs led to a number of complications in largely undernourished and anemic populations, which had little access to proper medical care. Consequently, there was little acceptance, and populations continued to rise. Efforts shifted to male sterilization in the 1970s, with significant government abuses as attempts were made to achieve regional sterilization quotas.[37] More recently, the aspirations of the people to lead a better life (after being exposed to widely accepted electronic media) have led more people to realize the virtue of a small family. In October of 1997, the focus shifted to a more broad approach to address the need to limit family size.[23] Literacy campaigns particularly among women combined with education about the advantages of smaller families are playing a greater role in causing this trend.[22] India conducts a census every 10 years. The most recent reports available suggest that India has succeeded in slowing its rate of growth from 2.1% annually to less than 1.8%. However, this rate is still sufficient to double the population in less than 30 years. Some parts of India continue to have a high fertility rate. Son preference has led couples who have girls to continue to have children until a son is born.[38] Girls are often seen as an economic liability. There have also been sex-selective abortions and unreported female infanticide.[38] This continued oppression of females in India has led to high rates of illiteracy and less health care.[38]

If a stabilized population growth is to be achieved, India must improve the status of Indian women by further expanding access to education, increasing opportunity for employment, and supporting programs that encourage later marriage.[25] All states in India have made notable progress in improving education and literacy among women over time, but equality has yet to be achieved.[38] The average age of marriage has increased considerably, and there has also been an increased use of effective family planning methods.[38] Government efforts to achieve birth reduction quotas through sterilization, abortion, or heavy socioeconomic penalties may result in regional and/or temporary successes. However, these efforts are contrary to the popular will and may lead to abuses and failures. In 1977, Prime Minister Indira Ghandi led her Congress Party to a crushing electoral defeat that was associated with these efforts to force a sterilization program on men during her previous 2-year emergency rule.[22]

Peru

Critics of a sterilization program that began in Peru in 1995 charge that government health workers are taking advantage of poor rural women who speak only local dialect. They report that the health workers are rewarded by promotions and money and so vigorously pursue government-imposed sterilization quotas by offering clothes and food to women in exchange for agreeing to become sterilized.[39] The TFR for Peruvian women who are uneducated and live in rural areas is 7.0, while the average TFR for Peruvian women with some college education is 1.7. The nation's poor women are, therefore, likely targets for sterilization. Critics further argue that the women are not being told about alternative methods of contraception or that tubal ligation is nearly always reversible. The opportunity for making informed decisions by these women is often minimized. An official of the U.S. Congressional Subcommittee on International and Human Rights Operations questioned whether the Peruvian Government is more interested in family planning or population control.[39]

China

China continues to enforce a one-child policy in the nation's largest cities such as Beijing and Shanghai. The one-child policy has successfully reduced China's population by 250 million.[40] There are severe penalties in much of China for having a second child that may include job loss and fines equivalent to 3 years' salary for each parent.[41] It was not long ago (1980s) that family planning goals were accomplished through forced abortions and sterilization. The one-child policy, which was announced in 1979, is gradually yielding to a healthier economic growth. The improved economy has permitted couples in some medium-sized cities to have additional children by simply paying the steep fines without concern for retribution. However, economic growth also appears to be reducing the number of births as families find new economic prosperity. Also, there are financial incentives to encourage couples to abide by the one-child policy.[40] The emerging government policy of China is moving away from enforcement to education and encouragement. There is an emphasis on later marriages, longer intervals between children, and smaller families.[40]

Some officials have come to realize that persuasion is more effective than threatening in effective long-term family planning goals.[40,41] Improvements in quality and choice of contraceptives in recent years has increased the effectiveness of family planning. But the one-child policy has led to increased sex discrimination against women. Threats have led to underreported births of both males and females, with the birth of a girl being twice as likely to be ignored.[40] It is believed that one half to one third of the difference in infant sex ratios is due to underreporting. Unreported daughters may be adopted out, left with relatives, or abandoned in orphanages. Sex ratios are further skewed by illegal use of ultrasound to determine fetal sex, which has led to widespread abortion. The girls who survive often have less health care and education.[40]

Mexico

The population of Mexico is currently 96 million, 21 million of which live in Mexico City alone.[42] If it were not for the success of sex education and family planning practices, Mexico's population could have been over 134 million today. The population growth rate has dropped from 3.5% in 1974 to 1.88 today.[42] Twenty-five years ago, the average number of children in a Mexican family was seven; today it is down to 2.5. The emphasis has been on promoting smaller families and better relationships. Some areas are even offering free birth control. Many families are realizing that large families are not really possible with Mexico's poor economy. Today, more than one in three women work outside the home, whereas 20 years ago, less than one in five did. Also, educators are handing out leaflets about medical care, offering courses on sexuality, and providing free HIV testing.

Iran

It is estimated that Iran will double its population from 65 million to 115 million by 2050.[2] However, Iran's family planning operation reaches almost the entire population and is one of the most successful operations worldwide.[2] There are advocates that disseminate information, and there are family planning messages advertised on television. Thirty-five thousand health volunteers operate mobile clinics, offering free contraceptives. There is one catch: Condoms are only offered to married couples since premarital sex is not acknowledged. Iranians who plan on getting married must first have a day of counseling, where family planning is discussed. Many people in Iran are waiting until they are 22 or 23 years of age before they start a family, and most of them only want two children.[2] Contraceptive use by Iranian women has increased from 26% to 75% since 1979.[2] Although fertility rates are still high, they continue to drop rapidly.

Family Planning Versus Population Control

The government-directed programs that set a policy for establishing an optimum population size are referred to as *population control*. In the countries described above, the policies were directed at reducing population. In a number of developed countries such as in Western Europe, policies may be directed at promoting a higher TFR by providing state financial support for hospital expenses related to childbirth, for monthly support for the parents during the first year or two of the child's life (as in Austria), or even tax deductions or special bonuses.

Population control is in contrast to family planning programs that are directed at assisting couples in having the number of children they desire regardless of how many. Family planning is parent centered and places an individual's or parent's desires for children before that of a state or national policy. Some policy makers argue that this style of family planning may not be enough to stem the tide of population growth in those countries where many children are considered an important resource for gathering wood and water, harvesting crops, or performing other forms of manual labor. The simple act of getting water may take several hours a day, and it takes longer and longer as resources diminish. Children are also viewed as a source of security in old age in those countries without a financial support system for the elderly.[12] The persistent desire for large families in many LDCs is estimated to add nearly 660 million people in the next 50 years.[6]

An additional problem of funding needs exists for family planning services in foreign countries. The ICPD in Cairo (1994) agreed that it would allocate at least $17 billion annually to provide services for population and reproductive health programs and the prevention of sexually transmitted diseases (STDs), including HIV. Unfortunately, annual world expenditures are much less than half of this amount, with the contributed dollars appearing to have leveled off.[6] One issue contributing to the decrease in funds for family planning efforts has been the anti-abortion lobbyists in the Republican-controlled congress.[43] Such shortfalls in LDCs are expected to result in as many as 220 million unintended pregnancies. As many as 88 million of these will likely be terminated by abortions. The pregnancies and abortions may cause as many as 117,000 additional maternal deaths and up to 1.5 million women with some form of chronic health problem. This unmet demand for family planning services may cause an increase of 1.2 billion in global population.[6] The major elements necessary for a better quality of life for the world's billions may depend on: (1) the willingness of governments to set reasonable policies with adequate funding; (2) develop greater motivation for small families through improved economies; and (3) universal access to knowledge and use of family planning services.

In July 1999, delegates from 180 countries approved proposals to slow world population growth, including offering women greater access to abortions and sex education to adolescents.[44] This was a significant advance from what was agreed to at the ICPD in Cairo because it forces governments to increase and improve access to reproductive health programs. The belief is that women who are educated and have access to reproductive health care have fewer children. There are new goals in place to increase access to universal health care, increase education for women, and reduce maternal mortality and HIV infection.[44]

METHODS OF FERTILITY CONTROL

Introduction

The process of conception requires that a viable sperm have access to a healthy egg. When the sperm enters the egg, the tail is released and the ovum secretes a protective chemical barrier that prohibits other sperm from entering. Methods that prevent egg fertilization are called *contraception*, and there are hundreds of different procedures and formulations available today. The enormous variety of contraceptive tools attests to the interest of society in preventing unwanted pregnancies. These methods vary in their risks to health, their efficacy in preventing pregnancies, ease of use, acceptance, and costs. Some of the methods and characteristics are described below.

Contraceptive Methods that are Reversible

Natural Birth Control and Family Planning

This process is sometimes referred to as the *rhythm method* and relies on voluntary changes in sexual behavior. The method requires that the female predict her fertile periods by examining cervical mucous, tracking basal temperature, or charting her menstrual cycle and ovulation times on a calendar. She then abstains from sexual intercourse during times of the month when she is fertile. This method has proven to be very unreliable because menstrual cycles vary, and the narrow window for sexual opportunity is not always observed.

Abstinence

The term abstinence has come to mean the exclusion of sexual intimacy including oral–genital kissing and mutual masturbation. These activities are often referred to as *outercourse*. Either of these methods can be 100% effective in prevention of pregnancies as long as there is not an opportunity for sperm getting near the vagina. The problem with outercourse is the increase in the probability of spreading STDs through oral–genital contact. Also, the motivation and opportunity for intercourse is increased.

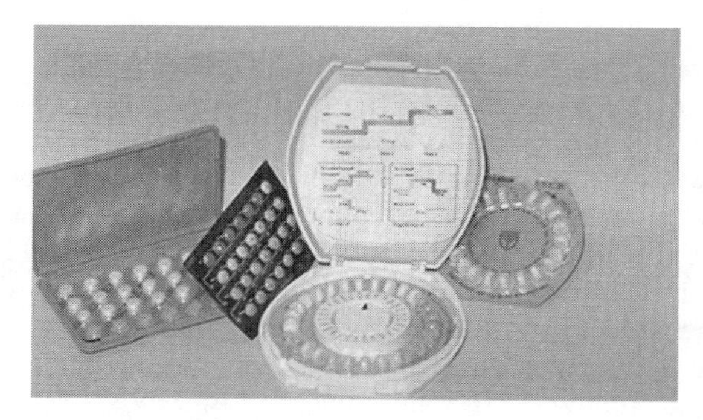

Figure 2.23 Various forms of the oral contraceptive pill designed to be taken daily.

Hormonal

Oral Contraceptives

Since the introduction of the pill in 1960, it has become widely accepted and is now used by nearly 40% of women who wish to avoid pregnancy. The combined effects of the synthetic hormones estrogen and progesterone in the pill prohibit the maturation of ova, prevent the development of the uterine lining, and form cervical mucous, creating a barrier against sperm. The amount of estrogen supplied in the pill has been decreased to reduce health risks. There remains a limited risk of increased blood pressure, the formation of blood clots, strokes and heart attacks for some women. These risks are low in most healthy women. The risks are further reduced for women less than 35 years of age and for those who do not smoke. A physician's prescription is required in most western countries, including a complete gynecological exam. When taken according to directions, the pill approaches a 97 to 99% effectiveness in preventing pregnancy (Figure 2.23). Oral contraception offers no protection against the transmission of sexually transmitted diseases.[45]

A progestin-only pill known as a *minipill* is also available for women who do not want to use estrogen. The minipill is 96% effective and has few side effects except for the possibility of irregular menstrual bleeding.

Depo-Provera

This is a long-acting synthetic hormone (progesterone) that is injected into muscles of the arm or buttocks every 3 months. It has an efficacy of 99% and is especially desirable for women who do not want to keep track of taking a daily pill or inserting a device or chemical into the vagina. Most women lose their menstrual period within a year on this hormone. The adverse side effects are minimal and are usually limited to a slight weight gain, occasional dizziness, and headaches.[45]

Norplant

Norplant consists of five or six silicon rubber capsules (Figure 2.24) containing progestin that are surgically inserted under the skin of the upper arm of the female.

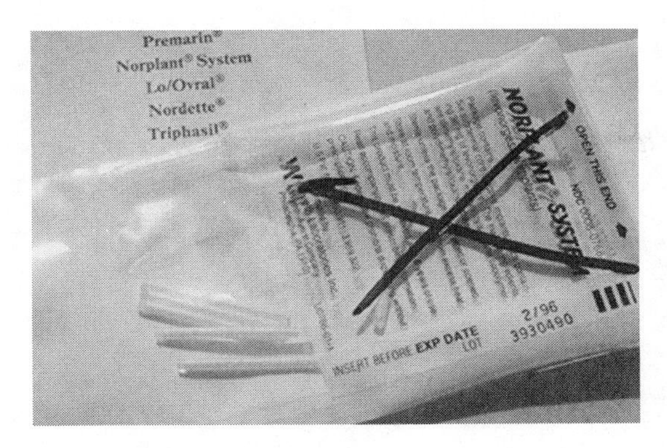

Figure 2.24 Norplant® showing elongated capsules.

Small amounts of progestin are time-released over a 5-year period and work by suppressing ovulation, thickening the cervical mucous, and inhibiting the growth of the uterine lining. Norplant has been successfully tested in more than 40 countries, demonstrating minimal health risks with nearly 100% effectiveness in preventing pregnancies. It has been used more frequently in the U.S. since it entered the market in 1991, but the up-front costs exceeding $550 may be a deterrent, even though the cost of the pill over the same time period is more than twice that amount.

Spermicides

Chemicals that are designed to kill sperm are called spermicides. The substance most commonly used is nonoxynol-9, and it comes in a variety of forms including jellies, creams, foams, and waxy suppositories (Figure 2.25). These are all meant to be inserted far enough to cover the cervix and provide a chemical and physical obstacle to sperm. Spermicides are most useful when used with a barrier method such as a condom. Together they provide a 98% effectiveness.

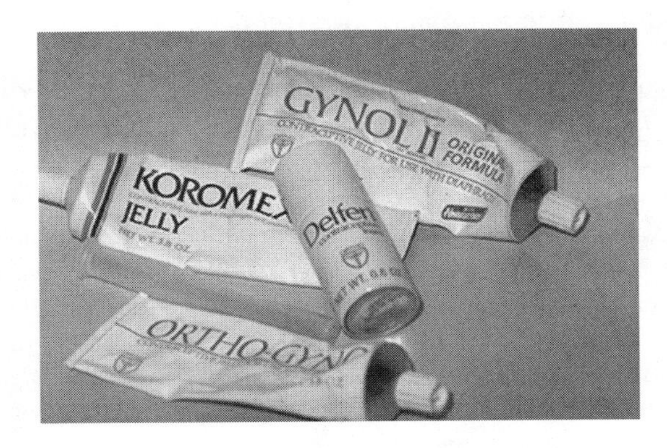

Figure 2.25 Samples of spermicidal jellies, creams and lubricants.

Figure 2.26 Male condoms appear in a variety of colors, shapes, and textures.

Barrier Methods of Contraception

Male Condom

Condoms are typically elastic material of latex rubber or polyurethane designed to be rolled onto the erect penis before intercourse (Figure 2.26). They appear in a variety of colors and textures, with or without lubricants, and with or without a reservoir that is designed to capture ejaculate. The effectiveness of condoms to prevent pregnancies is reduced to 88% because of improper use or storage, which can result in permitting sperm to enter the vagina through tears, rips or spillage. Formerly, it was believed a greater effectiveness could be achieved if the condom were used along with a spermicidal lubricant. A current study found that the spermicide nonoxynol-9 might actually irritate the vaginal walls, making infection in users more likely. Due to this study the CDC has changed its recommendations, advocating for use of a lubricated condom alone until this finding can be researched further.[46] Condoms are one of the top three methods used in the U.S. and provide the additional benefit of protecting against the transfer of HIV.

Female Condom, Diaphragm, and Cervical Caps

The female condom is a soft elastic polyurethane conical bag with an elastic ring (Figure 2.27). The narrow end is inserted into the vagina, leaving the wider end to extend out over the labia. It has a reported failure rate of around 12% based on use in the U.S.

The diaphragm is a thin latex rubber cup with a flexible rubber-coated ring intended to cover the cervix. Diaphragms come in a variety of sizes and must be fitted to the female by a trained health professional. Moreover, the user must be familiar with the correct method to insert and position the diaphragm. It is intended to be used with a spermicidal cream or jelly on the concave or inside surface before insertion, and then left in place for up to 8 hours after intercourse. This permits sufficient contact time to kill any remaining sperm.

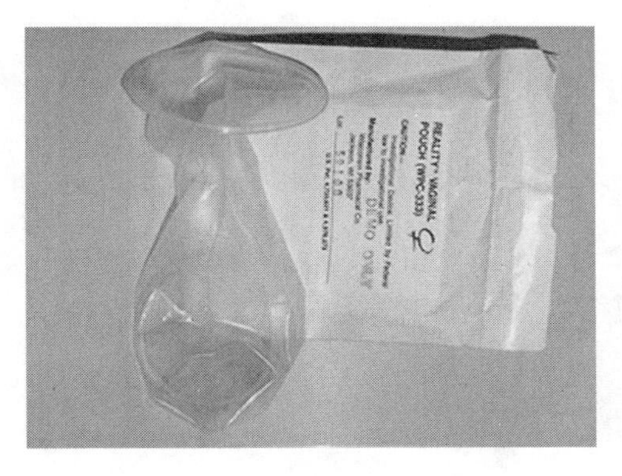

Figure 2.27 Female condom is inserted into the vagina with the wider open end extending over the labia.

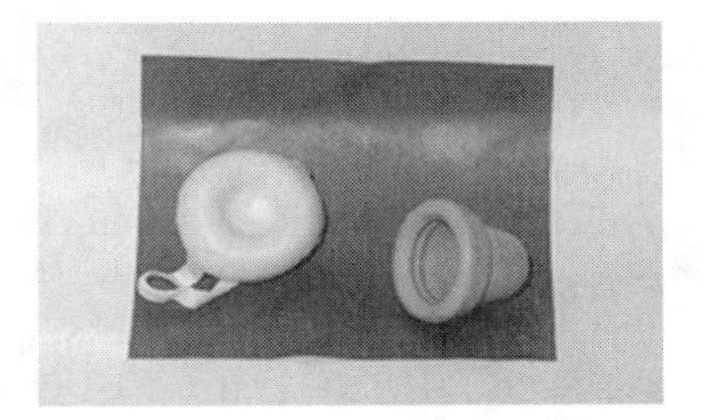

Figure 2.28 Sponge (left) and cervical cap (right) designed to be placed over the cervix with spermicide.

The cervical cap is similar in design but smaller than the diaphragm (Figure 2.28) and is also fitted by a health professional. It is designed to be used with a spermicide and left in place for several hours after intercourse. Its effectiveness is somewhat less than a diaphragm because it is smaller and more difficult to properly insert.

Intrauterine Devices (IUDs)

IUDs are plastic, T-shaped devices that are folded with a special instrument and inserted by a trained health professional into the uterus such that the arms of the "T" open up and cross the uterus where it is braced. Some brands come with a slow-releasing synthetic progesterone, while others are copper-wrapped. The specific action of IUDs is unknown, but it is believed they interfere with egg fertilization by sperm. The IUDs must be removed or replaced periodically — every 1 to 4 years depending on the type used. The effectiveness in preventing pregnancies can reach 95%. Risks include pelvic inflammatory disease, severe cramps, heavy menstrual flow, uterine perforation, tubal infections, and infertility. These are among some of the reasons for the negative publicity that prompted the manufacturer to remove the Dalkon shield from the market in the 1970s.

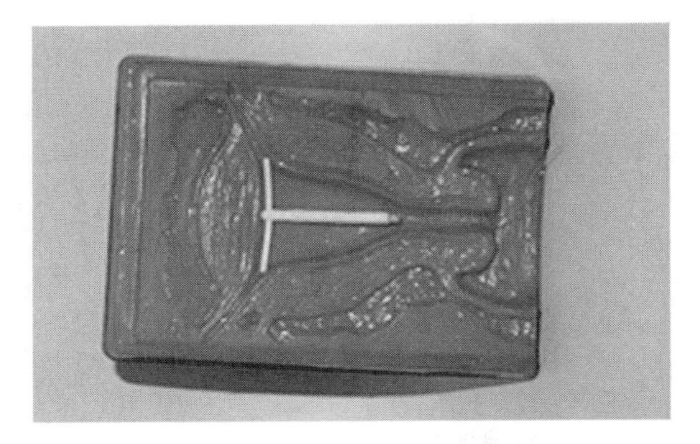

Figure 2.29 Intrauterine device showing approximate placement in the uterus.

Contraceptive Methods that are Permanent

Sterilization has become one of the most popular methods for contraception in the U.S. among married couples who have achieved their desired level of parenthood. Male sterilization by vasectomy is usually an outpatient procedure accomplished by making an incision on either side of the scrotum and snipping out a piece of the *vas deferens*. The two ends of the *vas* are tied off or cauterized. There is some discomfort in this area for about a week. The risks are minimal, and effectiveness in preventing unwanted pregnancies in partners is better than 99%.

Women may be sterilized by tubal ligation, which is a slightly more involved surgical procedure requiring small incisions near the navel and just above the pubic area. The abdomen is inflated with a gas so the physician can easily locate the fallopian tubes with the fiber-optic light on the laparoscope and tie off or cauterize the tubes. This procedure blocks the entry of eggs into the uterus, and eggs released from the ovaries dissolve and are reabsorbed into the body. Tubal ligations are largely reversible, while vasectomies have generally not met with similar success at reversal. However, developments in microsurgical techniques have resulted in a reversal approaching 50% depending on the elapsed time since the vasectomy and the skill of the surgeon.[28]

Abortion

Abortion refers to the medical means of terminating a pregnancy. In the U.S., this is considered to be a decision that a woman and her physician may make within the first trimester without legal restriction. This right was given under the 1973 *Roe vs. Wade* Supreme Court decision.[47] However, individual states were given the right under later court actions to impose restrictions on abortions. Despite these restrictions, about 1.6 million abortions have been legally performed each year in the U.S. since 1980.[28] Nearly 60 million abortions occur annually on a worldwide basis. Three quarters of the world's population live in countries where it is legal to induce abortion medically. There are different forms of abortion depending on how many weeks pregnant the woman is at the time of the abortion. The number of weeks of pregnancy is determined from the first day of a woman's last menstrual period. Most pregnancies are terminated in the first trimester when legal and health consequences are minimal. These are usually referred to as therapeutic abortions and are normally accomplished by vacuum aspiration. Abortion can also be safely induced within the first 9 weeks of pregnancy by administering the drug RU-486. This steroid blocks the action of progesterone and causes the uterine lining to break down, expelling the lining and the fetus, which terminates the pregnancy. The drug RU-486 was developed in the late 1970s by a French researcher consulting for Roussel-Uclaf, a French pharmaceutical company. There continue to be fears by many, including anti-abortion groups, that RU-486 would popularize abortion by making it available within a doctor's office and therefore remove the need for abortion clinics, which are frequent targets of protestors. Despite many setbacks, including halted production and distribution, RU-486 is now widely used in France and was approved by the FDA for sale in the U.S. on September 28, 2000. The approval of this drug gives women an alternative to surgical abortion for the cost of a standard abortion.[48]

REFERENCES

1. Kaplan, R., The coming anarchy, *The Atlantic Monthly,* 44, Feb., 1994.
2. Motavalli, J., Now we are six, *Environ. Mag.*, July/Aug. 1999.
3. United Nations Population Fund, *The State of the World Population 1999: 6 Billion, A Time for Choices,* 1999.
4. Population Reference Bureau, 1997 World Population Data Sheet, 1997.
5. United Nations Population Fund, the right to choose reproductive rights and reproductive health, *The State of the World Population 1997*, New York, 1997. http://www.unfpa.org/swp97e/toc.html
6. Fornos, W., 1997 World population overview, The Population Institute, www.population institute.org/overview97.html Dec. 30, 1997.
7. Odum, E., *Fundamentals of Ecology,* W.B. Saunders, Philadelphia, 1959.
8. Meyers, J. and Krebs, C., Population cycles in rodents, *Sci. Am.,* 38, June, 1974.
9. Turk, J. and Turk, A., *Environmental Science,* Saunders College Publishing, Philadelphia, 1988, Chap. 7.
10. Perlez, J., Birth control making inroads in populous Kenya, *The New York Times,* September 10, 1989.
11. Zero Population Growth, Frequently asked questions, http://www.zpg.org/q-a.htm, 2/10/98.
12. Dasgupta, P.S., Population, poverty and the local environment, *Sci. Am.,* Feb. 42, 1995.
13. Freed, S.A. and Freed, R.S., One son is no sons, *Nat. Hist.,* Jan. 1985.
14. Bongaarts, J., Demographic consequences of declining fertility, *Science*, 282, 419, Oct. 16, 1998.
15. Wattenburg, B.J., The population explosion is over, *The New York Times Magazine,* Nov. 23, Section 6, 60, 1997.
16. Deen, T., Development: Population Explosion Not Over, Warns U.N., Inter Press Service, March 1998. http://www.oneworld.org/ips2/mar98/population.html.

17. Kristof, N. D., Baby makes 3, but in Japan that's not enough, *New York Times International,* Oct. 6, 1, 1996.
18. Stanley, A., Blissful bachelorhood and the shrinking village, *New York Times,* Nov. 16, 1999. http://www.nytimes.com/news/world.
19. Funabashi, K., Reassessing the value of children, *Jpn Echo*, 26, Feb. 1999. http://www.japanecho.co.jp/docs/html/260111.html.
20. Sadik, N., United Nations Fund for Population Activities, 1985 Report.
21. Tilove, J., Racial generation gap growing, *Sunday Republican,* Feb. 22, B2, 1998.
22. Kataria, S., India succeeding in slowing population growth, Reuters, 02, 02, 1997. http://www3.elibrary.com/id/2525/getdoc.c…Q002D029&Form+RL&pubname=Reuters&pubur=0.
23. Visaria, L., Jejeebhoy, S., and Merrick, T., From family planning to reproductive health: Challenges facing India, *Int. Fam. Plann. Perspect.*, 25, 1999.
24. Population nears 1 billion in poverty-ravaged India, *Daily Hampshire Gazette,* 213 No. 294, 1, Aug. 16, 1999.
25. United Nations Population Fund, The State of the World Population 2000. http://www.unfpa.org.
26. 10 million orphans, *Newsweek,* Jan. 17, 2000.
27. Brown, L and Halweil, B., Breaking out or breaking down, *World Watch,* 13(5), Sept./Oct. 1999.
28. Doyle, R., Global fertility and population, *Sci. Am.,* March 26, 1997.
29. Motavalli, J., Contents under pressure, *E Mag.,* Nov./Dec. 28, 1996.
30. The Environment, A new positive image, *The Weekly Review,* Kenya, Nov. 10, 24, 1995.
31. Lynn, Walter, Megacities: Sweet dreams or environmental nightmares?, *Environ. Sci. Technol.,* 33, 1999.
32. Nelan, B.W., Taking on the thugs, *Time,* Dec. 14, 27, 1992.
33. Church, G., Anatomy of disaster, *Time,* Oct. 18, 40, 1993.
34. Gibbs, N., Cry the forsaken country, *Time,* Aug. 1, 28, 1994.
35. Xin, W., Population vs. development: challenge of the new century, *Beijing Review,* May 1-7, 12, 1995.
36. Elliot, M. and Dickey, C., Population wars: body politics. *Newsweek,* Sept. 12, 22, 1994.
37. Conly, S.B. and Camp, S.L., India's Family-Planning Challenge: From Rhetoric to Action, Country Study Series #2, The Population Crises Committee, 1992.
38. Arnold, F., Choe, M.K., and Roy, T.K., Son preference, the family-preference process and child mortality in India, *Population Stud.*, 52, 301, 1998.
39. Sims, C., Using gifts as bait, Peru sterilizes poor women, *New York Times,* Feb. 14, 1, 1998.
40. Choi, C.Y. and Kane, P., China's one child family policy, *Br. Med. J.,* 319 i7215, 992, Oct. 9, 1999.
41. Faison, R., China's strict 1-baby policy giving way to economic changes, *Springfield Sunday Republican,* Aug. 17, 1, 1997.
42. McConahay, M.J., A smaller but better future, *Sierra,* July/Aug, 1999.
43. Roach, John, Family Planning Urged for Developing Nations, October 6, 1998, http://www.enn.com/enn-news-archive/1998/10/1000689/famplan.asp.
44. U.N. moves to slow population growth, *Weekend Gazette,* A5, July 3, 1999.
45. Donatelle, R. and Davis, L., Pregancy, childbirth, and birth control, *Access to Health,* 3rd ed., Prentice Hall, Englewood Cliffs, N.J. 1988, Chap. 8.

46. http://www.unaids.org/whatsnew/press/eng/durban120700.html.
47. Boston Women's Health Collective, *The New Our Bodies, Ourselves,* New York, Simon & Schuster, 1984, p. 218.
48. Kaufman, Marc, Abortion pill wins FDA OK, *Union News,* Sept. 29, 2000.

CHAPTER 3

Environmental Degradation and Food Security

INTRODUCTION: THE DEBATE

The ability of our planet to sustain and feed the dramatic increases in human population growth has been an ongoing debate stretching back over 200 years. An English clergyman and economist by the name of Thomas Malthus published an article in 1789 titled *An Essay on the Principle of Population.* He concluded that human populations grow exponentially (2, 4, 8, 16…) and that plants grow arithmetically (1, 2, 3, 4) so that eventually uncontrolled human growth would lead to

OBJECTIVES FOR THIS CHAPTER

A student reading this chapter will be able to:

1. **Discuss the impact of population on resources and ecosystems**
2. **Define the following terms and explain their response to population growth: *retrogression*, *soil erosion*, *desertification*, *deforestation*, *wetlands destruction*, and *wildlife destruction***
3. **Define the term *food security* and discuss the reasons leading to food insecurity among many nations worldwide**
4. **List the suggested steps that might be taken to minimize global food insecurity**
5. **Explain the most likely reasons for a growing food insecurity in the U.S.**
6. **List and discuss the demographics of the populations in the U.S. at risk to food insecurity**

"misery and death," as growth surpasses the capability of the land to feed the expanding population. The predictions of Malthus were not widely accepted initially; they were even scorned as the opportunities for migration relieved the population pressures and technological advances in agriculture increased the available food supply.

It was not until the post-World War II period when growth rates sharply escalated that world leaders and demographic scientists pressed the population alarm, proclaiming that overpopulation was the greatest threat to continued habitation of the planet by humans. Government reports and published papers exclaimed that the world in the year 2000 would become more polluted, greatly crowded, ecologically unstable, and vulnerable to economic and political disruption.

Those holding these views are often referred to as neo-Malthusians. There would likely be more regional water shortages and major reductions in forested land. Significant deterioration in agricultural soils has been reported through the processes of erosion, desertification, waterlogging and salinization. World climate is predicted to change from the influence of greenhouse gases such as carbon dioxide that are produced in ever greater amounts by human activities such as combustion of fossil fuels. These gases are then thought to absorb greater amounts of heat energy from the sun, producing global warming. The combustion of fossil fuels is also thought to increase the formation of acid compounds in the air leading to increased acid deposition in lakes, streams, and rivers that become intolerant to many forms of aquatic life. There are predictions of increasing numbers of animal and plant species becoming extinct as pollutants increase and habitats are destroyed for agriculture and development. There is no shortage of professionals and lay people who predict worldwide calamities as a direct consequence of unbridled human growth.[1–5]

Technology and Policy will Save the Day

A real concern exists among many political leaders and the scientific community that the rapidly expanding human population cannot continue unabated without creating dramatic environmental crises. The human population may have already become so large, and the momentum for further growth so great, that any technologies that might mitigate the pressures cannot be developed or applied soon enough to prevent disasters from occurring.

Many counter these views, arguing that overpopulation is not the cause of a degraded environment but rather develops from poverty and unequal distribution of wealth. It is suggested by some that the Earth has resources capable of feeding numbers as great as 16 billion.[1] Still others predict the Earth has an enormous productive potential capable of supplying food for 400 to 1000 billion people.[6] Such wide deviation in the predicted carrying capacity of the Earth highlights the uncertainty in making such predictions. The counterview to "population as a problem" is that the real threat to global stability is the failure of nations to pursue economic trade and research policies that increase food production, more evenly distribute food and resources, and limit environmental pollution. These views usually recognize that more people are not a problem but rather an asset that drives the need for greater technology and innovation to sustain larger numbers of people. People who adhere to these views are sometimes called "Cornucopians" after the mythical horn-of-plenty. It may have been the pressures of population that resulted in the development of new genetically superior crops with dramatically higher yields, which became known as the *green revolution*. Preceding the green revolution was the development of more efficient and economical energy sources.

The Green Revolution

In the mid-1960s an interdisciplinary team of scientists from a worldwide network of agricultural centers began to operate under the Consultative Group on International Agricultural Research (CGIAR). Research from these groups resulted in the development of new varieties of important food crops such as rice, wheat, and corn.[7] The International Rice Research Institute in the Philippines was responsible for developing genetically improved strains that doubled the yields of rice per acre. Strains are being developed that resist diseases, pests, drought and flooding. Corn production per acre has continually risen since the 1960s, with the world average rising from 1 to 1.75 tons per acre. Corn is one crop that has shown increasing yields per acre of 1 to 2% even without the contribution of genetic engineering or other technologies. Using the latest technologies and best seeds, the state of Iowa now averages in excess of 4 tons per acre of corn, or five times the world average;

and Iowa master farmers now average nearly 8 tons per acre.[7] The new varieties have been shortened with thicker stalks to support the heavier weights, while the leaves are also shorter and more upright to reduce the problem of shading and crowding of nearby plants. These plants respond to increased fertilization by producing much greater yields of corn, rice or wheat. Wheat crops in Mexico exploded by a factor of 4.3 times the yield per acre when genetically enhanced seeds were used.[7] Dramatic increases in yields per acre of these crops, ranging from 50 to 200 or even 400%, have been seen in many areas around the globe. The International Rice Research Institute in the Philippines developed genetically improved strains of rice that have increased yields by two-fold.[6] There are now varieties being addressed that put more energy into developing grain-bearing seed heads, and other characteristics are being added that include resistance to drought, flood, salt and pests in order to further boost yields. So striking has been the increased production that the incorporation of these new seed varieties and processes became known as the Green Revolution.

Historically, improving yields in crops was a relatively slow process resulting from the selection of seeds that were better performers within existing crops. These seeds varied slightly in genetic makeup from the others and possessed the characteristics necessary to produce somewhat greater yields per acre. The investment by governments into international agricultural research allowed scientists to rapidly accelerate this process through crossbreeding and mutation by ultraviolet or gamma radiation. The technique of crossbreeding involved the selection of genetic and

Figure 3.1 The world markets and the "green revolution" may promote monoculture techniques that could prove to be ecologically unstable. (Photo from Discimagery.)

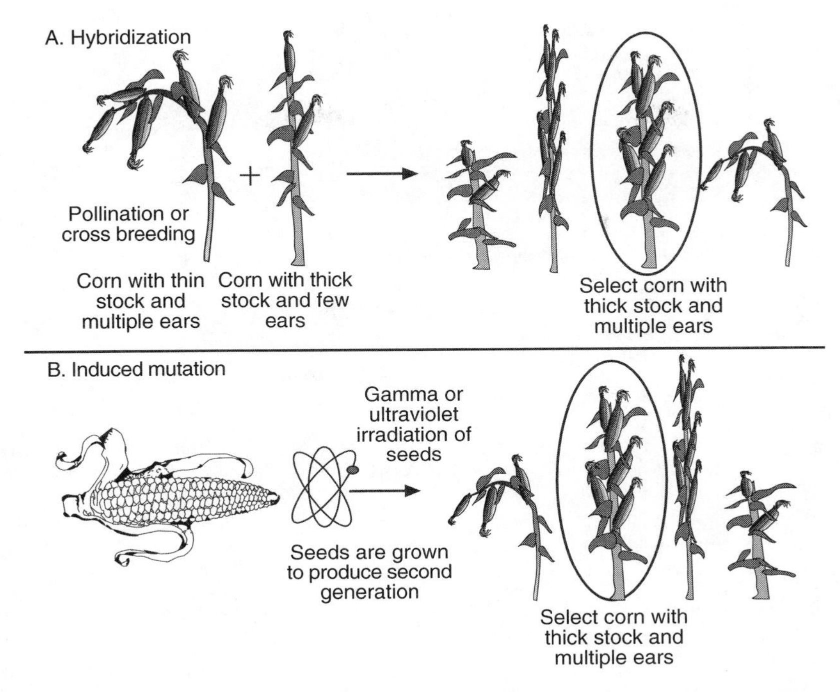

Figure 3.2 Historically, plant improvements were achieved with crossbreeding and artifical mutation. (Adapted from Turk, J., and Turk, A., *Environmental Science,* 4th ed., Saunders College Publishing, Philadelphia, 1988.)

morphologic variants that possess one or more of the desired characteristics and then crossing the plants in the hope that the most desired characteristics of the different plants will appear in the new crossbred plants. As an example, corn with a heavy stalk but few ears might be crossed with corn that has many ears but a thin stalk that would not ordinarily support the additional ears of corn. The desired product would be corn that has a thick stalk and many ears (Figure 3.2a). The genetic makeup of plants can be artificially altered by using ultraviolet or gamma radiation and producing mutations. The mutations will produce many variations of the plants, but some of these could be beneficial and produce the desired characteristics of many ears and a thick stalk, as described in this example (Figure 3.2b). More recently these processes have been improved with the advent of genetic engineering, which uses various artificial techniques to alter genetic information and produce new and unique plants. One of these techniques involves gene transfer. Gene transfer techniques were initially limited to bacteria that might be beneficial to certain plants because they were much simpler to treat in this way. Most plants are difficult to manipulate with gene therapy, but newly developed techniques use a "gene gun" that incorporates the desirable DNA fragments into a plastic bullet. This bullet is then fired by a gene gun into a barrier that shatters the bullet, creating a genetic spray that pierces the plant target at over 1000 mph.[7] Leaf cells are then cultured

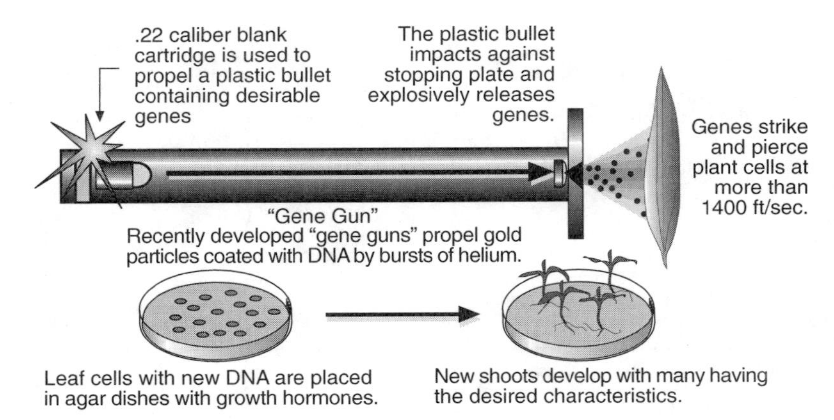

Figure 3.3 Methods of gene transfer in plants may use the "gene gun" that blasts the desirable genetic fragments into the plant. (Adapted from Budiansky, S., 10 billion for dinner please, U.S. News & World Report, Sept. 12, 57, 1997.)

with growth hormones to produce plant shoots in a few weeks. The plant shoots exhibit characteristics associated with the newly incorporated genetic materials (Figure 3.3). Viral-resistant barley was created this way, and these techniques are used to develop corn that can play host to nitrogen-fixing bacteria and thus require less fertilizer.[6] On March 3, 1988, the USDA and Delta and Pine Land Company announced that they had a patent on agricultural genetic technology. This technology, referred to as the terminator seed, has an embryo-killing toxin genetically spliced into the gene sequence before a gene promoter sequence can begin. To allow the parent plant to produce seeds, a genetic spacer is inserted between the toxin and promoter sequences. This way, in early germination, seeds are produced because the toxin and promoter are not touching. At this time, however, a protein recombinase that recognizes the spacer as unnecessary is activated and it removes the spacer sequence. The toxin and promoter are now touching, and the resulting toxic protein kills the embryo in late embryogenesis. The seed sits in the ground, producing no off-spring.[1]

Other agricultural techniques being developed involve the improved application of fertilizers using global positioning satellite (GPS) systems, and new soil tilling methods that reduce erosion (Figure 3.4). Precision farming depends on the use of GPS to exactly match the rates of application of seeds and fertilizer to existing soil conditions. Signals from the satellite are sent to the tractor's on-board computers which determine the coordinates and adjust the amount of fertilizer dispensed in each area of the field. This process reduces waste and increases yields. There are now "no-till" planters that cut a narrow trough or slit through the crop residues into the sod, drop the seed in place, and then cover the seed. This substantially reduces soil erosion.[6] New techniques are being developed all of the time.

However, there remain concerns about the application of new agricultural technologies. The new breeds of plants that are selected or genetically altered to produce higher yields often require productive soils along with extensive application of fertilizers, water, and insecticides to promote the potential growth of the plants and

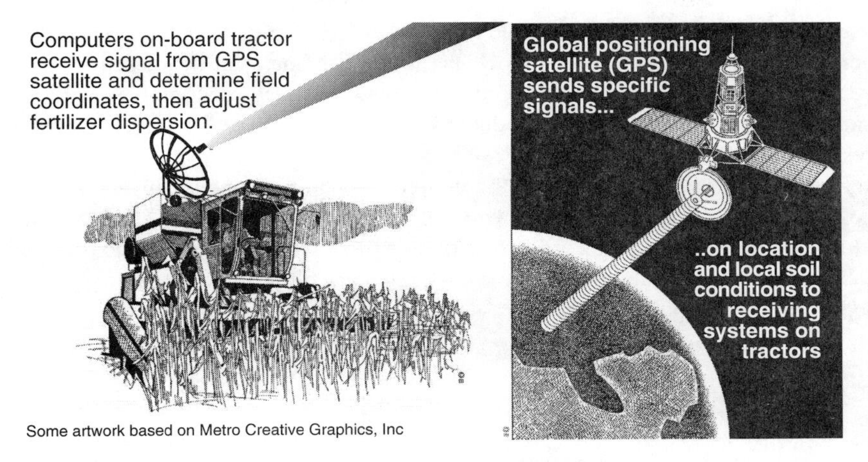

Computers on-board tractor receive signal from GPS satellite and determine field coordinates, then adjust fertilizer dispersion.
Global positioning satellite (GPS) sends specific signals...
..on location and local soil conditions to receiving systems on tractors
Some artwork based on Metro Creative Graphics, Inc

Figure 3.4 "Precision farming" techniques use satellite maps and onboard computers to adjust control rate of seed and fertilizer spread. (Adapted from Budiansky, S., 10 billion for dinner please, U.S. News & World Report, Sept. 12, 57, 1997.)

protect against insect invasion.[8] The higher yields that may be achieved often prompt farmers to plant single crops or monocultures because they may be profitable in the short term; but they could result in extensive ecological damage through the depletion of soil nutrients, erosion, and the attraction of pests. The requirement of these high-yield plants for more water, pesticides, fertilizers, and production equipment demands ever greater needs for energy as these components all require fuels to be produced or delivered. Many areas of the world likely to experience drought or poor water control (necessary for rice production) have not widely adopted these crops. Diffusion of high-yield crops is slow in areas that have inadequate infrastructure or market access such as Sub-Sarahan Africa or the hillside systems of Latin America and Asia.[8]

There is little debate that these advances in agricultural technologies have contributed significantly to reducing hunger in millions of people. Many LDCs have adopted the seeds and technologies produced via the Green Revolution. China's rice and maize crops are almost entirely planted in modern varieties. Although some exceptions exist, most of the growth in production in developing countries is a result of the higher yields generated by the Green Revolution.[8] Nearly three fourths of rice and wheat crop areas and more than half of all maize croplands were planted in modern varieties.[9] However, the growth of the human population in many of these lesser developed countries has exceeded the capacity of even these technological wonders in agricultural production,[6] while the energy to support high-yield cropping techniques is becoming more expensive.[8]

Agricultural researchers claim that we could expect more increases in crop yields, provided that money continues to flow into agricultural research and into credit and technical assistance for farmers who wish to apply the technologies. However, funding for CGIAR centers has been falling, causing a reduction in more than 100 scientists, while U.S. aid for foreign agricultural programs has sharply fallen by more than 75% since the 1980s.[7] Further, grain harvests have leveled off or seen a reduction in growth rate such that global cereal reserves will continue to remain below minimum safe levels for world food security.[10] Therefore, as global population increases and per capita food supply has leveled off or even decreased in critical areas, it is difficult to share the optimistic view that new technologies will cause another quantum burst of food productivity in the next 20 to 50 years. Decreasing fresh water supplies, reductions in fertile croplands, less effectiveness of fertilizers and pesticides, and the increasing likelihood of major climactic shifts in some of the more productive areas of the world all join to make it more difficult to meet future food demands at affordable prices. Efforts will need to be expanded in stabilizing world population and in sustainable agriculture if we are to avoid catastrophes to human populations.

Energy

Although there have been major technological advances in fuel use, LDCs continue to use wood as a primary fuel for cooking and heat. Wood is being used at such a rapid pace in some LDCs that forested regions have been decimated, and the collection of wood for fuel may require several hours each day or as much as 25% of average income.[11]

On the other hand, the history of fuel use in the developed nations moved from wood to more efficient fuels. As the industrial revolution entered the 17th and 18th centuries, a process for making charcoal from wood was developed; charcoal became the main fuel for iron smelting and virtually depleted much of the forested land of the U.S. Technology again entered the picture as wood resources became a limiting factor, and a process was developed that converted coal to a useable coke form necessary for powering iron refineries. As the need developed for economical fuels to power automobiles, trucks, tractors, and eventually planes, petroleum was refined into gasoline and oil. The requirements for petroleum increased, and new technologies were developed that permitted the detection and extraction of previously unrecoverable sources, driving the cost of oil down so that it has become an inexpensive fuel for much of the world when compared to total wages.

A few of the developed nations appear on the verge of entering yet another energy transition as the use of petroleum is reduced in favor of less polluting alternatives. Humankind continues to respond with the development of new, more efficient energy sources. As technologies expand, they create the possibility of feeding, clothing, and housing even greater numbers of people. Authors such as Julian Simon argue that human adaptation and change will avert disaster as more technologies are developed in response to growing population pressures.[12,13] Although such authors argue that technology will support ever-increasing populations, there is a practical limit to the maximum number of people this planet can sustain with a quality of life. The tragedy of the human condition is already evident in many parts of the world. Tens of thousands die each day from malnutrition and diseases in countries where resources have been exhausted and the land laid to waste. This is contrasted by those countries that have brought population under control, have enormous wealth, and enjoy a very high standard of living, such as the U.S., Western Europe, Japan, Australia, New Zealand, and other developed countries. Poor developing countries are generally blamed for world crises because of their high growth rates and subsequent environmental damage. However, Paul Erlich, professor of biological science at Stanford University, thinks the impact of humanity on the Earth's life-supporting systems are not merely decided by the numbers of people but also by how people behave. The impact of human activity on environments can be summarized by the following relationship:[14]

$$I = P * A * T$$

where

I is the impact of human energy-related activity on the globe
P is the population size
A is the affluence in terms of per capita consumption
T is the technology to supply each unit of consumption

According to Ehrlich, it is energy-intensive activities that cause most of the environmental degradation. The average energy consumption per person in the developed "rich" countries is 7.4 kw, while the average person in the LDCs uses less than 1 kw per year. There were 1.78 billion people in the "rich" nations in 1997, so their total environmental impact as measured by energy use was 8.7 trillion watts. Some 4.67 billion people lived in poor nations in 1997, hence their total impact was 4.67 trillion watts. The relatively smaller population of those in the developed countries accounts for nearly two thirds of the environmental destruction as measured by energy use. About 5% of the world's population lives in the U.S., where 35% of all raw materials are consumed. Moreover, the gap between the rich and poor countries is growing. A United Nations report released in September of 1997 shows a widening economic gulf occurring among the world's poorest countries driven by wars and political infighting.[15] Such unrest is thought responsible for reductions in per capita gross domestic product in the Congo, Burindi, Somalia, Sudan, and the Central African Republic, as examples.

Attitude and Behavior

The question posed by many is whether human society has the capacity to adjust to the dizzying pace of population increases without major political upheavals, where war and environmental destruction result in human calamity. We have seen in this chapter that neo-Malthusian arguments predict the inevitable misery of humankind as the population momentum outstrips the carrying capacity of the Earth.[16] We have also presented the optimistic "Cornucopian" outlook that more people will drive the technology to support the world's growing billions. Which of these views will ultimately be realized? Will we progress in a smooth transition to a world of global stability and health through carefully planned and thoughtful processes, or will national and personal interests prevail at the expense of the larger global community, resulting in the continued destruction of our environment and depletion of world resources? What are the attitudes and behaviors that may impact this outcome?

The Tragedy of the Common

If we are to believe the arguments proposed by Garrett Harden, the individual or nation that follows a healthy environmental protection ethic may be doomed unless everybody else follows the same path.[17] Where there is common land, air, or water, some people will take advantage of this situation and overgraze, overfish, or pollute that environment to their own advantage and at the expense of others. The deforestation and soil erosion in ancient Greece,[18] the overgrazing in the Sahel of Northern Africa,[11] or the polluting of air in New England by plumes of sulfur dioxide emissions from the Midwest are all examples of this behavior. As one person or segment of society emits its toxic brew into the environment while avoiding costly controls, the profit margin is enhanced in the short term. Others, seeing the advantage, are likely to follow suit, provided no adverse consequences are immediately applied. It is probably axiomatic that many members of any society will likely pass on the consequences of their destructive actions if they will benefit in the short term and receive little or no negative consequences from that action. Harden refers to this as the tragedy of the commons.[17]

The Pioneer

The consequences of laying waste to a land in the past were minimized by the ability of the population to emigrate, just as the pioneers in America could move west. There were enormous resources that permitted the early pioneers to cut trees, harvest the land, and trap or shoot wildlife with ecological impunity. The pioneer mentality cannot be continued indefinitely in the presence of massive population increases. The opportunity of mass emigration has been largely eliminated throughout the world as most habitable areas are now populated. The consequences of unrestricted use of resources well beyond their replacement level will eventually result in ecological and human disaster. We must seek a sustainable development where resource use comes into balance with replacement and the discharge of pollutants is minimized.

Declining Investment in Technologies

The optimistic attitude of the Cornucopians toward population increases demands an investment in new technologies that will help to sustain these populations. Unfortunately, the major gains in food crops experienced as part of the Green Revolution are unlikely to continue in the absence of investment in research and development. Further, these technologies, along with financial support, must reach the farmers in the developing countries. Just as it would be impossible to feed today's population with 50-year-old technology, it will be impossible to feed the population of the year 2050 with today's technology. However, government funding for such organizations as the CGIAR centers, which are largely responsible for the Green Revolution, has been falling. Despite contributions from the World Bank to shore up the losses, more than 100 scientists left the organization in 1993 to 1994 alone.[6] Further, the U.S. aid for agricultural programs has sharply declined by more than 75% since 1980.[6]

Family Planning Cuts

The 104th Congress of the U.S. reduced foreign assistance overall in 1996 with a 25% decrease in USAID funds and a 35% cut in the family planning/population assistance budget. The allowance was spread over a 15-month period that reduced the USAID 1996 population funding by nearly 80%.[1] Much of this reduced spending by the U.S. is intertwined with the debate over abortion. The 1998 budget for population assistance survived the extreme congressional cuts experienced in 1996, and $385 million was awarded for international population assistance. This was still $58 million below the administration's request, and there will remain severe shortfalls in planning assistance in LDCs resulting in up to: (1) 220 million unintended pregnancies; (2) 117,000 additional maternal deaths and 1.5 million women who experience permanent impairment; and (3) 9.3 million additional deaths of infants and young children.[4] Because the U.S. has normally contributed the most to family planning, the effect worldwide will be very limiting to programs in LDCs, and it is predicted that unavailability of family planning services will likely increase abortions. This is the very opposite effect that was intended by those leading the effort to cut population assistance funds.[1]

What Now?

We are now faced with two very different views on the effect of world population on the human condition and global ecosystems. The pessimistic view warns that population momentum will hurl humankind beyond the carrying capacity of the Earth, creating enormous human misery and suffering and dramatic destruction of environmental support systems. Others view people as an asset, finally prevailing with the development of new technologies and policies that will avoid a cataclysm. As with most polarities in life, the reality is probably somewhere in between and is already being played out in regions around the world. Many of the LDCs are showing a downward trend in birth rates associated with policies that empower women through education and economic opportunity. Overall world growth rates are declining, and the public news media boldly broadcasts that the population problem is over. Certainly, these are very substantial reasons to be optimistic about the future. However, just as there are positive examples, so are there dramatic failures that give reason to be apprehensive about the future.[19] Just as we can believe in technology as rescuing humankind, we can also point to declining investments in agricultural research, family planning, and the preservation of ecosystems. Social and political programs may not be capable of making the necessary adjustments, and ecological damage may already be sufficiently severe that it is irreversible in many parts of the world.[20] The ability of people to make predictions about the consequences of increasing populations depends on too many variables to be accurate. There are no simple answers to these issues. We can, however, examine the current state of human affairs and global ecosystems to capture a glimpse of where we have been and where we are now. This may give the students of environmental health enough information to personally develop their own visions of the future world.

IMPACTS ON THE ENVIRONMENT

We have all seen a spider web, sometimes glistening in the early sunlight still wet from the morning dew. Touch one strand of the web and all will move and vibrate. Our environment is so intertwined that each human action ripples across the environment, causing imperceptible changes throughout the system that become magnified by billions of people. As the population increases, the need for food increases. Land is cleared or grassland is cut and uprooted. Topsoils are lost, nutrients depleted, and balanced ecologies of insects, birds, and mammals are upset as habitats are destroyed. Fertilizers and pesticides are applied that wash into streams and seep into groundwater, causing soil and water contamination that biomagnifies up the food chain and causes reproductive abnormalities and disease in wildlife and humans. As insects develop resistance and natural soil nutrients are depleted, more chemicals are applied to croplands. These chemicals require energy in their manufacture that is drawn from the combustion of fossil fuels. These combusted fuels release pollutants into the air that produce acidic deposits and climate. Climate change and acidification of soil and water may cause further ecological damage, contributing to a downward spiral from a sustainable environment. The circle of influence moves from more people, greater deforestation, poorer soil, more deserts, fewer species of animals and plants, and less food. This is the sequence of events we will explore in the following pages.

Deforestation

Although life is highly adaptable, most species require a source of energy such as sunlight (required by most plants), nutrients, and moisture. Organisms are also affected by the presence of oxygen and the availability of warmth. Life exists in a relatively narrow shell, extending about the globe a few kilometers into the air and below the sea. The richest and most diverse forms of life tend to be in the tropical rainforests, while harsher conditions at the top of the Himalayan mountains or in the frozen, bleak landscape of the polar ice caps harbor very few life forms. Just as conditions of temperature, moisture, and nutrients vary in regions around the world, so do the types and diversity of vegetation and animal life change. Several major regions exist around the world that are identified by the major types of vegetation. These are called biomes, and include tropical rainforests, temperate forests, prairies, deserts, and arctic tundra. Since our present topic is deforestation, we will focus on tropical rainforests because the majority of them exist in areas of the world at risk from overpopulation, and many are threatened by slash-and-burn techniques to make room for croplands.

Rain Forests

The tropical rainforests (Figure 3.5) are located predominantly in the LDCs in a latitudinal band forming a belt around the globe from 20¼° South to 30¼° North. Rainforests are located in many countries along this band including Mexico, Belize, El Salvador, Guatemala, Nicaragua, Colombia, Ecuador, and Brazil, eastward to Cote

Figure 3.5 Typical lowland tropical rainforest. (NASA, The Earth Observing System Educator's Visual Materials, 1998, http://eospso.esfc.nasa.gov/cos-edu.pack/p24.html)

d'Ivoire, Nigeria, Congo, the Central African Republic, and further east to Malaysia, Indonesia, the Philippines, New Guinea, and the Solomon Islands (Figure 3.6).

The benefits of the tropical rainforests are that they: (1) are a major producer of oxygen for the global atmosphere; (2) are the major carbon dioxide sink; (3) are a potential source of new pharmaceuticals useful in the treatment of human disease; and (4) are an important source of species diversity. Rainforests cover less than 10% of the Earth's surface but contain more than one third of the Earth's biomass, which are instrumental in absorbing large amounts of carbon dioxide and releasing great quantities of oxygen and so serving as the "lungs of the planet."[21] In 1995, a team of British, Australian, and Brazilian ecologists found proof that the Amazon rainforest absorbs tons of carbon dioxide produced by burning fuels. The globe's one billion acres of rainforests may absorb one sixth of the amount of carbon dioxide produced by burning fuels each year.[22]

Some of the most important and historical drugs isolated from plants include those used for treating circulatory disorders and high blood pressure. Some of these come from the Indian Smoke Root, Rauwolfia Serpentina. Caloniea, from the Malaysian tropical rainforest tree, is presently being developed and tested for the treatment of HIV-infected persons in the U.S.[23] Plants from the Peruvian rainforests could help doctors in their fight against deadly tuberculosis. In a study of 1250 plant extracts from Peru, 46% showed an ability to inhibit the growth of Mycobacterium tuberculosis (M. tuberculosis), the bacterium that causes tuberculosis. There are myriad other pharmaceuticals being used and investigated in the treatments of cancer, blood pressure, HIV and other disorders. There are extensive efforts under way to investigate the possible pharmaceutical benefits from tropical rainforest plants before these plants disappear as a result of ecosystem destruction.

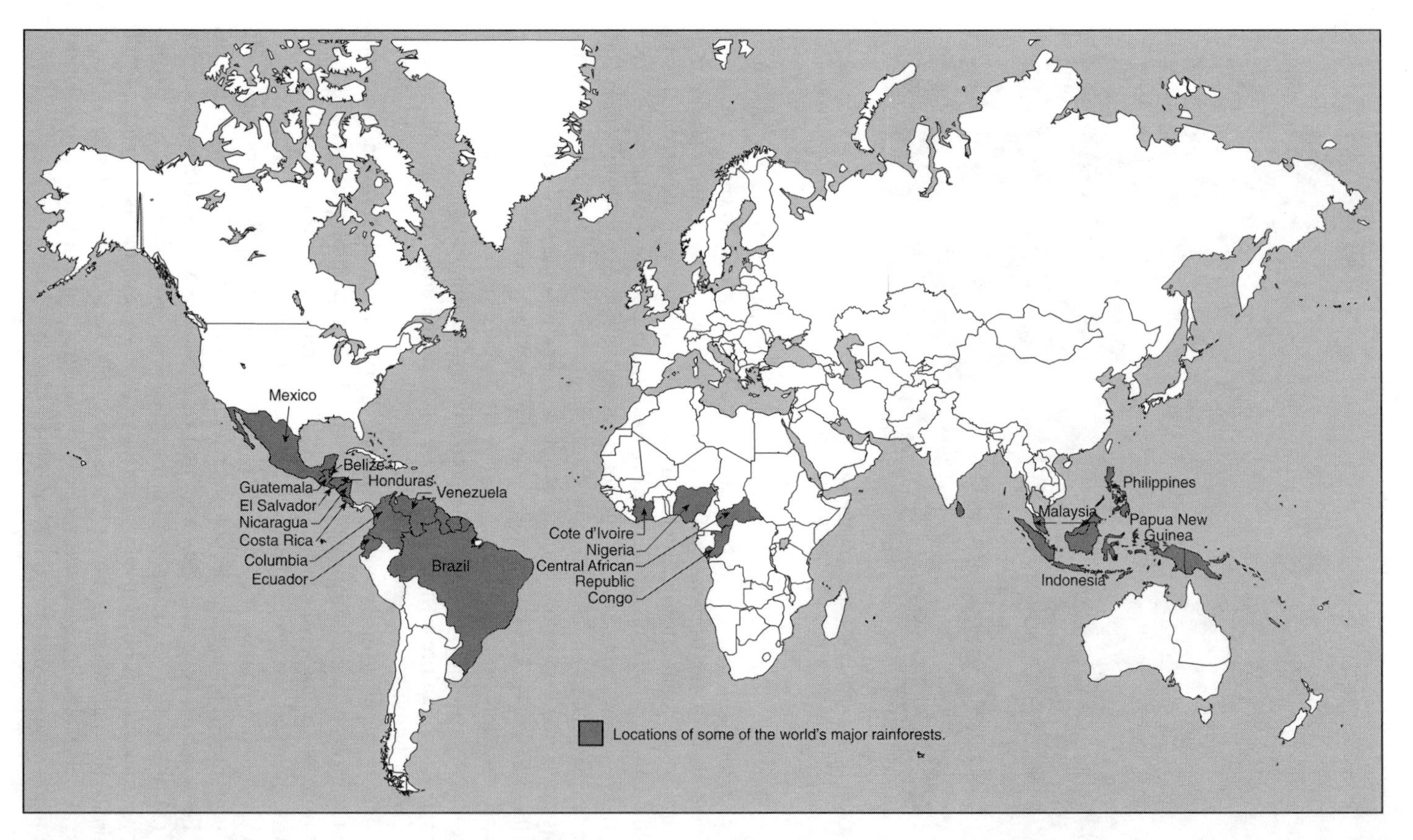

Figure 3.6 World map showing regions of major tropical rainforests. (NASA, The Earth Observing System Educator's Visual Materials, 1998, http://eospso.esfc.nasa.gov/cos-edu.pack/p24.html)

In spite of the numerous benefits from rainforests, they are disappearing at an alarming rate. This process is leading to *deforestation*, which has been defined as the permanent decline in crown cover of trees to a level that is less than 10% of the original cover. Prior to the human population explosion, by some estimates the tropical rainforests covered 16 million kilometers or nearly 11% of the Earth's land area. By 1987, tropical rainforests were disappearing at the rate of 42 million acres each year, representing a loss of 115,000 acres each day. The United Nations Environmental Program reports that half of the world's original forests have been lost and only one fifth of the remaining forests are "frontier forests" — pristine areas that have not been disturbed or degraded by human activities.[2] Tropical deforestation contributes significantly to the net movement of carbon (carbon dioxide) to the atmosphere because the concentration of carbon in forests is higher than in the nonforested areas that replace them. Additionally, the tropical rainforests occupy less than 7% of the land surface but contain more than half of all plant and animal species. The greatest problem associated with tropical deforestation may be related to the massive extinction of species. The diversity of plant and animal species is affected by: (1) destruction of habitat; (2) the formation of fragments of forested area less than 100 square kilometers; and (3) edge effects occurring as far as 1 km into the forest from adjacent deforested areas.[24] This boundary zone has greater exposure to winds, easier access by livestock and humans, and many other effects that result in loss of plant and animal species at the edge. Edge effects greatly magnify the adverse effects of deforested areas.

The Brazilian Amazon is the largest region of tropical forest in the world, containing 31% of the world's total.[24] NASA's (National Aeronautics and Space Administration) Global Inventory Monitoring and Modeling Study showed that, in 1987, fires burned 77,000 square miles of the Brazilian part of the Amazon Rainforest, or an area greater than New York by 1.5 times.[25] During the months between July and October 1987, 4% of Brazil's Amazon jungle was destroyed — most of this destruction followed the route of the World Bank-financed road through Rondonia known as BR-363. Brazil's poor and destitute have flocked to this area in search of arable land, and they clear the land by setting fire to it. NASAs heat-sensitive film taken from satellites shows nearly 350,000 fires were started in 1987, perhaps even contributing to the destruction of the Earth's ozone layer. However, the soil is so thin and of such poor quality that the settlers have to clear new plots every year or two. With no canopy to break the intense downpours, the thin topsoil is easily washed away. The birds and other natural predators to insects are eliminated through habitat destruction, and mosquitoes often abound, creating greater opportunities for vector-borne disease transmission.

The consequences of tropical deforestation are well documented and well understood globally. There are, in fact, 219 international groups concerned with saving the Amazon forests.[26] Unfortunately, reforestation efforts have met with dismal failure in the wake of the continued rapid growth of human populations. According to the Brazilian government's Institute for Space Research, weather satellites detected 39,900 fires in July of 1995, a nearly five-fold increase over the 8503 fires reported in July of 1994. These fires are not limited to Brazil but are spanning over a number of countries including Uruguay, Bolivia, and Paraguay, producing a plume of smoke covering over 6 million square kilometers. The results are respiratory irritation and airport closures on many days.[27] Environmental experts contend that Brazil's failure to analyze the extent of deforestation since 1991, along with stepped-up levels of burning in the Amazon region, raise doubts about the claims that the destruction of the Amazon rainforest has slowed as a consequence of government action. This situation has worsened, with LandSat images revealing that it was burning at an alarming rate within the last 4 years. Scientists working with the Washington-based Environmental Defense Fund report they "have never seen it so bad."[28] Further, the areas revealed by high-resolution LandSat images only showed cleared areas and did not reveal the almost equal areas that were impoverished beneath the canopy.[2]

Developing countries are mainly responsible for degrading nearly 14% of the world's remaining frontier forests due to their massive consumption of forest sources to meet their people's needs.[2]

Central America has been experiencing tremendous losses of its forests to industrial felling and the practice of slash-and-burn agriculture by the rural, landless poor. The felling of native forests has been leading to serious soil erosion in these parts. Panama, El Salvador, and Guatemala have been found to have the worst soil erosion in the world according to a study conducted by the Center of Studies and Social Action of Panama. Nearly 98% of El Salvador's natural forest cover has vanished, with Panama and Guatemala close behind.[29] Dramatic oil price increases by OPEC in the early 1970s placed kerosene and other petroleum fuels out of reach for millions of the world's poorest, leaving them no alternative but wood. Poor families in El Salvador and Guatemala are spending an entire day every week looking for firewood and traveling up to 12 kilometers from their homes to find it.[29]

The tropical forest of Africa is 18% of the world total and covers 3.6 million square kilometers of land in West, East, and Central Africa. Annual rates of deforestation in the region varies from 38,000 acres in Gabon to nearly 742,000 acres in Cote d'Ivoire. West African forests have been reduced to patches. African forests are the most depleted of all the tropical regions with only 30% of historical stands remaining.[30] Just as in part of South America, many Africans must travel miles away from home gathering firewood or expending 25 to 30% of their income for wood to burn.

A study published by the Worldwatch Institute, a nonprofit research organization dedicated to the analysis of global issues, concludes that nearly three quarters of the middle mountain range in Nepal has been deforested in the last 40 years. India has lost 40% of its forest cover within the last 30 years, and Pakistan has only a fraction of its original forested land. The deforestation contributes to extensive flooding, costing Pakistan about $1 billion each year in crop, building, and infrastructure losses.[31]

The rate of deforestation in the Pacific and Asia could decimate some 450 million acres before the year 2000. The greatest rates of deforestation are occurring in Indonesia, followed by Thailand, Malaysia, India and Laos.[31] Hundreds of thousands of acres are also being lost in Nepal and Vietnam.

Although 70 tropical countries have agreed to support the International Tropical Forests Action Program (TFAP) initiated in 1993, the full implementation of this program will require substantial investments of time and resources.[32] According to the Global Environmental Outlook Report released by the United Nations Environment Program (UNEP) at Nairobi, Kenya, "International and national funds and political action remain insufficient to halt further global environmental degradation."[20]

Forests in Developed Countries

The area under forest cover in developed countries has remained fairly unchanged during the last decade.[20] The forested areas in these countries are stable and increasing in the temperate regions where alternative fuels and forest management programs have sustained timber harvests. The major issues surrounding temperate forests include: (1) damage by air pollution; (2) the loss of biological variety; and (3) fragmentation with the loss of forest edge. Many Western European countries have actually experienced an increase in forest cover. Since the mid-1960s, Sweden, France, and the U.K. have experienced increases of forested area of 22, 34, and 37%, respectively. The U.S. has seen a significant return of forests compared to the 18th and 19th centuries, when farmers cut down all but 2% of the trees in the northeast.[33] However, in several important ways, the forests are fundamentally different from that of previous centuries. At 90 years old, the forest ecosystems are not yet mature. Ironically, the oldest trees are preserved in cities and suburbs. Exotic species from other continents have taken hold. Many native species are threatened by diseases and pests. Expanding suburbs have partitioned or fragmented the forest into patches, creating lots of forest "edges." Maine, Vermont and New Hampshire, as well as Northern New York, contain 26 million acres of forest known as the Northern Forest. This precious resource is in serious danger according to the Northern Forest Alliance, a coalition of about 25 local and national environmental groups. The danger stems from the process of clear-cutting by lumber companies, by recreational pressures from millions of campers,[33] and by the constant threat of acid deposition from air pollution.

Nevertheless, developed countries also bear inescapable responsibility for the rapid loss of forest sources. It is indicated that "North America, Europe and Japan, with just 16% of the global population, consume two thirds of the world's paper and paperboard and half of its industrial wood.[2]

Protecting the Rainforest

The World Bank has invested in sustainable forest development and offered incentives to companies to harvest a variety of tree species and plant replacement trees. The Brazilian government has begun talks with representatives of the major industrialized nations on the progress of an ambitious program to save the Amazon rainforest. The project, which was set up in 1991 and is funded mainly by the industrialized countries, aims to reduce deforestation, protect the Amazon's unique biodiversity, and encourage sustainable development. Brazil has lifted a ban on new logging permits in the Amazon rainforest after landowners and loggers agreed to slow their rates of forest destruction. However, Brazil's authorities refused to issue new logging permits in February after they discovered more than 15,500 square kilometers — an area the size of Belgium — was cleared in 1998.[3]

Soil Degradation

What is Soil?

Do not make the mistake of telling a farmer that his "dirt" must be really good to grow such excellent crops. Normally, he (or she) will acidly remind you that this is "soil," not dirt. What is the difference? The difference is that you can grow things in soil. Soil consists of small particles of rock and minerals mixed with a major proportion of plant and animal matter in various stages of decay. This organic matter is filled with living organisms including bacteria, algae, fungi, protozoa, insects, mites, and worms. A handful of soil is a complex, living ecosystem that provides the nutrients and conditions for plant growth. As plants grow and ultimately die, they serve as the nutrient source for a living mass of microorganisms that effectively recycle the organic materials into fertile soils. This organic material supports the growth of plants that may in turn become the nutrients for higher forms of animal life.[7] Plants are called autotrophic because they synthesize their own food from inorganic substances. Plants derive carbon, oxygen, and hydrogen from carbon dioxide and water. Plants also require nutrients they derive from the soil including nitrogen, phosphorous, calcium, magnesium, and potassium. These are called macronutrients because plants use significant amounts of them. There are a number of minerals that plants require in only trace amounts. These are known as micronutrients and include such substances as zinc and molybdenum.

Soils vary in physical nature. Some soils are sandy and contain larger, gritty particles that allow water to flow through rapidly and carry away or leach nutrients. Clay soils consist of extremely small or "greasy" particles that are bound together very tightly by chemical attraction. These soils tend to be impermeable to water, allowing rain to puddle on the saturated surface. Silt tends to be composed of fine particles that are larger than those producing clay but smaller than sand particles. Soils best suited to agriculture consist of sand, silt, and some clay in a homogeneous mixture referred to as *loam*. Fertile soils contain complex organic matter that has been biologically broken down so that original plant and animal matter is unrecognizable. This organic matter is *humus* and it serves to: (1) retain moisture much as a sponge; (2) serve as an insulator to heat and cold; and (3) bind and release nutrients to plants in useable forms. Soils that lack humus have low or poor fertility.[7]

Soil Biomes

Not all soils are equally fertile, and soils also vary substantially about the globe in their ability to sustain intensive agriculture. The most fertile soils for the growth of foods such as corn, wheat, and rice are located predominantly in the global temperate zones of Northern America, Europe and Asia. A smaller belt of the temperate zone climate may be found in Argentina, Southern Africa, Australia, and New Zealand (Figure 3.7). Such temperate zones of forest, woods, and grassland tend to have fertile and deep soils of one to two meters because the moisture and temperature of these regions permit the growth of substantial plant cover while permitting organic matter to accumulate more rapidly as it decomposes. In higher

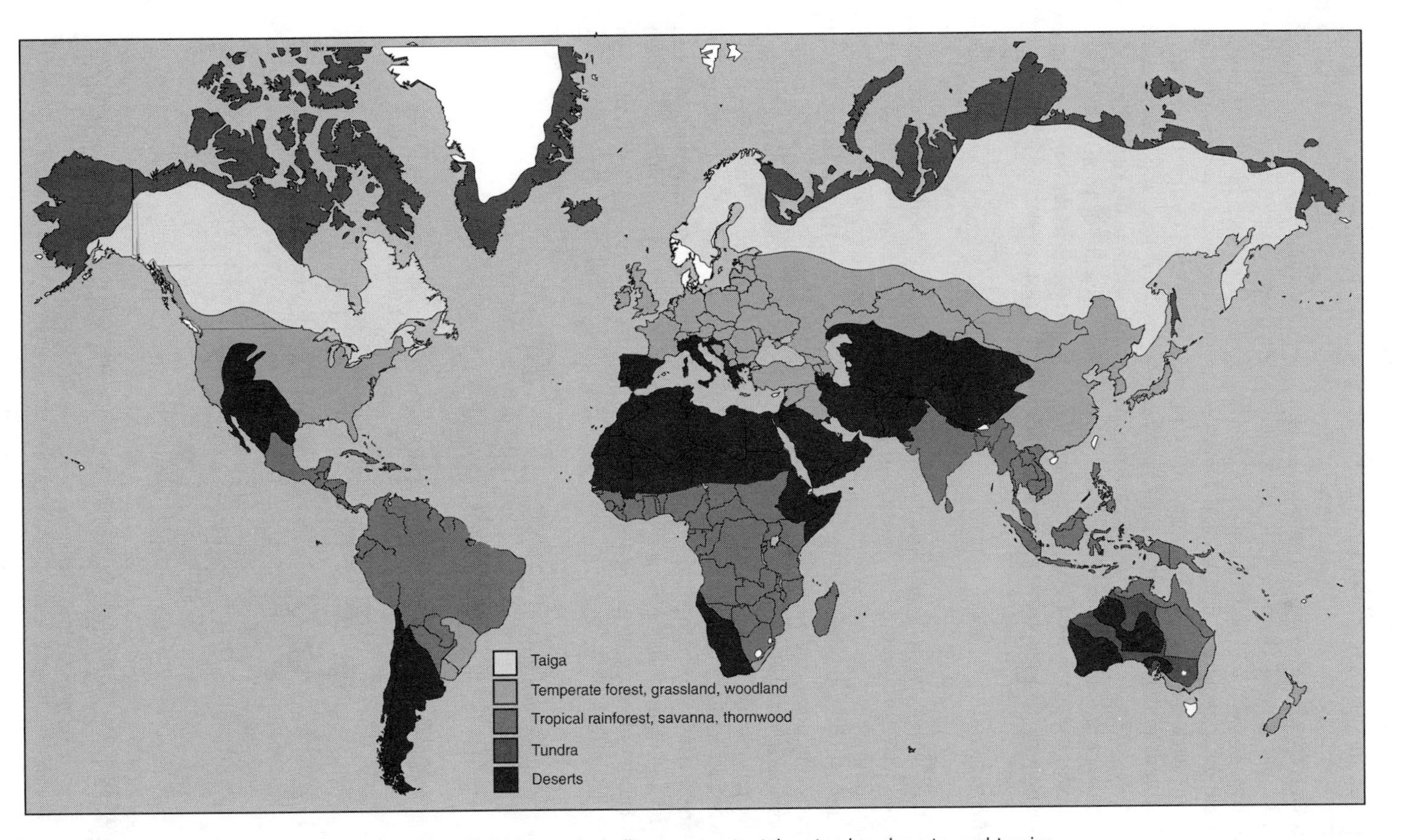

Figure 3.7 World map showing regions of major soil biomes including temperate, taiga, tundra, deserts, and tropics.

latitudes, the growing seasons become very short and substantially less organic matter is produced. The boreal forest or taiga is an example of this biome and is represented by a great circle of spruce and pine trees stretching from Alaska through Canada and Russia. Such organic matter is decomposed very slowly, resulting in thin soils. In even higher latitudes is the tundra. This is an area where there are no trees but great expanses of mosses, lichens, sedges and occasional low-lying bushes. The soils here are very thin because the weather is extremely cold, restricting plant growth and inhibiting decomposition. Further, the thin soil lies atop a permanent layer of ice or frozen soil known as permafrost. Any disruption to the sensitive tundra takes many years to recover. Southern or "horse" latitudes at 20° to 30° North and South exhibit some of the Earth's driest and largest deserts such as the Sahara because the regions receive a constant supply of warm, high pressure.

Soil Erosion

Consider that the soils that sustain life around the globe are really extraordinarily thin and vary from one to two meters to just a few centimeters. The critical layer of soil is in constant flux as new organic matter decomposes and is added to the soil, and some of it is carried or swept away by a process called erosion. In most natural conditions, the replacement of soil is equal or greater than erosion. This allows soil depth to be maintained or to increase. However, poorly managed agriculture has caused erosion to proceed more rapidly, resulting in impoverished soils. As woods are cut and fields are plowed to plant crops, soils are lost to the effects of wind and runoff water (Figure 3.8). Techniques to prevent soil erosion have been practiced in many parts of the world for centuries. Traditionally, such practices have included the planting of nitrogen-using (corn) and nitrogen-fixing (beans) crops in

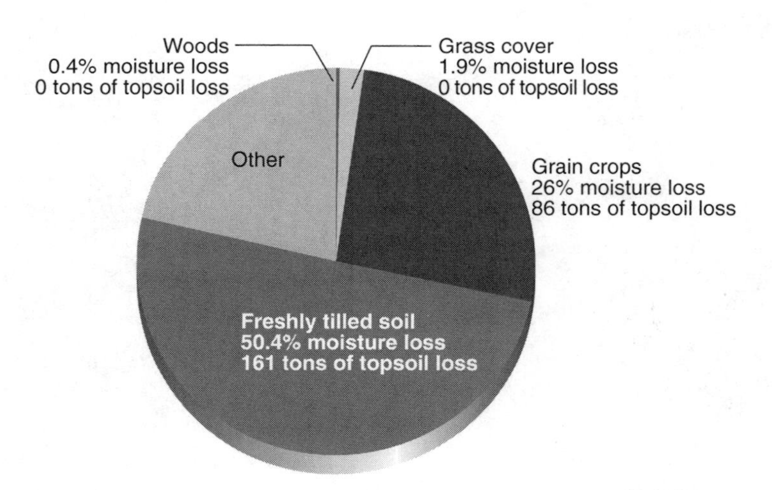

Figure 3.8 Loss of soil and moisture in a season per acre under various ground cover conditions. (Adapted form Turk, J., and Turk, A., *Environmental Science,* 4th ed., Saunders College Publishing, Philadelphia, 1988.)

alternative years in the same space or planting them in alternating bands in the same year. This is termed *rotation*. When fields remain unplanted every few years, this resting permits moisture and nutrients to return in a process called *fallowing*. When farmers are driven to plant on hillsides, they must create level fields along the contour of the slope in a process called *terracing*, or erosion will occur at an enormous rate and wash away the topsoil in a generation or two.[7] However, these techniques are not widely used, and soil erosion looms as one of the greatest potentially devastating problems facing the world today even as most countries fail to recognize it as such.[33]

Driven by logarithmically expanding populations, the developing nations of the world are forced to continuously cultivate their best farmland while moving agriculture into areas with decreasing agricultural sustainability. Many of these countries look to the U.S. for food, and that creates a high demand and high commodity prices. Farmlands in the U.S. have little time to lie fallow or be rotated as the land is pressed into service to grow food for the world's billions. North America now exports more than 100 million tons of grain per year, which is a four-fold increase over exports in the 1950s, and three times the combined exports of Australia, New Zealand, and Western Europe.[7] According to the U.S. Department of Agriculture (USDA) soil surveys, more than 30% of natural topsoils in North America has been lost since farming began in the U.S.[34] The average rate of topsoil erosion in the U.S. in the early 1980s was 4.7 tons per acre, with many of the agriculturally intensive states showing levels of 8.7 tons per acre. At this higher rate, and assuming an average topsoil of 12 inches, there would be a complete loss of topsoil in less than 150 years.[34] Nearly half of the cropland in the U.S. is losing soil to erosion, about 3 million acres are destroyed every year. In areas of moderate rainfall and temperate climate, it takes 30 years to form an inch of topsoil, which can be lost in 10 years in the presence of excessive grazing, monocropping, and ground cover destruction.[35]

Wind erosion devastated croplands in the Midwest during the 1930s, leading to a "dust bowl" environment and the subsequent creation of the Soil Conservation Service, which was renamed in the 1990s to the Natural Resources Conservation Service (NRCS). The NRCS has inventoried soil erosion in the major croplands of the U.S. and has shown that through crop rotation, improved tilling methods, and the Conservation Reserve Program (CRP) there has been either an improvement or no increase in average erosion rates since the 1980s.[35] The CRP was started under the Food Security Act of 1985 and offers farmers financial incentives to remove eroding land from production for periods of 10 years so they may recover. Despite these improvements, excess erosion still occurs in southern Iowa, northern Missouri, parts of western and southern Texas, and eastern Tennessee. A large section of the Great Plains stretching from Nebraska to northeastern Texas is unstable, and a prolonged drought could cause the entire area to become a Sahara-like desert.[35] A significant reduction in the U.S. food production would have a devastating effect on the world food supply.

When viewed globally, soil erosion claims over one billion acres every year, and 1.2 billion acres of global cropland is losing topsoil so rapidly that these acres are expected to become unproductive in the next few decades.[36] India is averaging topsoil losses at the rate of 9 tons per acre and is losing about 5 billion tons of topsoil a year at a rate double that of the U.S. Food security is a major issue in Africa, West Asia, and Latin America. The major concern in these areas is the availability of land and the prevention of degradation. More than one billion acres of land in Africa is moderately or severely degraded. Nearly 47% of the grazing lands in Latin America have lost their fertility because of erosion, overgrazing, and salinization.[20] As the lesser developed countries move towards another 5 billion people in the next decades, continued soil erosion in these countries could result in harvest reductions of 30% or more during the same period. Upwardly spiraling populations may make the political will, and funds to establish soil conservation programs, unlikely.[20]

The Process of Desertification

Piecing fences together by plywood, corrugated cardboard, and rusted scrap metal represent the futile efforts of the residents in Komsomolky to slow down the inevitable advancing desert from finally consuming their town. The Kalmyks, who inhabited the region for nearly 400 years, understood the fragility of the black lands in Southern Russia, which were once the bed of a larger Caspian Sea. They never tilled the soil. They raised sheep with flat-bottomed feet and did not let cattle graze in the area. These careful custodians of the grassy steppes of Central Kalmykia were expelled by Josef Stalin in 1943, and in their place were brought irrigation projects to develop cropland and sharp-hooved sheep. Because the topsoil lay over the former bed of the Caspian Sea, the irrigated land became a salt marsh and the sheep destroyed the thin topsoil, causing the desert to begin advancing in the 1950s.[37] The encroaching sands buried everything in sight and forced 25 towns to be abandoned while threatening 40 others. State-financed plans to reverse the trend have been halted as money ran out in the mid-1990s.[37] The agricultural mismanagement of this land has promoted the desertification of this delicate ecosystem, and it will take generations for the land to be restored even if the practices of mismanagement are halted.

What is Desertification?

The above scenario is being repeated in a number of areas throughout the world, but it was not until the late 1960s that desertification became a matter of international concern. Worldwide, with more than 10 million acres of farm land becoming unproductive each year, dust bowls are multiplying and raising legitimate concern about our planet's capacity to feed its rapidly growing population. During a period of severe drought in the Sahelian region of Africa from 1968 to 1973, the people suffered enormous tragedy through starvation. This drew worldwide attention to the problems of trying to survive on arid and semiarid land, especially on the perimeters of deserts. As a consequence, there have been a number of initiatives including a United Nations Conference on Environment and Development that was held in Rio de Janiero in 1992 (Rio Earth Summit). Here, more than 100 countries began the process of developing a legally binding international contract to combat desertification. This resulted in the United Nations Convention to combat desertification

Table 3.1 Types of Land Areas at Risk of Desertification by World Region [in square kilometers × 1000]. (Adapted from the brazil Embassy.[38])

Climate Type	Africa	Asia	Australia	Europe	North America	South America	TOTAL
Hyperarid	6,720	2,777	0	0	30	260	9,780
Arid	5,040	6,260	3,030	110	820	450	15,710
Semiarid	5,140	6,930	3,090	1,050	4,190	2,650	23,050
Dry Subhumid	2,690	3,530	510	1,840	2,320	2,070	61,500
TOTAL AREA	30,335	43,508	8,923	10,498	25,349	17,611	136,224

Brazil Embassy, What is desertification?, Science and Technology Section, National Research Network's Brazilian Regional Center, Feb. 30, 1997, http://brasil.emb.nw,dc.us./nuna/deserts/ingles/se/~edesert/desertmi.html#dm.

(known as the Convention). According to the Convention, desertification was defined as "land degradation in arid, semiarid and dry, subhumid areas resulting from various factors, including climactic variations and human activities."[38] The definition was adopted by the United Nations Environment Programme (UNEP), and areas susceptible to desertification were defined as those subjected to arid climates (arid, semiarid, and dry, subhumid). About 15 billion acres or one third of the Earth is dry land, and 2.5 billion (or 16% of the Earth's surface) of these dryland acres are hyperarid deserts where there is little or no growth. This leaves 12.5 billion acres, or one third, of the Earth's surface that falls into the categories of arid, semiarid, and dry, subhumid climates (Table 3.1). These areas include 51,720,000 square kilometers in nearly 100 countries that may be directly or indirectly affected by desertification. Some 900 million people live in these areas susceptible to drought and desertification, and 200 million of these people, or about 22%, are presently affected by this process.[38] Even more disturbing from a socioeconomic view is that the majority of high-risk areas to desertification are also home to some of the world's most impoverished people in LDCs. Poverty is globally recognized as one of the major reasons for degradation of arid, semiarid, and dry, subhumid regions.[38] The expanding population forces people to cultivate and overuse more and more marginal lands that are increasingly vulnerable to human and climactic disturbances. Growth conditions fluctuate from season to season and even within seasons. Dryland areas may receive an entire season precipitation in a few days, or experience prolonged droughts, although average rainfall over a decade may be normal. The people who inhabit these regions must adjust to the conditions. Drought years may require land to remain fallow or to be used in another way. However, poverty and the need for food are enormous pressures that defy flexible land use response and lead to desertification[39] (Figure 3.9).

The Costs of Desertification

The process of desertification is ongoing, and about 14.8 million acres (60,000 square kilometers) are removed from production annually because of poor land

Figure 3.9 Land degradation in arid, semiarid, and dry, subhumid areas resulting from various factors, including climatic variations and human activities, defines the process of desertification. This was once completely forested land in Africa.

Table 3.2 Estimated On-Site or Immediate Losses from Land Degradation with Loss in Productivity

	Estimated Direct Loss from Land Degradation (in Billion US Dollars)
Africa	9.3
Asia	21.0
Australia	3.1
North America	4.8
South America	2.7

Brazil Embassy, What is desertification?, Science and Technology Section, National Research Network's Brazilian Regional Center, Feb. 30, 1997, http://brasil.emb.nw,dc.us./nuna/deserts/ingles/se/~edesert/desertmi.html#dm.

management including overgrazing, improper irrigation, and intensive use. Economic losses are calculated to be $40 billion, while the cost of recovering these lands worldwide is estimated at $10 billion annually.[38] This $40 billion in losses from land degradation, is considered to be on-site or immediate losses from a reduction in productivity, are distributed as shown in Table 3.2. Costs may be many times higher if off-site or indirect costs were included relating to the loss of wells from silt formation, illiteracy among the young as they are forced to scavenge, and the cost of refugees and emergency aid.[38] There are also costs in human misery since the world's largest population increases are occurring in areas at the greatest risk to desertification.

Another loss difficult to measure is the reduction in biodiversity of important food crops. Many of our important food crops such as wheat, barley, and millet originate in drylands. The species that have adapted to drylands are few in number, and their loss to nonsustainable agriculture methods is a great threat to biodiversity. This is further exaggerated by the Green Revolution, which has developed high-yield crops but at the cost of fewer and fewer varieties. Overall, we are seeing a reduction in the genetic pool of important food crops. Consequently, the lack of genetic variability in the important food crops places us all at food security risks from changes in climate, emerging or reemerging pests, changing soil conditions, or other unpredictable events.

Another inestimable cost derives from the plight of environmental refugees. There is no accurate way to estimate the numbers of environmental refugees since they are not officially protected through the United Nations, although the figure is thought to greatly exceed 10 million.[38] The flight of refugees from degraded land has caused extensive disruption in developing countries resulting in economic, social, and political difficulties. These problems may worsen as populations increase in LDCs, causing greater poverty, increased desertification, and reduced food production.

Wetlands

What Are They?

Wetlands are found everywhere from tropical forests to the furthest reaches of the tundra. Wetlands appear virtually in every biome except the frozen expanses of Antarctica. Wetlands are areas of land where water saturation is the major factor influencing the nature of soil development and the communities of plants and animals that live in the soil and on the surface.[40] The National Research Council defines a wetland as "an ecosystem that depends on constant or recurrent shallow inundation or saturation at or near the surface of the substrate."[41] There is a somewhat different definition of wetlands used for regulatory purposes and written under the Clean Water Act (40 CFR 230,3[t]) as "those areas that are inundated or saturated by surface or ground water at a frequency and duration sufficient to support, and that under normal circumstances do support, a prevalence of vegetation typically adapted for life in saturated soil conditions." Because acreage that is classified as wetlands falls within the guidelines of a protected environment under federal law, the ultimate decision in this matter is affected by USEPA's definition of a wetland. Some ambiguity in delineating wetlands still exists. For example, the number of consecutive days that are required for a land area to have surface water can vary from 14 to 21 days depending on the scientific or legislative source. Other factors such as growing season, temperature changes, or the existence of records of environmental activity for the specified land site all must be taken into account. The key wetland features that are in question include the hydrology, anaerobic substrate, and biota. Policy makers and scientists dispute the number of wetland features that must be proven present.[43]

Wetlands are zones that fall in between areas of land and water and both separate and join the land and open water. Wetlands are covered or soaked by water for long periods during the year. They provide habitat for numerous species of plants and animals, including some that have specifically adapted to wetlands and appear nowhere else, such as cattails, mangrove trees and wood ducks.[42] Wetlands appear in a variety of forms including swamps, bogs, prairie potholes, bottomland hardwood forests, and estuaries. Swamps are are often transitional environments between large, open water and drier land. They are home to species from both extremes and are marked by flood-tolerant trees and shrubs.[40] Bogs are usually a stagnant, acidic, nutrient-poor ecosystem filled with peat and sphagnum moss. Its abundance of decaying material consumes enormous amounts of carbon dioxide. Prairie potholes are freshwater marshes found primarily in the upper midwestern U.S. This glacially formed landscape is pockmarked with potholes that fill with snowmelt and spring rains. They are among the most important wetlands in the world, supporting more than 50% of North American migratory waterfowl and wide varieties of invertebrates and plants (Figure 3.10).[41] River swamps or bottomland hardwood forests are found along streams and rivers in the Southeast and Southcentral flood plains of the U.S. These are deciduous forested wetlands consisting of gum and oak trees adapted to areas of seasonal flooding or covered with water most of the year. The trees in these wetland systems often have fluted or flaring trunks and knees or aerial roots as seen in the Bald Cypress.[40] Estuaries are inland marine waterways found worldwide. They are dominated by grass or grass-like plants and lie mostly between barrier islands and beaches. Nutrient-rich tides make estuaries among the most naturally fertile habitats in the world.[41] The USEPA lists a number of wetland categories including bogs, bottomland hardwoods, fens, mangrove swamps, marshes (tidal, saltwater, freshwater), playa lakes, prairie potholes, swamps, vernal ponds, and wet meadows.[40]

Figure 3.10 Various types of wetlands. (Source: USEPA, Office of Wetlands, Oceans, and Watersheds, *Facts about Wetlands,* Aug. 25, 1997, http://www.epa.gov/owow/wetlands/facts/types.html.)

Benefits of Wetlands

Wetlands have not historically received much positive press because they are often inaccessible, have boring landscapes of gray and muddy brown, and are identified with insects and snakes. However, wetland areas are very beneficial. For example:[40–42]

- Wetlands purify and replenish water supplies. They serve as natural recharge areas to groundwater. They act as natural water filters because the rich biomass of plants and microscopic animals strain pollutants from water flowing toward streams or soaking into the ground.
- Wetlands are extremely rich in biomass (the amount of plant and animal life). They are equivalent in biomass, acre for acre, to a tropical rainforest. Wetlands in the U.S. are habitat for 5000 species of plants, 190 kinds of amphibians, and one third of all bird species.
- Wetlands are an important source of food. Many species of fish and shellfish, such as shrimp, crabs, and crayfish, spawn in salt marsh wetlands.
- Wetlands absorb large amounts of carbon dioxide from the air taken in by plants or stored in peat moss.
- Wetlands control flooding in low-lying areas because they work like sponges, soaking up floodwaters from overflowing rivers. Vegetation slows the water flow, thereby sparing farms and communities downstream. The mammoth flooding of the Mississippi in 1993 worsened because the farm-filled states of Iowa, Indiana, and Missouri drained all but a small percentage of farmlands.
- Wetlands protect coastal areas from storms. Fierce coastal storms can create huge waves that wetlands are able to absorb and so buffer communities further inland.
- Wetlands provide recreation and beauty. Hunters and sport fisherman find enjoyment and often dinner in wetlands.

Wetlands Losses

The U.S. has lost more than half of its wetlands since the early days of colonization. Wetlands were considered a nuisance that made farming difficult. As settlements moved westward, these areas were filled. The great farm areas of Ohio, Iowa, and California have seen 90% of the wetlands disappear. There were no federal laws to protect wetlands until 1972 when Congress enacted amendments to the Clean Water Act, which prohibited discharge of fill material into "waters" of the U.S. without a permit from the U.S. Army Corps of Engineers. A federal district court ruled that this included wetlands and unleashed a bitter controversy over issues of property rights and the role of government.[42] Unfortunately, the Army Corps of Engineers has been granting quick building permits in environmentally-sensitive areas under the Clean Water Act. Permit 26, one of the 30 different quick permits under this Act, allows alteration of up to 10 acres of wetlands if they are isolated from other bodies of water. The regulations allow for small lots to be developed without forcing extensive documentation about the environmental impact of the project. Acquiring such permits may take as little time as one month.[44] Florida has lost an estimated 2000 acres in the past 4 years from this process, and nationally the estimate is tens of thousands of wetland acres lost since the program started.

From the mid-1970s to the mid-1980s, 2.6 million acres (1.05 million hectares) of U.S. wetlands were destroyed. The most recent and greatest losses have been in Louisiana, Mississippi, Arkansas, North Dakota, South Dakota, Nebraska, and Florida. An estimated 300,000 acres (120,000 hectares) of wetlands are drained or filled every year. About 87% is lost to agriculture, with urban development taking the rest.[42] More than three quarters of U.S. wetlands are privately owned, so preservation attempts are difficult to mandate. One successful approach incorporated in the 1985 Farm Bill removed farm program benefits from farmers who drain wetlands for agriculture. Enacted under the same legislation was the Conservation Reserve Program (CRP) which provides financial assistance to farmers who remove lands from cultivation and thereby promote the restoration of wetlands habitat. With wetlands in the U.S. rapidly vanishing, ecologists are seeking methods of restoring or replacing them. A symposium at the annual meeting of the American Association for the Advancement of Science in February 2000 offered evidence that man-made wetlands may be just as good at water filtration as untouched wetlands, but they may not offer the same rich resources for wildlife.[4] These efforts have seen a return of thousands of acres to wetlands along with significant increases in waterfowl populations.[45] Despite these positive approaches, the net annual loss of 300,000 acres per year represents an ongoing concern.

Wetlands are being threatened in many other parts of the world. As an example, despite concern for the destruction of the Amazon rainforest, another part of the Amazon is also suffering from rapid population and development. Brazil's Panatanal, the largest wetland in the world, is being threatened by human development and jeopardizing many of the rare 2000 species of animals and birds.[46]

The Loss of Biodiversity and Extinction of Species

Background

The extinction of species is not a new phenomenon. There have been many catastrophes on Earth, but none so destructive and overwhelming as the "mother of mass extinctions" some 250 million years ago at the end the Permian period. This mass extinction was far more severe than those events that occurred 65 million years ago and caused the extinction of the dinosaurs and nearly half of all species.[47] The Permian extinction caused 90% of all species in the oceans to disappear, two thirds of reptiles and amphibian families perished, and up to 30% of insect orders were lost. The events leading to this mass extinction are hypothesized to involve a huge and prolonged series of volcanic eruptions that released large volumes of carbon dioxide into the air and increased climactic instability.[47] The return of life may have taken 5 million years and resulted in dramatic changes in the forms of life that emerged. Very likely, most of the species (perhaps 99%) that have ever existed on this planet are now extinct. The diversity of species has ultimately recovered from each of these cataclysms as evolution has been somewhat greater than the extinction rate. Today, more than 1.8 million currently living species have been identified. There is general consensus, however, that the planet is now losing more species than are being created and that the activities of humans are the reasons for a rapidly growing species extinction and loss in biodiversity. Biodiversity refers to the range of animal and plant species and the genetic variability among those species. The greater the genetic variation, the more likely there will be survivor species in the event of major catastrophes.

Since the planet has experienced a series of cataclysms over many millions of years and has recovered each time with abundant and diverse life, what is the major concern about the loss of a relatively few species? In order to respond to this question, we need to remember that the forms of life that emerged after cataclysms were much different and that humans are a relatively recent phenomenon in the history of the Earth. A reduction of species diversity will often result in an unstable ecosystem that is in danger of collapse if key species in that ecosystem disappear. Since genetic variation is the basis for evolution, the diversity of genes and species "affects the ability of ecological communities to resist or recover from environmental change, including long-term climatic change."[48] A human-induced cataclysm resulting in major losses of species could lead to a retrogression in the global ecosystem incapable of supporting large populations of complex creatures such as humans.

Previous extinctions have been associated with major climatic changes arising from such forces as volcanism, or geologic changes that forced continents to clash and seas to rise or fall. Records of fossils show that entire groups of organisms including fish, reptiles, birds, and mammals have replaced one another over long periods of time. Birds and mammals, better able to respond to changes, began to proliferate as the numbers and species of reptiles declined (Figure 3.11). These changes are measured in millions of years. Even the cataclysm of volcanic eruptions

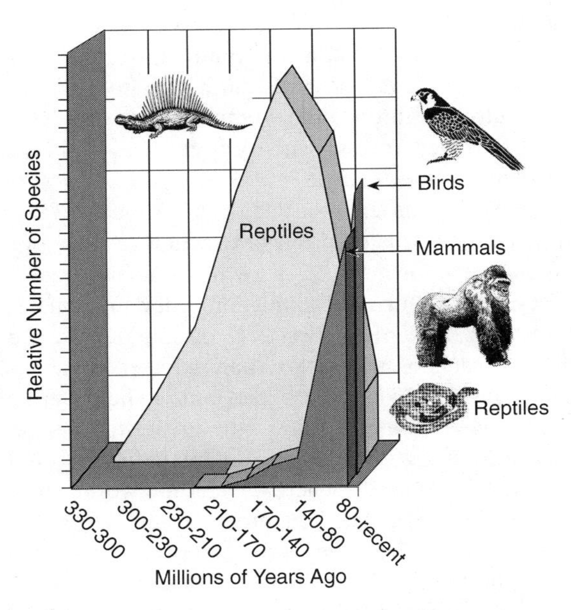

Figure 3.11 Comparative evolution of birds, mammals and reptiles over geologic time. (Adapted from Turk, J., and Turk, A., *Environmental Science,* 4th ed., Saunders College Publishing, Philadelphia, 1988.)

in the Permian period is thought to have occurred over several hundred thousand years. This is in sharp contrast to the sudden and precipitous global changes caused by humans. The massive increase in human population with corresponding pollution of land, air, and water is resulting in such rapid ecosystem changes that many species are unable to adapt. The extinction of species is thought to be occurring more rapidly than at any other time in recorded history.

More recently, a loss in biodiversity has also become a public health concern. One of the better known examples is the potential loss of medicines derived from plants. It has been estimated that there is one useful pharmaceutical out of every 125 plant species. Unfortunately, 3000 of the world's 300,000 catalogued plant species may become extinct in the next 25 years.[48] Disease epidemics may also be associated with a loss in biodiversity as ecosystems are disturbed and disease-producing organisms or conditions emerge. As an example, the deforestation of tropical rainforests eliminates the natural predators to mosquitoes, which may then proliferate along with many vector-borne diseases including malaria (see Chapter 6).

Loss in Biodiversity

A report by the United Nations Environmental Programme titled "The Global Biodiversity Assessment," was among the first comprehensive, peer-reviewed reports ever to examine the Earth's biodiversity. The report concludes that: (1) 4000 plants and 5400 animal species are threatened with extinction; (2) species have recently become extinct at 50 to 100 times the average expected natural rate; and (3) 1.75 million or about 13% of the 13 to 14 million species on Earth have been scientifically identified.[49] Background extinction rates through geological time have been roughly estimated at the rate of one mammal and two bird species every 400 years. Documented extinctions for mammals and birds during the last 400 years are already at 50 times that rate, and this figure is probably underestimated. The highest levels of biodiversity are in the tropical forests, where current rates of species extinction have been estimated at 1 to 11% per decade.[50]

A study by the World Conservation Union in October, 1996, showed that the world is entering a period of major species extinction rivaling five other periods in the past half billion years. Of the 4327 known mammal species, 1096 are at risk and 169 are in extremely high risk of extinction in the immediate future.[51] Besides mammals, 11% of birds, 20% of reptiles, 25% of amphibians, and 34% of fish are threatened. Of the 26 orders of mammals, 24 are threatened: most at risk are certain species of elephants, primates, rhinoceros, and tapirs. The pattern of species extinction and threatened species closely follows the greatest population densities. The countries with the most threatened mammals are Indonesia (128 species), China (75 species), and India (75 species) where 43% of the world's population lives and which are the most densely populated countries (Figure 3.12).

The U.S. has the potential for losing more plant varieties than any other country. Statistics show that 4669 of all existing plant species are threatened with extinction. Australia follows with approximately 2245 species threatened, followed by South Africa with 2215.[5]

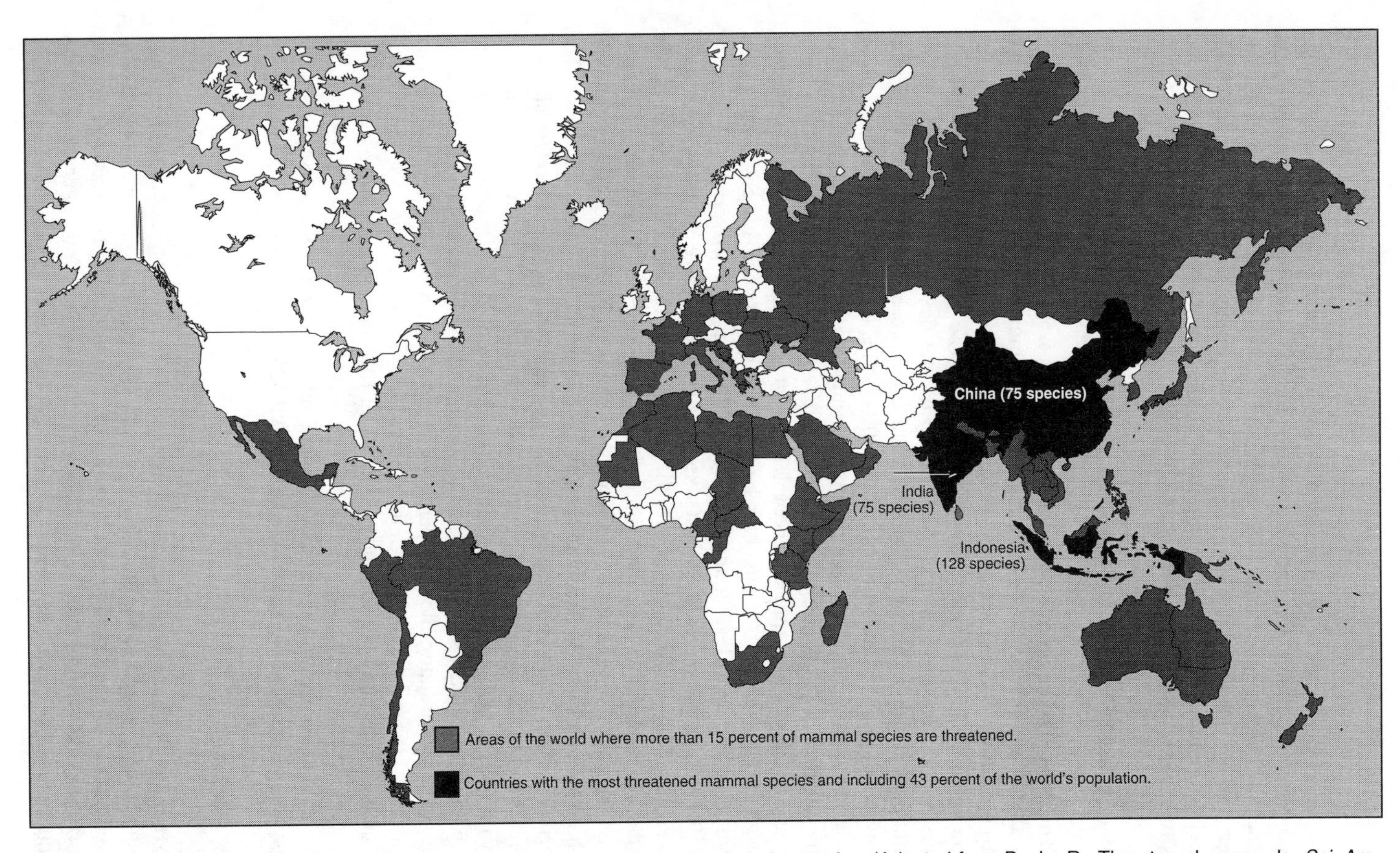

Figure 3.12 The majority of threatened mammal species occur in lesser developed countries. (Adapted from Doyle, R., Threatened mammals, *Sci. Am.*, Jan., 32, 1997.)

Threats to Biodiversity

Some of the threats to biodiversity include: (1) the destruction and fragmentation of habitat; (2) the harvesting of wild species for food, lumber, and other products; (3) the introduction of nonnative animals and plants; (4) the pollution of the air, water, and soil with toxic chemicals and waste products; and (5) the increased ultraviolet radiation and global warming resulting from atmospheric changes and possible human activities.[48]

Loss of Habitat

The most significant threat to biodiversity today is the elimination of habitat for agricultural and habitat purposes. As expanding populations assimilate geography by cutting down forests, filling in wetlands, and plowing fields, most species are unable to adjust to these rapid changes and perish. The greatest biological diversity occurs in the rainforests where much of the habitat destruction is occurring. But habitat destruction occurs on many levels and even threatens the mass extinction of freshwater mussels. More than half of the 300 species of mussels in the U.S. are in trouble to some degree, and most of this is brought on by human intervention such as the building of dams, gravel mining, sewage, and polluted runoff.[52]

For a lesson in how quickly ecosystems fragment across international borders, a report in December 2000 suggests looking no further than Lake Victoria. Lake Victoria is the world's second-largest freshwater lake covering an area of 67,850 square kilometers. This vast expanse, about the size of the Republic of Ireland, forms the headwaters of the Nile River. Prior to 1970, Lake Victoria, the world's largest tropical lake, had more than 350 species of fish from the cichlid family. Ninety percent of these were unique to the lake. Fishing for tilapia and Nile perch provides a living for many of the Luo people who live along the lakeside. The fishering of Nile perch and tilapia caused a collapse in the lake's biodiversity and deforestation in the three countries that border the lake — Uganda, Tanzania, and Kenya. Wood is needed to dry the oily perch, but resulting deforestation is jeopardizing the lake and its fish. Forest clearing has increased siltation and eutrophication in the lake, jeopardizing the Nile perch and tilapia fishery — initially the cause of the problem. (Often the result of human intervention, eutrophication is the process by which a water body is overwhelmed by plant life and nutrients at the expense of oxygen.)

Lake Victoria illustrates how profound and unpredictable trade-offs can be when management decisions do not take into account how the ecosystem will react.[4]

Overharvesting

Many species have been harvested to the point of extinction. For example, many fish stocks such as cod in the North Sea off New England are heavily exploited with as much 60% of the fishable stock removed annually. Despite warnings by the International Council for the Exploration of the Sea (ICES) and government action to reduce allowable catches or close fisheries completely, data suggest that management efforts have not been successful.[53] The Northern "right" whales are estimated to be less than 300 in number. Their decline is attributed to the devastation by turn-of-the-century whalers who preferred to hunt these whales because of their slow-moving and docile nature. Hence, they were called the "right" whale to hunt. Deaths are attributed today to the slowness of the beasts in which ship collision results in one third of the whale deaths per year.[54] The small number of remaining whales leads to a loss of genetic variability through inbreeding, which further jeopardizes the species. People along the coast of the Philippines use dynamite and cyanide to stun fish and then pound the reefs with crowbars to scare them out into waiting nets. Some fish are eaten; others are sold as exotic aquarium fish. The natives see this as necessary for daily survival. The process, however, is devastating the coastline coral reefs. Although efforts are under way through the Coastal Resources Management Project to empower local communities to act in their own self-interest, it is difficult to protect the environment from impoverished people seeking survival.[55]

Organized crime in countries such as the U.S., Japan, and Russia has discovered a new high-paying commodity — wildlife. The species they trade include parrots, elephants, whales, bear, tigers snakes, and rhino. Body parts from many of these animals are used in some cultures to stimulate sexual prowess. The shift in interest for criminals is that they used to use boa constrictors to carry cocaine across borders. Now they sell the snakes for faster and easier money.[56]

Nonnative Species

In the aftermath of World War II, no one paid much attention to the invasion of the tiny Pacific island of Guam. The opportunistic brown tree snake, which grows up to 2 meters in length, hitched rides from the island of Manus to Guam, slithering into the machinery used by the U.S. military. Without natural enemies, the population of brown snakes grew quickly within 30 years, eliminated eight of the island's natural bird species, and sharply reduced several others.[57] The freshwater mussels, already under attack by changing habitat and pollution, are also being displaced by Zebra mussels inadvertently transferred from the Black, Caspian, and Azov Seas. Many nonnative species of fish including large-mouth bass and carp have been implicated in the extinction of as many as two thirds of the native species, and now threaten many others.[58] Rainbow trout never encountered "whirling disease" before the parasite was unknowingly transplanted from Europe. In some reaches of the Colorado River, the disease has wiped out all but 1% of the population of young rainbows and has infected trout in many streams in the Rocky Mountains on both sides of the Continental Divide.[59]

Pollution

Acid deposition, promoted by the release of nitrogen oxides and sulfur oxides from combustive processes, has caused dramatic ecologic changes in the U.S., Europe, and in many cities of the lesser developed countries. The acidification of lakes and streams has led to juvenile recruitment failure among fish, resulting in the disappearance of many species in a number of industrialized countries. The acidification may also be leading to forest decline in many parts of the world, and it is thought to contribute to the loss of 40 to 50% of the mycorrhizal fungi in parts of Western Europe. These fungi are considered key to many ecosystems because they live on plant roots symbiotically and facilitate the uptake of nutrients to the plants.[60]

A significant decrease in some species of fish and birds on the Southern California coast has been attributed to an 80% decline in the population of zooplankton. This decline is explained by some researchers as a consequence of rising ocean surface temperatures of 1.2° to 1.6°C. Some scientists suggest this warming is a result of the greenhouse effect. A steeper density gradient, resulting from ocean surface warming, has slowed the upwelling to the surface of nitrates, phosphates, and other nutrients used by phytoplankton. The phytoplankton have diminished sharply from the lack of these nutrients. Because zooplankton feed on the phytoplankton, their numbers have also been markedly reduced; and consequently this effect is carried up the food chain to larger fish. If the ocean surfaces continue to warm over the next few decades, the biological impacts could be devastating.[61]

Deformed frogs have been reported in Minnesota, Wisconsin, South Dakota, Vermont, and even Quebec. The frogs have limb abnormalities. Some have stumps, others have four front limbs, sunken or missing eyes, and small or missing sex organs. Aquatic frogs appear to have the most severe abnormalities. Amphibians without these deformities are difficult to find.[62] The cause of the abnormalities is uncertain but has been hypothesized to be due to increased ultraviolet radiation resulting from stratospheric ozone depletion, heavy metals, pesticides, or a combination of these factors. Studies are under way to determine the cause, and there is concern that these effects may signal potential health effects to human populations.[62]

Protecting Endangered and Threatened Species

Legislation first aimed at protecting wildlife in the U.S. was introduced as a bill in 1926. It had little authority to protect. In 1969, the Department of the Interior added species to the list that were threatened with extinction worldwide. This international initiative helped to form the Convention on International Trade in Endangered Species (CITES) which attempts to reduce the trafficking of rare and exotic animals and/or plants.

In 1973, the Endangered Species Act (ESA) was promulgated in the U.S. Species are constantly evaluated by the Fish and Wildlife Service (FWS) and the National Marine Fisheries (NMF) to determine if they should be listed as threatened or endangered based on scientific data alone. Consideration is not given to economic or political concerns, and no branch of the federal government may proceed with a project that might threaten an endangered species or its habitat without first contacting the Fisheries and Wildlife Service.[63] Because projects may be stalled or blocked when threatened species are encountered, there are many critics of the ESA. However, 99.5% of the 100,000 projects going to consultation proceeded, usually with some modification. Less than 0.5% of the projects were prohibited. Even so, some isolated cases received national attention and thrust the ESA into the spotlight. The snail darter, a 3-inch fish, became a celebrity in 1978 when it was listed under

Figure 3.13 Endangered or threatened species whose numbers have increased include the whooping crane, peregrine falcon, American bald eagle, grizzly bear, and red wolf. (Source: U.S. Fish & Wildlife Service (USFWS) http://www.fws.gov/. Whooping crane photo by Steve Hillebrand; grizzly bear photo by Don Redfern; bald eagle photo by Robert Fields; gray wolf photo by USFWS.)

ESA as endangered and thereby temporarily halted the construction of the multimillion dollar Telico dam project on the Little Tennessee River. Thousands of jobs were expected to be lost in the Pacific Northwest when the spotted owl was listed as threatened in 1990, causing a fury among the loggers in the area.[63] The ESA seeks to protect the presently listed 632 species and subspecies of mammals, mollusks and plants classified as endangered in the 50 states. The ESA also attempts to protect some 200 additional species classified as threatened and, using trade sanctions, intends to protect 500 foreign and oceanic species outside the U.S. Many environmentalists praise the ESA for reducing the extinction rate of some animal species in the U.S. and even increasing numbers in as many as 65 species. Numbers have increased for the American bald eagle, whooping cranes, grizzly bears, red wolf, and peregrine falcons, and many other species (Figure 3.13). Former U.S. Secretary of the Interior, Bruce Babbett, announced in May 1998 that 29 species would be delisted from the government's endangered species list, including the bald eagle, the peregrine falcon, and the little-known Pahrump pool fish.[64] This marks the first time in the 25-year history of the ESA that such a large number of species has been targeted for removal from the list. Some groups, such as the Environmental Defense Fund (EDF), hope that politics do not drive the process, as efforts are made to show Congress that the ESA is working and deserves less criticism.[64] However, many opponents of ESA argue its powers are excessive and override the property rights of the individual, threaten jobs, and interfere with economic progress.

ESA is the only federal law that addresses biodiversity. It expired in 1992 and has not been reauthorized as of June 1998 because of the controversy. Bills submitted before Congress propose major changes in ESA that include: (1) compensation to landowners if federal actions lower property values; (2) better scientific justification for listing species; (3) tax incentives to property owners for habitat conservation planning; and (4) protection removal of endangered species habitat.[48] The idea of providing property owners incentives as part of the ESA reauthorization has gained new support based on a recent study finding populations of most endangered species in the U.S. live in a relatively few critical "hot spots" covering less than 2% of the land area, and most of the critical tracts are on private land.[65] These hot spots contain endemic species in that they are limited to a small area, and they are disturbed by urban development and agriculture. These hot spots are concentrated mostly in Hawaii, California, and the southeast, particularly Florida. Spending money on these limited areas would be very cost effective.[65]

Scientists are constantly working on new ways of protecting and rebuilding populations of endangered species. Captive attempts have been made to increase the reproductive rates among many species. However, these techniques have varying degrees of success. The cloning of Dolly the sheep 3 years ago presented scientists with new options for repopulating endangered species. At China's National Academy of Sciences, zoologist Chen Dayuan and his team used adult panda cells to produce a cloned embryo. Some teams are even working on cloning cells taken decades ago from species that are now extinct.[6]

Experts now generally agree that it is important to preserve large areas of landscape to protect species and keep them off the endangered species list, while permitting greater flexibility in balancing development and species protection.[65] Scientists and policy makers are also learning that absolute bans on wildlife destruction has led to disaster in LDCs because: (1) large areas of land have been destroyed by the uncontrolled growth of wildlife; (2) the wildlife is eliminated as a natural resource for the native human populations; and (3) these policies require very costly enforcement. The solution, as seen in some wildlife markets in the U.S., is to allow and regulate markets for use of these species in ways that will protect the native peoples and balance the use of natural resources.[66] In some programs, like one called "Campfire" in Zimbabwe, the villagers actually are the owners of the wildlife and are allowed to sell permits to hunters, under some governmental control, which allows the villagers to profit while maintaining a sustainable population.[67]

FOOD SECURITY

One of the biggest debates for the 21st century concerns whether the world can produce enough food for another few billion people. In the American Association for the Advancement of Science (AAAS) meetings, experts discussed how the world's growing population would be fed in the 21st century. It is said that about two thirds of preschool children in Southeast Asia have been ill nourished. By 2010, about 680 million people in the developing countries will not have enough food; and by 2020, about 135 million preschool children will be malnourished. Population

WEB PICK Go to http://www.unix.oit.umass.edu/~envhl565/index.html. Click on "Chapter Web Links" to find Earthwatch home page under Chapter 3	
Chapter 3	Environmental Degradation and Food Security
Whose Site?	Prepared and managed by the U.N. System-Wide Earthwatch Coordination, United Nations Environment Programme, Geneva, Switzerland.
URL	http://www.unep.ch/earthw.html
What's Here?	Earthwatch Coordination, based in Geneva, provides leadership and direction to the U.N. System-Wide Earthwatch established at the U.N. Conference on the Human Environment in Stockholm in 1972. It provides environmental and socioeconomic information for national and international decision making on sustainable development and early warning of emerging problems requiring international action. This site provides information on emerging environmental issues, near-real-time data on the state of the planetary environment, major environmental assessments, Earthwatch strategies, G3OS global observing systems, and more. Visit this site to gain access to the Global Fire Monitoring Center to see near-real-time reports on the status of forests and wildfires around the world. Visit and use interactive graphic viewing of global oceanographic data sets, including sea surface temperature and precipitation anomalies.

activists as well as some social scientists believe the answer is no, and that the only solution is to limit the numbers of mouths to feed. However, many agronomists and other scientists say yes, provided there are ample funds for research into improved plant varieties and agricultural technologies that will increase yields.[68] Depending on whether you live in an impoverished nation or a wealthy one, there is ample room for pessimism or optimism.

Food Production

For nearly 40 years the world production of grain has risen by more than 2% a year, but it declined to scarcely 1% a year in the 1990s. This is a level that is markedly below the 1.5% annual increase in the world's population.[4] The 1997 world carryover of grain stocks remained at a low level of 55 days of consumption, which is only four days more than the record low of 51 days experienced in 1996. A 70-day supply in carryover stocks of grain is thought important to maintain a minimal level of food security in the world.[4] Food security is said to occur when all people have physical and economic access to the basic food they need to work and function normally.[69] This 1996 low grain surplus drove up prices of wheat and corn, resulting in a 2.5% increase in grain harvest to 1.71 billion acres and a record 1997 grain harvest of 1.84 billion tons. Even so, the world was unable to rebuild depleted stocks to a secure level.[4] The 1998 cereal production is forecast to be just slightly below this record; it would be sufficient to meet 1998/1999 consumption requirements but would remain below minimum safe levels for world security. Malnourishment affects one in very three people on Earth, and about 18 million people, mostly children, die each year from starvation.[4] Thirty-seven countries are expected to have food emergencies, with Africa standing out as the continent with the most serious food shortages.[10] Countries with critical or low food security are shown in Figure 3.14.

Reasons for Regional Food Shortages

The main reasons for food shortages in Eastern Africa derive mainly from recent droughts followed by floods. Somalia, Kenya, Uganda, Burundi, and Tanzania all experienced serious flooding with crop losses. Harvests have been poor in several African countries near the Sahel desert such a Mauritania, Niger, and Senegal. In southern Africa, there has been localized crop damage with flooding in some areas and drought in others. Droughts associated with the effects of El Niño have caused a decline in cereal production in Indonesia, China, the Philippines, Papua New Guinea, and Thailand. Economic reforms are contributing to insufficient food for many in Mongolia, while Security Council Resolution 986 is causing continuing food shortages in Iraq.[10]

Distribution of Population and Food

Today there is a huge food gap between the developed countries and most developing nations. Already this gap poses moral, political, and economic problems of major dimensions; and if it widens, these problems will become increasingly serious.

The majority of the global population already lives in developing countries, including China, which faces the greatest food problems. Approximately one billion people live in the poorest thirty to forty countries — roughly the same number as in the well-fed developed nations. Moreover, the proportion of the total world population living in the developing nations will unquestionably continue to grow; 98% of the annual increase in the world's population is occurring in the developing countries.

The growing food gap between rich and poor nations has been, and will continue to be, due to the differences in rates of population growth, since food output has actually been increasing somewhat more rapidly in the rich than in the poor. However, because of a more rapid rate of population increase in the poor countries, their average rate of *per capita* increase in food output was less than one third the rate in the developed countries. Although there has been generally steady growth in the

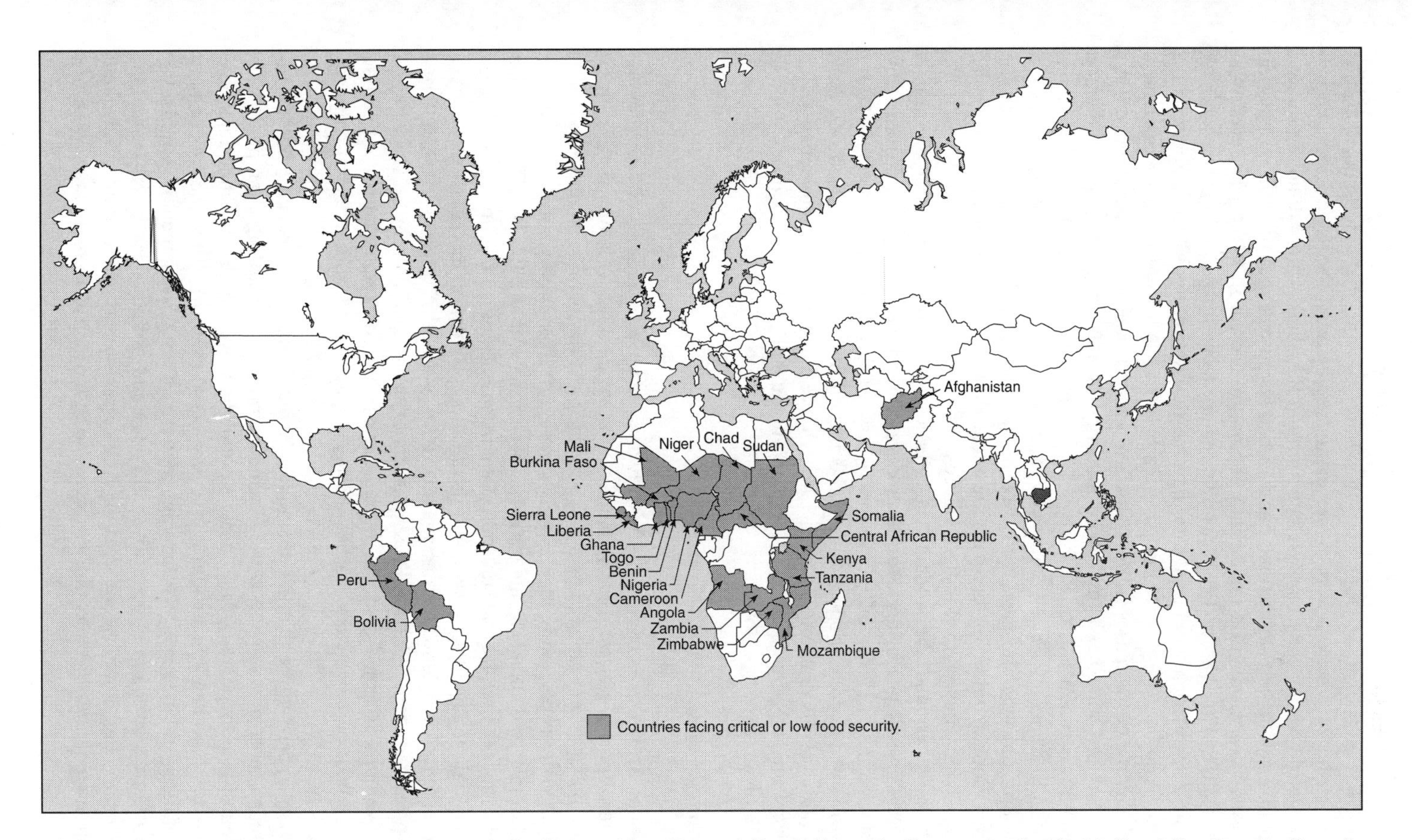

Figure 3.14 Countries with low and critical food security. (Adapted from Brown, L.R. and Kane, H., *Reassessing the Earth's Population Carrying Capacity,* W.W. Norton, New York, 1994.)

world production of most crops from the 1960s forward, there are unequal results among countries when viewed as per capita production. The food production in Africa has shown a steady increase beginning in the 1960s with a doubling in food production, but the per capita trend has declined because of the enormous population increases. Food production fell behind population growth in 64 of 105 developing countries between 1985 to 1995. If countries are to feed the 9 billion expected by the year 2050, Africa would have to increase production by 300%, Latin America by 80%, Asia by 70%, and North America by 30%.[4] Asia and Latin America have shown significant increases in both absolute and per capita food increases, but the former Soviet Union has experienced a decline of 40% in agricultural production since 1989 (Figure 3.15).[10]

In a fundamental sense, the world food problem, viewed globally, is less a problem of absolute shortages than a problem of unequal distribution between the rich and poor countries. Moreover, the inequality of food distribution is a basic problem within many developing countries.[7]

On the other hand, the production trends do not support a completely optimistic view. The growth rate of agricultural production worldwide has slowed from a 3% growth rate in the 1960s to 2.3% in the 1970s and less than 2% in the mid-1990s.[9] Growth rates in cereal production have been even more disturbing, declining from 2.8% in the 1960s to nearly 2.1% in 1992 (Figure 3.16).[10] So, though actual yields in agriculture and cereal may have increased in the 93 developing countries studied, the rate of increase has slowed.

Pessimistic View

Lester Brown, president of the Worldwatch Institute, expresses dire predictions for future food production. He maintains that agricultural technologies are going unused, fisheries and rangelands are approaching the limits of biological productivity, fresh water is becoming increasingly scarce, fertilizer applications are becoming less effective, and social disintegration is occurring in many LDCs.[69] There is also concern that the tropical and subtropical environments are so fragile in many LDCs that they will be unable to sustain significant food production increases. About 45% of the potential cropland in sub-Saharan Africa and Latin America is under forest or in protected areas.[70] Nearly three quarters of the potential cropland is so fragile that development would result in dramatic losses in biodiversity, increased carbon dioxide emissions, decreased carbon storage capacity, and high financial and ecological costs.[71]

Many experts agree that the potential to expand cropland area is quickly disappearing in most areas of the world.[72] China represents a startling example of this problem. More than 21% of the world's population lives in China, while its arable land covers less than 7%, resulting in a per-person share of 0.086 hectares or only 25% of the world average.[73] The acreage for agricultural development is diminishing while existing farmland is being overcultivated and polluted, and nearly one third of the land is threatened by soil erosion.[73] Overgrazing and reclamation of farmland has turned a quarter of the country's grassland into desert with more than 300 cities

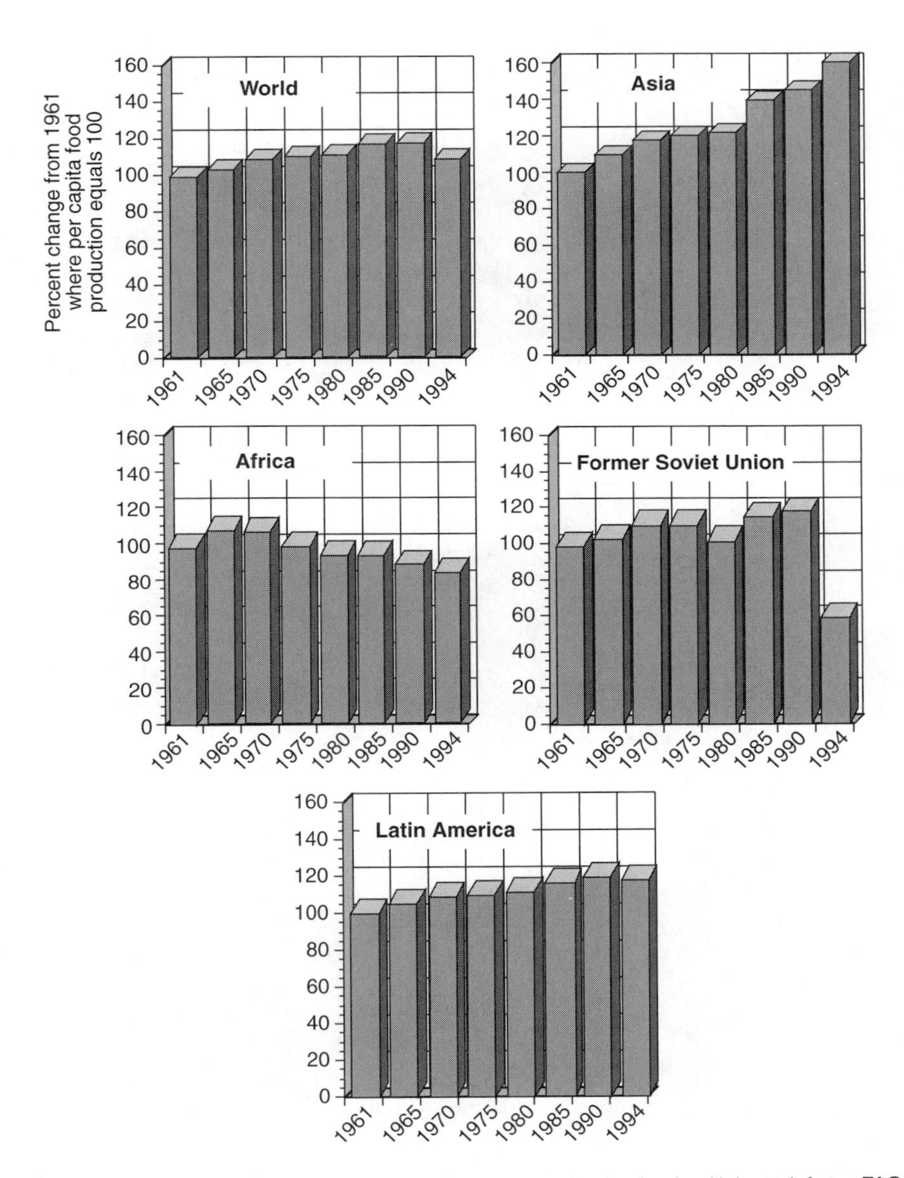

Figure 3.15 Some trends in food productin on a per capita basis. (Adapted from FAO, http://www.fao.org (WAICENT/faoinfo/econ-tie/giews/english/fo/fo9804/httoc.htm.)

or towns having a shortage of water.[73] In many of the LDCs, agriculture is hampered by unfavorable growing conditions. There is a continuing threat of civil strife, the impoverished cannot afford to pay for food, and governments cannot provide the policies or infrastructure to the rural poor that promote domestic agriculture production.

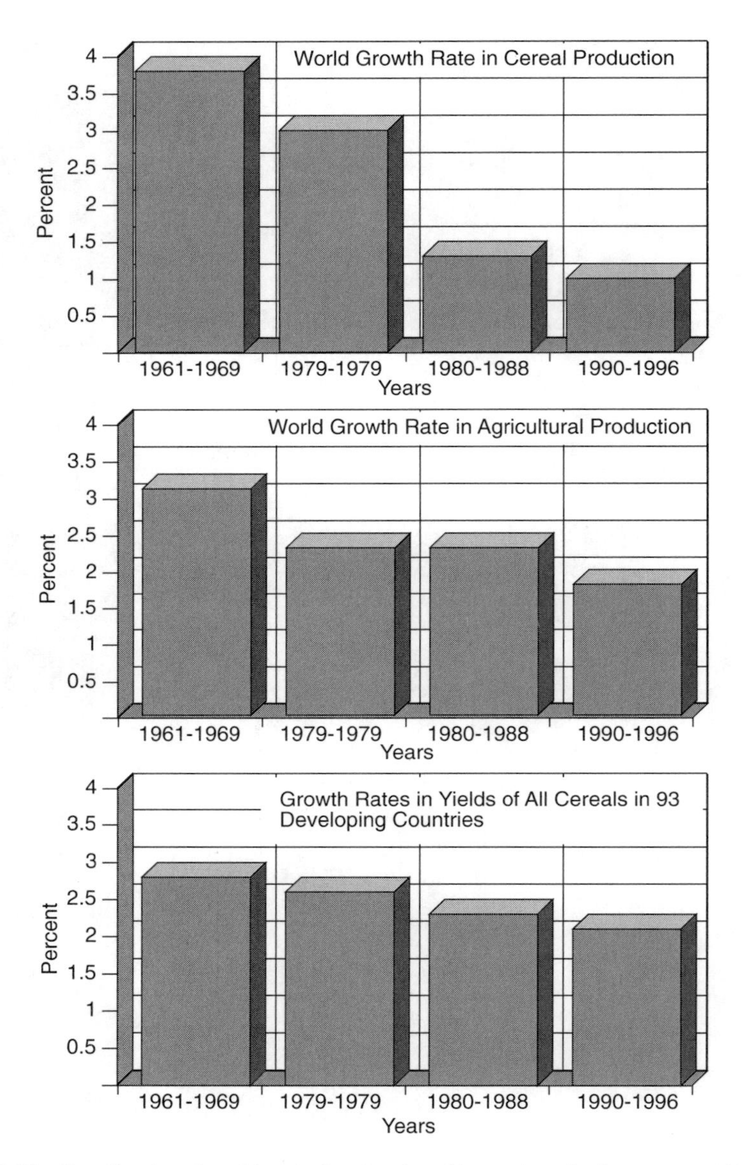

Figure 3.16 Growth rates of world agricultural and world cereal production compared to trends in actual total cereal yields in 93 developing countries. (Adapted from FAO, http://www.fao.org (WAICENT/faoinfo/econ-tie/giews/english/fo/fo9804/httoc.htm.)

Optimistic View

A number of organizations including The Food and Agriculture Organization (FAO) of the United Nations, The World Bank, and the International Food Policy Research Institute (IFPRA), which is a component of the Consultative Group of International Agricultural Research (CGIAR), have published studies looking at food production estimates from 2010 through 2030.[74–76] The studies and projections point to a number of hurdles with respect to worldwide food production, but they are similar to the FAO prediction that expected production increases will be able to meet the effective demands of a rising world population providing there are substantial investments in agricultural research.[10,74] Effective demand refers to that "demand level" corresponding to purchasing power without regard to food requirements. Unfortunately, the sharp rise in number of food-related crises has shifted priorities. In 1986, the World Food Programme directed 75% of its available monies to development projects. By the mid-1990s, more than 85% of its resources were going to meet emergency and refugee needs as those affected by natural and political disasters rose from 44 to 175 million people in the same years.[77] Admittedly, this last revelation is not optimistic, but must be included to realistically balance the view that we can meet the world's effective demand for food.

Sources

Where will the food come from to feed the world's billions? Increases in food supply must come from one or more of the following sources: (1) increases in yield (tons per acre); (2) increases in arable land placed under cultivation; and (3) cropping intensity (fewer fallow periods or more than one crop per year or field). The FAO projects that 66% of food production will come from increased yields, about 21% from arable land expansion, and the remaining 13% from increasing cropping intensity (Figure 3.17).[10]

Worldwide gross agricultural production is expected to increase in the next 20 years. Its growth will likely be at a slower rate of 1.8% per year compared to the previous decades because population growth rates and food demand have slowed in developing countries, and there is inadequate income growth in many countries to purchase large supplies of food.[78]

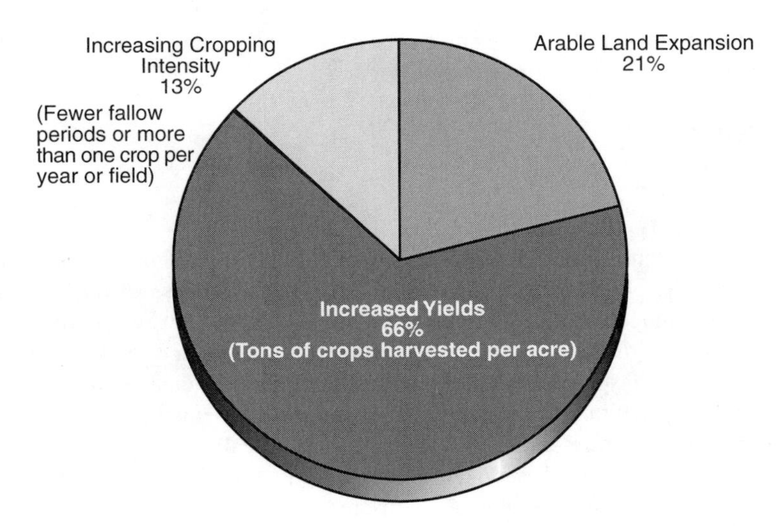

Figure 3.17 Projected sources of food increases. (Adapted from FAO, http://www.fao.org (WAICENT/faoinfo/econ-tie/giews/english/fo/fo9804/httoc.htm.)

Some scientists believe that the ability to expand cropland is limited and that it is disappearing in many areas of the world. The potential for increasing agricultural land is limited by: (1) the significant costs of developing an infrastructure in remote areas; (2) the lesser productivity of these alternative areas; and (3) the trade-offs in environmental destruction of sensitive ecosystems.[72]

The FAO projection of increased yields is based on assessments that there is room for improvement by farmers who are not achieving peak yields. Many countries are showing only half the yield (3.7 metric tons) of the best performing countries (6.7 metric tons).[72] The FAO projects that South Asia will make the strongest contribution of increased yields in crops of about 82%, while Latin America and Africa will experience less than 53% yield increases.[10]

This optimism is not shared uniformly. It is argued that increasing numbers of farmers in developing countries are already using crops with the highest genetic yield potential and that rice crops at experiment stations in Asia have leveled off for a number of years.[72] Further, the high input approach of using genetically programmed, high-yielding varieties has required large amounts of water, pesticides, and synthetic fertilizers. This approach can increase yields but may also cause the loss of genetic stocks and reduction in soil quality. Further, many LDCs do not have the funds necessary to support the increased costs of fertilizers, pesticides, and seeds. Alternative models are being evaluated and promoted that are more friendly to the environment. Integrated farming systems are being proposed that employ sustainable crop intensification strategies. These strategies include: (1) improved irrigation systems; (2) structured water pricing to reduce overuse; (3) alternative rotation of crops; (4) selective pesticide use; (5) use of pest-resistant varieties; (6) improved soil testing and fertilizer application; (7) regional crop breeding programs; and (8) more education of farmers.[10] High input and crop intensification strategies may not be viable in fragile ecosystems because of drought risk, inadequate support structures, and lower yield response. In these areas, alternative strategies might involve: (1) erosion and moisture control; (2) recycling nutrients; (3) and integrating annual and perennial crops in what is known as mixed farming.

Food Security

Worldwide

Temporary deficiencies in food and famine may occur in the presence of droughts, floods, political strife wars, or poor harvests. These events are on the increase and produce significant suffering in the short term. However, chronic undernutrition is a more difficult and pervasive problem resulting in a food security crisis in many LDCs. In such areas, more and more poor are unable to afford food, conditions for agriculture are less favorable, and governments have been unable to provide the infrastructure or policies needed to promote domestic production.[10] The lack of food security means people are unable to acquire the basic food they need to function normally because they are restricted physically and economically. Civil strife, inadequate marketing, poor storage facilities, and inadequate delivery capabilities reduce physical access to food.[10] Food insecurity is most pronounced in sub-Saharan Africa, where more than 300 million people may be undernourished by the year 2010 with a per capita food supply of 2170 calories per day or about 60% of the per-person calories available to people in developed countries. In other areas of the developing world, there is reason to be more optimistic. The growth rates and undernutrition are declining in East Asia, which is rapidly becoming industrialized. It is estimated that by the year 2010 the number of undernourished in South Asia will be 4%. Latin America is seeing progress in that its undernourished population is projected to decline by more than half to a low of 6% by 2010.[10] Overall, chronic undernutrition is expected to decline in the 93 developing countries studied from 20% in 1988 to 11% in 2010.[10] However, many countries will experience shortfalls in food production. The World Bank estimates that it will be importing 15% of its grain by 2101, and net imports to LDCs are expected to increase from 90 to 160 million tons in the years from 1990 to 2010.[10]

Although there is a hypothetical potential to meet the worldwide demand for food, the reality is that food crises in Africa, South Asia, and other countries are likely to persist. The International Food Policy Research Institute (IFPRI) has made recommendations that it hopes will lead to sustainable land use, efficient agricultural systems to produce low-cost food, and sufficient food that is physically and economically accessible to every person.[79] In order for these goals to be met, IFPRI recommended that: (1) governments of LDCs need to be strengthened; (2) agricultural and extension services need expansion; (3) agricultural yields need to be intensified with methods that sustain natural resources; (4) low-cost, efficient agricultural input and output markets need to be developed; and (5) international assistance needs to be expanded.[79]

Hunger in America

It seems unlikely, even absurd, but the American economy is surging forward and the stock market is breaking records just as a hidden crisis in poverty and hunger threatens millions of Americans. The rate of hunger in America grows steadily worse in the face of this prosperity, creating our own food insecurity problems as the gap widens between the rich and the poor, the secure and the threatened.[80] In 1994, Second Harvest, a nationwide network of food banks, released a study that revealed more than 25 million Americans, almost 50% of them under 17, resort to using food distribution programs such as soup kitchens and food pantries. This represents nearly one of every ten Americans who at least occasionally make use of these services.[81] In 1998 a report from the Tufts University Center on Hunger and Poverty and Nutrition Policy declared nearly 35 million Americans live in hungry or food-insecure households.[80] The U.S. Conference of Mayors is expecting the demand for food aid to rise in 92% of American cities, with requests for emergency assistance rising 16% in 1997 alone. Nearly 20% of these requests were unmet.[80]

The problem of feeding the rising number of hungry Americans is revealed by a number of charitable and government organizations. The U.S. Conference of Mayors reported a 16% increase in food emergency requests based on a survey of

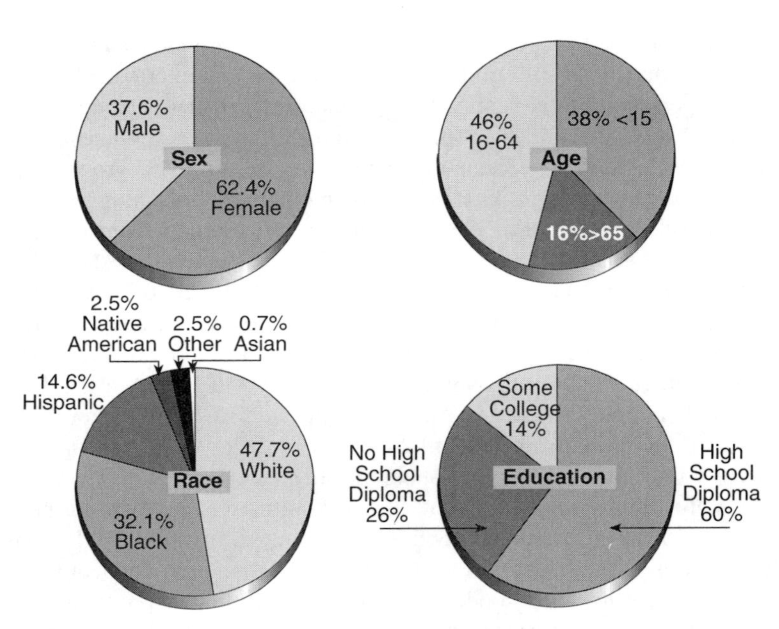

Figure 3.18 Demographic profile of Americans using food banks and pantries. (Adapted from Roberts, L., and Roberts, S.V., Hunger: a startling crisis, *USA Weekend,* March 27–29, 4, 1998.)

29 cities. Catholic Charities reported a 14% rise in people fed from 1995 to 1996 as they served 5.6 million people. The Congressional Hunger Center suggests the problem is underestimated and that some areas of the country are reporting greater than 42% increases in food aid demand.[80]

What is the profile of a person who makes use of food banks? Do you imagine a single man? Perhaps you think of an uneducated person, or a substance abuser. The face of hunger has changed to reveal that the food-insecure are children, mothers, the newly unemployed, the underemployed, and the working poor.[80,81] The Second Harvest report shows that 26 million Americans received food from food banks in 1997. The majority were female (62.4%), with 38% being under 17 years of age and 16% over 65 years of age (Figure 3.18). Most were white (47.1%); the majority had a high school diploma (60%); and nearly 19% had some college.[80] Two out of the five households using charity have at least one person employed, and in 50% of these cases, the employment is full time.[81] The number of working families requiring food aid is increasing as a percentage, but the majority of people requiring aid still includes those with disabilities, single mothers, and seniors on fixed incomes.

There is probably no single reason why more and more people are seeking food aid in the presence of a strong American economy. Efforts at welfare reform at the state and federal levels may represent one contribution to the problem. Many people who are fighting hunger in America have taken their efforts to Congress in order to restore food stamp benefits to nearly one million legal immigrants who were removed in 1997. Additionally, these advocates want to further increase minimum wage and child care subsidies. This will be a difficult task in the face of a conservative view that the answer to hunger lies in a stronger economy and additional charity from the private sector. Policy analysts in the Heritage Foundation, a conservative think tank, argue that the more programs you have to hand out food, the more people will use them. They also maintain that there is no evidence for widespread hunger or undernutrition in America.[81] This view is in contrast to that of J. Larry Brown, who in 1994, as director of the Center on Hunger, Poverty, and Nutrition at Tufts University in Medford, MA, claimed that hunger in America could be ended in 6 to 8 months by expanding the food stamp program and increasing funding for targeted programs in WIC (Special Supplemental Food Program for Women, Infants, and Children).[81] Unfortunately, the study conducted for Second Harvest in the summer of 1997 by Brown and Dr. J.T. Cook from the Tufts University National Center on Hunger, Poverty, and Nutrition revealed that food stamp cuts in the welfare bill from 1997 through 2020 will amount to nearly $28 billion or approximately 24 billion pounds of food.[82] In order to meet this shortfall, programs such as Second Harvest would have to increase capacity by 425%. Their most optimistic scenario projects a 10% growth rate per year over the next 6 years that would meet less than one-third of the food aid required. Food and grocery donations flattened in 1996 and 1997, making projections for growth in donations even more unfavorable. Although federal officials suggest that the private sector will be able to replace the food stamp safety net, this is unrealistic. Food insecurity in America is already a growing problem, and additional government cuts are not likely to result in increased personal giving. Christine Vlademiroff, president and CEO of Second Harvest, states that private donations are highest in areas where state and local governments give generously to the same issues that charities address.[82]

Food Security in China

With over one fifth of the world's population and only 7% of the globe's cultivated land, China faces a formidable challenge in meeting its food needs.

At the beginning of the 1950s, shortly after the founding of the People's Republic, China's annual grain output was only about 100 million tons. In 1978, grain output reached 304.8 million tons. Since 1995, China has witnessed bumper harvests for 4 consecutive years, which was rare in its history. The grain output reached 465 million tons in 1995 and jumped to 495 million tons in 1997.

Although per-capita cultivated area decreased from 0.18 hectares in the 1950s to 0.11 hectares in 1995, the country's per-capita consumption of grain grew from 209 kg to 380 kg. As the world's largest grain producer, China has basically solved the issue of feeding its huge population.

Nevertheless, there are several problems with food production in China:

1. Inadequate investment in grain production. A shortfall in investment for agriculture made it difficult for China to increase agricultural productivity. Inadequate investment in water conservancy projects, irrigation, and drainage systems led to a worsening condition for grain production.
2. Shrinking per-capita farmland devoted to grain production.

Absolute Decline in China's Farmland, 1990-1997 (thousand hectares)

Year	1990	1991	1992	1993	1994	1995	1996	1997
Decline in farmland	350	450	710	630	790	800	1070	460

3. Unfavorable natural conditions. There are 33 million hectares of cultivated land subject to continual natural disasters, and the perennial drought hit of flooded areas covers 6 million hectares. A considerable amount of farmland has low yields and needs to be transformed urgently.
4. Farmer's burden. The annual growth rate of farmers' net incomes dropped markedly during 1982–1990. Overwhelming economic burdens made farming less profitable, thus dampening farmers' enthusiasm.[8]

REFERENCES

1. Motava, W.J., Contents under pressure, *E. Mag.,* Nov/Dec 27, 1996.
2. United Nations Environment Program, Global environment outlook, *Environmental Health,* April 22, 1997.
3. Kaplan, R., The coming anarchy, *The Atlantic Monthly,* Feb., 44, 1994.
4. Fornos, W., 1997 World population overview, The Population Institute, Dec. 30, 1997, www.population institute.orgloverview97.html.
5. Barney, G.O., *The Global 2000 Report,* Pergamon Press, New York, 1980. (Vol. 1, Summary Report; Vol. 11, Technical Report; Vol. III, Government Global Model), 360 pp.
6. Budiansky, S., 10 billion for dinner please, *U.S. News & World Report,* Sept. 12, 57, 1997.
7. Turk, J. and Turk, A., *Environmental Science,* 4th ed., Saunders College Publishing, Philadelphia, 1988, chap. 9.
8. World Resources Institute (WRI), chap. 10: Food and agriculture, a guide to the global environment, *World Resources 1996–1997,* http://www.wri.org/wri/wr-96-97/fa-txt2.html.
9. Alexandratos, N., Ed., *World Agriculture: Towards 2010, An FAO Study,* John Wiley and Sons, Chichester, U.K., and Food and Agriculture Organization of the United States, Rome, 1995, p. 38.
10. Food and Agricultural Organization (FAO) of the United Nations, *Food Outlook: Global Information and Early Warning System (GIEWS) on Food and Agriculture:* Rome, April, No. 2, 1998, http://www.fao.org(WAICENT/faoinfo/econon-tie/giews/english/fo/fo9804/httoc.htm.
11. Dasgupta, P.S., Population, poverty and the local environment, *Sci. Am.,* Feb., 42, 1995.
12. Simon, J., *The Ultimate Resource,* Princeton University Press, Princeton, NJ, 1981, 495 pp.
13. Simon, J. and Kahn, H., *The Resourceful Earth,* Basil Blackwell Ltd., Oxford, 1984, p. 585.
14. Ehrlich, P., Too many rich people: weighting relative burdens on the planet, *Our Planet,* 6, 31, 1994.
15. AP (Associated Press): Geneva, U.N. gap widening between countries, *Daily Hampshire Gazaette,* Sept. 25, 1, 1997.

16. Glantz, M.H., Is the stork outrunning the plow?, *United Nations Environment Programme: Network Newsletter,* Vol. 9, No. 2, 1994.
17. Harden, G., The tragedy of the commons, *Science,* 162, 243, 1968.
18. Runnels, C., Environmental degradation in ancient Greece, *Sci. Am.,* March, 96, 1995.
19. Repetto, R. (Ed.), *The Global Possible Resources: Development and the New Century,* Yale University Press, New Haven, CT, 1985, p. 538.
20. United Nations Environmental Program (UNEP), Global environmental outlook, *Env. Health,* April 22, 1997.
21. Briefings, Satellite provides dim view of Amazon, *Greenpeace,* Nov/Dec, 22, 1988.
22. AP, Scientists hopes are borne out: rainforests and carbon dioxide, *New York Times,* Nov. 8, A4, 1995.
23. Farnsworth, N.R., New drugs from the rainforest, *World Health,* Mar/Apr, No. 2, 49, 1996.
24. National Aeronautics and Space Administration (NASA), Tropical deforestation and habitat degradation in the Brazilian Amazon Basin, The Earth Observing System Educator's Visual Materials, 1998, http://eospso.esfc.nasa.gov/cos-edu.pack/p24.html.
25. Moeller, D.W., *Environmental Health,* Harvard University Press, Cambridge, MA, 1997, chap. 19.
26. Fogel, L., Brazil suspicious of moves to protect forests, *The Washington Post,* Oct. 11, Al, 1993.
27. Schenio, D.J., Amazon is burning again, as furiously as ever, *New York Times,* Oct. 12, 1, 1995.
28. Benkoil, D., Fires worse than ever, *ABC News.com,* Oct. 9, 1997, http://www.abc-news.com/sections/word/amazon/I 09index.html.
29. Hernandez, S., Environment: Afro-Asian trees to the rescue in Central America, *InterPress Service English Newswire,* 01, 08, 1996, http://www2/efibrary.com/id/2525/ getdoc.cgi?id=53502173XOX634&Oqoo2D60&fores=R L.
30. Brevik, J.J., Global environmental outlook, *Environ. Health,* April, 22, 1997.
31. Haniffa, A., Degradation hits human security, *India Abroad,* 12, 106, 1996, http://www2. elibrary.com/id/2525/getdoc.c.cos&form=ri.&pubname=lndia-Abroad&puburl=O.
32. Depth News Asia, Action for tropical rainforests, *World Press Review,* Aug., 13, 1993.
33. Brown, L.R. and Woff, E.C., Soil erosion: a quiet crisis in the world economy, *Worldwatch Paper,* Worldwatch Institute, Washington, D.C., 1984, 49.
34. United States Dept. of Agriculture (USDA): Analysis of policies to conserve soil and reduce surplus crop production. *Agricultural Economic Report,* Washington, D.C., No. 534, 21, 1985.
35. Doyle, R., Soil erosion of cropland in the U.S., 1982–1992, *Sci. Am.,* Oct., 34, 1996.
36. Stutz, B., The landscape of hunger, *Audubon,* Mar/Apr, No. 2, 95, 1993.
37. Filipov, D., Desert's drift buries Russian region's dreams, *Boston Globe,* Sept. 27, A26, 1997.
38. Brazil Embassy, What is desertification?, Science and Technology Section, National Research Network's Brazilian Regional Center, Feb. 30, 1997, http://brasil.emb. nw,dc.us./nuna/deserts/ingles/se/~edesert/desertmi.html#dm.
39. National Aeronautics and Space Administration (NASA), Desertification Atlas (Africa) Maps I - 1, *Global Change Master Directory,* Oct. 24, 1997, http://gcmd.gsfc. gov/cgibin/detdy?=display+dif&entry=NBld0270/101.
40. EPA-OWOW (Environmental Protection Agency, Office of Wetlands, Oceans, and Watersheds), *Facts About Wetlands,* Aug., 25, 1997, http://www.epa.gov/owow/wetlands/ facts/types.html.
41. Secraig, B., What is a wetland? *Sierra,* May/June, 44, 1996.

42. Enteractive, Inc., Disappearing wetlands, *Earth Explorer,* Feb. 1, 1995, http://www3.elibrary.com/id/2525/getdoe.c ... lorer&puburl = http*c*//www.enteractive.com.
43. Adler, T., Scientist dispute legislator's take on wetlands, *Sci. News,* Vol., 148, 56, 1995.
44. Cushman, J.H., From wetlands to asphalt, a parcel at a time, *New York Times,* Oct. 27, 22, 1996.
45. Stevens, W.K., Prairie ducks return in record numbers, *New York Times,* Oct. 11, 1, 1994.
46. Marina, M., Progress threatens Brazil's vast wetlands: Foes say canal would drain Panatanal, CNN Interactive, Online Internet, April 23, 1997, http://www.cnn.com/Earth/9704/23/brazil.wetlands.
47. Erwin, D.H., The mother of mass extinctions, *Sci. Am.,* July, 72, 1996.
48. Callahan, J.R., Vanishing biodiversity, *Env. Health Perspect.,* 104:4, 386, 1996.
49. Hilchey, T., Biodiversity study sees more species in danger, *New York Times,* Nov. 14, Science section 1, 1995.
50. Worldwide Fund for Nature, Climate Change and Biodiversity Conservation: Worldwide Loss of Biodiversity, Executive Summary, 1996, chap. 3, http://www.panda.org/ida/climate/cIimate.3/Pagel.html.
51. Doyle, R., Threatened mammals, *Sci. Am.,* Jan., 32, 1997.
52. Cushman, J., Freshwater mussels facing mass extinction, *The New York Times,* Oct. 3, Cl, 1995.
53. Cook, R.M., Sinclair, A. and Stefarmson, O., Potential collapse of North Sea cod stocks, *Nature,* Feb. 6, 358, 1997.
54. Allen, S., Trying to do the right thing. *The Boston Globe,* Sept. 26, 27, 1995.
55. Butler, S., Rod? reel? dynamite? a tough-love program takes aim at the devastation of the coral reefs, *US News & World Report,* Nov. 25, 56, 1996.
56. Lemonick, M.D., Animal genocide, mob-style, *Time Mag.,* Nov. 14, 77, 1995.
57. Jones, A., Stowaway snakes deplete bird populations on Guam, CNN Interactive, 1, 9, 1996.
58. Luoma, J., Boom to anglers turns into disaster for lakes and streams, *The New York Times,* Nov. 17, 3, 1992.
59. Stevens, W., Bizarre parasite invades trout streams, devastating young rainbows, *The New York Times,* Sept. 19, C9, 1995.
60. Wilson, E.O., *The Diversity of Life,* Harvard University Press, Cambridge, chap. 1.
61. Kaiser, J., Vanishing zooplankton, *Sci. News,* March 11, 157, 1995.
62. AP (Associated Press), Many reports of deformities among frogs are puzzling, *The New York Times,* Oct. 13, 33, 1996.
63. Chadwick, D.H., Dead or alive: the Endangered Species Act, *Natl. Geogr.,* Mar. 21, 1995.
64. AP (Associated Press), Animals to come off list, *Daily Hampshire Gazaette,* May 6, p. 1, 1998.
65. Stevens, W.K., "Hot spots" for American endangered species cover surprisingly little land, *The New York Times,* Jan. 1, 1, 1997.
66. Budiansky, S., Killing with kindness, *US News & World Report,* Nov. 25, 48, 1996.
67. Satchell, M., Save the elephants: start shooting them, *US News & World Report,* Nov. 25, 52, 1996.
68. AP (Associated Press), Experts debate limits of Earth's resources, CNN Interactive, Oct 27, 1996.
69. Brown, L.R. and Kane, H., *Reassessing the Earth's Population Carrying Capacity,* W.W. Norton, New York, 1994.

70. The World Bank, *World Development Report 1992,* The World Bank, Washington D.C., 1992, p. 135.
71. Ehrlich, P.R. et al., Food security, population and environment, *Population Dev. Rev.,* 19:1, 7, 1993.
72. Pinstrup-Andersen, P. and Pandya-Lorch, R., Alternating Poverty, Intensifying Agriculture, and Effectively Managing Natural Resources, Food Agriculture and the Environment, Discussion Paper, No. 1, International Food Policy Research Institute, Washington, D.C., 1994, p. 6.
73. Xin, W., Population vs. development: challenge of the new century, *Beijing Rev.,* May 1–7, 12, 1995.
74. Food and Agriculture Organization of the United States (FAO), *The State of Food and Agriculture 1994,* FAO, Rome, 1994, p. 11.
75. Mitchell, D.O. and Ingco, M.D., *World Food Outlook,* The World Bank, Washington, D.C., 1993.
76. Rosegrant, M.W., Agcaoili-Sombilla, M. and Perez, N.D., Global Food Projections to *2020:* Implications for Investment, *International Food Policy Research Institute (IFPRI) Food, Agriculture and the Environment,* Draft Discussion Paper, IFPRI, Washington, D.C., 1.
77. Webb, P., A Time of Plenty, A World of Need: The Role of Food Aid in 2020, International Food Policy Research Institute (IFPRI) 2020, Brief No. 10, Washington, D.C., 1995, p. 2.
78. Pimental, D. et al., Environmental and economic costs of soil erosion and conservation benefits, *Science,* 267:5201, 1117, Feb. 24, 1995.
79. International Food Policy Research Institute (IFPRI), A 2020 Vision for Food, Agriculture, and the Environment, Draft Paper (117PRI), Washington, D.C., June, 1, 1995.
80. Roberts, L. and Roberts, S.V., Hunger: a startling crisis, *USA Weekend,* March 27–29, 4, 1998.
81. Shapiro, L. and Rosenberg, D. How hungry is America?, *Newsweek,* March 14, 59, 1994.
82. Brown, L.J. and Cook, J.T., Food stamp cuts bring grim forecast, *Second Harvest Update,* Winter, 1997, http://www.secondharvest.org/websecha/e-uw97tuft.htm.
83. RAFI, The terminator technology: new genetic technology aims to prevent farmers from saving seed, *(Rural Advancement Foundation International) Communiqué,* Mar/Apr 1998.
84. Earth's Green Lungs Begin to Fade, *Int. Wild.,* 25, Sept./Oct. 1999.
85. BBC Online Network, Oct. 27, 1999.
86. To restore or not to restore: That is the wetlands question, *Environ. News Service* http://ens.lycos.com/ens/ Feb., 2000; Dec. 2000.
87. Briscoe, D., Farmers worldwide losing seed varieties, *Springfield Sunday Republican,* Sept. 19, 1999.
88. Mzurek, R., Back from the dead, *New Scientist,* 150, Oct. 9, 1999.
89. Barnett, A.D., China and the World Food System, Monograph No. 12, ASIN 0686286834, 1982.
90. Baviera, A.S.P. and Shaolian, L., Food Security in China & Southeast Asia, 1999.

CHAPTER 4

Environmental Disease

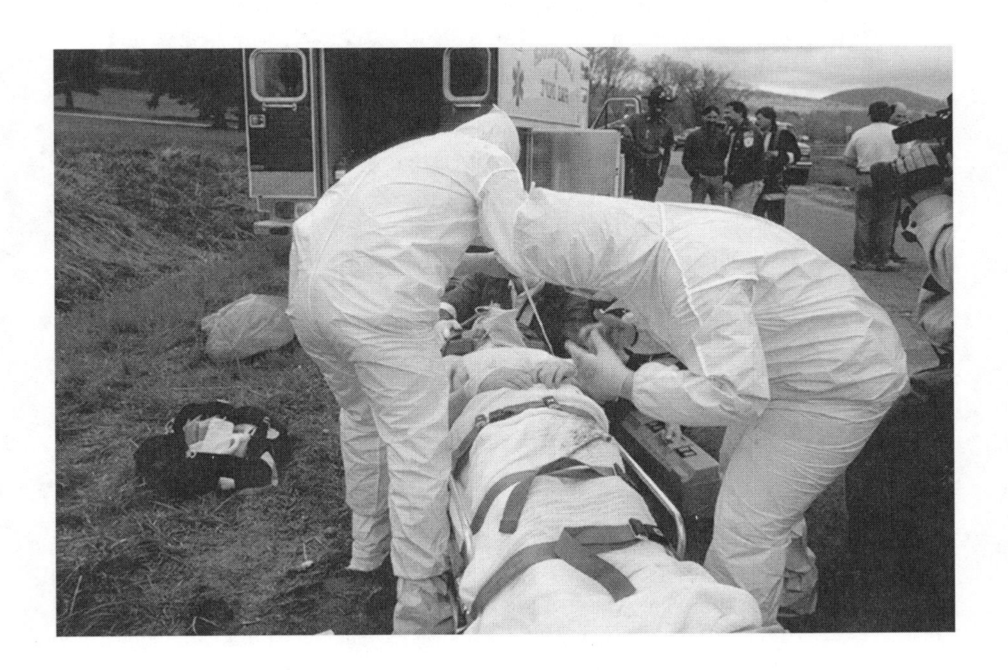

INTRODUCTION

There is hardly a day that passes without a news media report on the hazards of chemicals in our environment producing a possible cornucopia of diseases ranging from cancer, reproductive disorders, failed pregnancies, birth defects, asthma, emphysema, and any number of other diseases causing premature death or chronic illness. Some scientists report that chlorinated organics and heavy metals already abound in our bodies and are so ubiquitous that even the most remote life forms in the arctic oceans demonstrate measurable levels in their flesh. Some of these chemicals

OBJECTIVES FOR THIS CHAPTER

A student reading this chapter will be able to:

1. **Discuss and define the concepts of environment and disease**
2. **List and explain the factors influencing environmental disease including infectious disease, physical and chemical injury, ionizing radiation, developmental disease, neoplastic disease, and nutritional disease**
3. **Explain the process of genetic replication in mammalian cells; list the major genetic components and discuss their primary actions and features**
4. **Describe and discuss the major mechanisms of developmental and genetic diseases**
5. **Describe and provide a schematic of the new processes in recombinant-DNA and genetic screening techniques**
6. **Discuss and describe the new concepts on the origins of cancer, risk factors, and current trends**

have been implicated in developmental diseases in amphibians. Chemicals such as certain types of pesticides can mimic natural regulatory chemicals in animals. Mimics of the chemical retinoic acid have been linked to deformities in amphibians and fish, resulting in an increasing rate of the development of extra and missing limbs.[1] We are, according to some, placing our lives at jeopardy, risking our health, and threatening the life-supporting ecosystems on which humanity depends by the careless use and disposal of toxic chemicals.[1] We do all of this, it is reported, because we continue to manufacture and disperse billions of tons of more than 70,000 potentially toxic chemicals into the biosphere without regard for nature or an appreciation of the danger that these chemicals can bring to our Earth or to us.

Exposure to chemicals common in pesticides such as MPTP has been associated with the development of Parkinson's disease. Parkinson's disease had been previously thought of as a purely genetic disease.[2] This new association with environmental pollution may be the first in a series of diseases previously thought of as genetic but which may have an environmental component. Diseases such as heart disease, coronary artery disease, certain types of cancer, obsessive–compulsive disorder, and Tourette's Syndrome may all be implicated with an environmental cause.[3] The possibility that manufactured and discarded chemicals in our environment could potentially produce harm erupted into the public consciousness in 1978 when then-New York State Commissioner of Health, Robert Whalan, declared Love Canal in New York a health emergency with his document, *Love Canal: Public Health Time Bomb*.[4] A firestorm of activity lasting 2 years revolved around this revelation including EPA (Environmental Protection Agency) reports, nongovernment studies and reports, and news media events. The resulting emotional intensity accelerated the condition to a point where the federal government was forced in 1980 to relocate more than 700 Love Canal homeowners at a cost exceeding $30 million.

A sensitized public, now critical of government and environmentally astute, wanted answers and immediate resolution in December, 1982, when local flooding spread dioxin throughout Times Beach, Missouri. The dioxin was a contaminant in waste oil sprayed as duct control at a horse arena and onto unpaved roads in the mid-1970s. After flooding, soil levels were up to 1000 times higher than federally recommended safe levels. In an effort to protect the public health and calm angry, frightened citizens, the federal government ordered an evacuation of Times Beach at a cost of $33 million. Alarmed, frightened, and on guard, the public became responsive to revelations in the news media about chemicals and radionuclides in our midst that could cause potential harm.

Since the 1980s, popular magazines such as *Time*, major newspapers led by the *New York Times*, myriad television shows, and numerous books have warned against the evils of polluting our land. Widely recognized people such as Ralph Nader, Lester Brown, and Sam Epstein have authored an array of books vigorously challenging humanity to stop polluting our environment with synthetic chemicals. These and other books speak of "chemical violence … approaching epidemic proportions,"[5] "the merchandising of poison … a huge global growth industry,"[6] and "toxic chemicals (that) threaten every man, woman, and child."[7] Environmental advocacy has found its way into the television airwaves in documentaries such as the "The Poisoning of America" narrated by Hugh Downs.[8] Even cartoons have taken up the environmental protection flag while children watch as corporate villains threaten the health of innocent citizens with flagrant disregard for the proper disposal of festering and foul-smelling ooze in their bid for higher profits. What are the messages given through the public media about toxic chemicals? Among the more prominent messages are that: (1) exposure to toxic chemicals has dramatically increased the risk of cancer; (2) common household and agricultural chemicals are causing many human diseases and death; (3) polluted air and water are major sources of disease risk; and (4) environmental chemicals are interfering with the reproductive process in humans and producing harmful effects in the fetus and young children.[4] The major health concerns among many Americans revolve around what they hear about on television or talk radio shows or read in newspapers and in magazines. What Americans have read is that cancer is on the increase,[9,10] that cancer-causing chemicals or carcinogens pervade our food and drinking water[11] and are even spread onto lawns in the form of pesticides and fertilizers.[12]

Sometimes chemicals introduced to help the environment have unforeseen detrimental effects on the environment. MBTE was mandated to be added to automobile fuel to aid in more efficient burning. This was meant to reduce the amount of toxic chemicals released as well as lessen the amount of fuel burned altogether. Recently,

MBTE has been found in increasing amounts in drinking water supplies throughout the country, including Long Island, New York. MBTE has been classified as a possible carcinogen by the U.S. Environmental Protection Agency.[13]

Since the 1970s, both the popular media and the professional peer-reviewed articles claim that nearly 90% of cancers can be traced to an environmental cause and so may be largely preventable. Genetic factors (heredity) are said to explain only about 5% of cancers, while the remainder can be attributed to external or environmental factors that act in conjunction with acquired and genetic susceptibility.[14] The message is said to be optimistic in that exposure to environmental carcinogens is theoretically preventable, and therefore cancer risk can be modifiable and subsequently reduced. Based on what the public hears and reads in the news media and in some professional writing, it is easy to understand why many people associate the major cancer risks with synthetic chemicals discarded to the environment. This, however, would be largely incorrect. In order to explain why this so, we need to begin with definitions of the terms *environment* and *disease* — the subject of this chapter.

Defining the Term *Environment* in Relation to Disease (Cancer)

Koren defines environment as the "sum of all surrounding physical conditions in which an organism or cell lives, including the available energy in living and on living materials,"[15] while defined elsewhere as "the surrounding conditions or influences that affect an organism or the cells within an organism."[16] Based on such definitions and the pervasive use of the words "environmental toxins or chemicals" in the media, it is little wonder that most people interpret this to mean cancer-causing chemicals in our air, food, water, and workplace. However, the use of the word *environment* in relation to cancer causation was originally intended to be much broader and is thought to have originated with Dr. John Higginson, the founder and director of the World Health Organization's (WHO) International Agency of Research on Cancer. When he reported that nearly two thirds of cancers had an environmental cause, he was "… considering the total environment, cultural as well chemical … mode of life … what surrounds people and impinges on them."[17] Epidemiologists and other cancer researchers use the word *environment* to include almost everything that is not inherited. The scientific community has come to learn that many factors contribute to the risks of cancer, including personal habits and cultural behaviors. The term *environment*, in its broadest definition, refers to personal and cultural behavior including smoking, diet, alcohol consumption, sexual and reproductive patterns, workplace, infections, food additives, and pollution, along with the strictly physical environment.[1]

Some of the major factors associated with cancer risk are presented in Figure 4.1.[1] Diet represents one of the highest risks associated with cancer. Diet refers to those food-eating patterns high in fat and low in fruits, vegetables, and grains. This dietary pattern increases risks of colon, uterine, and breast cancers. An unhealthy diet may contribute to 35% of all cancers. Diet and other risk factors are discussed in more detail later in this chapter. The use of tobacco products is thought to contribute to 30% of cancers (Figure 4.1).[1] High risks of cervical cancer are

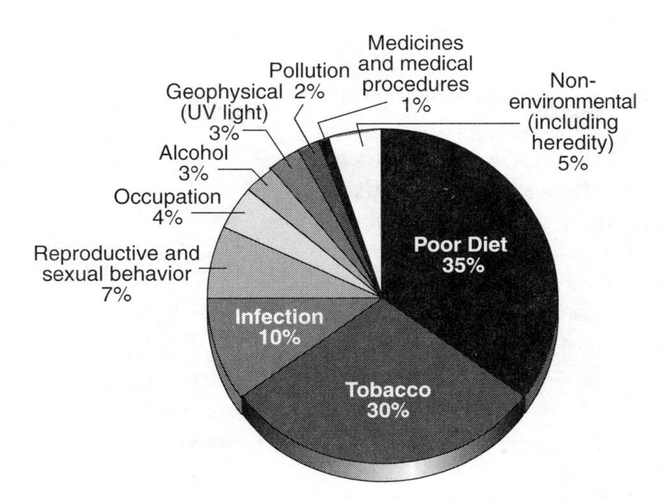

Figure 4.1 Proportion of cancer deaths associated with various environmental factors. These are best-estimate figures. (Adapted from *Morbidity and Mortality Weekly Report,* June 19, 1992, p. 47.)

associated with certain sexual behaviors such as having sexual intercourse earlier in life, having multiple sexual partners, or contracting specific sexually transmitted diseases.[18] Estimates of cancer risk associated with reproductive behavior suggest that such risks account for 1 to 13% of all cancer deaths, with the best estimate being 7%. Medicines and some medical procedures are also included as cancer risks such as diethylstibestrol (DES), which caused reproductive cancers in the offspring of women who used the drug during pregnancy to prevent premature births. DES is a nonsteroid estrogen that was used as a substitute for natural estrogenic hormones. The best estimate for the contribution of pollution to cancer risk is around 2% of the total risks. Pollution may be defined as "the presence of a foreign substance — organic, inorganic, radiological, or biological — that tends to degrade the quality of the environment so as to create a health hazard."[15] The vision presented to us in the public media and in a number of published books is that chemical pollutants contribute substantially to cancer risk. This term *environmental* has been corrupted to mean *chemical*. Consequently, many people believe that when they read the words "90% of cancers are environmentally caused," these cancers are necessarily produced from exposure to industrially produced toxic chemicals. In fact, such chemicals contribute to a very small percentage of cancer deaths, while personal habits including cigarette smoke, alcohol abuse, and unhealthy diet contribute to a combined risk of 70% or more to cancer deaths.[19] These behaviors are well within the control of the individual to modify, but translating belief or knowledge to changes in behavior is not always easy. Although 82% of Americans believe improving their diet can lower cancer risk, and 61% are aware that obesity is a risk for cancer, more than half of those surveyed in the American Cancer Society study believe there was little they could do to reduce their risk. The less than 15% that made dietary changes within the last year did so for weight loss or to reduce risk from heart disease, not to reduce cancer risk.[20]

DEFINING DISEASE

Disease may be defined as "a definite pathological process having a characteristic set of signs and symptoms which are detrimental to the well-being of the individual."[15] There is no shortage of diseases within the human population, and they take on an array of symptoms, are produced by a variety of sources, and are resolved in a number of ways and differing times. Diseases that already exist in a community where it is maintained in a low but constant incidence are said to be *endemic*. When there is marked increase in the incidence of disease within a limited area affecting ever-increasing numbers of people, this is referred to as an *epidemic*. Should the epidemic spread throughout the world, as in the case of various outbreaks of influenza, it is called a *pandemic*. Many diseases such as measles are acute and have a rapid onset, are usually self-limiting, and are of relatively short duration. Diseases having a slow onset and lasting for extended periods are said to be chronic and may include cancer, emphysema, some forms of heart disease, or AIDs. Diseases may be further classified by the form of the disease (Table 4.1). These diseases may be infectious in nature or the result of chemical or physical injury. They may be developmental diseases involving defects occurring during embryogenesis or the consequence of genetic errors. Neoplastic diseases such as cancer develop from multiple genetic defects acting in accord with a number of endogenous or environmental factors producing susceptibility. Nutritional diseases are often the consequence of poverty and a weakened society, but these characteristics may be be intimately bound with the environmental conditions.

Table 4.1 Forms of Human Disease

HUMAN DISEASE FORMS
• Infectious Diseases
• Physical and Chemical Injury
Mechanical
Thermal
Ionizing Radiation
Chemical
• Developmental Diseases
• Neoplastic Diseases
• Nutritional Diseases

Infectious Disease

The greatest mass of humanity suffers from infectious diseases that result from the pathologic process occurring when a microbial agent invades the body. The agents of such disease that are always inside the cell, or intracellular, include viruses (HIV, measles, mumps, human papilloma virus), chlamydia (often associated with nongonococcal urethritis), and rickettsia (epidemic typhus, Rocky Mountain Spotted Fever). Organisms that may be intracellular or extracellular include bacteria (strep throat, botulism, staph infections, gonorrhea), mycobacteria (tuberculosis), fungi (histoplasmosis, ringworm, jock itch [candidiasis]), and protozoa (Giardiasis, amebiasis). Metazoans are multiple-celled animals that are always extracellular because of their significant size. They include parasitic nematodes and roundworms (Ascaris, Trichinella), the cestodes or segmented flatworms (tapeworms), and the trematodes including the liver, lung, and intestinal flukes. Insects may also be included in the category of extracellular parasites and include mites, ticks, and lice (Table 4.2). In

Table 4.2 Agents of Infectious Disease, Reservoirs of Infection, and Mechanisms by which Diseases Are Spread

AGENTS OF INFECTIOUS DISEASE		RESERVOIRS OF INFECTION	THE SPREAD OF INFECTION
Virus **Chlamydia** **Rickettsia**	Always intracellular	**Human** • Exhibit signs of disease • Does not exhibit signs, but transmits disease This person is known as a **carrier**	**Contact** • Direct – no intermediate object involved in transmission; also known as person-to-person such as kissing or sexual intercourse • Indirect – transmitted from reservoir to host by nonliving object sometimes called a fomite; may include tissues, cups, utensils, etc. • Droplet – Organisms are in droplets produced by sneezing, coughing, and talking, and travel less than a meter
Bacteria **Mycobacteria** **Fungi** **Protozoa**	Intracellular	**Animal** • Domestic • Wild or sylvatic Diseases transmitted by animals to humans are called **zoonoses**	**Vehicles** Transmission of disease agents by a medium such as water, food, or air • Waterborne • Foodborne • Airborne – travel more than a meter in air
Metazoa	Always extracellular	**Nonliving** • Soil • Water	**Vectors** Animals that carry pathogens from one host to another; arthropods most important group • Mechanical – passive transport on feet or other body parts • Biological – active process in which agent reproduces in vector and then is transmitted by bite of insect, or in feces or vomit

order for a disease to perpetuate itself, the disease organism must be in constant supply. Such sources can be living organisms or inanimate objects that provide the conditions where the organisms may survive, multiply, and provide the conditions necessary for transmission. These sources are called reservoirs. Reservoirs may include humans, animals, and nonliving objects. As human populations grow and spread into previously unsettled regions, it is likely that they will be exposed to new reservoirs and potentially new diseases never experienced before. Diseases such as Yellow Fever, Ebola, and possibly HIV may have been introduced to human populations in this manner.[21] Most human diseases are harbored within the living bodies of humans and are transmitted directly or indirectly to others. Diseases may be transmitted from persons who exhibit signs of the disease, or they may be transmitted by persons known as carriers who exhibit no symptoms but harbor and transmit the disease organisms. Some people may never develop symptoms while others may pass the disease during a latent or symptom-free stage. Humans serve as reservoirs for such diseases as AIDs, hepatitis, and tuberculosis.

A number of wild (sylvatic) and domestic animals serve as reservoirs of disease agents that infect humans. Diseases that are spread mostly from animals to humans are called *zoonoses* and may include Lyme disease, eastern equine encephalitis, Rocky Mountain Spotted Fever, rabies, bubonic plague, salmonellosis, toxoplasmosis, trichinosis, and more than 140 other known zoonotic diseases. Some diseases that have animal reservoirs mutate in these animals such that they are able to change their characteristics enough that they continue to cause epidemics year after year.

One such disease is influenza. In 1997, a strain of influenza that had been previously thought to only infect birds was linked to deaths in Hong Kong. It was a particularly virulent strain, and it was feared that a pandemic similar to the large influenza pandemic of 1918 might ensue. The 1918 pandemic infected nearly every person on Earth and was linked to an estimated 30 million deaths.[22] The major nonliving reservoirs of most human diseases are soil and water, where disease-producing fungi and bacteria such as *Clostridium perfringens* or *C. botulinum* occur. Many organisms can survive in water, including parasites and bacteria that produce gastrointestinal disease and may be passed to the water from human feces or animal wastes.

The major routes of infections for most infectious disease agents include: (1) contact; (2) vehicles; and (3) vectors (Figure 4.2). The types of contact involved in transmission include direct, indirect, and droplet contact. *Direct contact transmission* occurs when the disease is passed directly from a source to the host by physical contact without an intervening object. Also known as person-to-person contact, this type of transmission may occur through kissing, touching, or sexual

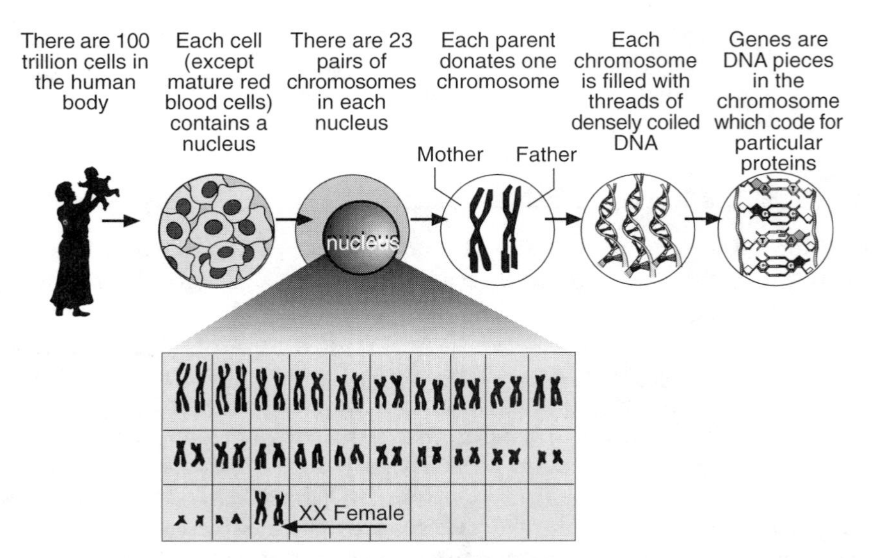

Figure 4.2 A perspective of DNA in the human cell. (Adapted from Elmer-Dewitt, P., The genetic revolution, *Time,* Jan. 17, 46, 1994.)

intercourse. AIDs, infectious mononucleosis, influenza, the common cold, and measles are examples of diseases transmitted this way. When the disease agent is transmitted from a reservoir to a host by a nonliving object, the process is called *indirect contact transmission*. The object to which the disease is transmitted is often referred to as a *fomite* and may commonly include eating utensils, drinking cups, shared toothbrushes, tissues, toys, syringes, and bedding. As examples, hepatitis and AIDs may be transmitted by contaminated syringes. Organisms such as the viruses involved with the common cold may be transmitted by droplet infections, whereby the organisms are contained within mucous droplets that are expelled to the air in sneezing, coughing, or even talking, and travel short distances to be inhaled or ingested by persons very near by (usually less than one meter).

When disease agents are transmitted by various media such as air, water, food, intravenous fluids or blood, and drugs, this is referred to as *vehicle transmission* (Table 4.2). Airborne transmission refers to disease agents that travel some distance greater than one meter in the air and are usually carried by dust particles or extremely small droplets. Fungal spores, staphylococci, and the measles virus can be transmitted by particles in the air. Diseases spread by water have usually been contaminated by human sewage or by animal waste that contains disease organisms. Foodborne disease usually occurs when food is improperly refrigerated or insufficiently heated and permits the disease organism to grow to numbers capable of producing toxins or invasive numbers of bacteria. In many cases, the disease organisms are carried in the flesh of a food product that has been previously infected (i.e., tapeworm or trichinosis) or has come into contact with the offending organism and serves to simply transport the organism (as in some forms of hepatitis). Diseases transmitted by air, water, and food are discussed in subsequent chapters.

Vectors are animals that carry pathogens from one host to another, either from another human who is infected or from an infected animal. Most vectors of significance to humans are arthropods (joint-legged invertebrates) including insects, spiders, and crabs. Insects represent the largest and most significant vectors of human disease. When the arthropods transmit the disease by simply carrying the organism(s) on the feet or other body part to a person's food or an open sore on the person, then the transmission is passive and the process is referred to as *mechanical transmission*. Most vector-borne diseases have worldwide importance in that they produce disease in tens of millions of people while influencing the social, economic, and political climates of many countries. In such diseases, the disease agent goes through some reproductive stage in the vector and is transmitted by a bite, fecal matter, or vomit of the vector in an active process referred to as *biological transmission*. Lyme disease, and an increasing number of other diseases such as Ehrlichiosis, are spread by deer ticks and Lone Star ticks, depending on the region of the U.S. Some of these new diseases are not easily detectable and are therefore more difficult to treat than Lyme disease.[23] Malaria is spread by the female Anophelene mosquito in which the malarial parasite (*Plasmodium* sp.) undergoes a sexual cycle before it can reinfect another human by the bite of the mosquito. Since the sexual cycle is in the mosquito, the mosquito is known as the definitive host, and the asexual reproduction takes place in the human who is the intermediate host.

Physical and Chemical Injury

Physical injury includes mechanical, thermal, and radiation injuries. Mechanical injury results from a transfer of a damaging excess of kinetic injury to tissues resulting in abrasions, lacerations, punctures, contusions, broken bones, and projectile wounds.[24] Automobile accidents, falls, shootings, stabbings, and other forms of trauma to the human body continue to be among the most important forms of death and injury in the world. The leading cause of injury-related deaths in the U.S. involves motor vehicles, accounting for nearly 40,000 deaths per year with more than 40% of those deaths in the 16- to 19-year age group, with alcohol involved in nearly half of all motor vehicle deaths.[25] Falls are the leading causes of accidental deaths in the home. Nearly 40,000 children are injured each year from cribs, high chairs, and bunk beds. Another 27,000 are injured in infant walkers, especially from falling downstairs. More than 750,000 falls and injuries occur from accidents on stairs, while falls from ladders account for another 100,000 injuries each year in the U.S. Those above 75 years of age are 12 times as likely to die from falls as all other groups combined.[24] Violence within the home and the community has become a significant public health problem. Firearm-related deaths in the U.S. include homicides, suicides, and unintentional shootings. Firearm-related deaths in the U.S. escalated sharply in the 30 years from 1968.[21] Deaths from firearms now approach 38,000 per year with most deaths involving males 15 to 19 years of age.

Thermal injury results when tissues are exposed to extremely low or extremely high temperatures. Low temperatures can produce hypothermia or frostbite. Frostbite occurs when extreme cold freezes body water and forms intracellular crystals that burst cell membranes to kill affected cells. Very high temperatures can produce hypothermia or burn injuries. Burn injuries claim about 500 people a year in the U.S. These deaths are generally from residential fires and most commonly occur among the very young or the very old, the black or Hispanic populations. Major reasons for the fires include children playing with matches, faults in electrical devices including heating sources, and careless smoking.[25]

When radiant energy interacts with matter to form charged particles, that energy is referred to as ionizing radiation and may include electromagnetic radiation (gamma and x-radiation) or particle radiation (alpha, beta, and neutron). Exposure to radiation may come from radon, diagnostic use of x-rays, use of radionuclides in medicine, and emissions from catastrophically failed nuclear power plants such as Chernobyl. Injury results from the radiolysis of water molecules in tissue that form free radicals (ion pairs) and react immediately with components of the cell, causing crosslinking of macromolecules and changing the tertiary or three-dimensional structure of the molecule. These changes may impair the function of proteins, deoxyribonucleic acid (DNA), and other molecules, resulting in reactions from subtle interference to cellular death.[24]

Chemical injuries result from exposure to chemicals that produce a deleterious effect on organ systems by stimulating or depressing metabolic function. Some chemicals may function at trace levels to promote or stimulate organ response, while larger doses may retard or destroy activity entirely. The major mechanisms for chemical injury may be categorized as: (1) interfering with enzyme activity; (2) directly combining with some cell component other than enzymes; or (3) producing a secondary action in which a chemical causes the release or formation of a more harmful substance. In this last case, allergens represent a good example in that otherwise harmless substances such as pollen, animal dander, or even certain foods can cause life-threatening allergic reactions in some people. Some chemicals produce a systemic disease. Systemic refers to conditions or a process that affects the whole body. Many environmental pollutants such as lead, mercury, cadmium, and chlorinated hydrocarbons are systemic poisons that affect a broad array of organ systems. Systemic pollutants are normally ingested, though inhalation and skin absorption may also occur. Once in the body they are distributed through the bloodstream to various organs, where common manifestations include central nervous system disorders, stomach and bowel disorders, and disruptions to the blood-forming tissues.

Developmental Disease

Every individual within each species is a representation of the information contained in the genome together with the expression of those genes in the development of that individual. Developmental disease occurs when faults or mistakes occur within the genes (or chromosomes) or when stages in development of the fetus are disturbed. Faults or mistakes occur within the genes when the genetic information is disturbed by chemical or radioactive insult, inherited errors, or random (spontaneous) errors in transcription. Major categories of genetic disease include single-gene defects and cytogenetic defects (abnormalities in the number or structure of chromosomes).[24] When defects arise during the embryonic period of development, these are said to be *developmental* or *teratologic defects*. Some teratologic diseases are inherited, some are produced by cytogenetic defects, while most occur from interference with embryologic development in an individual whose genetic makeup would ordinarily produce a normal individual. The causative factors in this case are extragenetic or extrinsic factors, and the agents producing these factors are referred to as teratogens. This is most often an environmental disease in that external factors are involved including maternal behaviors, tobacco and alcohol use, inadequate nutrition, use of various medications such as accutane or DES, stress, infectious diseases including toxoplasmosis or German measles, and exposure to chemical and physical agents in the home or workplace environment. The defects occurring during the embryonic stages of development are said to be congenital defects. More detailed information on developmental disorders are given later in this chapter.

Neoplastic Disease

The new and uncontrolled growth of abnormal tissue from the transformation of normal body cells is referred to as neoplasia. Such growths are referred to as tumors or neoplasms. Tumors may be benign or malignant and vary dramatically in form, rate of growth, potential injury to host, and appearance. Malignancies may derive from a number of exogenous factors (arising from the environment) or endogenous factors (originating from within the body). The exogenous factors may include habits (tobacco use, poor nutrition, alcohol use, sexual and reproductive activities), ionizing radiation, chemical exposure, environment (socioeconomic, geographical, and occupational differences), and oncogenic viruses. Endogenous factors contributing to cancer risk may include gender, age, hormonal imbalance, impaired immune system, and genetic predisposition.[24] This topic is covered in more detail later in this chapter.

Nutritional Disease

The consequences of overpopulation including an inadequate food supply in many parts of the world were discussed in Chapter 3. Inadequate diets also occur in developed nations such as the U.S. Fast food comprises an increasing percentage of the American diet. Critics of this trend point to the over-processing of the food for the purpose of homogeneity. This processing removes much of the nutritional value from the foods, often while increasing the sugar and fat content of the food. Obesity in the U.S. has risen from 8% of the population in 1980 to 20% currently.[26] In LDCs, most nutritional diseases are coupled with political, cultural, or emotional problems that make solutions more difficult than simply increasing caloric intake. Malnutrition, as a consequence of inadequate diet, is often associated with the ignorance and poverty that are manifestations of a weakened society. In the broadest definition of environment, malnutrition may be considered an environmental disease, but it must be known that in most nutritional diseases there is a social or emotional status that contributes to the problem. The consequences of an inadequate diet are far more disturbing to the child than the adult because the child needs food for growing in addition to functioning. Deprivation of nutrients at vital stages in development may result in delayed maturation, permanent damage, or even death. In the LDCs, such as certain African nations, many children 1 to 3 years old display a disease of dietary insufficiency known as kwashiorkor. These children are severely protein deprived and may exhibit anemia, diarrhea, graying or reddening of hair, edema, ascites (distended belly), fatty degeneration of the liver, failure to grow, and apathy with extreme weakness. Nutritional marasmus may appear in children one year of age who lack sufficient food and show signs of wasting, failure to grow, and complications of infections with multiple vitamin deficiencies. Infantile scurvy develops from a lack of vitamin C (ascorbic acid) in the diet and consequently the synthesis of collagen and mucopolysaccharides is disturbed. This results in widespread hemorrhages, anemia, failure to heal, tenderness in limbs, and possible cardiac failure with sudden death.[24]

Environmental Disease

We now have a better understanding of the terms environment and disease, and it is appropriate here to combine these terms and derive a more concise definition of the term *environmental disease*. Environmental disease refers to any pathologic process having characteristic signs and symptoms that are detrimental to the well-being of the individual and are the consequence of external factors including exposure

WEB PICK Go to http://www-unix.oit.umass.edu/~envhl565/index.html Click on "Chapter Web Links" to find Welcome to PubMed under Chapter 4	
Chapter 4	Environmental Disease
Whose Site?	National Library of Medicine
URL	http://www.ncbi.nlm.nih.gov/PubMed/
What's Here?	PubMed, a service of the National Library of Medicine, provides access to over 11 million citations from MEDLINE and additional life science journals. PubMed includes links to many sites, providing full text articles and other related resources. *Coffee Break* demonstrates online bioinformatic tools available at the NCBI through live searches and interactive tutorials. The PubMed Interactive Tutorial enables first-time PubMed searchers and experienced PubMed searchers to learn more about using PubMed.

to physical or chemical agents, poor nutrition, and social or cultural behaviors. Factors involved in the development of environmental disease include polluted air and water, excess noise, sunshine (UV radiation), ionizing radiation, food contamination, occupational exposure, toxic wastes, poor nutrition or diet, stress, and tobacco use. There is a rising public awareness concerning environmental disease, and these concerns have driven extensive research efforts into understanding the mechanisms producing cancers and developmental abnormalities. The current scientific efforts are directed at the cellular level, where injury to the genetic makeup of the cell can have ominous implications for the individual and possibly for several generations of progeny that will carry the genetic errors as inherited traits. A study of mutations, birth defects, or cancer must start with a knowledge of the genetic code.

THE ROLE OF GENETICS IN DISEASE

Structure and Function

A thread of life courses through all living things that defines and programs the commonality and differences among and between species. This thread of life is deoxyribonucleic acid (DNA), a spiral, staircase-shaped molecule over 3 billion steps long, compressed in the nucleus of living cells. DNA is the basic informational macromolecule that is the basis of heredity and whose chemical structure has been known for more than 45 years. The storage and transcription of biological information occurs within DNA, which has the power to express that biological information in the form of proteins. When viewed in a concentrated and pure form outside the cell, it is a clear, thick, viscous liquid that looks like water but has the consistency of mucous and is thread-like. DNA is located within the cell nucleus of each of the human body's 100 trillion cells (except mature red blood cells) (Figure 4.2). Each nucleus contains 46 chromosomes arranged in 23 pairs. Each parent contributes one of the chromosomes in very pair. The chromosomes in eucaryotic cells are tightly coiled (condensed) around a group of proteins known as histones. Genes are segments of DNA that code for a particular protein. The DNA of each species or cell type is specific for the species and reproduces that species precisely with the exception of mutations. When unfolded, DNA is a very long, thread-like molecule that is formed by four recurring subunits called mononucleotides. The mononucleotides are made up of one purine or pyrimidine base, one molecule of phosphoric acid, and one molecule of a sugar called 2-deoxy-D-ribose. Except for the base component, the nucleotides are identical and attached to each other in a long chain by phosphodiester links between carbon atoms in each molecule (Figure 4.3). The base components include adenine and guanine (purines) plus thymine and cytosine (pyrimidines). These

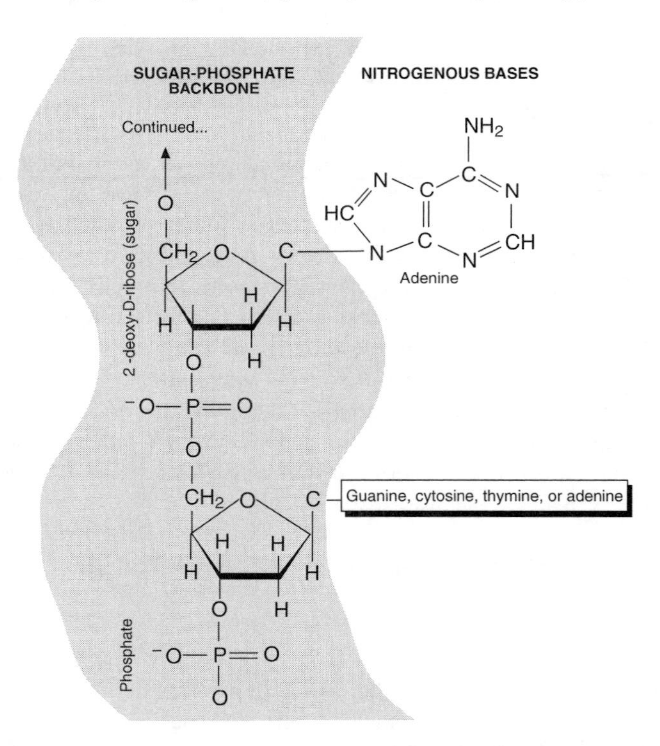

Figure 4.3 The chemical backbone structure of DNA. (Adapted from Lehninger, A.L., *Bioenergetics: The Molecular Basis of Biological Energy Transformations,* W.A. Benjamin, Inc., New York, 1965, Chapter 11.)

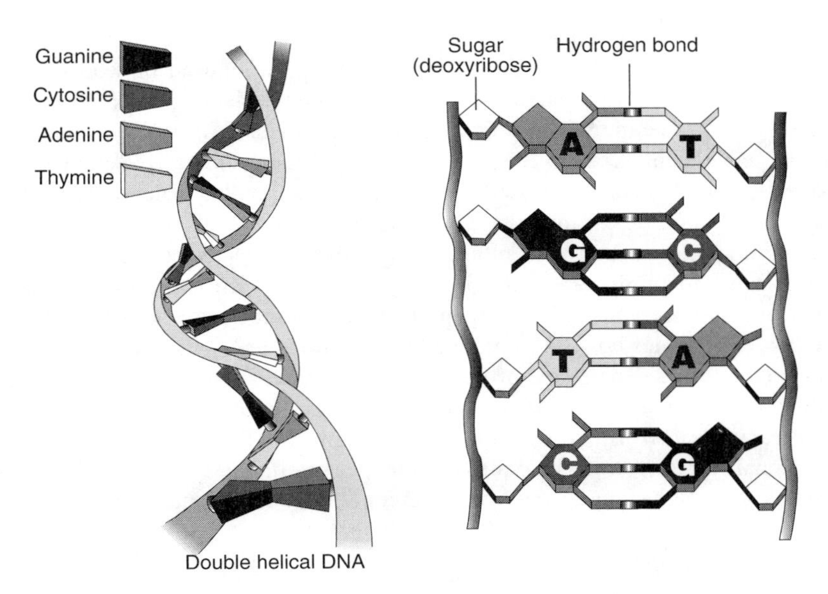

Figure 4.4 The double helical structure of DNA and the mechanism of hydrogen bonding.

four different recurring mononucleotides are the base elements of the DNA coding system from which all proteins are ultimately formed including over 1000 enzymes, numerous structural proteins, lens protein, myosin in striated muscle, hemoglobin in blood, nucleoproteins, and mucous. The proteins are essential for growth, the building of new tissue, and the repair of injured or broken-down tissue. The genetic message for the production of protein is imparted initially by the specific sequence of the four basic mononucleotides forming the long DNA strand. As an example, starting along one section of the DNA strand you might find a sequence that reads A_C_G_A_A_C_G_T_. Chromosomes are gene sequences along strands of compressed DNA and floating in the fluid of the cell nucleus. One thousand genes may be carried in a single chromosome, and it is the precise replication (copying) of this genetic information that genes are duplicated and transferred to daughter cells. The chromosomes do not exist as single structures in higher plants and animals. They are homologous pairs such that every chromosome has a near-twin counterpart and the two strands intertwine to form a double-helical structure so that the two molecules run in opposite directions (Figure 4.4). The pair of nucleotide bases fit together in a precise way held together by hydrogen bonds between the base pairs and by hydrophobic bonding (Figure 4.4). Each strand of DNA contains the same information because the base pairs are complementary in that adenine is always paired with thymine, and guanine is always paired with cytosine. Dividing cells provide each daughter cell with one complementary strand of the parent, then synthesize a complementary strand of the parent, and then synthesize a complementary strand to produce double-stranded DNA that is a duplicate of the parent DNA. This process is referred to as replication. Consider the human genome of 23 chromosomes pairs as 23 chapters in a book that describe the complete set of instructions for making a person. Within each chapter are sentences (genes) composed of words that are each three letters in length. There are only four letters available to make these words and they are A (adenine), G (guanine), C (cytosine), and T (thymine). There about 100,000 genes (sentences) in this book, so tens of thousands of different proteins are dictated by the sequences stored in the genetic code. In this genetic code, a group of three nucleotide bases (three letters) is called a codon and specifies the position of one amino acid in a protein molecule that will ultimately accumulate with all other proteins to produce a living individual. Proteins are composed of specific sequences of amino acids linked by peptide bonds, and the sequence of amino acids ultimately determines how the completed protein folds or layers into specific three-dimensional shapes. The shape of the completed proteins are usually critical to its function. Enzymes are proteins that interact with a substrate much like a key in a lock to catalyze the formation of a new molecule or product. Changes in the three-dimensional structure of the enzyme could reduce or eliminate this ability and cause outcomes that might range from barely perceptible to severe, including death.

Protein Biosynthesis

The DNA molecule is locked in the nucleus and cannot reach the ribosomes outside the nucleus in the cell cytoplasm where protein synthesis occurs. The DNA must first be transcribed to another molecule called messenger RNA or mRNA. Messenger RNA is quite similar to its DNA counterpart except it is single-stranded, contains the nucleotide base uracil instead of thymine, and the sugar D-ribose instead of 2-deoxyribose in the four mononucleotides. The complementary mRNA diffuses through the nuclear membrane into the cytoplasm and attaches to the surface of cellular ribosomes where protein synthesis occurs. The mRNA now carries the code that will direct the specific sequence of amino acid molecules to be inserted into the peptide chain(s) that will form the protein. The coding is accomplished through the sequence of three consecutive (triplet) mononucleotide units (codons) in the mRNA that each represents a single, specific amino acid molecule (Figure 4.5). The consecutive nucleotide codon triplets (i.e., A-U-G) are read out by the protein-synthesizing machinery of the ribosome to form peptide bonds among a specific sequence of amino acids in order to form a complete protein. Amino acids are first prepared for this process by an activation enzyme that forms a high-energy link between the amino acid and an amino acid carrier molecule called transfer or soluble RNA (sRNA) that is specific to each of the 20 amino acids. The sRNA molecule has a sequence of nucleotides that recognize and attach to a specific codon triplet. The recognition site for nucleotides on the sRNA is called an anticodon and is thought to be a complementary triplet of bases to each mRNA codon. The activated amino acid plus its anticodon forms a specific bond with the corresponding codon of mRNA. Thus, the amino acid tyrosine with its specific anticodon will attach only to the triplet codon sequence of A-U-U appearing as a sequence on the mRNA attached to the surface of the ribosome. Once fixed in position, enzymes form peptide bonds between this amino acid and the end of the peptide under construction, releasing sRNA while other enzymes peel the peptide chain from the ribosome. A

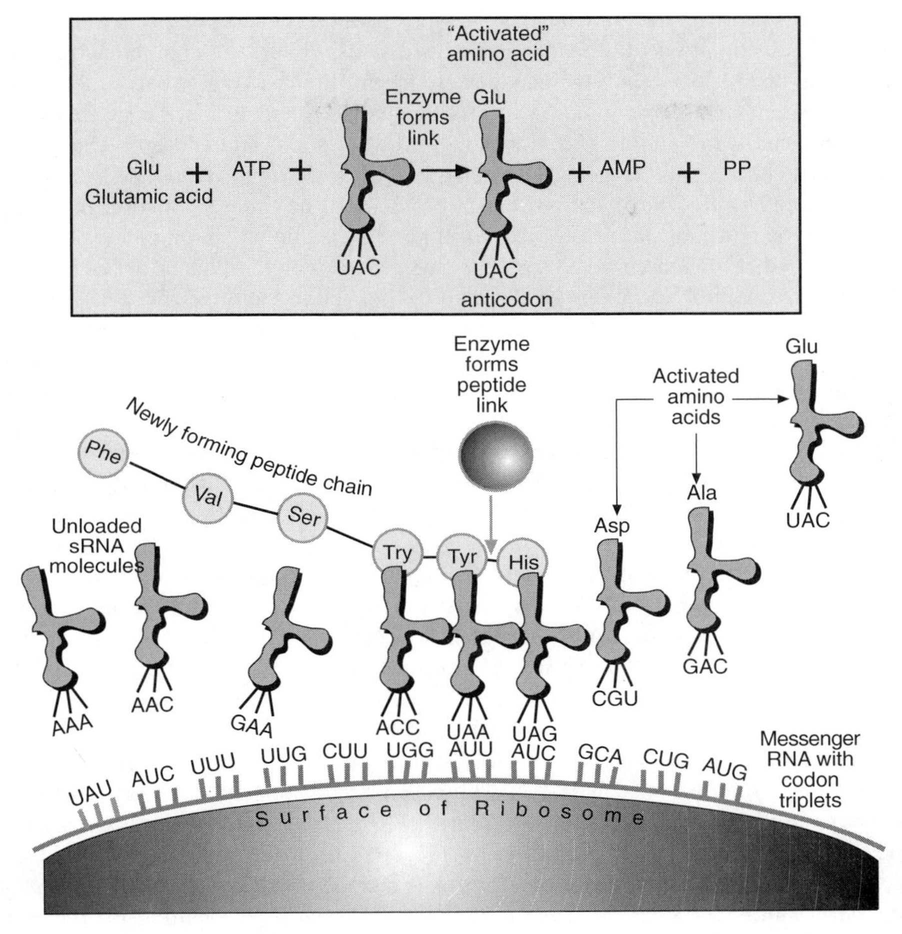

Figure 4.5 A schematic representation of protein synthesis on the ribosome. (Adapted from Lehninger, A.L., *Bioenergetics: The Molecular Basis of Biological Energy Transformations,* W.A. Benjamin, Inc., New York, 1965, Chapter 11.)

great deal of energy is input into this reaction through the formation and breaking of peptide bonds via adenosine triphosphate (ATP). The accuracy of replication is therefore insured thermodynamically by the creation of very stable attachments between the triplet codon and its anticodon. However, mistakes do occur, and if a single amino acid is out of sequence, the coding is disrupted, an amino acid may be eliminated or replaced, and the corresponding protein will be altered.

Diseases of Genetics and Development

Genetic Abnormalities

A change in the nucleotide base sequence of DNA is called a mutation. When such changes occur, there is often a resulting change in the protein encoded by that gene. If this protein is an enzyme, that enzyme may become less active or completely inactive as its amino acid sequence is changed, causing alterations in the active site of the enzyme. Most genotypic changes are harmful or damaging, but some mutations are beneficial. That is how species change and compete in an ever-changing environment. Many simple mutations are neutral or silent and cause no change in the product of the encoded gene. This can occur when the amino acid that is changed is in a nonvital portion of the protein molecule and the protein function remains. However, many changes cause variations in protein function, resulting in a spectrum of decreased or increased activity. These are genetic polymorphisms reflecting the existence of several variations of the same gene. As an example, the P450 genes produce enzymes that detoxify chemicals such as certain polyaromatic hydrocarbons (PAH) that may occur in tobacco smoke. Variations in P450 genes result in decreased activity of the resulting enzymes and are associated with an increased risk of lung cancer in smokers in about 10% of the Caucasian population.[14] Other genes, such as GSTM1, also have been shown to have a natural function of inactivating some of the carcinogens in tobacco smoke. This gene is missing in 50% of Caucasions, which may increase the risk of lung cancer from second-hand smoke by 2.6 to 6 times.[27]

Unfortunately, many human diseases are associated with genetic defects. These defects can occur as single-gene defects known as point mutations or base substitutions, or cytogenetic defects in which there are abnormalities in the number or structure of chromosomes. A point mutation occurs when a single nucleotide base is replaced at one point in the DNA molecule with a different base. The DNA replicates with a substituted base pair such that A_T might be substituted for C_G. This base substitution could result in an incorrect amino acid in the synthesized protein. The effect may be imperceptible or neutral, or it may result in one of many serious human ailments such as cystic fibrosis, phenylketonuria, hereditary spherocytosis, alpha-1-antitrypsin deficiency, Gaucher's disease, achondroplastic dwarfism, hemophilia, or certain forms of diabetes (Figure 4.6). Point mutations may occur spontaneously because of occasional mistakes created in the process of DNA replication, resulting in a spontaneous mutation without any intervention of external factors. There are also agents in the environment such as chemicals or radiation that promote such mutations and are said to be mutagens. When mutations occur, they become part of all cells of next-generation individuals. Defective genes may be inherited according to Mendelian laws just as are normal genes.

Some genes are inherited as autosomal dominant genes in that the genes are always expressed phenotypically (physical appearance) even when appearing on just one of the paired chromosomes. These are usually mild defects since they are always expressed and have occurred over many generations. The hereditary disease, spherocytosis, is caused by a point mutation that is autosomal dominant. Dominant genes that are lethal before reproductive age would clearly not survive in a population. A lethal dominant gene produces Huntington's disease, but this disease is normally expressed in the fourth and fifth decades.[24]

Codominant genes are partly expressed if present as a single allele (an allele is a pair of genes situated at the same site on paired chromosomes) and fully expressed if present on both. If the gene is defective on one chromosome, with a normal gene on its counterpart chromosome, the expression of the gene may be subdued. When

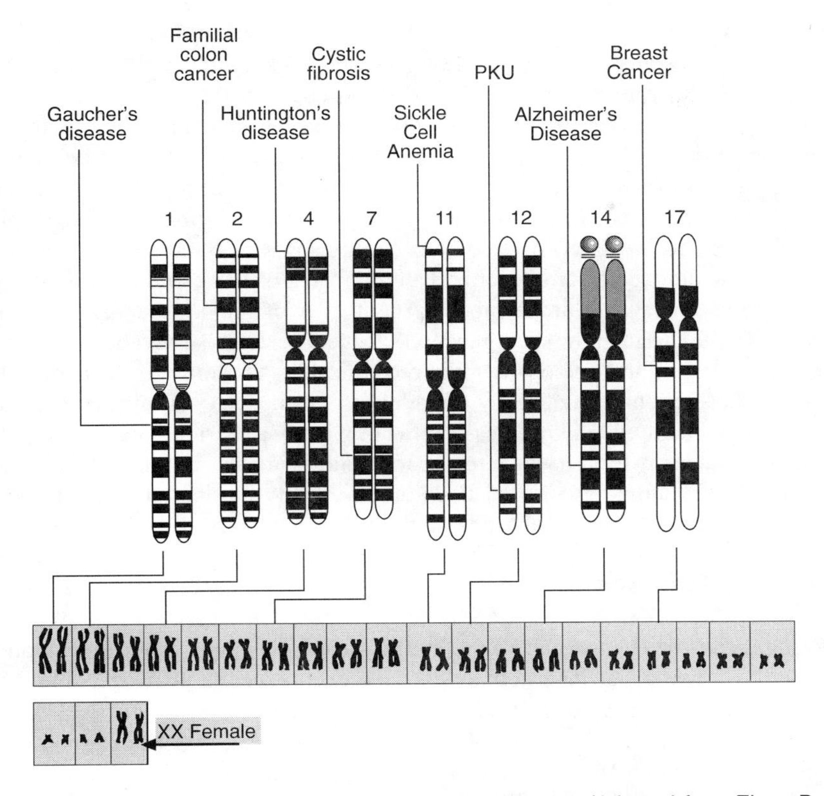

Figure 4.6 Chromosome locations of some genetic diseases. (Adapted from Elmer-Dewitt, P., The genetic revolution, *Time,* Jan. 17, 46, 1994.)

defective on both chromosomes, the disease may be fully expressed and possibly even fatal before adulthood. An example of a codominant gene is that which produces Sickle Cell disease. Sickle cell describes the distinctive crescent or sickle-shaped erythrocytes (red blood cells) that develop in deoxygenated erythrocytes carrying hemoglobin S. It is a disease seen primarily in the black race. Hemoglobin S differs from normal hemoglobin A by the substitution of a single amino acid in one of the three polypeptide chains that make up the protein portion of the molecule. The misshapen erythrocytes are quite fragile, become "sickled," and accumulate in the tiny capillaries where oxygen is released to the tissues. This causes vascular obstruction, tiny infarcts, and fibrosis. When the defective gene responsible for Sickle Cell disease is present on one chromosome, about 40% of the hemoglobin will be hemoglobin S. When defective in both genes, the Sickle Cell disease will be fully expressed, leading to severe anemia and probable death in early adulthood.

Some genetic diseases are inherited as autosomal recessive, meaning that a defective gene must occur in both chromosomes at complementary sites in order for the disease to be expressed. A person with this genetic disease is deprived of an essential protein, resulting in a potentially severe disease such as phenylketonuria

(PKU) or cystic fibrosis.[24] There may also be dominant and recessive defective genes in the sex chromosomes as well, producing a variety of sex-linked diseases including hemophilia-A, color blindness, and forms of muscular dystrophy.

Some genetic diseases are produced by abnormalities in the number or structure of the chromosomes and may be referred to as cytogenetic defects. Such defects tend not to be inherited, but they result from disorders occurring during mitosis when eggs or sperm are formed in the parent or when the embryo is developing. These defects involve large parts of the chromosome and most often appear as deletions or additions of chromosome material. Most of these defects are fatal and fail to survive embryonic development. About 20% of fetuses that spontaneously die while in the uterus have such genetic defects. There are survivable cytogenetic disorders, however, and one of the more recognizable is trisomy 21. Trisomy 21 or Down's disease is characterized by the addition of an extra chromosome to chromosome number 21, which is expressed phenotypically as various degrees of mental retardation, predisposition to leukemia, congenital heart disease, and characteristic facial features including epicanthal folds of the eyelids causing an oriental appearance of the eyes, flat nose or absent bridge, and low-set ears.

Teratologic Diseases

A number of physical defects arise during the period of embryonic development and may be the result of inherited and/or cytogenetic defects in the genome. Most result from factors that cause problems in embryonic development in a person otherwise genetically normal. Such diseases are teratogenic, and the external factors are teratogens. Teratogens may include ionizing radiation; chemicals such as accutane, dioxin, diethylstibestrol (DES), thalidomide, organic mercury, and alcohol; pathogenic infections including toxoplasmosis, German measles (Rubella), cytomegalovirus, and syphilis; and maternal conditions such as age, stress, nutritional adequacy, smoking, drinking, hormone levels, and previous births (Table 4.3). Most teratogens exert their effect during certain critical time windows in the early stages of tissue and organ formation known as organogenesis. This is a period occurring from about the 18th to 60th day after conception, with the most sensitive period around 30 days. Exposure prior to this time is very often fatal to the developing fetus, while exposure after 8 to 9 weeks occurs after the organs are developed and will not produce major physical defects but can produce blindness, mental retardation, or damage to sexual organs. As an example, the sedative thalidomide, synthesized and marketed in Western Europe in the 1950s, was eventually linked to an unusual birth defect in which the long bones of the arms and legs were missing so that hands and feet were attached directly to the torso. This condition was known as phocomelia or "seal limb" disease.[22] Most of the women giving birth to children with this defect had taken the mild sedative, thalidomide, during the first trimester of pregnancy, but more than 10,000 such births occurred before the drug was removed from the market. A similar critical window of exposure was demonstrated for German measles. It was discovered in the early 1940s that women who contracted this disease during the first trimester were at high risk for having babies with congenital cataracts, deafness, and heart abnormalities.

Table 4.3 Some Selected Teratogens, Sources, and Effects (Adapted from Whelan, E.M., How to reduce your risk of cancer, *Priorities,* Summer, 36, 1990, and Nader, R., Brownstein, R., and Richards, J., Eds., Who's Poisoning America: Corporate Polluters and Their Victims in the Chemical Age, Sierra Club Books, San Francisco, 1981.)

TERATOGENS	COMMENTS
Chemicals (including pharmaceuticals)	
Accutane (isotretinoine)	A derivative of vitamin A used in the treatment of cystic acne; when used during pregnancy, 25% of births result in defects
Alcohol	May cause fetal alcohol syndrome (FAS); associated with microcephaly, low IQ, memory problems, abnormalities of upper lip, teeth, hyperactivity
Anesthesia	Various anesthetics associated with miscarriages and structural deformities
Cigarette smoke	Associated with low birth weight, stillbirths, and miscarriages
Dilantin	Dilantin is a drug used to control seizures and produces a wide spectrum of birth defects
DBCP (1,2, dibromo-3-chloropropane)	Widely used nematocide between 1955–1977; produced infertility in factory workers
Dioxin	An unwanted byproduct in some forms of paper production, mass burn incineration, and production of defoliants; produces miscarriages and structural deformities
Methyl mercury	Often produced by soil microorganisms from organic mercury wastes that reach the bottom of streams and bays. Concentrated in fish flesh, eaten by humans; produces gross physical and mental abnormalities
Tegison	A pharmaceutical producing an effect similar to Accutane
Thalidomide	Used in the 1960s in western Europe as a mild sedative; resulted in phocomelia or absence of long bones in legs and arms
Pathogens	
Cytomegalovirus	A virus in the Herpes family transferred across the placenta from mothers with latent infection; about 1/4 of infants develop symptoms that result ultimately in death
German measles (Rubella)	Virus crosses placental barrier to reach fetus; most serious in first trimester causing congenital cataracts, deafness, and heart defects.
Syphilis	Sexually transmitted disease producing microcephaly and mental retardation
Toxoplasmosis	Disease caused by protozoan parasite, *Toxoplasma gondii* found in many mammals; the congenital form causes lesions in central nervous system, jaundice, and anemia
Ionizing Radiation	
Gamma or x-rays nuclear fallout	Causes eye problems, microcephaly, and mental retardation

Just as chemicals taken into the body during critical windows of development can cause physical abnormalities, so can the absence of certain important nutritional compounds. The absence of folic acid, a B vitamin, is associated with neural tube defects including spina bifida and anencephaly. Spina bifida is characterized by the protrusion of a portion of the spinal cord from the spinal column, while anencephaly is a disorder in which most of the brain is absent. Consequently, the simple addition of this vitamin to the diet of a woman about to conceive can prevent a seriously debilitating and often fatal disease. There is a growing list of physical and chemical agents that have the ability to produce birth defects. It should be remembered, however, that 600 to 800 of these agents are teratogenic in laboratory animals but not necessarily humans. About 30 chemicals are currently known to be responsible for human birth defects.[4,28,29] However, the science of teratology is relatively new; and the little that is known rests upon those substances and conditions that produce clear abnormalities. Very little is known about teratogenic substances that may produce mild deficiencies, lack of vigor, or slight losses in mental acuity. New studies have suggested a much stronger link between fetal development and many adult-onset diseases. A study in England found that women who were pregnant during times of severe food shortage gave birth to more children who were obese as adults. It is thought that as a fetus it is possible that factors such as food shortages may program the fetus to retain food, predisposing them to obesity. Other possible diseases that may have a fetal implication may include heart disease, breast cancer, as well as other physiological characteristics such as body size and metabolism.[30] Expectant mothers wishing to protect their developing fetus from teratogenic effects should be under a physician's care, pursue a healthy diet (including folic acid), avoid smoking and alcohol consumption, reduce stress, and limit exposure to known chemical teratogens and infections known to produce teratogenic effects.

New Approaches in Genetics

Methods of Studying Genes

The artificial techniques for creating recombinant DNA developed rapidly in the 1970s and 1980s, permitting the transfer of genes between unrelated species and creating a powerful new science called genetic engineering. These techniques relied upon the discovery of a number of critical tools and techniques including the discovery of restriction enzymes. These are microbial enzymes that are used to cut DNA into several predictable and reportable pieces. These fragments come together to form genetic "maps" that are unique to each individual, as we all contain many slightly different forms of some genes. Most of these polymorphisms are harmless in effect but invaluable in using DNA as forensic evidence in courtrooms as well as genetic screening tests.[31] These fragments of DNA from different sources can also be spliced together through the enzymatic action of DNA ligase, creating a recombinant DNA molecule (Figure 4.7).[32] Once created in a test tube, the recombinant DNA molecules can be inserted in cells where copies of the gene may be produced, or protein products may be created from the newly inserted gene. The protein product can then be harvested and used for many different purposes. One of the original uses is the production of insulin. The two small polypeptide chains of insulin are created by inserting the gene for each chain into two different bacteria, which are then cultured in large amounts to produce the polypeptide chains. The polypeptides are linked to another compound normally produced by bacteria, so they must be separated from the compound. Then the insulin polypeptide chains are joined together to form a complete insulin molecule. The genetic manipulation in insulin production requires methods for inserting the recombinant DNA molecules into cells.

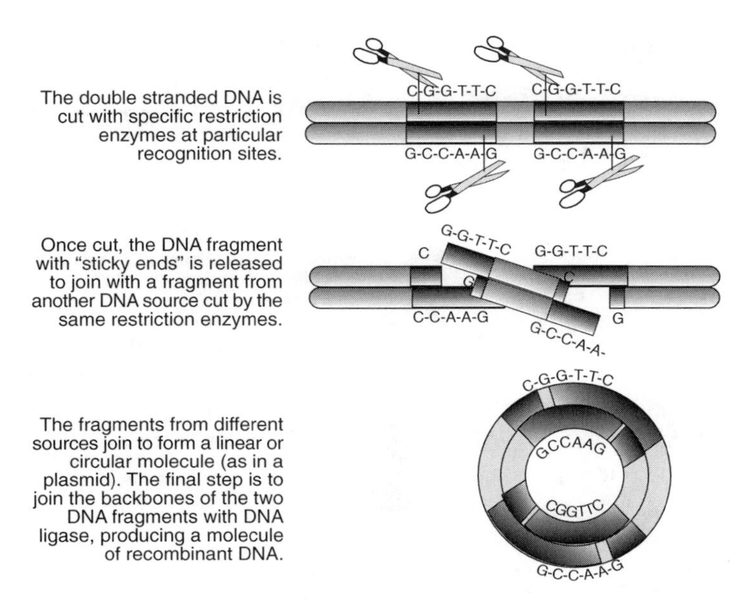

The double stranded DNA is cut with specific restriction enzymes at particular recognition sites.
C-G-G-T-T-C
G-C-C-A-A-G
Once cut, the DNA fragment with "sticky ends" is released to join with a fragment from another DNA source cut by the same restriction enzymes.
G-G-T-T-C
C-C-A-A-G
The fragments from different sources join to form a linear or circular molecule (as in a plasmid). The final step is to join the backbones of the two DNA fragments with DNA ligase, producing a molecule of recombinant DNA.
GCCAAG
CGGTTC
G-C-C-A-A-G

Figure 4.7 Restriction enzymes help create recombinant DNA. (Adapted from Tortora, G., Funke, B.R., and Case, C., *Microbiology: An Introduction,* 5th ed., Cummings Publishing, Redwood City, CA, 1995.)

The method developed relies on a fragment of self-replicating DNA molecules called cloning vectors, to which the target gene is attached, and transmits the gene from one organism to another (Figure 4.8). Once inside the bacterial or yeast cell, the cloning vector will replicate itself including the newly inserted target DNA sequence. The most commonly used vectors are plasmids which are circular, self-replicating, gene-containing DNA fragments found in most bacteria and some yeasts such as Saccharomyces cerevisiae. Restriction enzymes can be used to cut out a section of the plasmid and insert the desired DNA sequence into it. The plasmid must then be inserted into bacterial or yeast cells. One method involves placing bacterial cells in a solution of calcium chloride which makes them competent to take up the plasmids when placed together in a process called transformation. In addition to human insulin, many useful proteins are developed in this way, including erythropoetin, growth hormone, epidermal growth factor, bone morphogenic proteins, taxol, monoclonal antibodies, factors useful in treating hemophilia, and many others.

Another application of gene manipulation is already becoming commonplace in the food that we eat. New techniques are being employed that insert DNA from one species into crops. These genes encode proteins that reduce reliance on fertilizers, make crops more palatable in taste and appearance, and encode natural defenses against pests. The U.S. in particular has been at the forefront of this new technology. However, in the face of rising opposition from consumer groups, the increase in transgenic crop-planted acreage has leveled off. China, facing a burgeoning population, has embraced the new technologies to a larger extent.[33] These new technologies

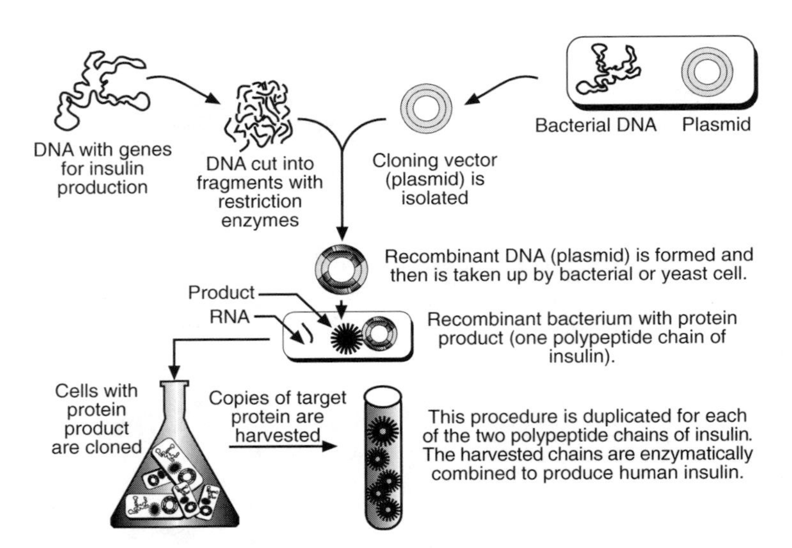

Figure 4.8 The genetic engineering of insulin. (Adapted from Tortora, G., Funke, B.R., and Case, C., *Microbiology: An Introduction,* 5th ed., Cummings Publishing, Redwood City, CA, 1995.)

are not without controversy. One possible consequence is the decline of the insects and other fauna that are dependent upon them, including the Monarch butterfly.[34]

In addition to making products, recombinant DNA technology can produce millions of exact copies (cloning) useful in analytic techniques. This technology enables researchers and clinicians to determine the exact sequence of nucleotide bases in DNA using techniques that are now largely automated. Such sequencing permits the identification and cloning of mutant genes such as those responsible for cystic fibrosis (CF). This cloned gene can then be used to diagnose CF and other genetic diseases by way of techniques based on hybridization. Once a gene is located and its sequence of nucleotide bases determined, a hybridization test can be devised to detect the presence of the gene even before the disease is manifested in a process called genetic screening. Such diagnostic tests exist for cystic fibrosis, Huntington's disease, breast cancer, Duchenne muscular dystrophy, retinoblastoma, neurofibromatosis, and many others.

One hybridization technique is called Southern Blotting, and a variation on this has been automated by manufacturers into a genechip that permits rapid genetic screening of a wide array of genetic disorders. In the original Southern Blotting technique, DNA is extracted from human cells and fragmented by restriction enzymes (Figure 4.9). The fragments are separated by gel electrophoresis and then transferred to a nitrocellulose filter by blotting. The DNA is transferred to the filter as an exact copy of the gel and is exposed to a radioactive probe. For cystic fibrosis screening, the probe consists of cloned DNA fragments with the defective gene, and these are radioactively labeled. Once exposed, the probe will pair up with complementary bases (hybridize) on the patient's separated DNA on the filter if a close match exists. When exposed to x-ray film, the gene fragment containing the hybridized

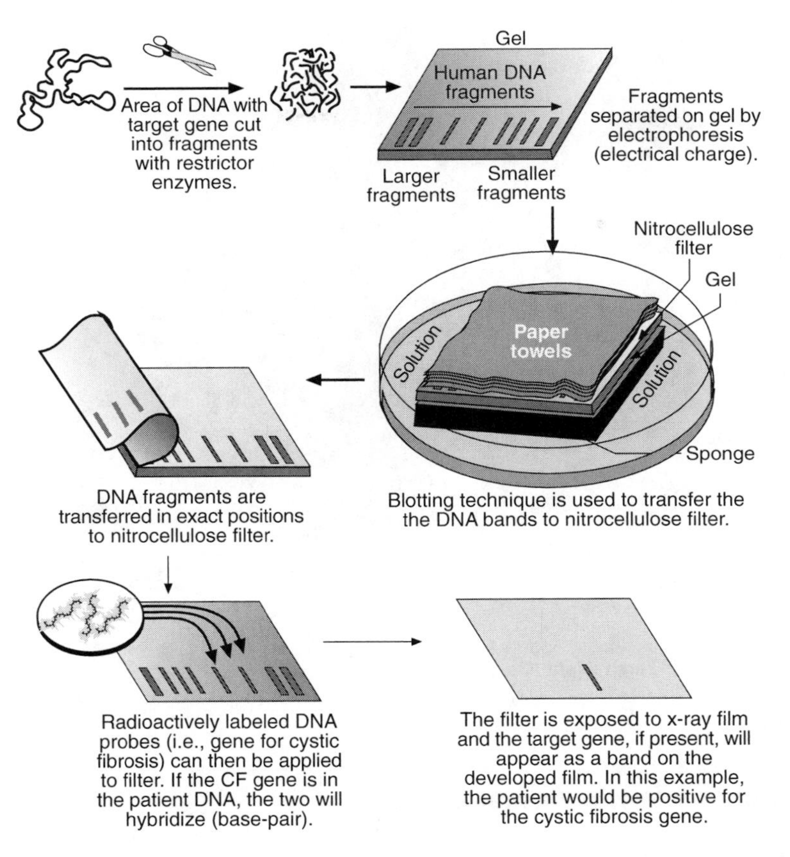

Figure 4.9 The hybridization technique called Southern Blotting. (Adapted from Tortora, G., Funke, B.R., and Case, C., *Microbiology: An Introduction,* 5th ed., Cummings Publishing, Redwood City, CA, 1995.)

pair will show up as a band in the developed film and confirm the presence of the abnormal CF disease gene fragment in the patient.[35]

The introduction of genechip technology has greatly simplified and automated this process. In a typical assay, the user extracts and purifies the patient's nucleic acid, prepares a fluorescent-labeled target including the genes of interest, and fragments the DNA for hybridization. The material is applied to a single-use probe array (genechip) containing hundreds of thousands of different oligonucleotide probes (Figure 4.10). Computer-controlled equipment incubates, washes, and processes this material automatically. The probe arrays are then laser scanned to detect any fluorescent-tagged probes, indicating a complementary match with the patient's DNA. In this way, genetic disease can be rapidly screened. Such devices are useful in sequence analysis, genotyping, and gene expression monitoring. Such technology is being used in the Human Genome Project. By 2003 the project leaders intend to develop a detailed map of every one of the 100,000 genes in human DNA and the sequences of the 3 billion nucleotide bases positioned along those genes.

Figure 4.10 A probe array (gene chip) used for genetic screening developed by Affymetrix, Inc. Santa Clara, California.

The Hunt for Environmental Genes

Similar techniques are also being proposed by the director of the National Institutes of Environmental Health (NIEHS) in the Environmental Genome Project. This effort is based on experiences of researchers who have implicated certain genes in enhancing environmental risk. There is, for example, gene coding for the enzyme paraoxonase, which produces various versions of the enzyme. These versions differ in detoxifying capability. Ten percent of Caucasians and 25% of Asians express a version of the enzyme paraoxonase that converts organophosphates, such as the nerve gas sarin, into a less toxic form nearly ten times faster than those who do not produce this version.[35] This became evident during the sarin nerve gas attack in a Tokyo subway by Aum Shinrikyo cult members in 1995. Variations in gene coding for cytochrome P450 and NAT families can increase bladder cancer risk six-fold in smokers compared to nonsmokers. The HLA-DPbet1 genetic marker is found in people who are much more susceptible to beryllium disease (lung disorder caused by exposure to beryllium) than those who do not have the marker. Based on decades of research demonstrating variations in genes that detoxify or activate offending chemicals, every person is suspected of having unique drug-metabolizing enzymes causing differences in how we handle them.[35] These variations in susceptibility may have broad implications for setting environmental policy and regulations. Current standard-setting methods set permissible exposure at some fraction (i.e, 1/10th) determined to be acceptable for the general population without regard for members of the population who may be at higher risk. The coming trend in environmental risk assessment is to include genetic susceptibility and to determine both distribution and ranges for risk. The Environmental Genome Project is proposed to assist in this effort. Substantial technical obstacles exist including reliability of the gene sequencing process, determining the relevance of polymorphisms to disease, and demonstrating how genotypes relate to phenotypes (physical manifestations).[35] Francis Collins, director of the National Human Genome Research Institute, has urged NIEHS to collaborate because of the substantial complexity of the task. The collaboration is expected to take place and begin a project that will attempt to identify genetic predisposition to environmental pollutants, which may ultimately be reflected in development of environmental policy and regulations. Although the technology is exciting, legal and ethical questions abound. For instance, will workers be obligated to submit to genetic screening before working with toxic chemicals? Will the results of the screening be a basis for job denial?

The Promise of Gene Therapy

Genetic screening will rapidly advance the discovery of genetically-based diseases while offering improved capabilities in gene therapy. Gene therapy may involve removing cells with defective genes from a person, replacing those genes with normal or healthy genes, and then placing the cells back into the patient where the cells will begin to produce normal product and "cure" the disease. As fantastic as this sounds, these procedures are being attempted by such people as Dr. W. French Anderson, director of gene-therapy programs at the University of Southern California/Norris Comprehensive Cancer Center in Los Angeles. Dr. French and colleagues performed the first federally approved gene therapy procedure in 1990 when they replaced a faulty gene in Ashanthi DeSilva, a 4-year-old girl with severe combined immunodeficiency disease (SCID) or "bubble-boy" disease. The disease results from a deficiency in the enzyme adenosine deaminase because of a rare and deadly mutation. Dr. French extracted some white blood cells from the girl and exposed them to viruses that were genetically engineered to carry the normal gene for the production of adenosine deaminase. When the white blood cells were returned to her bloodstream, her cells began producing the necessary enzyme that restored immunity. The white blood cells eventually die, requiring follow-up treatments. Although very promising, the success of this gene therapy story has not been easy to duplicate. Many technological problems in subsequent human trials have prompted the National Institutes of Health (NIH) Committee on Gene Therapy to suggest that NIH funding for such activities be confined to laboratory animals.[36] Ten years of gene therapy experimentation has produced a number of adverse reactions and at least one death. It is unclear if these adverse reactions were related to the gene therapy.[37] Technical problems revolve around methods of introducing DNA into the right place in human cells and then having the genes express themselves. Ideally, inserting the "corrected" genes into undifferentiated human stem cells would provide long-term or permanent cures as the stem cells would serve as an unending source for the differentiated cells throughout the the body. Unfortunately, stem cells are few in number and divide infrequently, making the introduction of vector-borne genes difficult. Even so, gene therapy sits on our horizon as a promising technology and the ultimate solution to curing gene-based diseases. Even infectious diseases such as AIDS may be cured via gene therapy. A new treatment under development is to introduce a gene into an HIV-infected person's white blood cells — the very cells that HIV targets. This introduced gene would produce a protein that prevents the HIV virus from replicating, thus denying the virus the reservoir it requires to reproduce.[38]

In the meantime, current practical approaches to using recombinant DNA technology is in genetic screening where errors in DNA can be detected. In some cases, the patient may be advised to follow certain activities and diets that will reduce the risk from the disease. In other cases, it may be possible to counteract the effect of faulty genes by replacing missing or abnormal proteins.

The Ethical Dilemma

The discovery of defective genes is progressing at a much faster pace then the development of other treatments and gives rise to many ethical and legal dilemmas. One of these dilemmas revolves around informing patients of a disease that is difficult or impossible to treat. As an example, Huntington's disease can now be identified through genetic screening. However, this disease is autosomal dominant and so is invariably expressed. The disease normally shows itself in the fourth and fifth decades of life as a relentlessly progressive disease leading to mental retardation, emotional disturbances, loss of self control, debilitation, and death. There is presently no known cure for the disease.

Another gene, BRCA1, has a 50 to 85% risk of breast cancer when present.[39] The knowledge of the presence or absence of this gene may cause some women to decide to prophylactically have one or both of her breasts removed before cancer develops. Other women have decided to have their breasts removed at the first indications of cancer.[40] Knowledge that one has the defective gene for this disease is not liberating and could lead to more immediate unwelcome effects, including severe depression and even suicide. However, the genetic knowledge gained by a person may be important to other family members. In the event that severe disease or early death is likely, some argue there must be an ethical obligation to communicate it to a spouse or a significant other because it influences their lives, such as the choices of whether to have children.[41]

The other ethical questions center on whether employers or insurance companies could use the information to identify people that may be insurance or occupational risks. What would potential employers do if they knew in advance a person was at increased risk to cancer or Alzheimer's disease? Prenatal testing could ultimately reveal a comprehensive disease profile of an individual potentially leading to a type of eugenics through abortion, or perhaps stigmatism by society because of revealed inborn genetic deficiencies. The ethical and legal questions are enormous and are only beginning to be realized as the weight of the technology settles upon the social consciousness. Perhaps there is no more pressing need for genetic information than in the area of cancer, and yet the information that comes with it often elicits fear.

CANCER

What Is It?

The clinicians and scientists treating and researching cancer now understand that it is a genetic disease caused by multiple mutations in genes responsible for controlling cell division.[42] Most cancers develop from the interaction of genetics with environment; perhaps less than 5% can be attributed to hereditary factors alone. The majority of the scientific community attributes most cancers to a combination of exogenous (external environmental) factors in combination with endogenous (internal) factors that facilitate the accumulation of genetic errors on the prolonged pathway to cancer.[14] Many endogenous and exogenous factors enhance susceptibility in this process (Figure 4.11). Endogenous factors occur from within the body and include gender and ethnicity, health status, age, and heredity (genetic polymorphism).

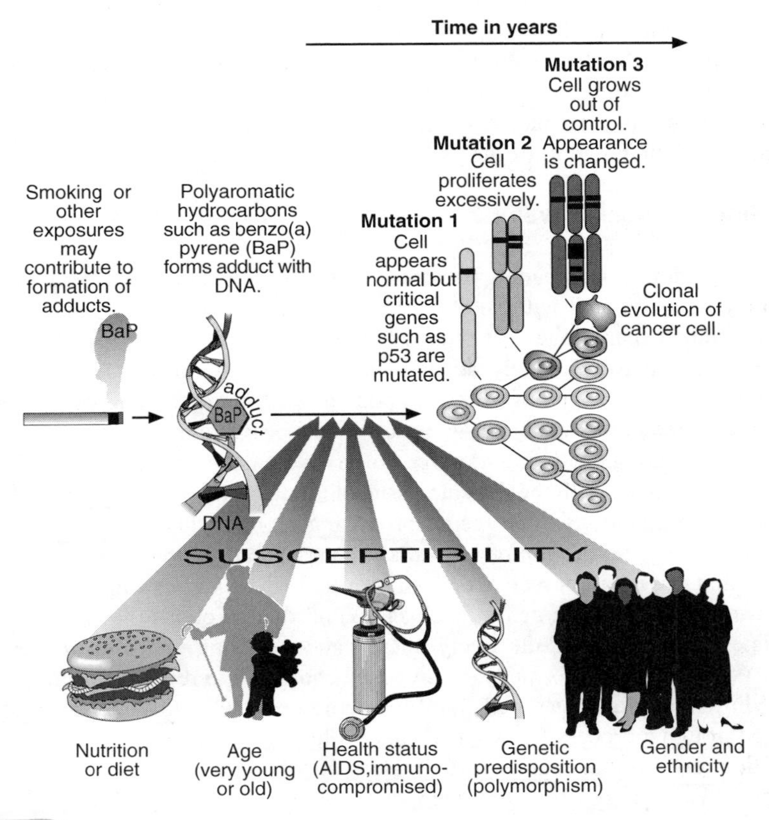

Figure 4.11 The evolution of cancer cells and role of susceptibility. (Adapted from Perera, F.P., Environment and Cancer: Who Is Susceptible?, *Science,* 278, 1068, Nov. 7, 1997.)

External factors include cigarette smoking, inappropriate nutrition, exposure to cancer-causing chemicals and physical agents (ionizing radiation), and oncogenic viruses. It has been hypothesized that frequent cellular telephone use may increase the risk of certain types of brain cancer. Studies have been inconclusive in establishing this relationship, however.[43] Eventually, a single abnormal cell develops that has accumulated enough mutations to bypass all of the regulatory mechanisms that normally control cell growth. It divides again and again, accumulating mutations that discard or silence protective genes that ordinarily monitor replicating DNA for errors. The malignant cells grow into a small mass that develops nutrient-bearing blood vessels in a process called angiogenesis. The blood vessels stimulate further growth of the tumor and establish a pathway for the slippery cancer cells to metastasize to other body sites.[44] Cancer produces local damaging effects, metastasis, and systemic effects. Local damaging effects include the obstruction of airways, urinary tract, and digestive tract, resulting in infections and stasis. They can erode blood vessels causing acute or chronic hemorrhaging or the perforation of intestines or the stomach, causing peritoneal infections. Adhesions may develop in the abdominal or thoracic organs. When bone marrow is affected, critical bone marrow is replaced leading to anemia and thrombocytopenia. Very often malignancies will travel, producing tumors at distant sites in a process called metastasis. Tumors may also produce systemic effects causing endocrine tissues to produce either too much or too little of essential hormones, alteration of intrinsic blood coagulation process, autoimmune inflammation of the skin, and wasting away.[24]

How Does Cancer Develop?

In a typical pathway leading to cancer, a chemical agent such as benzo(a)pyrene (BP) may be associated with smoking activity and combines with human DNA to form an adduct (carcinogenic residues bound to DNA), which leads to increased mutations that may eventually accumulate and lead to cancer. The smoke from cigarettes contain as many as 4000 compounds, with more than 40 of these known to be carcinogenic. Included among these are benzo(a)pyrene, benzene, and arsenic. The act of smoking (or being exposed to passive smoke) brings these substances into intimate contact with the delicate tissue of the respiratory tract. Cancer develops from multiple mutations in the genes producing proteins that control cell division. Some of these genes are tumor suppressor genes based on their ability to produce proteins that can repair defective cells or prevent cells from growing that are grossly defective. Other defective genes (such as ras) produce proteins that inappropriately stimulate cell division. Collectively, these cancer-causing genes are called oncogenes. When behaving normally, these genes cooperate in repairing damaged cells, destroying greatly damaged cells, and replacing dead cells. Cancers are thought to emerge through a process known as clonal evolution whereby a single cell develops a genetic mutation that enables it to divide even when normal cells do not replicate (Figure 4.11).[45] These cells copy their defective DNA to daughter cells that show similar unrestricted growth. At some point, a descendant of this line undergoes additional mutation(s) that removes the mutated cell further from regulation — allowing the progeny to pass through tissue, enter the bloodstream, travel to other

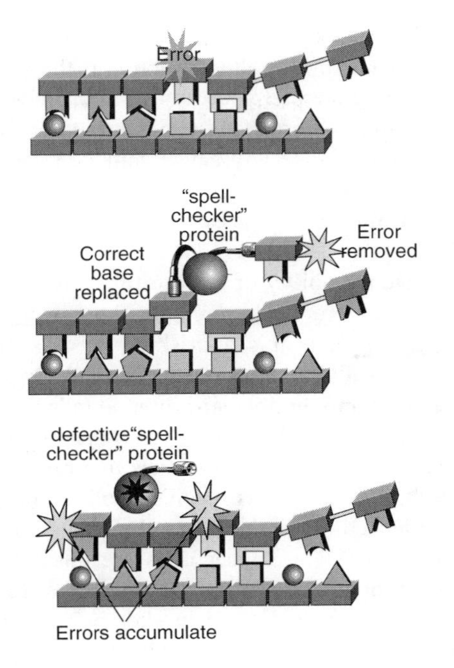

Genes such as hMLH1 and hMSH2 code for proteins that correct errors in the order of nucleotide bases during DNA replication. These proteins are known as "spell-checker" proteins, and defects in the gene will produce defective "spell-checker"proteins that fail to correct errors. When such errors occur in genes controlling cell growth, a major step forward in the development of a cancer cell occurs.

Figure 4.12 The action of "spell-checking" proteins created by hMLH1 and hMSH2 genes.

body locations, and colonize distant organs (metastasize). In support of this mechanism, a brain cancer known as astrocytoma follows a defined sequence of genetic changes involving inactivation of the p53 gene, loss of a gene on chromosome 9, enhancement of epidermal growth receptor by another altered gene, and loss of one copy of chromosome 10.[45] Certain colon cancers often result when defects develop: (1) in both alleles of a gene called APC (adenomatosis polyposis coli); (2) in certain ras genes that promote cell growth; and (3) in the p53 gene. The genetic changes occur early in APC while subsequent changes in the other genes appear much later. These cumulative genetic errors may be facilitated by defective genes such as hMLH1 and hMSH2, which code for proteins that correct errors in the order of nucleotide basis during DNA replication by scanning for errors, detecting them, and repairing them much like a spell checker on a computer (Figure 4.12). Defects in these genes would result in incomplete or nonfunctioning proteins that would permit errors to accumulate prior to becoming cancerous. Eventually, the most important step in the development of a cancer cell is its failure to die when genetic mutations render them cancerous. This process of cellular self-destruction is called apoptosis and occurs when cell damage accumulates to the point where cell integrity is compromised.[46] At this point the cell produces enzymes that cause the cell's DNA to disintegrate in an orderly way, causing cell death. One of the most important gene coding for enzymes involved in apoptosis is p53, a tumor-suppression gene. More than half of all human cancers involve a p53 gene either mutated by free radicals or inherited as a defect. The normally functioning enzyme halts cell division until DNA can be repaired, or turn on the mechanism for cell destruction if the DNA is severely damaged. Mutations occurring along the p53 gene appear at characteristic sites and with unique base sequences known as biomarkers depending on the damaging environmental agent at fault. Cigarette smoke will cause damage to specific locations along p53 whereas UV light will produce damage at different sites. Using such information may eventually be useful in risk assessment, or even legal proceedings.[47]

Certain p53 mutations may also reduce the ability of cells to limit blood vessel formation. Tumors may stimulate their own growth by creating their own blood supply in a process known as angiogenesis. It is this nourishing process that is believed to give cancer cells their enhanced ability to multiply and spread. Dr. Judah Folkman and his colleagues have been testing anti-angiogenesis compounds for years as a possible mechanism for cancer treatment. They have discovered that endostatin and angiostatin prevent and reverse the formation of blood vessels in a broad array of tumors growing in rodents.[44] In 1998, this news was received with enthusiasm by many, and some skepticism by experienced researchers and clinicians, but with the prospect of human trials to begin soon. Many cancer cells are unstable and discard both copies of the p53 gene while other cells retain mutant copies. It turns out that cells with normal p53 genes are quite sensitive to chemotherapy while those with mutant p53 genes tend to be very resistant. This sensitivity of cells to p53 actually is the cause of many of the difficult side effects of chemo and radiation therapies. Cells that are only slightly mutated are destroyed, which causes the common side effects of nausea, hair and weight loss, and a weakened immune system. Some researchers have found that by suppressing p53 during these therapies the side effects may be lessened.[48]

New therapies are being sought that may prove more effective against cancers with mutant p53 genes. Cancer is a complicated disease made up of different cells, often expressing many different genes, influenced by multiple growth factors in different ways, and variably responsive to pharmaceuticals. Consequently, combination treatments of chemotherapy with multiple drugs, gamma radiation, and other combinations of interventions are proving more successful.[49] With the use of DNA genechips to screen cancer cells by the proteins they produce, cancers will be named by the genes they express and not by the site such as colon or stomach cancer.[50] Eventually, treatment in the 21st century will rely upon determining the specific DNA damage and custom-fitting the medicinal cocktail to each tumor.[44] Some of the genes associated with cancers are listed in Table 4.4.

Routine testing for genetic errors is very likely 10 years off, although a long-range project called the Cancer Genome Anatomy Project is underway to create a DNA library of tumor types. Individual screening for cancer risk has been associated with technological problems to date. This difficulty, combined with the potential abuse of genetic information and the limited use of such information in preventing cancer, has prompted the National Advisory Council for Human Genome Research to declare such testing as premature in general medical practice.[14] However, advances in the knowledge of genetic and acquired susceptibility in the general population will likely emerge in future environmental health and regulatory policies as high-risk

Table 4.4 Some Known Cancer Genes (Adapted from Wallis, C., Molecular revolution, cancer: a special report, *Time,* 48, May 18, 1988; and ELSI Project, All About Breast Cancer Genes, LBL, June 1988, http://www.LBL.gov.education/ELSI/screening-main.html.)

NAME	ACTION OF DEFECTIVE GENE	COMMENTS
ras (rat sarcoma)	Causes cell to divide again and again	Plays role in 90% of pancreatic cancers, 50% of colon cancers, 25% of lung cancers
EGF (epidermal growth factor)	Takes in growth signals, relays them to ras protein; abnormally high concentrations in tumors	Found in high concentrations in 40% of tumor cells, some prostate tumors
HER-2/neu	Helps relay signal telling cells to divide; excess produced with defective gene	Rapidly fatal, hard-to-treat breast cancer
BRCA-1	Uncertain	Isolated on chromosome 17; appears in 5% of 182,000 breast cancers; 85% of women with altered BRCA-1 develop breast cancer
p53	Fails to block growth or destroy damaged cells	Found in 50% of all cancers; these cancers tend to be more aggressive and often fatal
p65	Linked to overproduction of hormones associated with breast and prostate cancers	p65 protein marker detectable in blood and decreases as tumors are destroyed during therapy
hMSH2/hMLH1	Fails to correct errors in DNA replication	Proteins normally correct order of nucleotide bases during DNA replication; errors linked to colorectal cancers; gene errors in 1 of every 200 people

Table 4.5 Modifiable Behaviors for Reducing Cancer Risk (Adapted from Perera, F.P., Environment and Cancer: Who Is Susceptible?, *Science,* 278, 1068, Nov. 7, 1997; and Nadakavukaren, A., *Our Global Environment,* 4th ed., Waveland Press, Inc., 1995, Chapter 6.)

MODIFIABLE BEHAVIORS
1. Quit smoking and use of other tobacco forms
2. Avoid excess alcohol consumption
3. Adopt a diet low in fat, high in vitamins A, C, and E. Include cruciferous vegetables such as broccoli, cabbage, brussel sprouts, and cauliflower
4. Adopt a high-fiber diet such as whole grain cereals, breads and pasta
5. Reduce total fat intake
6. Reduce intake of smoked, charred, and salt-cured foods
7. Avoid obesity through diet and exercise; maintain a sensible weight
8. Avoid psychosocial stress
9. Avoid known chemical carcinogens and overexposure to UV radiation (sunlight)

groups are defined by the genetic information that they carry.[14] The risk of cancer can be largely reduced by modifying our behaviors (Table 4.5). The science of epidemiology and genetics (molecular epidemiology) is producing more knowledge on the common factors that influence genetic susceptibility to cancer. Strategies in protecting the public health will be challenged to follow this information.

Major Cancer Risks

Smoking

Smoking cigarettes is recognized as one of the leading causes of cancer in the U.S., causing more than 30% of all cancers. In excess of 88% of the 120,000 people newly diagnosed with lung cancer each year are smokers. Lung cancer used to be a disease primarily of men, but the popularity of smoking among women has driven the cancer cases upward so that lung cancer surged past breast cancer as the leading cause of cancer deaths among women beginning in the 1980s.[29] Since the 1960s there has been a decline in smoking of more than 15% among American adults that is being reflected in a decline in lung cancer.[51] Smoking is showing some increase among teenage girls and blacks. There is also a disturbing trend of increased cigarette smoking outside of the U.S. with dramatic increases in countries of the former Soviet Union, China, South Korea, and Spain.[52] The disturbing proliferation of smoking abroad, developed in concert with some American cigarette manufacturers, comes at a time when cigarette smoking in the U.S. is under attack by public health officials. This battle was given impetus by EPA's designation of environmental tobacco smoke as a class A carcinogen. While the federal and state governments battle with the major cigarette manufacturers in court, many health departments and agencies are passing health regulations banning smoking in public places.

Diet

Dietary factors also play an important role in cancer risk. Diets that are limited to 20% fat or less of total calories have shown dramatic decreases in prostate-specific antigen (PSA) circulating in the blood from the high twenties (indicating dangerous cancer activity) to levels of 0.4 to 2.5 (suggesting the tumors were no longer growing).[46] High fat intake has been linked with increased risks to colon and prostate cancer, while obesity encourages the growth of endometrial cancer.[53] The precise mechanism for how low-fat diets lower risk to cancer is unknown, but studies on mice suggest that calorie restriction boosts apoptosis and gene repair.[46] Apoptosis may also be triggered by butryric acid produced when bacteria ferment fiber in the gut. Consumption of high-fiber foods instead of refined flour and processed foods also appears to reduce the risk of cancer.[53] Cruciferous vegetables including broccoli, cabbage, cauliflower, kale, and brussel sprouts all produce a powerful isothiocyanate anti-carcinogen known as sulforaphane. Sulforaphane appears to increase the production of detoxifying enzymes that quickly jettison cancer-promoting substances from the body.[54] Fruits and vegetables rich in antioxidants such as vitamins C and E protect against a variety of cancers by reducing free radical damage to DNA or to membranes that would otherwise increase the access of environmental toxins. Garlic and onions have a long history of medicinal use throughout the world. Both garlic and onions have strong antimicrobial activity and also increase production of detoxifying enzymes through the allylic sulfides (allium) in the plant tissues. Many fruits, vegetables, and nuts contain ellagic acid that binds to DNA, preventing the powerful carcinogen aflatoxin from forming a DNA adduct. Research in Sweden has shown that human breast milk contains a protein with anti-cancer activity. This protein seems to be a natural protection in infants from cancer and other diseases such as bacterial infections. In laboratory tests this protein seems effective against many different types of cancer cells.[55]

Although not a food, aspirin has been linked to reduced cancer risk. Persons who use aspirin regularly experience half the rate of colon cancer and lower rates of esophageal cancers than do persons who do not use aspirin.[46] An enzyme called cox-2 prevents DNA damaged cells from dying; aspirin inactivates this enzyme, allowing colon-cancer cells *in vitro* (in the test tube) to die. Aspirin has been associated with reduced risk of colorectal cancer in humans. It has been recommended by Dr. Aaron Marcus of the New York Veterans Affairs Medical Center and Cornell University Medical College that people at risk to colorectal cancer should take a single aspirin tablet (325 mg) every other day, provided there are no contraindications.[46]

Trends in Cancer

For the first time in nearly two decades, the rate of Americans diagnosed with cancer and dying from it has begun to decline. The results published in 1998 represent a collaboration of several agencies including the American Cancer Society (ACS), The National Cancer Institute (NCI), The Centers for Disease Control (CDC), including the National Center for Health Statistics (NCHS). The data on cancer incidence was generated from the NCI's Surveillance Epidemiology and End Results (SEER) program representing nine cancer registries around the country and 9.5% of the U.S. population.[51] The data on cancer deaths comes from the NCHS, which collects death certificates from every state and so represents 100% of the U.S. population. The data show that after increasing for 1.2% annually from 1973 to 1990, the rate of new cases (incidence) of 23 major cancers sites combined fell an average of 0.7% a year from 1990 to 1995. The overall cancer death rate, which had risen 0.4% annually from 1973 to 1990, fell by an average of 0.5% per year in the 1990s (Figure 4.13). A decline in adult smoking from 42% in 1965 to 25% in 1995 has played a major role in the declining incidence of cancer and overall cancer death rates. Another reason for the decline may be the increasing use of mammography to detect breast cancer early with follow-up treatment.

The decline in death rates for many other cancers is not easily explained. The overall cancer death rate for males declined 0.9% per year from 1990 to 1995, while women showed a decline of 0.1%. Blacks were more likely to be diagnosed with cancer between 1990 to 1995 than any other racial group for all categories except breast cancer. Black women diagnosed with the disease were more likely to die from it than were white women.[51] While the cancer experts are elated by the evidence that some progress is being made in the struggle against cancer, there remains the cold statistic that more than 550,000 people will die from cancer this year in the U.S. Further, the goal for reduction in death rates for most common cancers (causing most of the deaths) has not been met since President Nixon declared war against

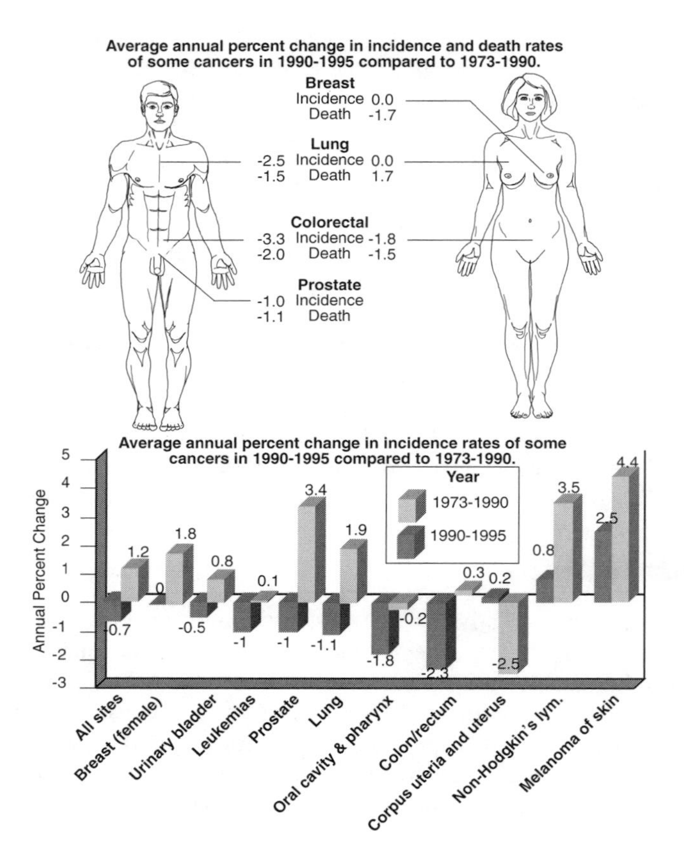

Figure 4.13 Average annual percent change in incidence and death rates for various cancer sites in the U.S. in 1990–1995 compared to 1973–1990. (Adapted from Wingo et al., Cancer incidence and mortality, 1973–1995: a report card for the U.S., *Cancer,* 1197, Mar. 15, 1998.)

cancer with a "campaign" that began in 1971 with the National Cancer Act.[56] The National Cancer Institute stated that its goal was to reduce cancer mortality by 50% by the year 2000. In the years 1973 to 1992, overall cancer mortality actually rose.[57] Carcinomas of the lung, breast, prostate, pancreas, and ovary still claim the lives of hundreds of thousands of people; and the prognosis for most diagnosed cancers that are advanced invasive or metastatic disease has not much improved in the last 25 years.[56]

Additionally, the NCI has reported that the rate of cancer among American children has been steadily rising nearly 1% per year, with an increase in incidence from the early 1980s to the early 1990s of 36.6% in children under 10 years old. The incidence rate in infants aged zero to one year has increased 53.6%.[58] Acute lymphoblastic leukemia in boys and girls increased 27% between 1973 and 1990.[42] This has resulted in double-digit increases over the last few decades. Cancer is still rare in children when compared with adults and is diagnosed each year in an estimated 8000 children below the age of 15. It is, however, the most common form of fatal childhood disease, accounting for nearly 10% of all childhood deaths, while one of every 600 newborns is likely to contract cancer by the age of ten. Acute lymphoblastic leukemia rose 27% in boys and girls from 1973 to 1990, but the rate has declined in boys since then, while still rising in girls. Glioma, a form of brain cancer, increased 40% in children from 1973 to 1994. Recent and spectacular advances in the treatment of these childhood diseases has produced a decrease in the percent of deaths in spite of the increased number of children diagnosed with cancer. The causes of the cancers are uncertain, but some experts point to toxins in air, food, dust, and drinking water as probable suspects. Carol Browner, the former director of the EPA, sought increased federal support to investigate the effects of environmental pollutants on children, and to look at children's total cumulative risk from all exposures to toxic chemicals.[58] Children tend to be more susceptible to environmental toxins because: (1) they take in more food, water, and air and accompanying carcinogens relative to body weight compared to adults; (2) nursing infants get higher exposures of substances such as dioxin in breast milk than adults exposed to background levels; (3) children have a higher internal dose of toxins and greater genetic damage than adults who have similar exposures to tobacco smoke and polyaromatic hydrocarbons (PAH); (4) children have reduced detoxification and repair systems; (5) they have a higher rate of cell proliferation during early developmental stages; and (6) children have increased absorption and retention of toxins.[14] Just as health and regulatory policies are likely to target more specific at-risk groups because of enhanced genetic knowledge, children may well become another more specific target for such policies and regulations. It may take a number of years to confirm that children require lower doses of carcinogens to promote disease; but if this is confirmed, it is expected that more rigid controls on pesticides, toxic wastes, and other chemicals will occur. Congress has already started to move in this direction by approving new laws that incorporate potential exposures to children, such as pesticide residues in food and contaminants in drinking water.[58]

REFERENCES

1. Sessions, S.K., et al. Morphological clues from multi-legged frogs: are retinoids to blame?, *Science*, 284, 800–802, Apr. 1999.
2. Tanner, C., Parkinson's disease in twins: an etiologic study, *J. Amer. Med. Assoc.*, Jan. 27, 1999.
3. Cowley, G., The real "hot zone," *Newsweek*, 136, 66-67, Nov. 27, 2000.
4. Whelan, E.M., *Toxic Terror: The Truth Behind the Cancer Scares,* 2nd ed., Prometheus Books, Buffalo, New York, 1993, chap. 1.
5. Nader, R., Brownstein, R., and Richards, J., Eds., *Who's Poisoning America: Corporate Polluters and Their Victims in the Chemical Age,* Sierra Club Books, San Francisco, 1981, chap. 1.
6. Grossman, K., *The Poison Conspiracy,* Sag Harbor, NY, The Permanent Press, 1983, chap. 1.
7. Naar, J., *Design for a Liveable Planet,* Harper and Row, New York, 1990.

8. Downs, H., *The Poisoning of America,* New York, ABC-TV, Sept. 8, 1988.
9. Waldholz, M., War on cancer: new study disputes claims of progress, *The Wall Street Journal,* May 6, 1, 1986.
10. Brown, D., Nonsmoking cancer is found to increase in baby boomers, *Boston Globe,* Feb. 9, 1, 1994.
11. Epstein, B., *The Politics of Cancer,* Sierra Club Books, San Francisco, 1987, chap. 1.
12. Skow, J., Can lawns be justified?, *Time,* Jun. 3, 63, 1991.
13. Rather, J., Contaminant from gas is found in water, *New York Times*, Aug. 29, 1999.
14. Perera, F.P., Environment and cancer: who is susceptible?, *Science,* 278, 1068, Nov. 7, 1997.
15. Koren, H., *Illustrated Dictionary of Environmental Health and Occupational Safety,* CRC/Lewis Publishers, Boca Raton, 1996, p. 407.
16. Thomas, C.L., Ed., *Taber's Cyclopedic Medical Dictionary,* 13th ed., F.A. Davis Co., Philadelphia, 1977.
17. Interview with John Higginson, *Science,* 1363, Sept. 28, 1979.
18. Whelan, E.M., How to reduce your risk of cancer, *Priorities,* Summer, 36, 1990.
19. Tlrichopoulis, D., and Hunter, D.J., What causes cancer? *Cancer Causes, Screening and Prevention,* April 22, 1998, http://oncolink.upenn.ed/cuaseprevent.html.
20. Associated Press (AP), Most link cancer to diet, but eating habits unchanged, *Daily Hampshire Gazette,* Northampton, MA, 2, Feb. 24, 1987.
21. Preston, R., What new things are going to kill me?, *Time*, 69, Nov. 1999.
22. Gadsby, P., Fear of flu, *Discover,* 82-89, Jan. 1999.
23. Gorman, C., Beyond lyme, *Time*, Jul. 26, 1999.
24. Bickley, H.C., *Practical Concepts in Human Disease,* The Williams and Wilkins Co., Baltimore, 1975, chap. 7.
25. Morgan, T.M. et al., *Environmental Health,* 2nd ed., Morton Publishing Co., Englewood, Co., 1997, chap. 15.
26. Cummings, C.H., Entertainment foods, *The Ecologist*, 29, 16-19, Jan/Feb 1999.
27. Associated Press (AP), Mutation May Raise Cancer Risk From Secondhand Smoke, Cable News Network, December 1, 1999, http://www.cnn.com/health/aids/.
28. Sheperd, T.H., Detection of human teratogenic agents, *J. Pediatr.,* 101, 810, 1982.
29. Nadakavukaren, A., *Our Global Environment,* 4th ed., Waveland Press, Inc., 1995, chap. 6.
30. Underhill, W., Shaped by life in the womb, *Newsweek*, Sept. 27, 1999.
31. Shreeve, J. Secrets of the gene, *National Geographic,* 43-75, Oct. 1999.
32. Tortora, G., Funke, B.R., and Case, C., *Microbiology, An Introduction,* 5th ed., The Benjamin Cummings Publishing Company, Inc., Redwood City, CA, 1995, chap. 9.
33. Smith, C.S., China rushes to adopt genetically modified crops, *New York Times*, Oct. 7, 2000.
34. Brower, L.P., Will biotechnology doom the monarch?, *Defenders of the Wildlife*, Fall, 1999.
35. Kaiser, J., Environmental institute lays plans for gene hunt, *Science,* 278, 569, Oct. 24, 1997.
36. Jaroff, L., Keys to the kingdom, *Time, Special Issue:Frontiers of Medicine,* 124, Fall, 1996.
37. The Washington Post, Belated gene therapy records reported deaths, *Daily Hampshire Gazette,* Vol. 214, No. 126, Jan. 31, 2000.
38. BBC Online, *Gene Therapy Advance For Kids*, British Broadcasting Company, Feb. 20, 2000, http://news2.thls.bbc.co.uk/hi/english/health/newsid%5f647000/6472820.stm.

39. Couzin, J., Quandaries in the Genes, *U.S. News and World Report*, 64-66, Nov. 1, 1999.
40. Hyman, D. and Huang J., The courage to heal cancer survivor doesn't regret her difficult decision, *Springfield Sunday Republican*, Oct. 24, 1999.
41. Berg, K., Ethical aspects of early diagnosis of genetic diseases, *World Health,* 5, 20, Sept/Oct, 1996.
42. Glasser, R.L., *The Greatest Battle,* Random House, 1976, chap. 1.
43. Center for Devices and Radiological Health, *Consumer Update on Mobile Phones*, Oct. 20, 1999, http://www.fda.gov/cdrh/ocd/mobilephone.html.
44. Gorman, C., The hope and the hype, *Time, Cancer: A Special Report,* 38, May 18, 1998.
45. Cavenee, W.K. and White, R.I., The genetic basis of cancer, *Sci. Amer.,* 72, Mar. 1995.
46. Kohitak, R., Cancer's nemesis: The suicide gene, *Chicago Tribune,* 257, 1, Sep. 14, 1997.
47. Harris, C.C., At the crossroads of molecular carcinogens and risk assessment, *Science,* 62, Dec. 24, 1993.
48. Recer, P., Drug said to aid cancer therapy toleration, *The Boston Globe*, Sept. 10, 1999.
49. Nash, J.M., The enemy within, *Time: Special Issue: The Frontiers of Medicine,* 14, Fall, 1999.
50. Saltus, R., Cancer cells' gene signature may aid in diagnosis, *The Boston Globe*, Oct. 15, 1999.
51. Wingo, P.A. et al., Cancer incidence and mortality, 1973–1995: A report card for the U.S., *Cancer,* 1197, Mar. 15, 1998.
52. Brown, L.R., Kane, H., and Ayres, E., *Vital Signs, 1993,* Norton, W.W., Worldwatch Institute, 1993.
53. Willett, W.C., Diet and health: what we should eat, *Science,* 264, Apr. 22, 1994.
54. Davis, D.L., Natural anticarcinogens: Can diet protect against cancer?, *Health Environ. Dig.,* No. 1, 4, Feb. 1990.
55. Radetsky, P., Got cancer killers?, *Discover*, June 1999.
56. Sporn, M.B., The war on cancer, *Lancet,* 347, 177, May 18, 1996.
57. Kosary, S.L. et al., Eds., *SEER Cancer Statistics Review, 1973-1992, Tables and Graphs,* National Cancer Institute NIH, Bethesda, Pub. No. 95, 2789, 1995.
58. Cushman, J.H., U.S. reshaping cancer strategy as incidence in children rises, *The New York Times,* 147, 1999.
59. *Morbidity and Mortality Weekly Report,* June 19, 1992, p. 47.
60. Elmer-Dewitt, P., The genetic revolution, *Time,* Jan. 17, 46, 1994.
61. Lehninger, A.L., *Bioenergetics: The Molecular Basis of Biological Energy Transformations,* W.A. Benjamin, Inc., New York, 1965, Chapter 11.
62. Wallis, C., Molecular revolution, cancer: a special report, *Time,* 48, May 18, 1988.
63. ELSI Project, All About Breast Cancer Genes, LBL, June 1988, http://www.LBL.gov.education/ELSI/screening-main.html.

CHAPTER 5

Toxicity and Toxins

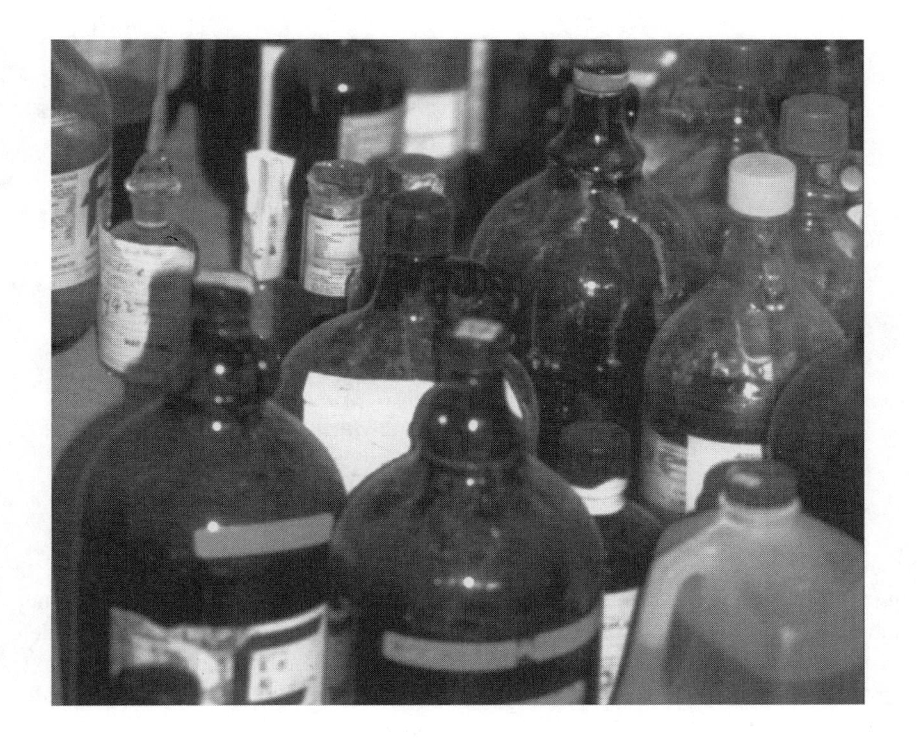

INTRODUCTION

In the U.S., there are more than 70,000 synthetic chemicals currently in commercial use, and for most of them, their toxicity is not widely known or understood. More than 60,000 of them are thought to be potentially harmful and their effects

OBJECTIVES FOR THIS CHAPTER

A student reading this chapter will be able to:

1. **Discuss and define the concepts of toxic triangle, poison, hazardous material, and hazardous waste**
2. **List and explain the various methods of absorption including diffusion, facilitated diffusion, active transport, and special processes**
3. **Explain the processes of endocytosis including phagocytosis, pinocytosis, and receptor-mediated endocytosis**
4. **Describe and discuss the major mechanisms by which toxic materials produce their adverse effects including: (1) inactivation of enzymes, (2) direct effect on cells and tissues, and (3) production of intermediate compounds or secondary action**
5. **Describe and provide an overview of the immune system, the cellular and humoral immune system, and allergic mechanisms**
6. **Discuss and describe the adverse health effects associated with endocrine disruptors, PCBs, dioxin, lead, mercury, asbestos, and organic solvents**

could be devastating, causing cancer, neurological and reproductive disorders, or other injuries.[1] Some of these chemicals are so widely distributed and of such potential harm to human health and/or ecology that they serve as sentinel models for understanding the potential harm for the indiscriminate use and disposal of toxic agents. The compounds to be discussed in this chapter include: (1) dioxin, polychlorinated biphenyls and other endocrine disruptors; (2) the heavy metals, lead and mercury; (3) asbestos; and (4) the organic solvents. These represent a wide divergence in chemical make-up, in mechanisms of toxicity, distribution, and patterns of exposure. Some substances such as lead have been recognized as a hazard since early civilization when it was used to store wine, to pipe water, and even as vessels in which to cook food.[2] Dioxin, PCBs, and other organochlorines have only recently been synthesized as products or unwanted byproducts and have been implicated as agents in reproductive disorders or other diseases. Exposure to many of these chemicals at critical points in human development may produce lifelong adverse effects — even effects that are passed on to subsequent generations.

The reason that toxicological information is incomplete is because animal testing requires significant time and money and deals with one chemical at a time. Such testing provides insufficient information on the interactive effects that may occur during or after exposure to multiple chemical exposures. Additionally, animal studies often fail to detect future or delayed conditions. Human exposure studies are seldom performed because of the ethical problems associated with exposure of humans to various chemicals and metals. In addition to these insufficiencies, the federal government is currently reducing support for further research, while private corporations are filling the gap — which may potentially lead to bias in both study designs and interpretations.[1]

Similar to the legal premise that a person is innocent until proven guilty, a chemical is often considered safe until proven otherwise by the appropriate federal agency. Conferring such a privilege on chemical substances is not the best approach to protecting the public health. Since 1000 to 2000 new chemicals are introduced each year into our society, there is significant opportunity for untested materials to enter our environment and expose humans, wildlife, and plants to toxic effects. Sometimes decisions on what substances are allowed to enter the environment are political and economic, with little regard for the effects on human health. In other instances, the health effects are simply unstudied or not recognized.[1]

Figure 5.1 The toxic triangle. (Adapted from Schroeder.[15])

A potentially toxic substance produces its adverse effect by interacting with humans (or organisms) and the environment in a relationship referred to as the toxic triangle (Figure 5.1). Consequently, a poison or toxic substance does not constitute a hazard unless contact is made with the organism in a form and quantity that can cause harm. Furthermore, the environmental conditions of heat, humidity, or even noise may aggravate the response to the poison.[3] Hazardous substances, or wastes, may be explicitly defined under federal law, and many of them are listed by name and regulated under the Clean Air Act (CAA), the Toxic Substances Control Act (TSCA), and the Clean Water Act (CWA). A hazardous substance is defined in the Comprehensive Environmental Response Compensation and Liability Act (CERCLA) as any chemical regulated under the previously mentioned acts. Hazardous materials are defined and regulated under the U.S. Department of Transportation's (DOT) regulations (49 CFR). The definition is quite broad and pertains to "a substance or material, including a hazardous substance, which has been determined by the Secretary of Transportation to be capable of posing an unreasonable risk to health, safety and property when transported in commerce …." Hazardous wastes are solid wastes defined under 40 CFR.261 as "any discarded material garbage, refuse, sludge or other discarded material, that is solid, liquid, or confined gas which is abandoned, inherently waste-like, or recycled." Listed wastes appear under 40 CFR 261.30. Wastes not listed may still constitute a hazard if they possess defined characteristics of toxicity, corrosiveness, reactivity, and flammability. Toxic substances are those that: (1) can produce reversible or irreversible bodily injury; (2) have the capacity to cause tumors, neoplastic effects, or cancer; (3) can cause reproductive errors including mutations and teratogenic effects; (4) produce irritation or sensitization of mucous membranes; (5) cause a reduction in motivation, mental alertness, or capability; (6) alter behavior; or cause death of the organism.[3]

People continue to be exposed to a wide spectrum of toxic and hazardous substances that potentially threaten their health. In most cases, the consequences of such exposures are unknown. The following sections in this chapter outline some of the principle mechanisms of toxicity and spotlight a few of the more widely distributed toxic and/or hazardous chemicals in our society.

EXPOSURE AND ENTRY ROUTES

Exposure

In order for a toxic substance to produce its harmful effects on the human body, a person must first be exposed to the chemical. The harmful effect may occur upon contact, or the substance may be absorbed after which the adverse effect is produced. Some toxic materials such as strong bases or acids can express their harmful effects by causing chemical lesions upon contact with the skin. Although no absorption of toxic substances takes place, the lesions produced are local and often result in killing areas of tissue in a process known as necrosis. Absorption refers to the passage of substances across the membranes through some body surfaces into body fluids and tissues by a variety of processes that may include diffusion, facilitated diffusion, active transport, or special processes.[3,4] Diffusion is a passive process that occurs when molecules move from an area of high concentration to one of low concentration. When this process involves water molecules across a semipermeable membrane, it is referred to as osmosis. The chemical nature (basic or acidic), electrical charge, and the size and shape of the molecules determine their ability to passively diffuse across the membrane.

Some molecules such as amino acids and sugars require specialized carrier proteins to be transported across a membrane. No high-energy phosphate bonds such as adenosine triphosphate (ATP) are required in this process and so it is referred to as facilitated diffusion. When ATP is required in conjunction with special carrier proteins to move molecules through a membrane against a concentration gradient (i.e., high concentration to low), this process is called active transport. Typically, about 40% of the ATP in a human body cell is consumed in this process of actively transporting molecules across the cell membrane. This process is important in nerve impulse conduction in which sodium and potassium are actively transported across the nerve cell membrane by the sodium-potassium ATPase pump. The energy from ATP is required because three sodium ions are actively transported out for every two potassium ions actively transported in.[3,4]

Particles and large molecules that might otherwise be restricted from crossing a plasma membrane can be brought in or removed by the process of endocytosis. There are three major types of endocytosis including phagocytosis, pinocytosis, and receptor-mediated endocytosis. Phagocytosis (phag = eat; cyto = cell) occurs when specialized cells such as phagocytes extend out streams of cytoplasm much like an amoeba and engulf solid particles to form sacs around the particles. The sacs detach from the plasma membrane inside the cell, and enzymes attack the particles. When the materials are liquid, a small drop adheres to the cell membrane surface where an indentation forms to surround the drop and form a vesicle that detaches from the membrane in a process called pinocytosis.[3]

Waste products, poisons, nutrients, and hormones that are located in interstitial fluids (between cells) at lower concentrations than in the cells are called ligands. The ligands bind to receptor sites on the cell surface, causing a vesicle to form around the substance in a process called receptor-mediated endocytosis. In a series of steps, the receptor is separated from the ligand, and the ligand is enzymatically digested.[3,4]

Routes of Entry

Toxic substances may be gases, liquids, solids, or vapors. There are several ways in which toxic substances can enter the the body. They may enter through the lungs by inhalation, through the skin, mucous membranes or eyes by absorption, or through the gastrointestinal tract by ingestion.

Respiratory System

The respiratory system is composed of the nose, pharynx, larynx, trachea, bronchi, and lungs (Figure 5.2). Its function is to supply oxygen to the body's cells and to expel carbon dioxide from the body in a process called respiration. The act of breathing or ventilation brings air into and out of the lungs. The exchange of oxygen and carbon dioxide between the atmosphere and blood is known as external respiration, while the exchange of gases between blood and individual cells is called internal respiration. Air is drawn into the nasopharyngeal area through nasal hairs at the vestibule and then over mucous membrane-covered bony plates called the nasal conchae. These structures combine to filter, warm, and moisten inhaled air. Additionally, nerve endings in this nasal area may be stimulated to cause a sneeze reflex, helping to eliminate the mucous and the trapped particles. The inhaled air then enters the trachea-bronchial area, where the bronchi branch into numerous bronchioles with even smaller diameters until they terminate into thin-walled, delicate alveoli where gas exchange takes place. The trachea, bronchi, and bronchioles are lined by a velvety layer of cilia with mucous cells throughout. Particles such as dust and pollen of 10 mm or larger are removed by the constant streaming of mucous propelled from the bronchial and tracheal passages by the cilia beating at over 1300 times per minute in a process known as mucociliary streaming. The tube-like structures of the bronchi and trachea are also largely surrounded by smooth muscle layers that contract in response to irritating substances and allergens (Figure 5.3). This bronchoconstriction narrows the lumen and restricts the flow of air, other gases, and

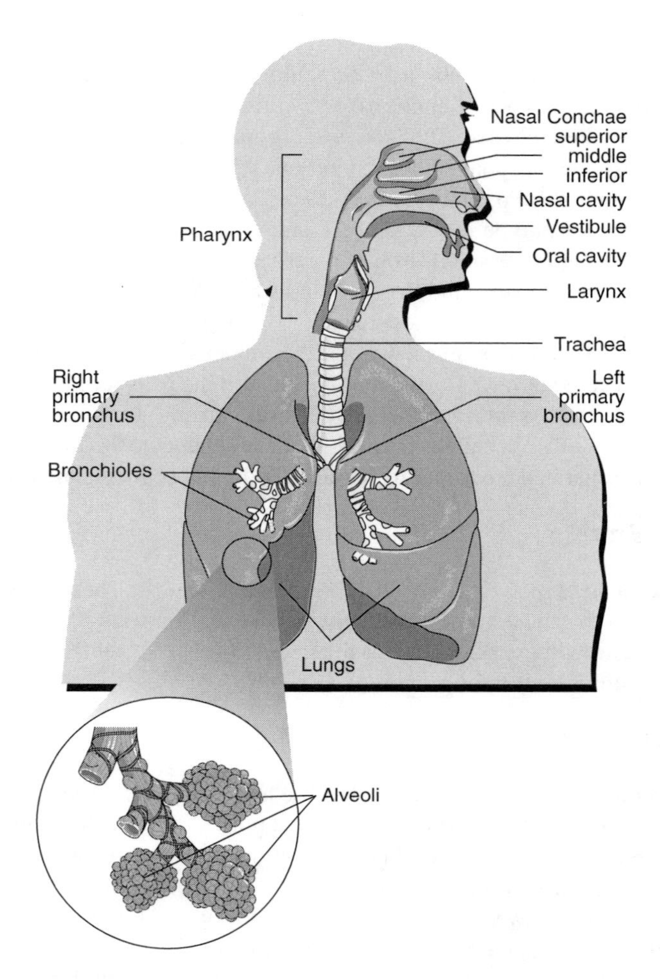

Figure 5.2 The respiratory system.

particles from reaching more delicate tissues deeper in the lung. This process may also be combined with excess mucous secretions or other fluids that make breathing difficult. When prolonged and severe, this process can be life threatening as in asthmatic attacks. A cough reflex may be initiated at the level of the bronchi to eliminate accumulated mucous in which foreign substances have become lodged.[2,3,5]

Inhaled air ultimately reaches the delicate, elastic and extremely thin alveolar sac at the terminus of the bronchioles (Figure 5.3). The alveoli are surrounded by tiny capillaries in which red blood cells give up their carbon dioxide and receive oxygen in exchange. Within the lumen of the alveoli are macrophages that serve as the final line of defense against microbes and a variety of organic and inorganic materials.[2]

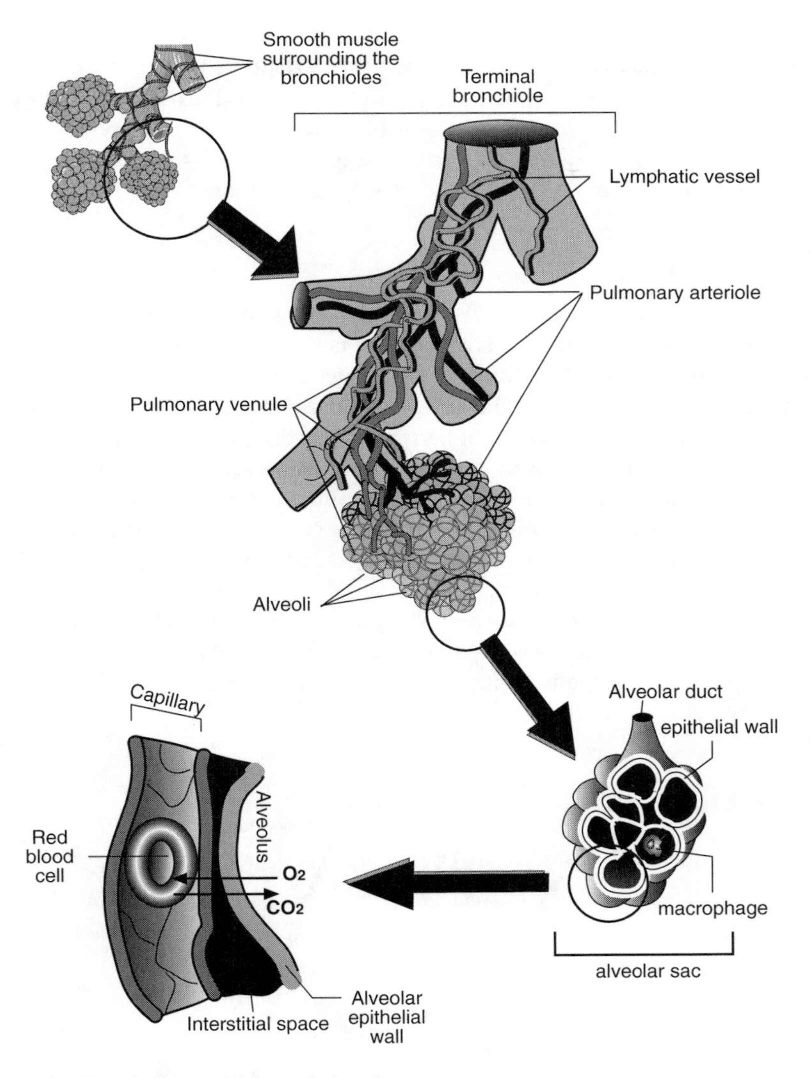

Figure 5.3 Terminal bronchiole and alveoli.

Particulates in the airstream greater than 5 mm are mostly removed within the nasopharyngeal or bronchial areas by mucociliary streaming and either swallowed, sneezed, or coughed up. Smaller particles in the range of 1 mm or less reach the alveoli, where they may be engulfed by the macrophages, moved up the respiratory tree in a mucous sheet, lodge in the interstitial tissues of the lung, or enter the lymphatic spaces and lymph glands to be carried to the bloodstream.[3] Some inorganic particles cannot be digested by the macrophages and prove to be lethal, resulting in the formation of scar tissue and loss of elasticity in the associated alveolar tissue. The size of particles largely determines where in the lungs they will be deposited, whereas the solubility of gases determines their fate in the lungs. Highly soluble gases such as sulfur oxides and ammonia tend to act in the upper moist airways of the respiratory tree, whereas less soluble nitrogen dioxide tends to reach deep into lung tissue where it can cause tissue destruction and more easily be absorbed into the bloodstream.

The Skin

The skin is normally thought of as a relatively impervious barrier to chemicals and biologic agents, and for the most part this is true. The skin is the body's largest organ, consisting of many interconnected tissues covering an area of nearly 3000 in.^{2} in the average adult. The skin helps to: (1) regulate body temperature through sweat glands; (2) provide a physical barrier to dehydration, microbial invasion, and some chemical insults; (3) excrete salts, water, and organic compounds; (4) serve as a sensory organ for touch, temperature, pressure, and pain; and (5) provide some important components of immunity.[6,7] The skin consists of an outer, thinner layer called the epidermis that is attached to an inner and much thicker layer called the dermis (Figure 5.4). Beneath this layer is a subcutaneous layer of glandular and adipose tissues. The skin is effective in keeping out water and water-soluble (hydrophilic) materials, although small amounts of substances may be absorbed through hair follicles or sweat glands. Materials may also enter breaks in the skin from cuts,

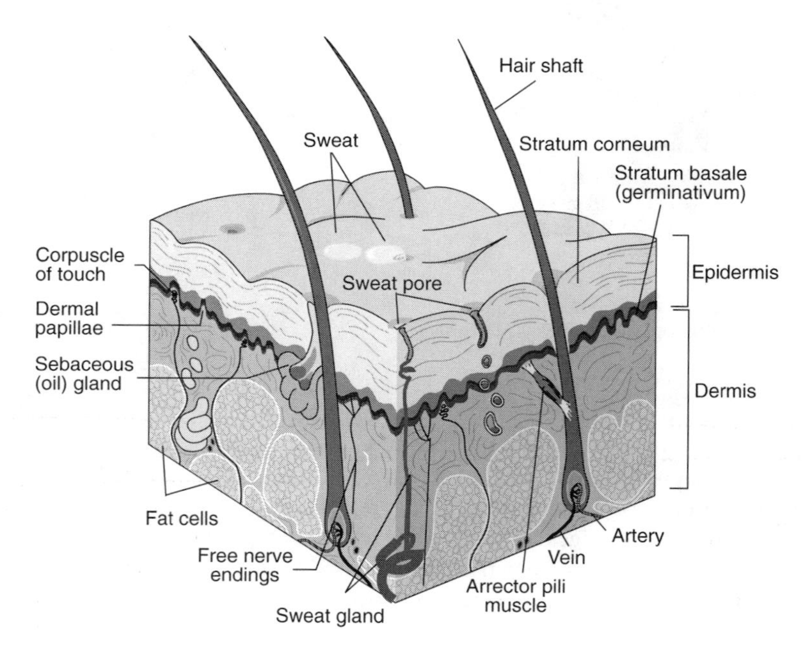

Figure 5.4 Cross-section of human skin.

injections, insect or snake bites, and high-pressure steam or liquids.[7] Absorption of hydrophilic materials can be increased as sweating increases, or in the presence of moisturizers or oily materials. Substances that are fat soluble (lipophilic), such as organic solvents, petroleum compounds, and organochlorine compounds, can be absorbed by passive diffusion through the skin. A variety of substances such as hexane, trichloroethylene, many insecticides, military nerve gases, and numerous other substances fall into this category.

The eye is also an important route of absorption, especially by way of the moist epithelial tissue surrounding the eye in the socket that normally facilitates movement.

The Gastrointestinal Tract

The gastrointestinal tract is a continuous tube of nearly 30 feet in length extending from the mouth to the anus (Figure 5.5). The components of the digestive tract include the mouth, pharynx, esophagus, stomach, small intestine, large intestine, and anus. The purpose of the digestive system is to ingest and move food for digestion and nutrient absorption. The system also functions to eliminate indigestible substances by defecation.[7,8] The gastrointestinal tract is a major route of absorption for many toxic agents including mercury, lead, and cadmium that appear in food and water. Most of this absorption occurs between the stomach and the upper portion of the intestine. Toxic substances can also be transferred from hands to mouth.

The inner layer of the small intestine or mucosa consists of a velvety layer of finger-like projections of 0.5 to 1.0 mm long called villi. There are up to 40 of these villi per mm^2, giving a tremendous absorptive capacity to this section of the intestinal tract. Within each villus is a venule, a capillary network, an arteriole, and a lymphatic vessel. Nutrients as well as toxic agents can penetrate through the epithelial cells of the villus, enter the blood and lymph vessels, and be carried to various parts of the body (Figure 5.6). Nutrients, including carbohydrates, proteins, electrolytes, sodium, and calcium are absorbed with a combination of facilitated and active transport. Toxic materials that are similar in configuration, electrical charge, and molecular size may also be transported in this way. Such substances include lead, mercury, cadmium, and fluorides. The absorptive process can be influenced by physical health such as the presence of parasites, viruses, infectious bacteria, ulcers, cancer, or other digestive disorders. Absorption is also affected by nutritional status, the chemical and physical nature of the substances in the intestinal tract, and level of physical activity and age. Both spastic movements and mucous secretion are defensive capabilities of the intestinal tract that serve to remove many noxious agents either through diarrhea, vomiting, or both. However, soluble toxic chemicals may be quickly absorbed into the bloodstream while less soluble ones are often carried into the lower bowels for excretion. It is also possible for particles up to 50 mm to penetrate the mucosa and enter the lymphatic channels and blood vessels to reach the internal organs.

Mechanisms of Action

Nearly 2500 years ago, Hippocrates described the toxic effects of numerous poisons including lead, mercury, and arsenic. In the Middle Ages, the art of poisoning

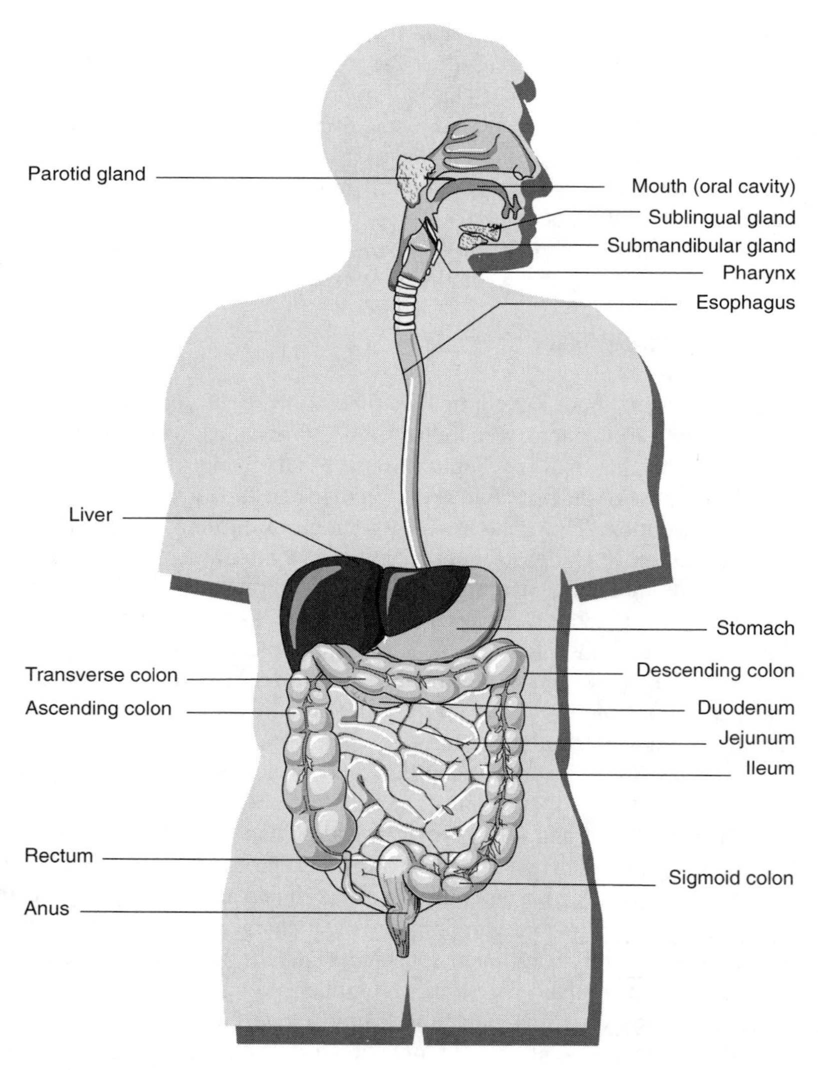

Figure 5.5 The gastrointestinal tract.

and the study of toxic effects became a formidable science in the hands of Catherine de Medici, who reported on the use of poisons, their clinical signs and symptoms, their potency, site of action, and how rapidly they produced their effects.[3] Today, we know that the toxic actions of pollutants vary dramatically based on dose, the length of exposure, and individual response. In fact, the symptoms that develop among a group of people exposed to the same toxic substance may vary so much as to appear caused by completely different substances. A toxin can produce a harmful effect upon an organ by either stimulating or depressing the normal metabolic actions

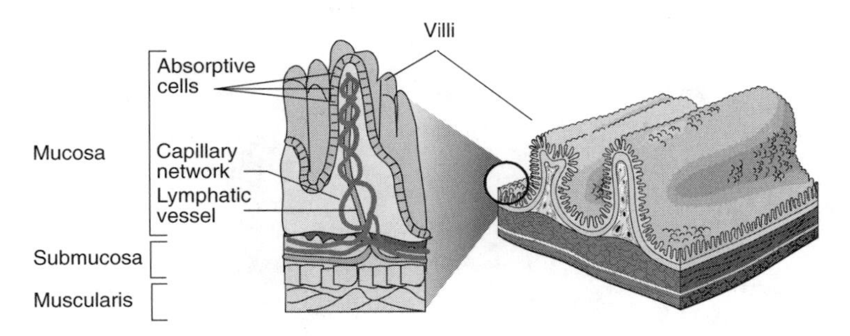

Figure 5.6 A cross-section of villi in the gastrointestinal tract.

of that particular organ. Very small amounts of toxic agents have the ability to stimulate function(s) in a variety of organs, whereas larger doses may impede or even destroy the activity of the organ involved. The body is able to detoxify foreign substances principally through the liver and other organs such as the kidneys, intestines, and lungs. Sometimes adverse effects occur when the quantity of the toxic substance in the body overwhelms the body's detoxification mechanisms or when injury or illness has compromised the body's ability to detoxify substances. The harmful effects of environmental toxins are dominated by three principal mechanisms: (1) the toxin's influence on enzymes; (2) direct chemical combination of the toxin with a cell constituent and; (3) secondary action as a result of the toxin's presence in the system.[12]

Effects of Toxic Agents on Enzymes

Enzymes are complex, three-dimensional proteins that have the unique ability of catalyzing (promoting or speeding up) metabolic activities in the body without undergoing any changes themselves. Most enzymes are highly specific and selective. The enzyme "prototype" may be envisioned in the following manner: an enzyme has the capacity to hold molecules together by binding to them, keeping them stabilized in order for a reaction to occur between them, and then moving on when the reaction has occurred. Many enzymes require a protein component (apoenzyme), and a non-protein component (cofactor) to become active. The apoenzyme and its cofactor combine to form a complete and active enzyme called a holoenzyme. Cofactors are usually metal ions such as manganese, magnesium, copper, or iron. They may also be an organic molecule called a coenzyme including vitamins, nicotine adenine dinucleotide (NAD+), and nicotine adenine dinucleotide phosphate (NADP+). Enzymes act on substrates to add or remove molecules of water, oxygen or hydrogen, amino or other functional groups. Enzymes may also rearrange atoms within a molecule, or join molecules (Figure 5.7). Enzymes are important in nearly all life functions including most aspects of energy production and biosynthesis. Many toxic substances have the ability to interfere with or block the active sites of the enzyme — inactivate or remove the cofactor, compete with the cofactor for a site on the enzyme, or alter enzyme structure directly, thereby changing the specific three-dimensional

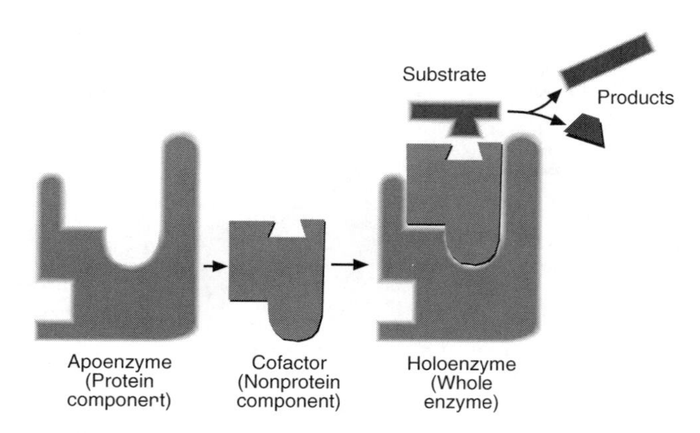

Figure 5.7 Components of an enzyme and mechanism of action.

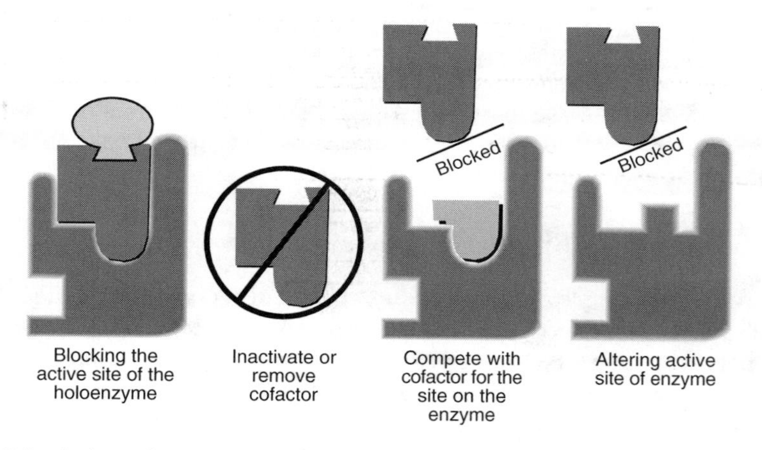

Figure 5.8 Actions of enzyme inhibitors.

nature of the active site (Figure 5.8). Hydrogen cyanide and hydrogen sulfide inactivate enzyme cofactors by binding to the iron present in the cytochrome oxidase system and inhibiting enzymes critical to the final steps in glycolysis and ATP production.[2,9] Similarly, dithiocarbamate will bind to copper and prevent the activity of copper-dependent enzymes important in the breakdown of ethanol. Consequently, alcohol consumption results in severe nausea and headaches. Substances similar to dithiocarbamate are used in alcohol anti-abuse drugs.[9] Cadmium and beryllium are believed to inactivate enzymes by blocking the sites on the enzyme where such cofactors as magnesium or zinc normally attach.[10]

The inhibition of enzymes may be reversible or irreversible. Usually, reversible enzyme inhibition occurs when the enzyme acts upon a substance similar to its normal substrate and so is unavailable to carry out its normal function. When a toxic chemical such as lead, arsenic, parathion, ozone, or fluorine covalently bonds to an enzyme, the bonding tends to be permanent for that particular enzyme molecule and is therefore considered an irreversible reaction. In most of these cases, the active group of the enzyme is blocked or destroyed.[9]

The Direct Action of Pollutants on Cell Components

Many substances affect cells directly. Strong acids, bases, and phenol will combine directly with cellular materials and also etch tissue, producing areas of necrosis. Nitrous and sulfuric oxides and ozone are powerful oxidizers that irritate mucous membranes. Inhalation of phosgene or chlorine gases can blister tissue and produce edema. Inhaled carbon monoxide reacts directly with hemoglobin to produce carboxyhemoglobin and prevents the attachment of oxygen which has a much lower affinity to hemoglobin than carbon monoxide. The greater the level of carboxyhemoglobin, the more severe the symptoms. Many sulfur-containing compounds including sulfonamides produce sulfhemoglobin in blood which is unable to carry oxygen. A number of aromatic amines, nitrites, and nitro compounds can oxidize hemoglobin to methemoglobin which is unable to carry oxygen, resulting in oxygen-poor blood and a syndrome in infants called "blue baby."

Pollutants that Cause Secondary Actions

Many substances may themselves be harmless but can cause the formation of other chemicals in the body that are harmful or potentially lethal. As an example, fluoroacetate (rodenticide 1080) is enzymatically treated as acetic acid in the respiratory metabolic chain, causing the formation of fluorocitric acid. This substance prevents the formation of NADP and ATP by inhibiting aconitase. The consequence is often lethal in relatively small doses.[3,9] Methanol is degraded to formaldehyde, which attacks the optic nerve and causes blindness. In most instances people are aware of the danger of these chemicals, but many are threatened by such innocuous substances as animal dander, pollen, milk products, shellfish, peanuts, and even sunlight. These exposures all trigger an exaggerated immune response in some people, leading to mild allergies associated with hay fever to rapidly fatal anaphylactic shock. The intermediary substances produced include histamine, serotonin, and similar products. Many pollutants also have an allergic component. Toluene diisocyanate (TDI) is a component of polyurethane foam and is reported to produce allergic reactions.[5] The air pollutants nitrogen dioxide and sulfur dioxide are also reported to have an allergic component. In fact, there is an allergic or immune component to many environmental diseases. Many parasites, bacteria, viruses, and toxic chemicals may have an allergic component. What are allergies? How do otherwise harmless substances cause the release, production, and final expression of intermediate substances in the body that can result in significant discomfort and even death? To understand this mechanism, a discussion of specific immunity and allergies is in order.

Immunity and Allergies

The immune system is critical to the defense of the human organism against invading microbes and abnormal or cancerous cells. An immune system debilitated by disease, chemotherapy, or inherited defects will permit the rapid onset of overwhelming infections or possibly cancer. The movie titled *The Boy in the Plastic Bubble*, starring John Travolta, was based on the true story about a boy who had inherited Severe Combined Immunodeficiency Syndrome (SCIDS). The absence of immune responses in this boy required him to live his life within an antiseptic, plastic-lined room in his home. Attempts to build an immune system in the boy through bone marrow transplants failed, and he died soon after. Just as threatening is an exaggerated response of the immune system as in anaphylactic shock from insect bites, or food allergies that can lead to unpleasant death within minutes. Chemical pollutants such as ozone can depress the immune response by inactivating alveolar macrophages. Others such as nitrogen dioxide or sulfur dioxide have an allergic component that makes the reactions to exposure more severe. Immunity is a double-edged sword.

Origins of the Immune System

Immunity is based on the premise that certain immune cells in the body can recognize microbes, tissues, and other substances that are "non-self" or foreign, and so destroy, encapsulate, or remove them. There are two separate but cooperating components of the immune system known as the humoral (antibody-mediated) immunity and the cellular (cell-mediated) immunity. Cell-mediated immunity (T cells) respond primarily to viruses, fungi, foreign tissue transplants, and cancer cells. Antibodies produced by B cells are directed against bacterial and viral infections and soluble antigens, while also being involved in anaphylactic or "Type I" allergic responses. The responses of cellular and humoral immunity are quite different. A rabbit injected with bovine serum albumin (BSA) and a purified protein derivative (PPD) from a modified tuberculosis bacterium will develop an amplified immune response to these materials within two weeks. When these substances are injected separately just under the skin (subcutaneously) in two separate areas, the immune response is quite different (Figure 5.9). The raised, hardened, warm area is caused by the cellular immune system to the tuberculin PPD. This area is filled with macrophages and lymphocytes and takes 2 to 3 days to develop. The subcutaneous bleeding is caused by humoral (antibody) response to subcutaneously-injected BSA, which causes localized vasodilation and leaking of blood components into the area. The mechanisms of these two components of the immune system are discussed in the following sections. Each component of the immune system is formed in the embryonic stages from lymphocytic stem cells that appear in bone marrow. During this development, some stem cells migrate to the thymus gland, where they are influenced to develop into T cells and then populate specific areas in lymph glands, spleen, bone marrow, and other lymphoid tissue in the gastrointestinal tract. The remaining lymphocytic stem cells come under the influence of tissues (possibly

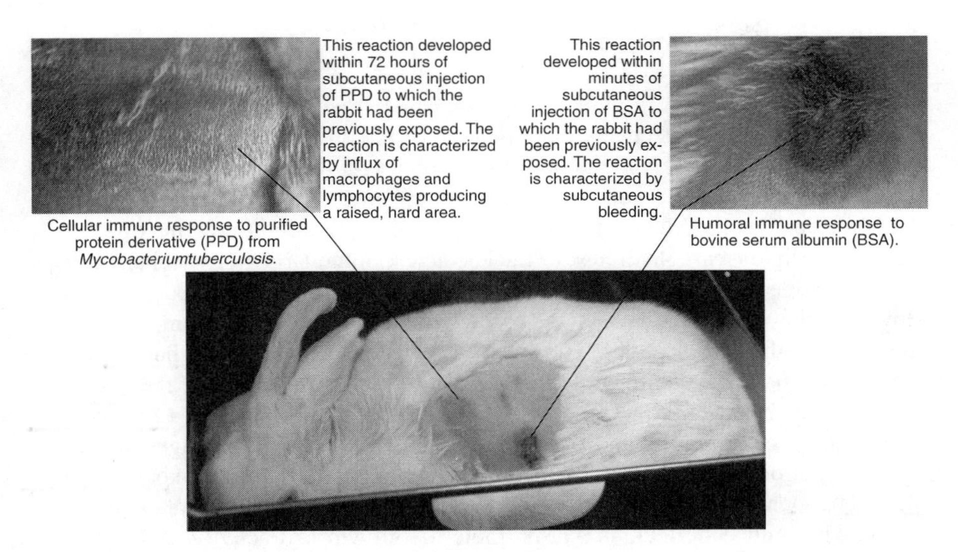

This reaction developed within 72 hours of subcutaneous injection of PPD to which the rabbit had been previously exposed. The reaction is characterized by influx of macrophages and lymphocytes producing a raised, hard area.
This reaction developed within minutes of subcutaneous injection of BSA to which the rabbit had been previously exposed. The reaction is characterized by subcutaneous bleeding.
Cellular immune response to purified protein derivative (PPD) from Mycobacteriumtuberculosis.
Humoral immune response to bovine serum albumin (BSA).

Figure 5.9 Demonstration in the rabbit of humoral and cellular components of the immune system.

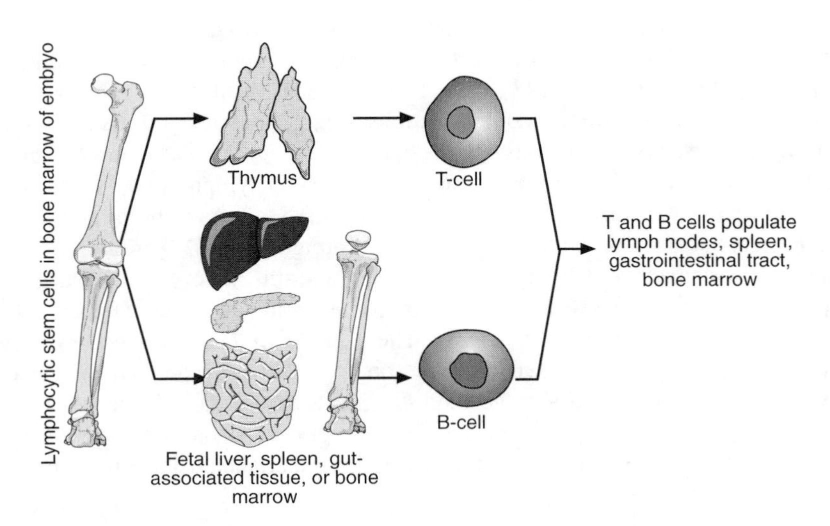

Figure 5.10 The origin of T cells and B cells.

gut-associated lymphoid tissue, bone marrow), to become antibody-producing cells or B cells that also occupy lymphoid tissue, but in different compartments other than T cells (Figure 5.10).

The Initial Immune Response

The immune system responds to agents, cells, or substances that are foreign or non-self. These are collectively called antigens. An antigen is a chemical, cell, or agent that causes the immune system to respond with the production of antibodies (humoral immunity) or T cells (cellular immunity) or both, and then react with the antibodies or T cells. Most antigens are high-molecular weight proteins often associated with lipids, carbohydrates, or nucleic acids. Some large polysaccharides may also be antigenic. Many molecules are too small to promote antibody formation. However, they may link to a larger carrier molecule such as a serum protein and function together as an antigen. The attached molecule is then called a hapten. Under most circumstances, an antigen, such as a bacterial cell, reaches one of the numerous lymph nodes in the body where a large white blood cell known as a macrophage (large eating cell) ingests and processes the antigen by partially digesting it and displaying it on the surface of the macrophage along with human leukocyte associated (HLA) antigens to B cells and T cells. The macrophage also produces lymphokines, and interleukins that promote the growth and activity of T cells and B cells.

Cellular Immunity

T cells respond to a particular antigen, and then enlarge, divide, and give rise to clones of several subpopulations of T cells including cytotoxic (killer T cells), helper T cells, memory T cells, suppressor T cells, delayed hypersensitivity T cells, and amplifier T cells (Figure 5.11).[11,12] Killer T cells are sensitized to recognize particular antigenic groups on the surfaces of cancer cells, foreign transplanted tissues, viruses, and fungi. As the cells migrate through the body and recognize such chemical groups, they respond by attaching to the cell and: (1) secreting lymphotoxins that can kill the cell directly; (2) releasing a series of chemicals that attract, hold, and activate macrophages to assist in the destruction of the foreign cell; (3) secreting a transfer factor that sensitizes nearby lymphocytes and recruits them to the battle; and (4) producing interferon, which inhibits multiplication of viruses. Helper T cells stimulate other immune cells to activity. These helper cells are sometimes termed CD_4 cells because of certain cell receptors on their surface. The CD_4, or helper T cells are among the primary targets of the AIDS virus, thereby depressing the immune response. There also appear to be memory cells that are long-lived and recognize the reentry of a particular foreign cell or material, causing a rapid immune response.

Humoral Immunity

B cells do not normally leave the protection of their locations in the lymph nodes, spleen, gastrointestinal tract, or bone marrow. Their response to foreign invaders is to produce liquid proteins (humoral) known as antibodies and secrete them into the bloodstream, where they can travel to the affected site and carry out their destructive action. A foreign substance or cell is processed by a B cell with antigenic receptors

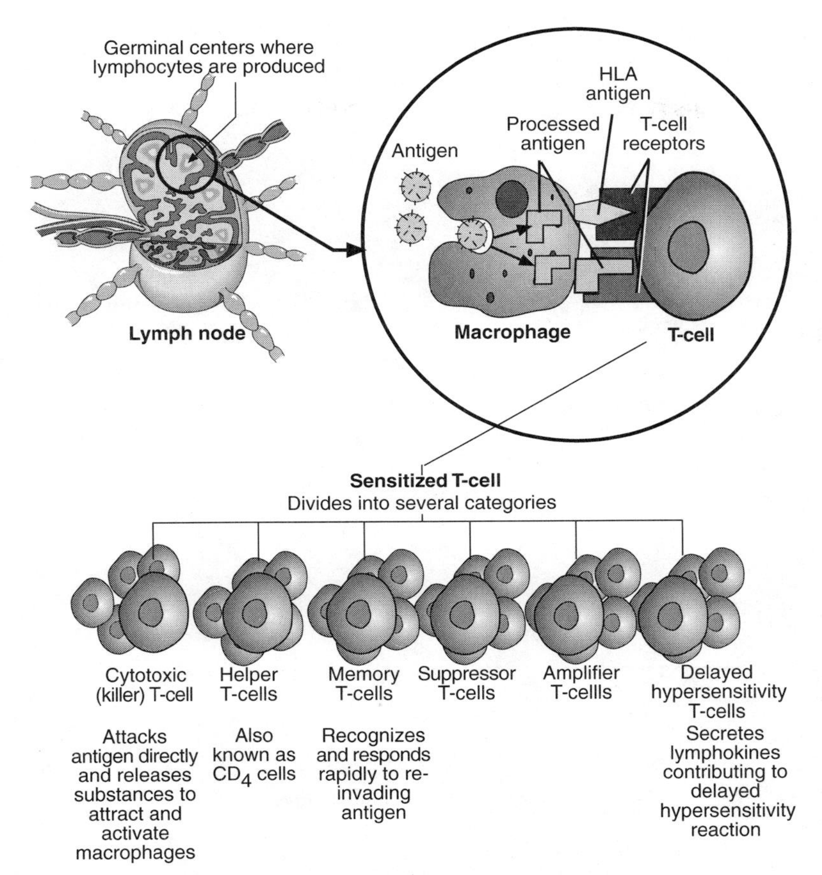

Figure 5.11 Action of killer T cells. (Adapted from Tortora, G.J. and Anagnostakos, N.P., *Principles of Anatomy and Physiology,* 6th ed., Harper and Row Publishers, New York, 1990, chap. 22; and Tortora, G.J., Funke, B.R., and Case, C.L., *Microbiology: An Introduction,* The Benjamin/Cummings Publishing Company, Inc., Redwood City, CA, 5th ed., 1995, chap. 17.)

that recognize specific chemical groups on the foreign material. Each B cell is specific to one particular antigenic group. An antigen selects a specific B cell and attaches to its antibody receptor on the B cell. The antigen is processed and presented with its HLA antigens on the B cell surface to helper T cells (Figure 5.12). The helper T cells produce substances that promote B cell proliferation into a clone of identical plasma cells. These plasma cells produce antibodies to this specific antigen. Other B cells are recruited as memory B cells that are long-lived and will direct a rapid and aggressive response if the identical foreign material should reappear at a later time. Antibodies to a particular antigen are then released from the B cells into the bloodstream, where they are carried to the affected site and released to the area. The antibodies will attach to the antigenic groups on the foreign substance, causing a number of possible actions including neutralization, precipitation or agglutination, or activation of the complement system. Prior to explaining these activities, it is necessary to describe a typical antibody molecule and its mode of action.

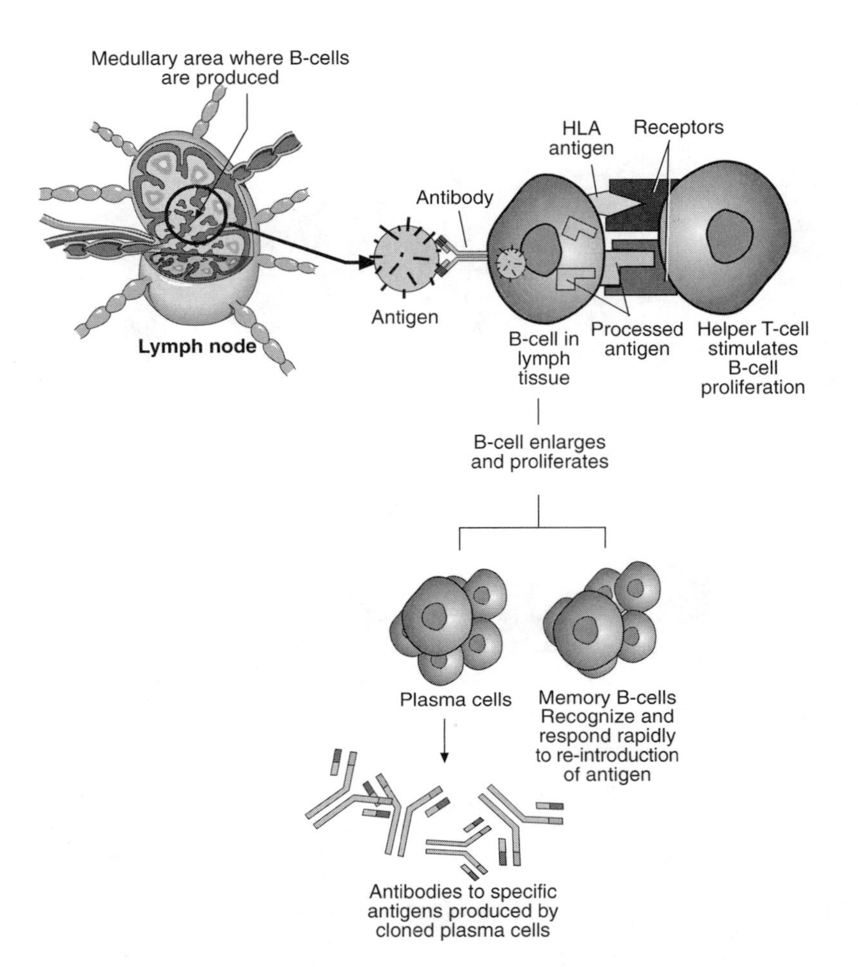

Figure 5.12 How B cells respond to antigens in humoral immunity. (Adapted from Tortora, G.J. and Anagnostakos, N.P., *Principles of Anatomy and Physiology,* 6th ed., Harper and Row Publishers, New York, 1990, chap. 22; and Tortora, G.J., Funke, B.R., and Case, C.L., *Microbiology: An Introduction,* The Benjamin/Cummings Publishing Company, Inc., Redwood City, CA, 5th ed., 1995, chap. 17.)

The Antibody Molecule

Antibodies (also called immunoglobulins) are proteins with molecular weights of 150,000 to 900,000 Kd.[80] The simplest from of an antibody appears as a bivalent or Y-shaped structure consisting of two identical light peptide chains and two heavy peptide chains joined together by disulfide linkages (Figure 5.13). Each light and heavy chain has a constant region composed of specific sequences of amino acids. These regions remain largely unchanged and differ only among the five major classes of antibodies known as immunoglobulins — IgG, IgA, IgM, IgD and IgE plus four subtypes of IgG (IgG1-4) and two of IgA (IgA1 and IgA2). Light chains exist in two classes, lambda and kappa. Each antibody molecule has either lambda or kappa

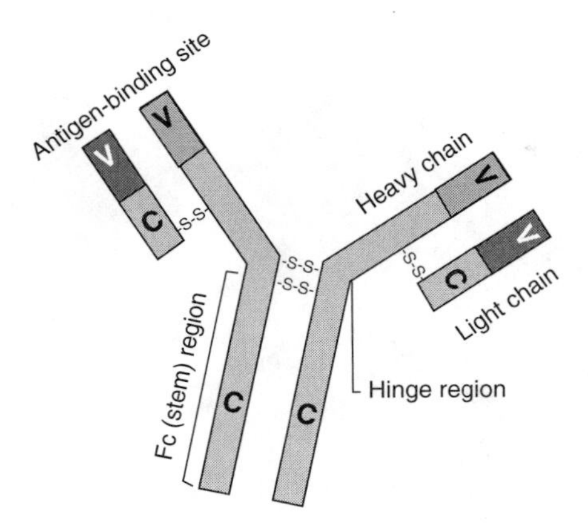

Figure 5.13 Typical bivalent antibody molecule. (Adapted from Tortora, G.J. and Anagnostakos, N.P., *Principles of Anatomy and Physiology,* 6th ed., Harper and Row Publishers, New York, 1990, chap. 22; and Tortora, G.J., Funke, B.R., and Case, C.L., *Microbiology: An Introduction,* The Benjamin/Cummings Publishing Company, Inc., Redwood City, CA, 5th ed., 1995, chap. 17.)

light chains, not both. Each of these classes has a separate function. The stem of the constant region of the heavy chain is also known as the Fc region and is capable of attaching to a number of host cells including mast cells. The opposite end of the antibody molecule, the fab portion, consists of variable regions in the heavy chains and the light chains. The combination of the variable regions on a single heavy chain and its attached light chain form a single antigen-binding site that will attach tightly to one very specific antigenic group.

Since there are two such sites on the simplest Y-form antibody, it is called bivalent. Further, the peptide arms of the Y are flexible and may form a T shape as well. Each possible antigenic group on a foreign substance must have its own unique antibody. Consequently, the variable regions of the antibody are created in a specific three-dimensional form that is preconfigured in the B cell clone to only one antigenic group (Figure 5.14). The largest percentage of antibodies belongs to the IgG class. These bivalent antibodies have a molecular weight of 150,000, can cross the placenta to provide passive immunity to the newborns, enhance phagocytosis, neutralize toxins and viruses, and activate the complement system. The role of IgE in allergic reactions will be discussed later.

Antibody Activities

Antibodies can neutralize viruses and organic toxins (as from snake bites or bacterial toxins) by attaching to receptors and preventing their attachment to host cells or neutralizing toxins by blocking active sites. Antibodies will combine with cells, soluble proteins, or haptens to promote agglutination or precipitation. This effectively enlarges the particle sizes and enhances the process of phagocytosis by macrophages. Yet another function of antibodies is opsonization in which the bacteria is coated by the fab region, thereby facilitating subsequent phagocytosis by cells possessing an Fc receptor. The binding of an antibody with its specific antigen can

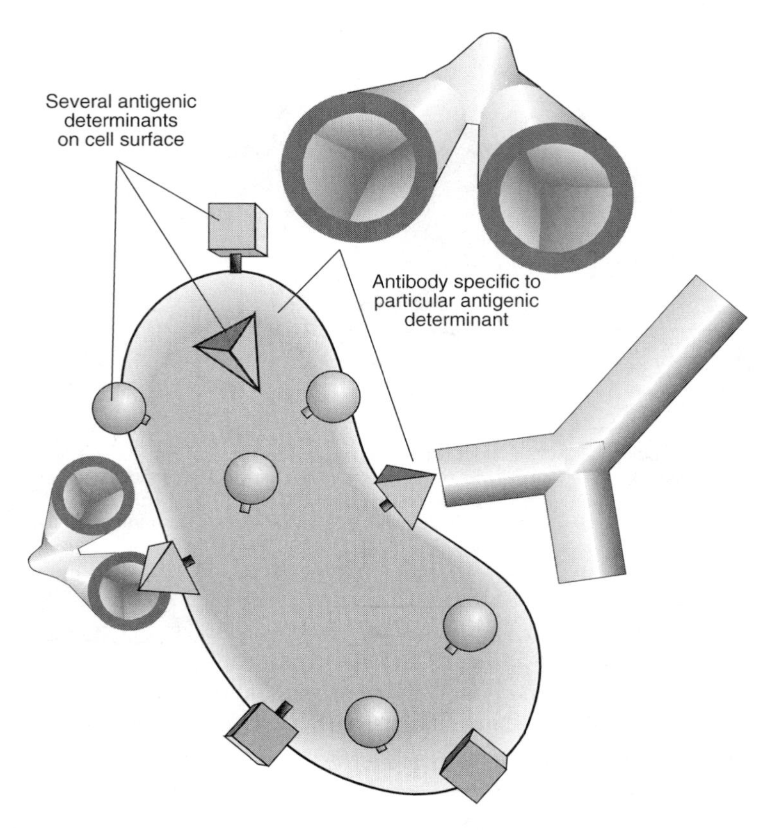

Figure 5.14 Specificity of antibody and antigenic determinants.

activate the complement system. The complement system is composed of several (at least 20) serum proteins that follow a sequence or cascade of actions beginning with the Ag–Ab binding. This sequence of action enhances phagocytosis, inflammation, and cell lysis (Figure 5.15). When exposed to an antigen, there are two types of responses. Primary response is slow in onset, low in magnitude, and short-lived. Secondary response is rapid in onset, high in magnitude, and long-lived. There is a lag of several days before a specific antibody becomes detectable. This antibody is IgM. After a short time, the antibody level declines. This is the primary response. If at a later date we are exposed to the same antigen, there is a far more rapid appearance of antibody and in greater amount. It is of the IgG class and remains detectable for months or years. These are the features of the secondary response.

Hypersensitivity

An exaggerated immune response to the presence of an antigen is termed hypersensitivity or allergy. Allergies are the other side of the two-edged sword of immunity. The immune reactions are usually unnecessary, harmful, and sometimes fatal. People

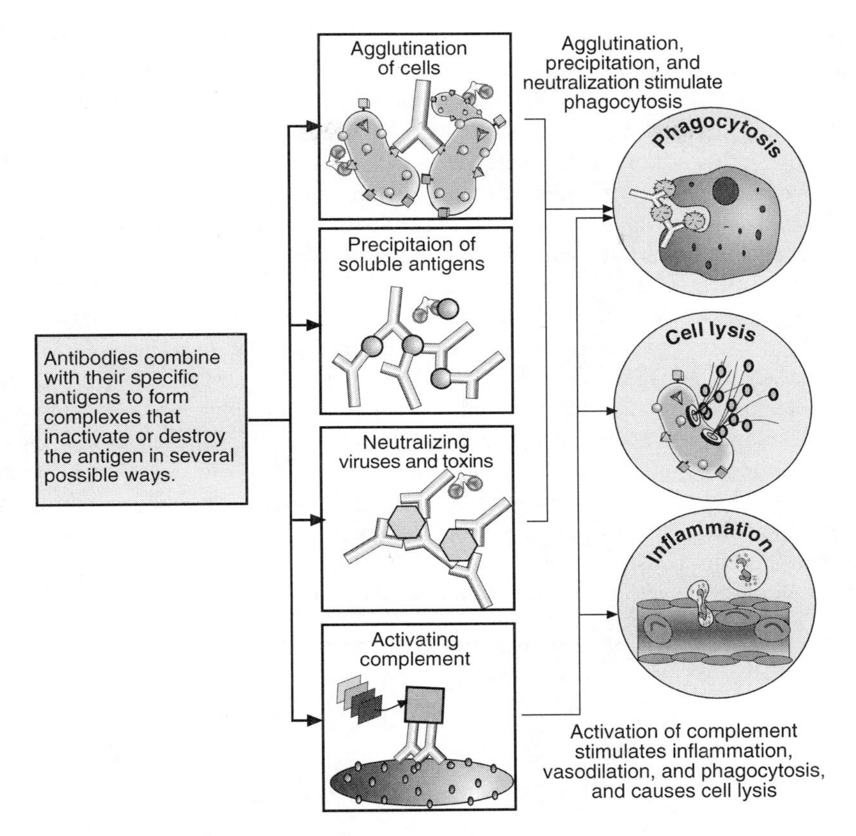

Figure 5.15 Antibody activities stimulated by Ag–Ab binding. (Adapted from Tortora, G.J. and Anagnostakos, N.P., *Principles of Anatomy and Physiology,* 6th ed., Harper and Row Publishers, New York, 1990, chap. 22; and Tortora, G.J., Funke, B.R., and Case, C.L., *Microbiology: An Introduction,* The Benjamin/Cummings Publishing Company, Inc., Redwood City, CA, 5th ed., 1995, chap. 17.)

have developed allergies to most everything including insect bites, pollen and other plant products, environmental toxins, a variety of foods, sunlight, human semen, and much more. Some people have become so sensitized to chemical substances they are unable to physically cope in routine surroundings. Such people are said to have multiple chemical sensitivities (MCS), and this is a highly controversial subject. There is also a very significant rise in the number and severity of asthma cases. The prevailing scientific thought is that environmental pollutants such as ozone and fine particulates contribute to this phenomenon.[13]

There are four major types of hypersensitivity reactions — cytotoxic, cell-mediated, immune complex, and anaphylactic types. Cytotoxic hypersensitivity involves complement activation and is most often seen in transfusion reactions when incompatible ABO and Rh antigens on blood cells are transfused to a person resulting in the donor cells being lysed. Cell-mediated reactions involve T cells that respond to certain antigens on transplanted tissues and organs, and to haptens associated with poison ivy or poison oak, cosmetics, latex gloves and condoms, nickel in coins and zippers, and bacterial extracts such as PPD from Mycobacterium tuberculosis. Immune-complex reactions involve the coupling of antibodies with soluble antigens circulating in the serum in specific ratios that permit the complex to become trapped in the basement membrane of blood vessel cells where they can activate complement, and cause inflammatory, responses. The anaphylactic or Type I hypersensitivity is associated with hay fever, asthma, and anaphylactic shock from insect bites, certain foods such as shellfish, and a variety of chemicals. The reactions are usually immediate, occurring within moments in persons previously sensitized by exposure to the offending antigen. The Type I hypersensitivity occurs when the antigen binds in specific ways to IgE antibodies attached to mast cells and basophiles. Mast cells occur in the connective tissue of the respiratory tract, the skin, and the blood vessels. Basophiles are a form of white blood cells circulating in the bloodstream. The IgE antibodies attach by their Fc fragment to the mast or basophile cells. Persons who have developed an excess of IgE antibodies run a higher risk of developing an anaphylactic type of allergic response because most of the sites on the mast cell or basophile cells will be covered by IgE antibodies. When an antigen such as animal dander or pollen attaches across two closely spaced IgE molecules specific to that antigen, this causes the mast or basophile cell to degranulate and release histamines that stimulate vasodilation and bronchoconstriction (Figure 5.16). Vasodilation

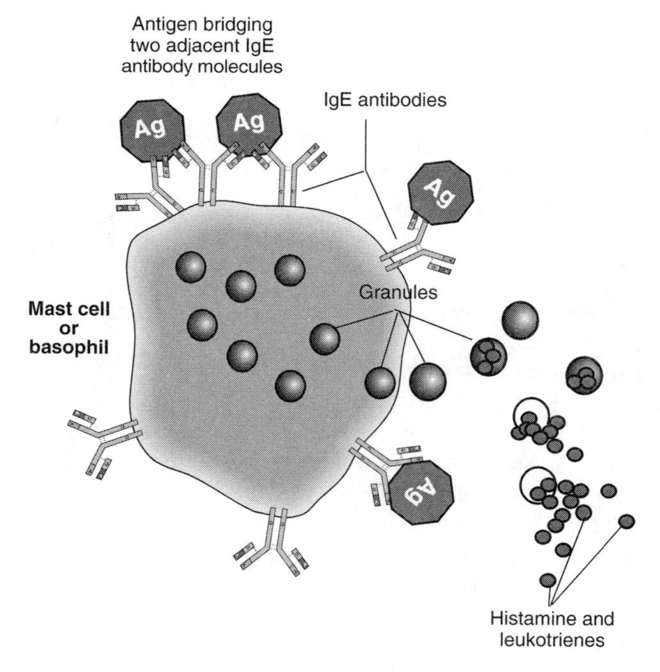

Figure 5.16 Mechanism of anaphylactic allergic responses involving IgE. (Adapted from Tortora, G.J. and Anagnostakos, N.P., *Principles of Anatomy and Physiology,* 6th ed., Harper and Row Publishers, New York, 1990, chap. 22; and Tortora, G.J., Funke, B.R., and Case, C.L., *Microbiology: An Introduction,* The Benjamin/Cummings Publishing Company, Inc., Redwood City, CA, 5th ed., 1995, chap. 17.)

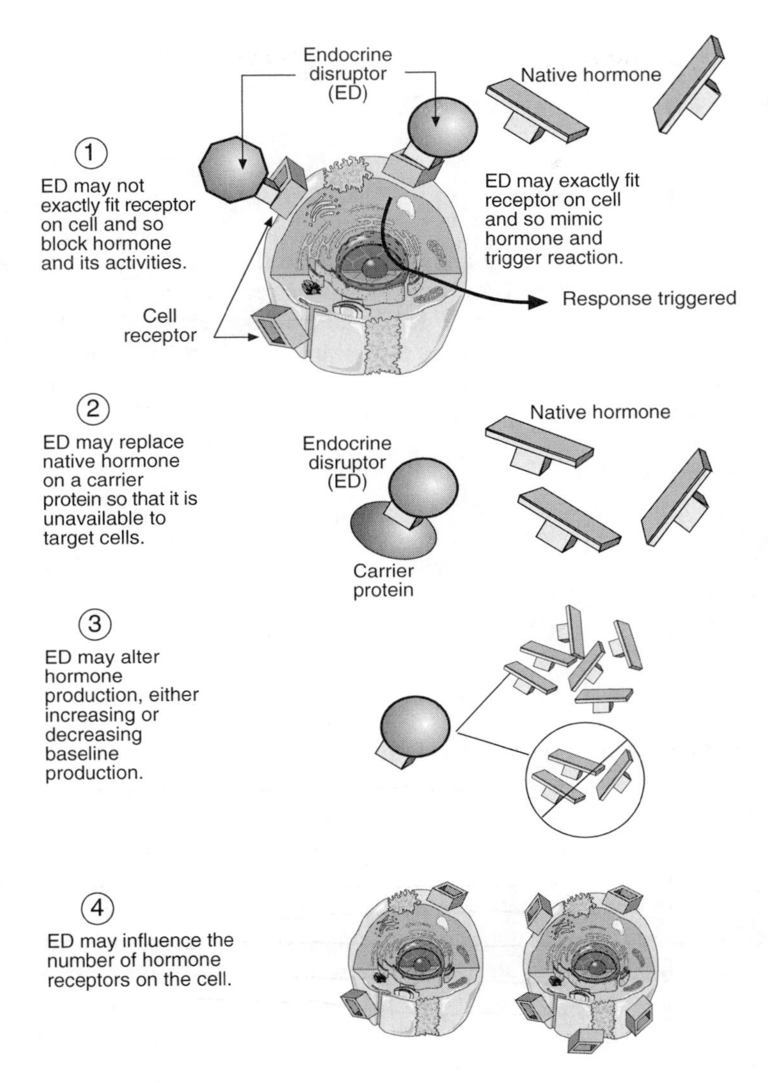

Figure 5.17 Mechanisms of endocrine disruption.

increases the permeability of blood vessels, allowing blood components to seep into the affected area (Figure 5.17). Swelling and redness of mucous membranes in the eyes, nose, and throat are also common symptoms of this type of allergy when the allergens are confined to the respiratory tract. The histamines also cause smooth muscle contraction that occurs in the bronchi and bronchioles of the lung. This bronchoconstriction decreases the lumens of the respiratory tract and makes breathing difficult. The mast or basophile cells, once triggered by an antigen, cause the synthesis of leukotrienes and prostoglandins that cause prolonged contractions of smooth muscle and increased mucous production, which further aggravates respiratory distress. Localized reaction may lead to hay fever, hives, and asthma. These allergic reactions can also be systemic (involving the whole body), producing breathing difficulties, a precipitous decline in blood pressure, cyanosis, and possibly death.[11,12]

Factors Governing Toxicity

The outcome of exposure to a toxin depends on a number of factors that may include: (1) the chemical properties of the substance; (2) the concentration of the substance; (3) the length of time to which the person is exposed; (4) the route of exposure; (5) whether or not the toxin is biotransformed in the body; (6) the interaction with other pollutants; (7) and a number of personal or internal factors including a person's age, level of activity or stress, and health status.

Chemical Properties

Chemicals vary in their toxicity to mammals. An oral dose of sodium chloride would require 3000 mg/kg to achieve an LD50 in rats, whereas the LD50 for strychnine is less than 2.5 mg/kg. The LD50 is defined as the dose of a chemical that will cause death in 50% of a defined animal population.

Concentration

The concentration of the substance to which one is exposed is also important because the concentration and the length of time of the exposure determine the dose. Each toxin and health effect has its own dose–response relationship. Adverse effects from exposure to toxic agents may range from a slight irritation, such as itchy, red eyes, to rapid death. Large doses may cause more harm than would the exposure to the same toxin in small amounts over a long period of time. In acute illness, effects are often witnessed immediately, with characteristic symptoms occurring over a limited time. Chronic effects are usually the result of long-term exposure to small amounts of toxins where illness develops slowly and insidiously (without obvious identifiable symptoms).[2] A sudden exposure to large amounts of carbon monoxide will lead to dangerous concentrations of carboxyhemoglobin in the blood with resulting unconsciousness and death. Persons exposed to the same toxin in smaller doses over longer periods may slowly develop headaches, dizziness, chest pains, and possibly produce damage to heart and blood vessels.[14]

Usually, only very small amounts of an absorbed chemical reach target tissue and attach to molecular receptors, becoming the site of action for the chemical. The effective dose is the amount that reaches the general circulation and ultimately the location where a particular effect is seen or felt. This dose, along with the speed with which it happens, largely determines the pharmacologic effect. Drinking five beers in an hour has a noticeably different effect than drinking five beers in four hours. Once in the general circulation, the toxic chemicals may be absorbed, stored, biotransformed by other tissues, and/or excreted. The route of entry is also critical to determining the toxic consequences of the agent. Some substances such as lead and mercury are more efficiently absorbed through the lungs than the intestines, making airborne forms of these substances of more concern.[15] When materials containing lead or mercury are incinerated or otherwise combusted, the potential consequences of even small amounts should be of greater concern.

The mechanisms of tissue absorption have been previously discussed and include passive and active processes. Some substances may be localized or stored in various body compartments characterized by having similar structures and functions. Lead tends to be stored in long bones, while organochlorines tend to be stored in fatty tissue. If the substance is stored at levels above what would be normally expected, then this is termed *bioaccumulation*. When metabolic processes alter the structure and characteristics of a chemical this is called *biotransformation*. A wide number of enzymatic pathways is capable of biotransforming toxic substances into non-toxic substances, and these pathways exist in the liver, kidneys, lungs, intestines, and other organs. Pesticides such as malathion are detoxified in humans and other mammals by the enzyme carboxyesterase. However, such biotransformations are not always beneficial, as discussed previously, when sodium fluoroacetate is readily biotransformed into fluorocitrate by mammalian cells resulting in chemical asphyxia.[2,9] Absorbed toxic substances may also be excreted. The major excretory organs are the kidney, liver, and lungs. Chemicals that are not readily soluble in blood or that may have high vapor pressures will likely be exhaled via the lungs. Urine and bile are often the recipients of water-soluble toxic materials. Substances not eliminated in this way are generally biotransformed or stored in some body compartment.

Interactions

Most of the information on toxicity of chemicals is based on studies that attempt to limit all variables except the effects of the single chemical under study. However, most exposures come with multiple chemicals; and these additional chemicals influence the outcome of the exposure. The effects of some substances are additive in that the two substances taken together are the sum of the two substances taken separately. If you are exposed to chemical A and it produces a measurable response of a 5% decrease in red blood cells (RBCs), and chemical B produces a reduction of 3% in RBCs, the two together (A+B) would be expected to produce an 8% reduction if their effect were additive. Generally, substances behaving in this way employ a similar set of toxic mechanisms. In our example, if the two substances produce a 20% reduction in RBCs the effect would be much greater than the sum of the two substances, producing an effect that is synergistic. This has great significance in environmental health since many toxic agents act in this way. Sulfur dioxide is far more toxic in the presence of small particulates or sulfuric acid mists which carry the pollutant deeper into lung tissue.[16] Again referring to our example, if substance A and substance B are combined and the effect is to produce a reduction of only 2% in RBCs, the effect is lesser than the sum of the two acting independently, and this is referred to as antagonistic. Although particulates enhance the toxicity of sulfur dioxide, they have quite the opposite effect when inhaled simultaneously with nitrogen dioxide by decreasing the toxicity.[17] Stimulants such as caffeine in some teas or coffee will tend to counteract the effects of sleeping pills.[9]

Age

There are a number of personal factors that influence the effects of toxic substances. The age of the person is a major factor. The very young and the elderly tend to be at increased risk to toxic substances. The young are more susceptible for a number of reasons including: (1) they inhale or ingest more toxic substance per pound of body weight than adults; (2) they are more likely to put dangerous substances into their mouths; (3) children often absorb materials into the tissues at a higher rate than adults; (4) they tend to retain larger proportions of toxic substances than adults; and (5) children's nervous systems and detoxifying systems are not fully developed, causing disproportionate effects from toxic materials.[18] The adverse effects of ozone, lead, mercury and sulfur dioxide are more pronounced in children than in adults.[18,19] The elderly are also at greater risk to the effects of toxic substances since they tend to have preexisting respiratory and cardiovascular diseases,[2] depressed immune systems, and higher levels of stored toxic agents in their bodies than do younger people.

Exercise and Physical Stress

Most athletes are aware that you should not run in hot, humid, heavily polluted air. The consequences are reduced performance and even pain in the lungs. There is a tendency when exercising heavily to take in air through the mouth in large amounts. Consequently, higher levels of pollutants bypass the nasopharyngeal area and enter the sensitive tissues of the lungs more directly. It has been known for some time that rats exposed to 1.0 ppm of ozone while being forced to exercise periodically in a rotating cage will die, while nonexercising rats will experience little discomfort with the same exposure.[19]

Health Status

The health status of a person is also important in the outcome of toxic chemical exposure. Preexisting cardiovascular disease will render a person more susceptible to a variety of air pollutants. Persons who have asthma or who smoke are also increasingly sensitive to the many pollutants. Alpha1-antitrypsin deficiency is a genetic defect in humans that markedly increases the risk of emphysema from a number of chemical exposures. Genetic polymorphisms, or variations of the same gene, can produce subtle or dramatic differences in response to toxic chemicals. The P450 genes produce enzymes that detoxify some agents in tobacco smoke. Variations in P450 genes result in enzymes of varying activity that may increase the risk of lung cancer for many people.[20] Unfortunately, many human diseases are associated with genetic defects. Debilitating diseases, poor nutrition, and lack of exercise all contribute to increasing the risk of adverse effects from toxic substances.[5]

SOME SPECIFIC EXAMPLES OF TOXIC AGENTS

Endocrine Disrupters and Reproductive Health

Hormone Function

Hormones are chemical messengers that originate in an organ, gland, or other body part, and then convey through the blood to a target organ or tissue that is stimulated by the hormone to increased functional activity or secretion. Hormones do not provide energy and are active in very small quantities. They are critical in the regulation of many life processes, including sexual development, metabolic functions, development of the brain, human growth, and stress response. As examples, androgens regulate the development and maintenance of male sexual characteristics, while estrogens stimulate the development of female sexual characteristics. The predominant and most active androgen is testosterone, which is produced by the male testes. The other androgens that support the functions of testosterone are produced mainly by the adrenal cortex — the outer substance of the adrenal glands and only in relatively small quantities.[82] Hormones function by attaching to specific receptors on the inside or outside of cells. These receptors have a three-dimensional complementary fit to the active site of the hormone. Once the hormone links with its specific receptor site, a series of events is initiated that may involve the production of other chemical products or specific biochemical reactions.[1,21] When the activity of such hormones is disrupted, a number of unwelcome outcomes may occur.

Adverse Effects of Endocrine Disruption

In a report issued in 1997, the EPA concluded that endocrine disrupters can produce cancer, developmental problems, and sterility in animals.[22] There are many examples of adverse effects on reproduction in wildlife and fish in many geographic areas including Florida, the Great Lakes, and California.[23,24] In those areas where fish and wildlife have been exposed to endocrine-disrupting hormones, the effects on reproduction and development have been marked and include the mixing of male and female characteristics, or indeterminate sex; the feminization of male species and the masculinization of female species; unsuccessful reproduction; thyroid dysfunction in birds and fish; decreased fertility in birds, shellfish and mammals; gross birth deformities in turtles, fish and birds (ie., twisted beaks in birds); and behavioral abnormalities.[25,26]

The effects of endocrine disruptors on humans is much more problematic. Significant differences in opinion have erupted concerning the potential threat. Some authorities emphasize that there is no strong evidence that environmental disruptors cause adverse health effects in humans at the current ambient exposure levels and that there is no causal link between such chemicals and perceived human illnesses.[21] In fact, the science of endocrine disruptors is highly controversial and there is little knowledge or agreement on how much of a problem there is.[27] Many people have pointed to endocrine disruptors as the cause of a number of emerging human ailments including reduced sperm counts, precocious puberty, increase in non-hodgkins lymphoma, marked increase in males having undescended testicles (cryptochordism), and testicular cancer.[27–29] Studies have shown a reduction in sperm counts within the general population of many different countries to levels that suggest increased infertility.[30] Other studies suggest there may be regional problems of sperm decline challenge the assertion that sperm counts are declining worldwide.[31] The frequency of undescended testicles has more than doubled in England and Wales[32] and has increased in other countries as well.[29]

More recent studies have not demonstrated an association between breast cancer and the endocrine disruptors DDE (a metabolite of DDT) and polychlorinated biphenyls (PCBs). In one study, blood levels of PCBs and DDE were measured in 600 pregnant women. When their babies were born, the researchers then measured levels in the mothers' breast milk. Finally, the team monitored the children as they grew and entered puberty. The most prominent effect reported was that boys exposed to DDE and girls exposed to PCBs were heavier than their unexposed peers at age 14. The study also noted that girls with high prenatal PCB exposure tended to hit the first stages of puberty a bit earlier than others.[83] The role of these chemicals in increasing testicular cancer has also not been demonstrated.[27] However, numerous other endocrine disruptors exist, and the level of concern among public health professionals is great. The combination of increasing testicular cancer, undescended testicles, and falling sperm counts is thought to be caused by fetal exposures to endocrine-disrupting chemicals.[21] Studies of rats exposed *in utero* have shown that alkylphenols disrupt Sertoli cell development in the fetuses, causing decreased testes size and a decrease in daily sperm production of the rats growing to adults.[33] Because humans are exposed from many pathways at levels similar to exposure levels in rats, there is a significant concern that these and other endocrine disruptors may be the reason for the increasing reproductive disorders seen in males.[21] Efforts continue by the EPA, working groups, and an advisory committee, together with industry, to develop screening strategies for environmental disruptors.[27]

What Are Endocrine Disruptors and How Do They Work?

Endocrine disruptors have been referred to as xenoestrogens, estrogenic, hormone-mimicking, and endocrine-disrupting chemicals. The synthetic chemicals presently identified include a number of pesticides such as DDT, lindane, and atrazine plasticizers such as phthalates and alkylphenols; bisphenol A; PCBs; dioxin; the food antioxidant, butylated hydroxyanisol (BHA); and several fungicides. A variety of naturally occurring plant compounds or phytoestrogens are also known to have estrogen compounds (Table 5.1). There are at least four different mechanisms by which endocrine disruptors can exert their adverse effects (Figure 5.17).

1. Some endocrine disrupters are similar enough to the native hormone that they can occupy its site on the receptor tissue causing a cascade of events, as if the native hormone had bound to the site. Such a chemical is called a hormone mimic. Some chemicals may bind to the site but are not similar enough to activate a response. They simply prevent the hormonal action through a hormone block.

Table 5.1 Endocrine-Disrupting Chemicals

CHEMICAL	USES
Phthalates	Used as plasticizers; found in medical and household products, clothing, toys, and packaging including plastic wraps, and liners of metal food and beverage cans; phthalates leach from food wraps and lining of containers into food and beverages where they are ingested; phthalates may also be absorbed through the skin and have been associated with testicular toxicity
Dioxin	An unwanted byproduct in the production of some chlorinated hydrocarbons, paper production, and incineration of plastics
PCBs	Used in electrical transformers and capacitors, hydraulic fluids, plasticizers, and adhesives; production was banned in the U.S. in 1977 because of bioaccumulation, persistence, and toxicity
Alkylphenols	Present in detergents, paints, pesticides, plastics, food wraps,and plastic laboratory ware; they are known to be estrogens
Pesticides	Methoxychlor, lindane, dicofol, atrazine, and DDT are pesticides that exhibit weak to moderate estrogenic activity, but mixtures of two or more pesticides have been shown to produce potent activities sometimes 1000 times that of individual pesticides
Natural sources	Grains, some vegetables, nuts, berries, grasses, and soybeans may contain phytoestrogens such as flavinoids and liganans with hormone-like properties

(Adapted from Crapo, L., *Hormones: Messengers of Life,* W.H. Freeman, New York, 1985, chap. 1; and Warhurst, M., Introduction to hormone disrupting chemicals, 1998, http://easyweb. easynet.uk/-mwarhurst/estrogenic.html.

2. Endocrine disruptors may cause artificial and unwelcome changes in the numbers of hormone receptor sites on tissues and organs.
3. The baseline production of hormones may be diminished or increased.
4. Endocrine disruptors may replace hormones on the carrier proteins that transport hormones via the bloodstream to various parts of the body. This makes some proportion of the population native functional hormone unavailable and produces adverse health consequences that may be barely noticeable to very severe.[21]

Pollutants originally described as estrogens or androgens were based on their ability to bind to two different types of steroid receptors, and not much thought was given to the possibility of one binding to both sites. However, recent studies *in vitro* reveal that some environmental pollutants can bind to estrogenic, androgenic receptors and binding proteins. If this phenomenon occurs in living organisms, then the effects of environmental disruptors may occur across several different receptor sites and magnify the effects, making even very low concentrations quite potent.[35]

Reducing Exposure

Although no definitive evidence exists on the role of endocrine disruptors in adversely affecting human health, there is sufficient concern that federal agencies, industry, and private organizations are attempting to respond to this complex dilemma with more answers through research and recommendations for risk reduction. Many of the endocrine-disrupting chemicals are chlorine-based and persistent in the environment. Furthermore, they are ubiquitous and find their way into the food supply through: (1) ingestion of contaminated grains and grasses by livestock which then store the lipophilic chemicals in their fatty tissues; (2) contamination of fruits and vegetables by spraying with pesticides; and (3) leaching of endocrine-disrupting chemicals from plastic wrappers, plastic liners of cans, and polystyrene containers. The majority of exposures may come from such contaminated food. Individuals who are concerned about the exposure to endocrine disruptors may reduce that exposure by:

1. Reducing or limiting ingestion of dairy products and meat high in fat where organochlorines tend to accumulate
2. Avoiding synthetic pesticides by purchasing foods low in pesticide residues and switching to herbal or scent-based repellents
3. Keeping children away from vinyl toys or teething rings
4. Using detergents and shampoos that do not contain alkylphenols such as nonoxynol and octoxynol[22]

Dioxin

Dioxins are not manufactured for direct use in any product. Dioxin is actually an unwanted by-product from heating mixtures of chlorine and organic compounds in industrial processes, such as the bleaching of paper pulp, the incineration of chlorine-containing materials, or in the production of pesticides, herbicides, and chlorophenol wood preservatives. They are formed naturally by the incomplete combustion of wood products and industrial and municipal wastes. Unfortunately, dioxin does not break down easily, therefore dioxin tends to bioaccumulate (the accumulation of toxic chemicals in living things through the consumption of food or water) in the fatty tissue of humans and animals. This substantially increases the exposure of nursing infants whose mothers may have concentrated dioxin in the milk and fatty tissues of the breast.[36]

There are 75 separate chemical compounds that make up the group referred to as dioxins. The most acutely toxic chemical of this group is known as 2,3,7,8-tetrachlorodibenzo-p-dioxin, or TCDD. The International Agency for Research on Cancer (IARC), part of the World Health Organization, announced on February 14, 1997, that TCCD is now considered a class 1 carcinogen, meaning a "known human carcinogen."[84] Dioxin slowly breaks down in the environment when it is exposed to the ultraviolet rays of the sun; otherwise, dioxin is a stable compound. Since it is highly insoluble in water, it is not readily able to migrate through the soil; and once in the soil, it persists.[37]

Dioxin is considered to be one of the most toxic man-made chemicals (Figure 5.18). It is not an estrogen, however, but it can block the action of estrogens.[38] Dioxin is capable of lowering androgens and affecting the levels of thyroid hormones. It effects both the endocrine system and the immune system.[29] Interestingly, dioxin does not attach to the estrogen receptor; rather, it attaches to a receptor called the Ah-receptor, whose function is unknown. Even though it does not attach to the estrogen receptor, it still exerts both estrogenic and antiestrogenic activity, indirectly influencing estrogen activity. Its antiestrogenic effects may result from (1) causing the DNA, which is attached via the receptor, to produce an enzyme that is capable of breaking down the body's normal estrogen; or (2) decreasing the number of estrogen receptors available for naturally occurring estrogen.[39]

Although dioxin is considered to be among the most potent synthetic toxic chemicals known, its effects are highly variable among species and even differ among age groups within the same species. Animal studies show wide variations in reproductive and developmental effects from dioxin in different species at different levels of exposures. Adult hamsters are several thousand times more resistant to the toxic effects of dioxin than adult guinea pigs, while there is much less difference among the fetuses.[40]

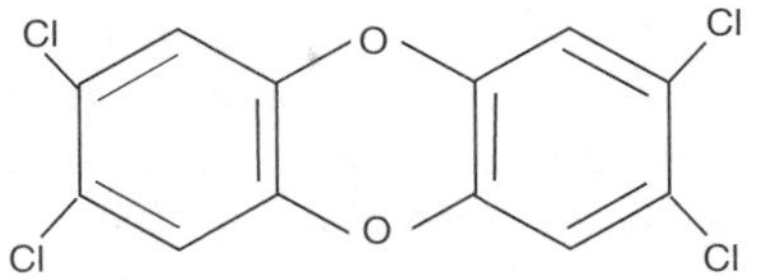

Figure 5.18 Chemical structure of dioxin.

Most of the effects of dioxin are based on animal studies, but there have been unintended or accidental human exposures exceeding normal ambient levels. The largest known dioxin contamination occurred between 1963 and 1971, when 12 million gallons of Agent Orange, a defoliant mixture consisting of 2,4,5,-T and 2,4-D contaminated with dioxin, was sprayed over parts of Vietnam. Anecdotal reports followed by media attention thrust the spectre of Agent Orange into the public eye as a possible cause for veterans' health complaints. Most studies to date show little or no adverse health effects from such exposures. A study by the Centers for Disease Control showed Vietnam veterans have slight losses in sperm concentrations and somewhat more hearing loss than other veterans.[41] The health of the Air Force unit responsible for aerial spraying of the herbicides (Ranch Hands) was compared with units handling other cargo. The two groups were similar in health except that Ranch Hands had slightly more basal cell carcinomas.[42] There were also no differences seen in major, minor, or suspected defects among children of Vietnam and non-Vietnam veterans.[43]

In July 1976, pressure built up in the reaction chamber of the ICMESA chemical company, which was producing trichlorophenol, the parent compound of hexachlorophene and 2,4,5-T. At noon the pressure became too great; up to 4 pounds of dioxin were released into the air, descending onto the town of Seveso, Italy, and the patios where meals were being served. Many of the animals in the town, including birds, rabbits, and chickens, died within a few hours of exposure. People developed chloracne (a disfiguring acne involving the sebaceous glands), nausea, and temporary nerve damage. Since that time, scientists have been monitoring the health of 2000 families in Seveso and have documented increases in cardiovascular disease and questionable increases in certain types of cancers. Research carried out by the same group of scientists has strengthened the evidence that dioxin may be a carcinogen in humans. They have found that people living in the second-most contaminated area, referred to as zone B, were nearly three times more likely to acquire liver cancer than the general population. In the same zone, the occurrence of a form of myeloma was noted 5.3 times more often among women, while among men, cancers of the blood were 5.7 times more likely. The scientists did not find a greater number of cancers in the most heavily polluted area due to the small group of people who were most affected and who moved immediately, thereby shortening the duration of their exposure. Those people in zone B had a lower and more prolonged exposure.[44]

Although these findings were not the first to associate dioxin with cancer in humans, the Seveso study has significance. This population was well monitored because new techniques had made blood levels of dioxin easy to measure, which enabled researchers to more accurately determine exposure levels. The investigators also based their findings on extrapolations from soil data, which corresponded to the analyses of the blood samples.[44]

In the 1970s, a waste hauler sprayed dioxin-contaminated oil on several towns in Missouri and at the Shenandoah Stables near Moscow Hills in Missouri. More than 50 quarterhorses died, along with chickens, dogs, cats, and birds. The owner's young daughter was hospitalized with aching joints, inflammation, headaches, and diarrhea.[45] In 1983, the Environmental Protection Agency announced that the riverside community of Times Beach, Missouri, should abandon its homes and property and evacuate. During the early 1970s, the roads were oiled to control the dust. Soil analyses revealed high levels of dioxin contamination; the oil had been scavenged from a trichlorophenol factory by a waste hauler.[37] Subsequent studies revealed no health problems associated with the former citizens of Times Beach.[37]

The EPA released a draft dioxin reassessment document in 1994 based on an extensive review of dioxin's toxicity. The document asserts that levels of dioxin present in the environment may be very close to those that can produce health consequences to humans. The report further states that dioxin poses increased risk for cancers, adverse reproductive and developmental effects, neurological damage, including both cognitive and behavioral effects from *in utero* exposure, endocrine disruption, and reproductive and development effects.[46]

In a continuing effort to provide a balanced view, it should be reported that: (1) there are no formal studies to date that any person has died from environmental exposure to dioxin in the U.S.; (2) dioxin has been reported to cause chloracne, and short-term reversible nerve dysfunction, and may be toxic to both the liver and the kidneys; (3) dioxin is highly toxic to some species of animals and minimally toxic to others; (4) the evidence that dioxin causes birth defects is inconclusive; and (5) occupational exposures to dioxin may increase the risk of soft-tissue sarcomas.[37]

Polychlorinated Biphenyls (PCBs)

PCBs (polychlorinated biphenyls) are chemically inert, nonflammable fluids with high plasticizing ability and a high dielectric constant (Figure 5.19). They were commercially synthesized in the U.S. from 1929 to 1977 and used in transformers, capacitors, hydraulic and heat transfer fluids, and solvents in adhesives and sealants. They were also used to coat electrical wires, and as a protective coating on lumber, concrete, and

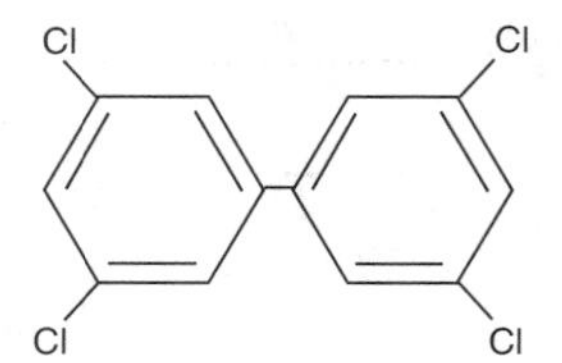

Figure 5.19 Chemical structure of polychlorinated biphenyl (PCBs).

metal surfaces. PCBs were used as an antioxidant in paints, varnishes, and rubber compounds.[46] The PCBs were found not only around us in air, food, soil, water, and in human and wildlife fat tissue, but also in plastics, food wraps, cosmetics, detergents, pesticides and even baby bottles.[1] PCBs are a family of approximately 209 chemical compounds whose characteristics vary considerably. Some of the PCBs are light and oily fluids, while others are heavy, greasy, or waxy. The properties that made PCBs desirable in electric equipment, such as their resistance to thermal degradation and their stability, are the very same properties that have led to their bioaccumulation in the environment.[29] PCBs were banned from most uses in the U.S. in the 1970s due to their persistence in the environment and their ability to bioaccumulate in animal tissue. There are various chemical alternatives to PCBs like silicone fluids, fluorocarbons, high-molecular weight hydrocarbons, low-molecular weight chlorinated hydrocarbons and high boiling oils, and esters, which work well in electrical equipment. PCBs[85] remain widely spread in the environment, including the fat tissue of humans and animals. More than 94% of fish collected in the U.S. show PCB residues at an average concentration of 0.53 ppm.[48]

Sources of PCB Exposure[86]

1. Using old fluoroscent lighting fixtures and old appliances such as television sets and refrigrators; these may leak small amounts of PCBs into the air when they get hot during operation
2. Eating food, including fish, meat, and dairy products containing PCBs
3. Breathing air near hazardous waste sites that contain PCBs
4. Drinking PCB-contaminated well water
5. Repairing or maintaining PCB transformers

While the U.S. stopped producing many of these PCBs, they continue to be made in other parts of the world. These PCBs are binding to dust particles and are transported through the air, depositing on plants that enter the food chain, or they land on water where they enter aquatic life and bioaccumulate.[29]

A few of the PCBs are dioxin-like, exerting similar biological activities. Some PCBs, or their metabolites, are estrogenic, while others can alter the levels of thyroid hormones in the bloodstream.[49] Some PCBs have the ability to bind to proteins in the blood that would otherwise bind to thyroid hormones, interfering with the transport of the thyroid hormone and altering normal growth and metabolism.[39] PCBs may also be altered within the body, so one portion of the PCB molecule resembles a portion of an estrogen molecule, allowing for the PCB molecule to bind to the estrogen receptor. This activity will cause either an estrogen or antiestrogen response.

Researchers are consistently finding delayed psychomotor development of children exposed to PCBs in the uterus. There is evidence of a connection between paternal dioxin exposure to birth defects in offspring. There are worldwide increases in diseases and conditions of the reproductive system in infants, children, and adults that may be linked to environmental chemical exposure.[1]

There have been two major PCB poisoning incidents in which the adverse effects have been clearly observed in children born to women exposed to high levels of

PCBs. The PCBs were also contaminated with dioxin analogues, known as polychlorinated dibenzofurans (PCDFs). In Japan in 1968, a large outbreak of what was thought to be PCB poisoning occurred, when 1300 Japanese developed chloracne, eye discharge, swelling in the joints, weakness, and other symptoms after consuming rice oil. Apparently the rice oil had been contaminated when a PCB-containing heat exchanger leaked fluid from a pipe during processing. This incident was referred to as "Yusho (rice oil) disease."[50] A similar incident occurred in Taiwan in 1979, dubbed the "Yu-Cheng disease."[51]

Children were studied over time after the Yu-Cheng incident. Children at birth were observed to have abnormalities in teeth, nails, and pigmentation. Many of the children exhibited developmental delays in addition to being smaller than average size. The symptoms of a decreased body size and a five-point IQ deficit was observed even after the children continued to mature.[51] Behavioral problems and cognitive deficiencies have also been evidenced in children born to mothers who have eaten PCB-contaminated fish.[51] Studies have also shown that *in utero* exposure to PCBs and related contaminants is associated with poorer short-term memory functions in early childhood.[52]

Although the environmental levels of PCBs have declined over recent years, the risk of exposure from industrial toxic waste continues due to the current use of PCBs in older electric equipment or landfill deposits. The PCBs in such landfills exceed the quantity that has escaped into the environment.[52] PCBs have been banned in this country since 1977, but the persistence and ubiquity of these compounds in the environment has continued to expose children, causing deficiencies in cognitive, behavioral, and motor functions of the brain.[1]

Lead

Lead has been in documented use since the days of Egyptian Pharaohs. The Romans used it for piping water in parts of the viaduct, in cooking vessels, and as a sweetener and preservative in wines. Its dangers in affecting the human mind have been known for more than 2000 years.[53] Despite its known dangers to human health, lead continues in wide use throughout the world and especially in North America, where it has had a history of use in agriculture in the form of lead arsenate for pesticidal use, as solder in pipes, as a solder in food containers, and as an antiknock compound in gasoline. It has also been used in paints, inks, and glazes for ceramics. Lead has been used so extensively, for such a long time, that it has been widely dispersed throughout our environment in substantial quantities and appears in the flesh of every human, nearly all living creatures, and virtually everywhere in the environment, including the most remote locations of Antarctica.[54] Lead has been placed at the forefront of environmental health concerns due to compelling evidence of a wide range of adverse effects. In the 1970s, both federal regulatory and legislative efforts were begun to reduce lead hazards, including the limitation of lead in paint and gasoline (Figure 5.20).[55] Lead was also reduced in evaporated milk and juices; and by 1989, less than 4% of food cans manufactured in the U.S. contained lead solder.[53] Since 1980, there have been intensive actions to further reduce lead

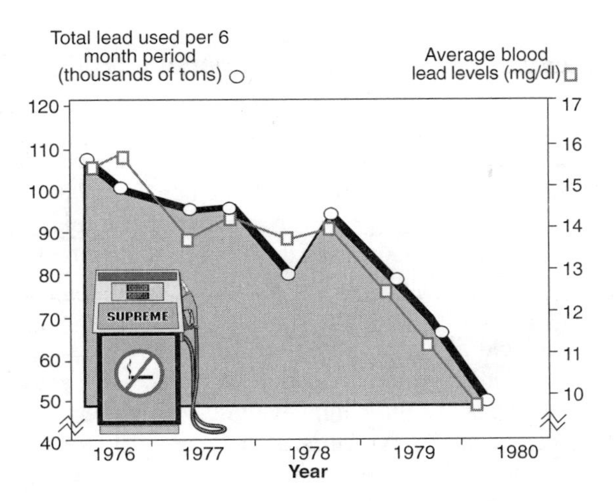

Figure 5.20 Comparison of lead in gasoline and average blood lead levels. (Source from Schwartz, H. et al., *Costs and Benefits of Reducing Lead in Gasoline: Final Regulatory Impact Analysis,* EPA-230-05-006, Washington, D.C., U.S. Government Printing Office, 1985.)

exposure from gasoline, paint, solder, and other sources. The EPA regulations requiring the reduction of leaded gasoline were highly successful in reducing the blood lead levels in people.[56] Secondary prevention programs, such as screening for early detection and lead education programs, have also been implemented.

The most significant sources of exposure to inorganic lead include food, water, soil, and lead-based paint.[53] The most current environmental exposures to lead in the U.S. are from lead paint in older homes, although people are exposed to lead in their drinking water supplies due to the leaching of lead from pipes. Additional sources of exposure may include lead-glazed pottery, medicinal and cosmetic preparations, and food grown in lead-contaminated soil. Since lead solder is no longer used in commercial food canning and due to the phaseout of lead in our gasoline since 1983, lead exposures in the U.S. have been reduced. However, leaded gasoline is currently being used in other parts of the world and will continue to expose pregnant women and children for many more years to come.[54]

Lead may enter the body through ingestion and inhalation, while organic lead may also be absorbed through the skin. Lead does not have any normal function in living organisms, but lead does produce toxic effects. After lead enters the body, it distributes to various organs, crosses the blood–brain barrier, and can cross the placenta. Blood lead levels in the fetus are up to 90% as high as blood lead levels in the mother.[57] Lead is also excreted into breast milk during lactation.

Whether lead enters the body via ingestion or inhalation, it first enters the bloodstream where its half-life is estimated at 36 days. Therefore, measurements of blood lead levels are calculated as micrograms of lead per deciliter of blood (ug/dl), which gives the most accurate indication of short-term lead exposure. While some lead will make its way to soft tissues such as the brain and kidneys, 50 to 60% of the lead that enters the body is excreted relatively quickly. Over time, the lead will be distributed, deposited, and stored in bones, especially the long bones of the arms and legs.[53] Lead can remain in the long bones for years. The lead stored in long bones can be released back into the bloodstream secondary to conditions such as high fever, osteoporosis, or pregnancy, which will result in acute lead poisoning.

Lead may cause numerous toxic effects on the body. Lead affects the formation of blood in two distinct mechanisms: (1) by slowing the normal maturation of red blood cells in the bone marrow, decreasing the number of red blood cells and possibly causing anemia; and (2) lead inhibits the synthesis of hemoglobin by interfering with two important substances that are necessary for the formation of hemoglobin, delta-aminolevulinic acid and coproporphyrin III. In lead poisoning, both of these substances are subsequently excreted in the urine in excessive amounts. The inhibition of the activity of the acid delta-aminolevulinic dehydrogenase in red blood cells (an enzyme involved with the synthesis of hemoglobin), serves as a diagnostic marker in the early stages of lead poisoning before the symptoms are pronounced.[58]

Over the past 10 years, there has been more and more evidence that lead may have serious health effects at lower exposure levels that were previously not thought to be harmful (Figure 5.21). In the U.S., women of childbearing age have an average blood lead level of about 2.0 ug/dl, while the average blood lead level for men of

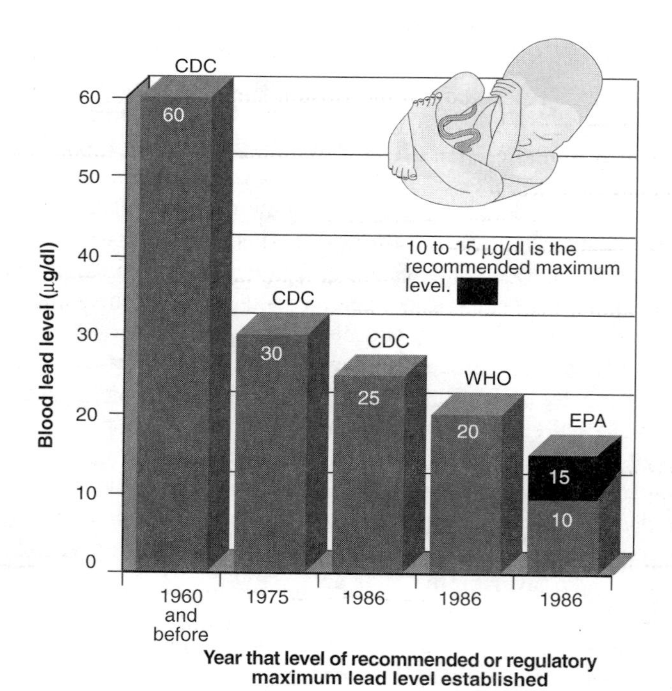

Figure 5.21 History of blood lead levels considered acceptable by various agencies over time. (Adapted from U.S. Congress, Office of Technology Assessment, *Neurotoxicity: Identifying and Controlling Poisons of the Nervous System,* OTA-BA-436, Washington, D.C., U.S. Government Printing Office, Apr. 1990, chap. 10.)

the same age is 4.2 ug/dl.[59] The EPA has listed 10 ug/dl as the maximum acceptable blood lead level for fetuses and young children. The CDC recommends treatment in children found to have blood lead levels over 10 ug/dl.[54] Nearly 22% of black children one to two years old currently have blood lead levels over 10 ug/dl.[59] In the workplace, employers are not required to remove workers from the exposure until their blood lead levels reach 50 ug/dl. This occupational health standard will only protect against acute lead poisoning.[54]

Lead impairs fertility in both men and women when blood lead levels approach 50 ug/dl. Lead may act directly on the testes, lowering the sperm count. At high blood lead levels, lead may cause spontaneous abortions and stillbirths. It is interesting to note that in the past lead was actually used to induce abortions.[60]

The public health threat posed by blood lead levels is worrisome. Young children are at a greater risk for elevated lead levels due to their increased oral activity, increased ability to absorb lead, higher retention of absorbed lead, and the incompletely developed nervous system.[53] Blood lead levels as low as 10 ug/dl, previously thought to be safe, have been associated with developmental delays, deficits in intellectual performance and neurobehavioral functioning, as well as decreased stature and diminished hearing activity.[55]

Over the past 30 years, we have justified the declining levels of blood lead which define a case of lead poisoning. The results of numerous studies carried out over the last several decades continue to show that at levels thought to have been safe, chronic, low-level exposure to lead can inhibit the normal development of a child's intellectual ability. A 1979 study in the Boston suburbs of Chelsea and Somerville was performed on second-grade students. The study concluded that those whose baby teeth had a higher lead level also exhibited a greater incidence of unruly classroom behavior, lower IQ scores, and a diminished capacity to follow instructions than did classmates with lower lead levels in their teeth (Figure 5.22).[61] A follow-up study was carried out in 1990.[62] All students in the 1979 study were now 18 years of age. In comparison to classmates whose baby teeth had minimal lead levels, the students with higher lead levels had a greater school drop-out rate, higher incidence of reading difficulties, and a lower class rank, all in part due to childhood lead exposure. Several studies conclude that lead exposure at levels even below 10 ug/dl is associated with aggressive and delinquent behavior, impaired neuro-behavioral development, and low birth weight.[62,63] Studies on rats confirmed that lead exposure can produce long-term cognitive deficits. Additionally, rats exposed to lead post-weaning did not show impairment in performance despite high body burdens of lead, whereas exposures during gestation and lactation (maternal exposure) resulted in significant impairment. The authors concluded that a window of vulnerability to lead levels exist that results in long-term cognitive impairment.[64]

The prevention of lead poisoning should be one of society's top priorities. Human exposure to lead had been reduced in recent years by the near-phaseout of lead in gasoline, paint, food containers, and home plumbing. Nevertheless, preschoolers are continually exhibiting elevated lead levels, possibly secondary to lead paint and dust still present in their homes. Other children are exposed by family members involved in furniture, refinishing, pottery, or stained glass businesses who unintentionally carry lead dust home with them on their clothes. Since blood lead levels as low as

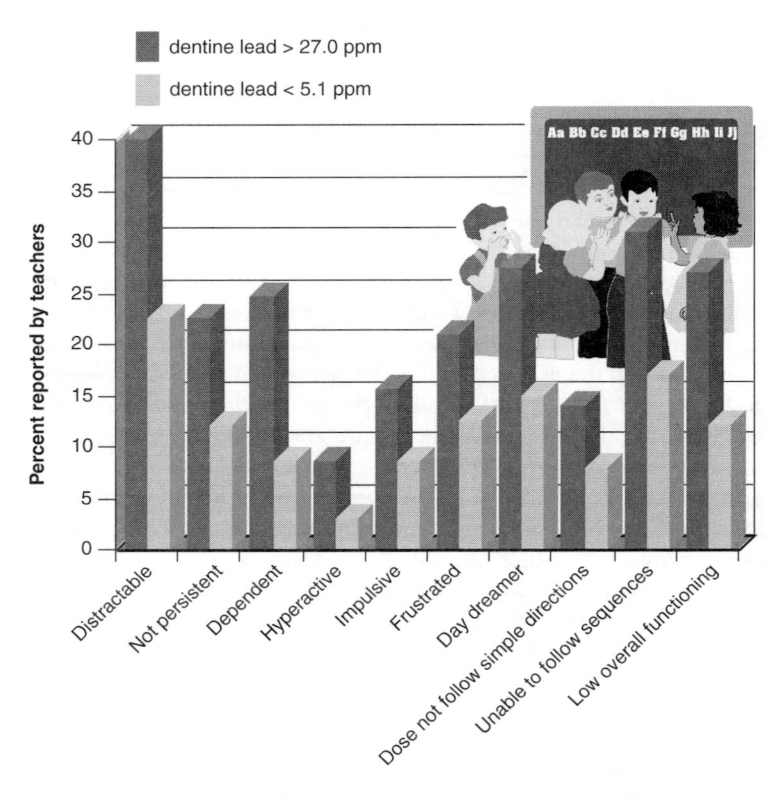

Figure 5.22 The negative effect of classroom behaviors associated with dentine lead levels. (Adapted from Needleman, H.L. et al., The relationship between prenatal exposure to lead and congenital anomalies, *JAMA,* 251, 2956, 1984.)

10 ug/dl are a cause for concern, there is a continuing need to identify and eliminate the source of lead exposure for every child.

Organic Solvents

Organic solvents are simple organic liquids that have the capacity to change from liquids to gases in the presence of air. Organic solvents contaminate our drinking water. Not only can solvents enter the body by ingestion from contaminated drinking water supplies, but they may enter the body via skin absorption and inhalation in the shower. The total exposure from taking a 10-minute shower in contaminated water is actually greater than the exposure after drinking 2 quarts of the same water.[65] Organic solvents are widely used in the U.S., both in the home and industry. Organic solvents are components of many products including paints, varnishes, paint removers, adhesives, glues, degreasing and cleaning agents, pharmaceuticals, plastics, and pesticides. Nearly 50 million tons of industrial organic solvents are manufactured yearly in the U.S.[53]

Individuals may be exposed to organic solvents if they work with electronics, in the health care field, auto repair, dry cleaning, laboratory work, or painting business. Solvents may enter the body through a variety of ways such as inhalation, skin contact, or ingestion. A major route of exposure is inhalation, since the chemicals are volatile. The amount of solvent entering the body strictly depends on factors such as the concentration of the solvent in the air and the route of exposure.[53] Organs such as the brain, due to its blood supply and high-fat content, tend to achieve high levels of solvents more quickly than organs with lower fat contents. All solvents are fat-soluble and will produce effects on the central nervous system.[66]

Researchers have examined the reproductive effects of solvents, in which many of the human studies have found harmful effects. The evidence remains consistent that maternal exposure to solvents during pregnancy increases the risk of spontaneous abortions by up to four-fold.[67] Birth defects, including cleft palate and lip, cardiovascular malformations, and CNS defects, have also been reported in association with organic solvents.[68]

Glycol ethers, in both animal and human studies, conclude that they damage testicular function, cause testicular atrophy, lower sperm counts, and may lead to infertility.[69] Solvents may also lead to acute leukemias in children of exposed parents as well as an increased risk of neurological and urinary cancers.

There is disturbing evidence that organic solvents, most notably the glycol ethers, cause spontaneous abortions, birth defects, and childhood cancers. It is estimated that nearly 10 million workers come in contact with organic solvents on a daily basis.[53] Workers and their families need protection through intensive education, informing workers about the health risks involved and mandatory training sessions to teach employees how to minimize exposure to hazardous substances.

Asbestos

Asbestos is a collective term for a group of six fibrous silicate materials: asmolite, chrysolite, tremolite, actinolite, anthophyllite, and crocidolite, which are found throughout most of the world. Nearly 95% of all asbestos in the U.S. comes from just one of the six types of commercially used asbestos: chrysolite. The "white" asbestos is used in wall insulation, ceilings, around pipes and vents, and for heat resistance in roofing shingles. The chrysolite-based asbestos is also widely used to reduce friction in automotive brake shoe linings. Asbestos has been used as an essential component in various products and processes, including building materials, brake linings, textiles and insulation, as well as floor tiles, cement, and potholders (Figure 5.23). There are approximately 3000 uses of asbestos in various industries.[70] Unfortunately, during 1950 to 1972, asbestos was used in spray insulation in buildings, which poses a significant environmental health hazard as those buildings are being demolished. Asbestos can enter our water supplies by airborne settling, dumping of asbestos waste, and by the dumping of effluent from mining operations. Asbestos may also leach from asbestos-cement pipes in municipal water systems. In cities where asbestos-cement pipes are leached heavily by the water, long fibers of asbestos have been known to clog household appliances.[71] Asbestos is a human carcinogen that can be found in air, water, and food throughout the U.S. Asbestos

Figure 5.23 Asbestos used in insulating pipes.

is naturally released into the environment, but there are also man-made emissions, from mining asbestos ores to the manufacture of asbestos products. Exposure to these fibers may occur throughout the environment, especially in our drinking water supply.

Asbestos, although extremely useful, represents an enormous health risk. It is estimated that of the 8 to 11 million current and retired workers who were exposed to vast amounts of asbestos, between 30 and 50% can be expected to die from asbestos-related cancers.[70] Roughly 5 to 10% of all workers who are employed in asbestos manufacturing or mining will die from mesothelioma (an asbestos-caused cancer of the lining of the lung or abdomen). Asbestos workers who smoke have a 90 times greater increase in the incidence of lung cancer over that of nonsmokers.[72] Their families are also at risk, since asbestos workers presumably carry dust on their clothes and in their hair into their homes.

Exposure to asbestos is primarily achieved through the inhalation of these tiny fibers that are suspended in the air, often getting trapped deep within the lungs. Most fibers are carried through the gastrointestinal tract and excreted within a few days, while the few that remain imbedded deep in the lungs lead to irreversible disease processes including asbestosis, lung cancer, and mesothelioma. Certain fibers such as crocidolite are more likely to produce disease than other forms of asbestos, but it is generally believed that the size of the fiber may be the more important characteristic in disease production than the fiber type. Asbestosis is a slowly developing disease normally occurring over 20 to 30 years, resulting in gradually worsening breathlessness on exertion and the development of a productive cough. The impoverished respiration leads to a barrel chest, bluish discoloration of the skin, and a restricted chest expansion.[70] The lungs become typically scarred with thickening of the lining, which is often replaced by calcium deposits. Mesothelioma is a rare malignant tumor involving the pleura or the abdominal wall and abdominal organs caused only by exposure to asbestos. The disease is a diffuse cancerous tumor that spreads rapidly, is usually fatal, and results in death within one to two years after diagnosis.[70] Cancer can also develop in the digestive tract as a result of ingesting asbestos in food, in beverages, or swallowing the fibers in contaminated saliva.

Asbestos is durable, flexible, strong, inflammable, and an excellent insulator. Therefore, it is of no surprise that it was widely used in construction. Buildings constructed between 1945 and 1978 are likely to contain asbestos material, while pipe insulation manufactured from 1920 to 1972 contained asbestos to prevent heat loss from pipes. In schools, the EPA estimates that 15 million students and 1.4 million teachers and other employees are in buildings that contain asbestos.[73]

Asbestos was determined by the EPA in 1971 to be a hazardous air pollutant as defined by the Clean Air Act of 1970. Although this regulation has helped to control asbestos emissions from manufacturing, roadway surfacing, and demolition projects, the main concern and target of asbestos regulations has been school systems. Congress passed the Asbestos School Hazard Abatement Act (ASHAA) in 1984 to provide financial assistance to schools having significant asbestos problems. The Act was reauthorized in 1990 and required additional training requirements for abatement along with revisions to accreditation plans. President Reagan authorized the Asbestos Hazard Emergency Response Act (AHERA) on October 22, 1986, which required schools to inspect for asbestos, develop asbestos management plans, and implement actions involving repair, encapsulation, or removal of the asbestos. This was all to take place prior to 1988 and be conducted by accredited asbestos removal personnel. Nearly $300 million has been distributed to schools under these various Acts for abatement, training, and to conduct other activities related to asbestos.[74] Despite this enormous cost and effort, the average level of asbestos in schools was less than 1/3000 the level (.0007 fibers per ml) than the federal limit permitted by OSHA in the workplace environment. Work on the removal of asbestos from schools continues today at very significant costs. The EPA estimated that the the cost of removing all asbestos from public and commercial buildings throughout the U.S. may exceed $51 billion.[75]

Mercury

There are three different forms in which mercury is found in the environment: elemental mercury vapor, inorganic mercury compounds, and organic (usually methyl) mercury. The three forms are significantly different from one another. They are each absorbed by the body differently, they are used and manufactured for different purposes, and they have different effects on reproduction and development.[65]

Examples of the uses of mercury are in the production of chlorine, for use in thermometers, batteries, and fluorescent light bulbs. For centuries, mercury was an essential part of many different medicines such as diuretics, antibacterial agents, antiseptics, and laxatives. More recently, these drugs have been substituted and drug-induced signs of mercury toxicity are rare.[88] Mercury is also a constituent of dental amalgams. Concerns regarding mercury stem from its volatile nature, resulting in global dispersal throughout the air and water. To make matters even worse, inorganic mercury may be methylated by microorganisms that are naturally present in the environment to an even more toxic form — methyl mercury — which tends to bioaccumulate in marine life.[76]

In the 1940s and 1950s, mercury became known as the product that caused acrodynia, also known as Pink disease. Manifestations of acrodynia included pain and erythema of the palms and soles, irritablity, insomnia, anorexia, diaphoresis, photophobia, and skin rashes.[88] Debate continues about the lowest levels at which mercury causes human health effects. Studies have found that the health effects of organic mercury are similar in both animal and human populations. When mice are exposed to organic mercury, studies have shown that the cells in the developing brain stop in the middle of cell division.[65] Mercury poisoning usually is misdiagnosed due to the insidious onset, nonspecific signs and symptoms, and lack of knowledge within the medical profession.

Organic mercury is the most toxic form of mercury due to its ease of oral absorption while circulating in the bloodstream, crossing the blood–brain barrier, destroying the very cells which control coordination. Symptoms of poisoning appear one to two months after exposure, showing initial symptoms like numbness of the tongue, lips, and fingertips. Speech gradually becomes slurred, and difficulty swallowing soon follows. The most infamous episode of mass poisoning occurred during the years 1953 and 1961 in the town of Minamata on the Japanese island of Kyushu. A local vinyl chloride manufacturing facility had been dumping inorganic mercury waste into the Minamata Bay, where bacteria in the sediment converted it into organic mercury. The mercury moved through the aquatic food chain, contaminating the fish. The fish showed concentrations of mercury 100,000 times that of the water levels. The population of Minamata relied heavily on a fish diet and vividly recall the day when the cats started acting weirdly and hurling themselves into the ocean to die. Then in 1956, people started showing the symptoms of central nervous system damage including tremors, convulsions, numbness of hands and feet, and losses in speech, hearing and vision.[77] In many cases of mercury poisoning, a pregnant mother will remain relatively free of symptoms while children born to her will demonstrate gross physical abnormalities and severe mental retardation.[76,77] More than 1000 Minamata Bay sufferers have since died. Minamata's population has shrunk since the time of the poisoning from 50,000 to less than 33,000. The chemical company (Chisso Corporation) implicated in this disaster did not stop dumping until 1968.[77] More than 40 years later, the prefectural government declared that the fish in Minamata Bay now meet safety standards and the net that confined the fish to the bay has been removed. However, the social stigma and discrimination for victims of Minamata will likely continue.[77] By 1991, more than 2000 cases of Minamata disease had been confirmed.[76] A similar epidemic occurred again in 1965, but this time in riverside villages along the Agano River, Niigata, Japan.[76]

Elemental mercury is found in dental amalgam fillings, thermometers, and batteries. It is hazardous only when inhaled. Due to its low vapor pressure, it can be inhaled at room temperature. Elemental mercury can easily pass the blood–brain barrier and cross the placenta to the fetus once it is in the body. Individuals can be exposed to mercury in the air from waste incinerators that burn medical waste or batteries, and also from oil and coal burning, since mercury is a contaminant of these fuels.[78] It is far more dangerous to inhale mercury fumes than to swallow inorganic mercury, such as the mercury found in an oral thermometer. If swallowed, the mercury passes through the gastrointestinal tract and is excreted in a few short days.

Inorganic mercury vaporizes at room temperature, and its volatility increases markedly with each degree increase in temperature. This makes it especially hazardous when spilled on floors or surfaces where dirt or grease may exist and combine with the mercury to form tiny drops that expose larger surface areas to air. Vaporized inorganic mercury can enter the body either by inhalation or absorption through the skin following prolonged contact. Symptoms may include damage to kidney tubules, liver damage, interference with coordination, generalized tremors, and possible hallucinations. These symptoms are described among hat makers in 17th century France who soaked animal hides in a solution of mercuric nitrate in order to soften the hairs. This practice allowed for the hat makers' arms and hands to be in contact with a solution of inorganic mercury, and they developed the typical symptoms of inorganic mercury poisoning referred to as "Mad Hatters' Disease" (featured in the children's story *Alice in Wonderland*).

Organic mercury is a developmental toxin implicated in some of the largest epidemics of environmental toxicity in recent history. Some of these outbreaks have occurred in Sweden, the Middle East, and Pakistan, where grain seeds coated with the organic mercury fungicide, Panogen or Granosan M, mistakenly were fed to food animals or were mistakenly used to make bread that was then consumed.[58] Infants studied in these outbreaks exhibited mental retardation, lack of coordination, muscle weakness, seizures, delayed development, and visual loss.[79]

Chronic and intense acute exposure causes cutaneous and neurological symptoms. The classic triad found in chronic toxicity is tremors, gingivitis, and erethism (i.e., a constellation of neuropsychiatric findings that includes insomnia, shyness, memory loss, emotional instability, depression, anorexia, vasomotor disturbance, uncontrolled perspiration, and blushing).

Arsenic

Since the Middle Ages, arsenic has been a popular way of poisoning people. Arsenic is a grey metal that is not very poisonous. Arsenic and related compounds occur naturally in granite and mountains as arsenopyrites or arsenic sulfide (*Arsenikon* is the Latin word for yellow pigment). Other sources include microbiological activity, pesticides, herbicides, fertilizers, treatments of human waste, volcanic eruptions, and industrial sources.[89] Another form is white arsenic or arsenic oxide. This is extremely poisonous, although the symptoms of arsenic poisoning could be confused with those of many other illnesses. It was also very difficult to detect arsenic after death. All in all, it is a handy way to get rid of someone. Politicians, Popes and parents were all victims at different times. Indeed, white arsenic became known as "Inheritance Powder."

Arsenic poisoning happens because arsenic binds very strongly to sulfur groups. Many proteins contain one or more sulfur atoms, and enzymes (also proteins) are

WEB PICK Go to http://www-unix.oit.umass.edu/~envhl565/index.html. Click on "Chapter Web Links" to find Envirofacts Querying Capability under Chapter 5	
Chapter 5	Toxicity and Toxins
Whose Site?	U.S. Environmental Protection Agency
URL	http://www.epa.gov/enviro/html/ef_query.html
What's Here?	Envirofacts provides tools to obtain information on the environment, including chemical releases, hazardous waste, Superfund, risk management plans, toxics release inventory, water discharge permits, preformatted query forms, and mapping applications through enviromapper, http://www.epa.gov/enviro/html/mod/index.html. You can type in search criteria in query forms (e.g., county and state) to retrieve information and obtain a report of environmental information that matches your criteria. The content varies according to the different forms available. The mapping applications generate maps that display environmental information for the entire U.S. You can go to either the mapping applications directly, or you can "Map the Results" of your query through tools such as "enviromapper." Information is available from a collection of databases formed from those facilities that are required to report on releases and other environmental activities. You can retrieve information about hazardous waste, toxic and air releases, Superfund sites, water discharge permits, releases to the air from power plants, universities, steel mills, and factories.

responsible for regulating the body's internal chemistry. An enzyme with an arsenic atom bound to sulfur will not function.

Bangladesh and parts of the Indian state of West Bengal are facing a disaster of astronomic proportions. Beginning in the 1970s, tubewells were sunk in Bangladesh and West Bengal to avoid cholera and other bacterial infections present in above-ground water supplies. However, arsenic has slowly crept in on unsuspecting villagers. Over 6 million people in West Bengal and 30 million Bangladeshis are drinking water containing arsenic over five times the WHO limit of 10 ppb.[90] The lowering of the water table resulted in exposure to air in the zone of aeration. This exposure resulted in the oxidation of arsenic minerals previously present below the water table in the Bengal sediments. The arsenic oxides migrated to the groundwater and were reduced to the poisonous forms in the reducing environments below the water table.

Arsenic poisoning causes a variety of systemic problems like diaphoresis, muscle spasms, nausea, vomitting, abdominal pain, garlic odor to the breath, diarrhea, anuria, dehydration, hypotension, cardiovascular collapse, aplastic anemia, and death.[91]

REFERENCES

1. Schettler, T. et al., *Generations at Risk: How Environmental Toxins May Affect Reproductive Health in Massachusetts,* Greater Boston Physicians for Social Responsibility, and Massachusetts Public Interest Research Group (MASSPIRG) Education Fund, July 1996, chap. 1.
2. Waldbott, G.L., *Health Effects of Environmental Pollutants,* 2nd ed., The CX Mosby Co., St. Louis, 1978, chap. 5.
3. Malachowski, M.J., *Health Effects of Toxic Substances,* Govt. Institute, Inc., Rockville, MD, 1995, chap. 1.
4. Tortora, G.J. and Anagnostakos, N.P., *Principles of Anatomy and Physiology,* 6th ed., Harper and Row Publishers, New York, 1990, chap. 3.
5. Tortora, G.J. and Anagnostakos, N.P., *Principles of Anatomy and Physiology,* 6th ed., Harper and Row Publishers, New York, 1990, chap. 23.
6. Tortora, G.J. and Anagnostakos, N.P., *Principles of Anatomy and Physiology,* 6th ed., Harper and Row Publishers, New York, 1990, chap. 5.
7. Malachowski, M.J., *Health Effects of Toxic Substances,* Govt. Institute, Inc., Rockville, MD, 1995, chap. 2.
8. Tortora, G.J. and Anagnostakos, N.P., *Principles of Anatomy and Physiology,* 6th ed., Harper and Row Publishers, New York, 1990, chap. 24.
9. Malachowski, M.J., *Health Effects of Toxic Substances,* Govt. Institute, Inc., Rockville, MD, 1995, chap. 6.
10. Simon, F.P., Potts, A.M., and Gerard, R.W., Action of cadmium and thiols on tissue enzymes, *Arch. Biochem. Biophys.,* 12, 283, 1947.
11. Tortora, G.J. and Anagnostakos, N.P., *Principles of Anatomy and Physiology,* 6th ed., Harper and Row Publishers, New York, 1990, chap. 22.
12. Tortora, G.J., Funke, B.R., and Case, C.L., *Microbiology: An Introduction,* The Benjamin/Cummings Publishing Company, Inc., Redwood City, CA, 5th ed., 1995, chap. 17.
13. Friebele, E., The attack of asthma, *Environ. Health Perspect.,* 104:1, 22, Jan. 1996.
14. Astrop, P., Kjeldsen, K., and Wanstrup, J., Enhancing influence of carbon monoxide on the development of atheromatosis in cholesterol-fed rabbits, *J. Atheroscler Res.,* 7343, 1967.
15. Schroeder, H.A., Metals in the air, *Environment,* 13, 18, 1971.
16. Amdur, M.O., The physiologic response of guinea pigs to atmospheric pollutants, *Int. J. Air Pollut,* 1, 170, 1969.
17. Labelle, C.W., Long, J.E., and Christifano, E.E., Synergistic effects of aerosols, *Arch. Ind. Health,* 11, 297, 1955.
18. U.S. Department of Health and Human Services, The Nature and Extent of Lead Poisoning in Children in the U.S.: A Report to Congress (Washington, D.C. Public Health Service, Agency for Toxic Substances and Disease Registry, 1988.

19. Stokinger, H.E., Ozone toxicology: a review of research and industrial experience, 1954–1964, *Arch. Environ. Health,* 10, 719, 1965.
20. Perera, F.P., Environment and cancer: who are susceptible?, *Science,* 278, 1068, Nov. 7, 1997.
21. Crapo, L., *Hormones: Messengers of Life,* W.H. Freeman, New York, 1985, chap. 1.
22. Biondo, B., Are common chemicals scrambling your hormones?, *USA Weekend,* Feb., 13, 15, 18, 1998.
23. Fry, D.M., Speich, S.M., and Peard, R.J., Sex ratio skew and breeding patterns of gulls: Demographic and toxicologic considerations, *Stud. Avian Biol.,* 10, 26, 1987.
24. Colburn, T., von Saal, F.S., and Soto, P.M., Developmental effects of endocrine-disrupting chemicals in wildlife and humans, *Env. Health Perspect.,* 101, 378, 1993.
25. Guillette, L.J., Gross, T.S., et al., Developmental abnormalities of the gonad and abnormal sex hormone concentrations in juvenile alligators from contaminated and control lakes in Fla., *Environ. Health Perspect.,* 102, 680, 1994.
26. Colburn, T. and Clement, C., *Advances in Modern Environmental Toxicology, Vol. XXI Chemically Induced Alterations in Serial and Functional Development: The* Wild-lifiefHuman *Connection,* Princeton Scientific Publishing, Princeton, 1992, chap. 1.
27. Ziegler, J., Environmental endocrine disruptors get a global look, *J. Natl. Cancer Inst.,* 88:16, Aug., 20, 1185, 1997.
28. McLachlan, J.A., Functional toxicology: a new approach to detect biologically active xenobiotics, *Envir. Health Perspect.,* 101, 386, 1993.
29. Birnbaum, L., Endocrine effects of prenatal exposure to PCBs, dioxins, and other xenobiotics: implications for policy and future research, *Env. Health Perspect.,* 102:8, 676, 1994.
30. Kimmee, C.A., Approaches to evaluating reproductive hazards and risks, *Environ. Health Perspect.,* 101 (Supp. 2), 137, 1993.
31. Forum, New evidence in sperm counts, *Environ. Health Perspect.,* 104:9, 919, 1996.
32. Jackson, M.B., et al., Cryptochordism: an apparent substantial increase since 1960, *BMJ,* 293, 1401, 1986.
33. Sharpe, R.M., Fisher, J.S., Miller, M.M., et al., Gestational and lactational exposure of rats to xenoestrogens results in decreased testicular size and sperm production, *EHP,* 103:12, 1136, 1995.
34. Warhurst, M., Introduction to hormone disrupting chemicals, 1998, http://easyweb.easynet.uk/-mwarhurst/estrogenic.html.
35. Raloff, J., Hormone mimics get harder to pigeonhole, *Sci. News,* 151, 254, Apr. 26, 1997.
36. Smith, A.H., Infant exposure assessment for breast milk dioxins and fumas derived from waste incineration emissions, *Risk Anal.,* 7:3, 347, 1987.
37. Whelan, E.M., *Toxic Terror,* Prometheus Books, New York, 1992, chap. 9.
38. Safe, S., Astroff, B., Harris, M., et al., 2,3,7,8-Tetrachlorodibenzo-p-dioxin (TCDD) and related compounds as antiestrogens: characterization and mechanism of action, *Pharmacol. Toxicol.,* 69, 400, 1991.
39. McKinney, J.D., and Waller, C.L., PCBs as hormonally active structural analogues, *Environ. Health Perspect.,* 102(3), 290, 1994.
40. Olson, J.R., Holscher, M.A., and Neal, P.L.A., Toxicity of 2,3,7,8-tetrachlorodibenzo-p-dioxin in the Golden Syrian hamster, *Toxicol. Appl. Pharmacol.,* 55, 67, 1980.
41. Centers for Disease Control Vietnam Experience Study, Health Status of Vietnam Veterans, Part 11, Physical Health, *JAMA,* 259, 2708, 1988.
42. Wolfe, W.H. et al., Health status of Air Force veterans occupationally exposed to herbicides in Vietnam, *JAMA,* 264:14, 1824, 1990.

43. Centers for Disease Control Vietnam Experience Study, Health status of Vietnam veterans, Part 111, reproductive outcomes and child health, *JAMA,* 259, 2715, 1988.
44. Holloway, M., Dioxin indictment: a growing body of research links the compound to cancer, *Sci. Am.,* Jan., 25, 1994.
45. Monks, V., The truth about dioxin, *Natl. Wildlife,* Aug./Sep. 4, 1994.
46. Schecter, A. et al., Agent Orange and the Vietnamese: The persistence of elevated dioxin levels in human tissues, *Am. J. Publ. Health,* 85:4, Apr., 516, 1995.
47. Jensen, S., The PCB story, *Ambio,* 1, 123, 1972.
48. Schmitt, C.J. et al., National pesticide monitoring program. Residues of organochlorine chemicals in freshwater fish, 1980–81, *Arch. Environ. Contam. Toxicol.,* 14, 225, 1985.
49. Safe, S.H., Polychlorinated biphenyls (PCBs): environmental impact, biochemical and toxic responses, and implications for risk assessment., *Crit. Rev. Toxicol.,* 24, 1, 1994.
50. Kuratsume, M., Yusho, a poisoning caused by rice oil contaminated with polychlorinated biphenyls, *H.S.H.M.A.S. Health Rep.,* 86, 1083, 1971.
51. Gun, Y.L., Ju, S.H., Chen, Y.C., and Elsu, C.C., Sexual developments and biological findings in Yucheng children, *Chemosphere,* 14, 235, 1993.
52. Jacobson, J.L., Jacobson, S.W., and Humphrey, H.E.B., Effects of *in utero* exposure to polychlorinated biphenyls and related contaminants on cognitive functioning in young children, *J. Pediatr.,* 116, Jan. 1990.
53. U.S. Congress, Office of Technology Assessment, *Neurotoxicity: Identifying and Controlling Poisons of the Nervous System,* OTA-BA-436, Washington, D.C., U.S. Government Printing Office, Apr. 1990, chap. 10.
54. Schettler, T., Solomom, G., Burns, P., and Valenti, *Generations at Risk: How Environmental Toxins May Affect Reproductive Health in Massachusetts,* Greater Boston Physicians for Social Responsibility, and Massachusetts Public Interest Research Group (MASSPIRG) Education Fund, Jul. 1996, chap. 4.
55. Pirkle, J.L., Brody D.J., Gunter, et al., The decline in blood lead levels in the United States: The National Health and Nutrition Surveys (NHANES), *JAMA,* 272, Jul. 1994.
56. Schwartz, H. et al., *Costs and Benefits of Reducing Lead in Gasoline: Final Regulatory Impact Analysis,* EPA-230-05-006, Washington, D.C., U.S. Government Printing Office, 1985.
57. Clarkson, T.W., Nordberg, G.E, and Safar, P.R., Reproductive and developmental toxicity of metals, *Scand. J. Work Env. Health,* 11, 145, 1985.
58. Waldbott, G.L., *Health Effects of Environmental Pollutants,* 2nd ed., The CX Mosby Co., St. Louis, 1978, chap. 12.
59. Brody, D.J. et al., Blood lead levels in the U.S. population: Phase I of the Third National Health and Nutrition Examination Survey (NHANES III), 1988-1991, *JAMA,* 272, 277, 1994.
60. Agency for Toxic Substances and Disease Registry (ATSDR), *Toxicological Profile for Lead,* U.S. Department of Health and Human Services, Apr. 1993.
61. Needleman, H.L. et al., The relationship between prenatal exposure to lead and congenital anomalies, *JAMA,* 251, 2956, 1984.
62. Needleman, H.L. et al., The long-term effects of exposure in low doses of lead in childhood: a 1-year follow-up, *N. Eng. J. Med,* 322, 83, 1990.
63. Needleman, H.L. et al., Bone lead levels and delinquent behavior, *JAMA,* 275, 363, 1996.
64. Kuhhnann, A., MeGlothan, L., and Guilarte, T.R., Developmental lead exposure causes spatial learning deficits in adult rats, *Neurosci. Lett.,* 223, 1010, 1997.

65. Schettler, T. et al., *Generations at Risk: How Environmental Toxins May Affect Reproductive Health in Massachusetts,* Greater Boston Physicians for Social Responsibility, and Massachusetts Public Interest Research Group (MASSPIRG) Education Fund, July 1996, chap. 5.
66. Dick, R.B., Short duration exposures to organic solvents: the relationship between neurobehavioral test results and other indicators, *Neurotoxicol. Teratol.,* 10, 39, 1988.
67. Taskinen, H., Kyyronen, P., et al., Laboratory work and pregnancy outcome, *JOM,* 36, 311, 1994.
68. Holmberg, P.C., Hernberg, S., et al., Oral clefts and organic solvent exposure during pregnancy. *Int. Arch. Occup. Environ. Health,* 50, 371, 1982.
69. Wess, J.A., Reproductive toxicity of ethylene glycol monomethyl ether, and their acetates, *Scand. J. Work Environ. Health Suppl.,* 2, 43, 1992.
70. Waldbott, G.L., *Health Effects of Environmental Pollutants,* 2nd ed., The CX Mosby Co., St. Louis, 1978, chap. 16.
71. Archer, S.R. and Blackwood, T.R, *Status Assessment of Toxic Chemicals: Asbestos,* U.S. Environmental Protection Agency, Monsanto Research Corp., Dayton, OH, Dec. 1979.
72. Selikoff, Q., Churg, J., and Hammond, E.C., Asbestos exposure and neoplasia, *JAMA,* 188, 22, 1964.
73. Koren, H. and Bisese, M., *Handbook of Environmental Health and Safety: Principles and Practices,* Vol. 1, 3rd ed., CRC Lewis Publishers, Boca Raton, 1995, chap. 7.
74. Whelan, E.M., *Toxic Terror,* Prometheus Books, New York, 1992, chap. 8.
75. EPA, Study of Asbestos-Containing Materials in Public Buildings: A Report to Congress, United States Environmental Protection Agency, Washington, D.C., Feb. 1988.
76. Ratcliffe, H.E., Swanson, G.M., and Fischer, L.J., Human exposure to mercury: A critical assessment to the evidence of adverse health effects, *J. Toxicol. Env. Health,* 49, 221, 1996.
77. Pollack, A., Mercury, mostly gone from bay in Japan, still poisons town's life, *Minamata J.,* Aug. 23, 1, 1997.
78. Agency for Toxic Substances and Disease Registry (ATSDR), Draft Toxicological Profile for Mercury. U.S. Department of Health and Human Services, Oct., 1992.
79. Cox, C., Clarkson, T.W., et al., Dose–response analysis of infants prenatally exposed to methyl mercury: an application of a single compartment model to single strand hair analysis, *Env. Res.,* 49, 318, 1989.
80. Sheldon, P., Ph.D., Department of Microbiology and Immunology, University of Leicester, www.micro.msb.le.ac.uk/MBChB/3b.html.
81. Huskey, J.R., www.people.virginia.edu/~rjh9u/imsys.html.
82. Androgens, *Encyclopaedia Britannica.*
83. Teens before their time, *Time,* Health, Oct. 30, 2000, Vol. 156 No. 18.
84. www.iarc.fr.
85. www.asch.org/publications/report/pcupdate2.html.
86. www.astdr.cdc.gov/tfacts17.html.
87. www.som.tulane.edu/ecme/leadhome/poison.html#Health.
88. Diner, B., Department of Emergency Medicine, The Brooklyn Hospital Center, mercury toxicity, www.emedicine.com/emerg/topic813.html.
89. Dey, N. and Clark, G., Department of Chemistry, University of Aberdeen, Arsenic poisoning through drinking water.
90. Phillips, W.M., Department of Geology, Colorado College, Arsenic Poisoning in Bangladesh and West Bengal, www.coloradocollege.edu/dept/gy/faculty/wphillips/JJ1.html.
91. www.crystal.biol.csufresno.edu:8080/projects97/122.html.

CHAPTER 6

The Trouble with Pests

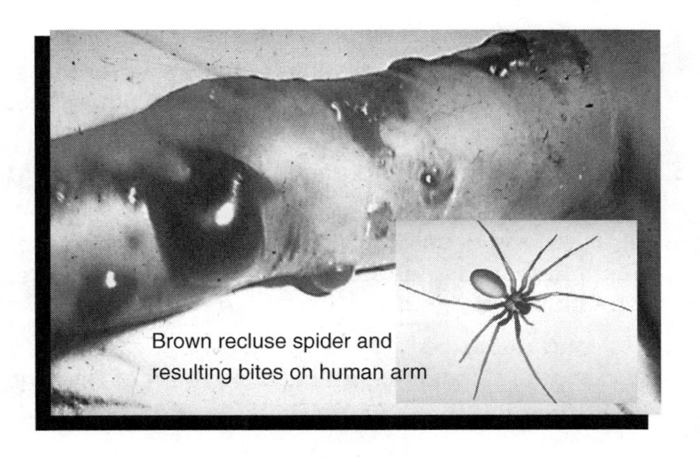

Brown recluse spider and resulting bites on human arm

INTRODUCTION

According to a report prepared by Bechtel, Parsons and Brinckerfoff, co-managers of the massive Central Artery/Third Harbor Tunnel project in Boston, rats were sighted at the rate of 500 per hour shortly after 9:00 p.m. on November 12, 1993, in the open-air market known as Haymarket in the downtown area.[1] Fears were so great that construction would displace the rats into nearby neighborhoods that the construction managers were required to develop plans and countermeasures for exterminating the rats. They were not the first. The blunt-nosed, muscular rats (*Rattus norvegicus*) arrived from Norway as stowaways on boats in the 1600s and numbered in the tens of thousands by the mid-1700s. Nearly 10,000 rats were killed in the last 4 months of 1742 as the town of Boston paid four pence for each rat killed. An all-out effort in 1917 to rid Boston of the more than 700,000 rats failed dismally, and all efforts throughout the following years have failed to eradicate them.[1]

OBJECTIVES FOR THIS CHAPTER

A student reading this chapter will be able to:

1. **Discuss and define the term *pest***
2. **List, differentiate and classify the major arthropod pests to the order level**
3. **Discuss and describe the general structure and development of insects and arthropod pests**
4. **List and describe the major arthropod and mammalian vectors of disease, including kissing bugs, flies, mosquitoes, fleas, sucking lice, ticks, mites, rats, and mice**
5. **Describe and provide specific methods for rodent control**
6. **Discuss and describe the purpose and function of pesticides and the issues of ecological damage, pesticide resistance, and adverse health effects**
7. **List and describe the major classes of arthropod and rodent pesticides, their mechanism of toxicity, and some alternatives to chemical pesticides**

Boston is not alone in these efforts. In one 60-foot deep pit in Bombay, India, there are 442,186 dead rats of the variety *Rattus rattus*. After more than half a century of relief from the plague, the return of the rat in huge numbers has promoted the return of the plague to many of India's cities.[2] In most areas of the world, rats are feared and detested and periodically become the targets of military-style campaigns to destroy them as pests. This is not so everywhere. In Mexico's impoverished northern desert region, rats are hunted as a favorite food, boiled in soup or fried. The government views the selling of rats for meat (about $1 each) as illegal.[3] The prospect of hunger and starvation is a powerful motivation in adjusting attitudes toward what is food and what is a pest.

Also on the comeback trail — after a pause in years of effort to destroy them — are mosquitoes. *Aedes aegypti* has returned in enormous numbers to Latin America after funds for mosquito spraying were diverted elsewhere, human populations exploded, sewage and waste were improperly handled, and sanitation suffered. The result has been a dramatic increase in dengue fever.[4] For similar reasons, increasingly larger epidemics of dengue fever and other vector-borne diseases have returned to India, Sri Lanka, Taiwan, China, and Africa.[4] Despite heroic attacks against the female Anophelene mosquito to control malaria, these programs have also proved ineffective; and once more malaria affects 500 million people worldwide and causes 2 million deaths per year.[5] Deer mice have proliferated in New Mexico and spread hanta virus to the human population in that area, while on the East Coast of the U.S., the deer tick, *Ixodes scapularis*, carries the disease agent for lyme disease in a complicated life cycle involving the tick, deer, and mice. Rats, mice, mosquitoes, and ticks all have something in common. They are largely perceived as undesirable because of their ability to transmit diseases and have been labeled as pests to be targeted for destruction.

WHAT ARE PESTS?

Pests are unwanted plants and animals.[6] Pests are not limited to organisms that transmit disease; they may be any living thing that negatively affects human interests. The interests usually include: (1) a loss of resources such as agricultural crops, food and property damage, and damage to lawns and gardens; (2) agents of disease; and (3) sources of annoyance and discomfort. Pests may be weeds that compete for agricultural space; freshwater mussels that clog intake pipes; rabbits that decimate field crops; moles that burrow through lawns; starlings that threaten collisions with aircraft; wasps that sting; moths that eat through clothes; cockroaches and flies that carry disease organisms onto food; or insects, fungi, and viruses that attack crops or people. Pests cover the entire spectrum of the plant and animal kingdoms ranging from small viruses to large mammals. Although well represented across the animal and plant kingdoms, the actual number of pests is only a small percent of the total members. The vast majority of plants and animals are viewed by most as beneficial, worthwhile, or at least neutral.

The most unwelcome pests from a public health perspective tend to include arthropods and rodents. The arthropods (phylum Arthropoda) are invertebrate animals with jointed and paired appendages, a chitinous exoskeleton, and segmented bodies. This is the largest animal phylum with over 700,000 species including insects, arachnids, myriapods, and crustaceans (Figure 6.1). Within the arthropods are included the insects and arachnids that are medically important as vectors of disease, disease agents themselves, or noxious pests. Certain arthropods and rodents have been associated with numerous disease outbreaks, crop damage, annoyance from bites, and contamination of food sources. Insects, ticks, and mites are involved in the majority of important human vector-borne diseases, and most of these diseases cannot be prevented by vaccines or chemotherapy. Consequently, control of vector-borne diseases often relies upon reducing the number of vectors.[8] Early success in the control of insect vectors has been met with reversals as: (1) insects have developed resistance to insecticides; (2) insect control programs have been halted or underfunded; (3) the use of less expensive pesticides such as DDT have been reduced because of environmental concerns and political pressures; (4) the combination of poverty and overpopulation has led to poor sanitation with greater opportunity for insect proliferation; (5) destruction of forested areas has eliminated natural insect

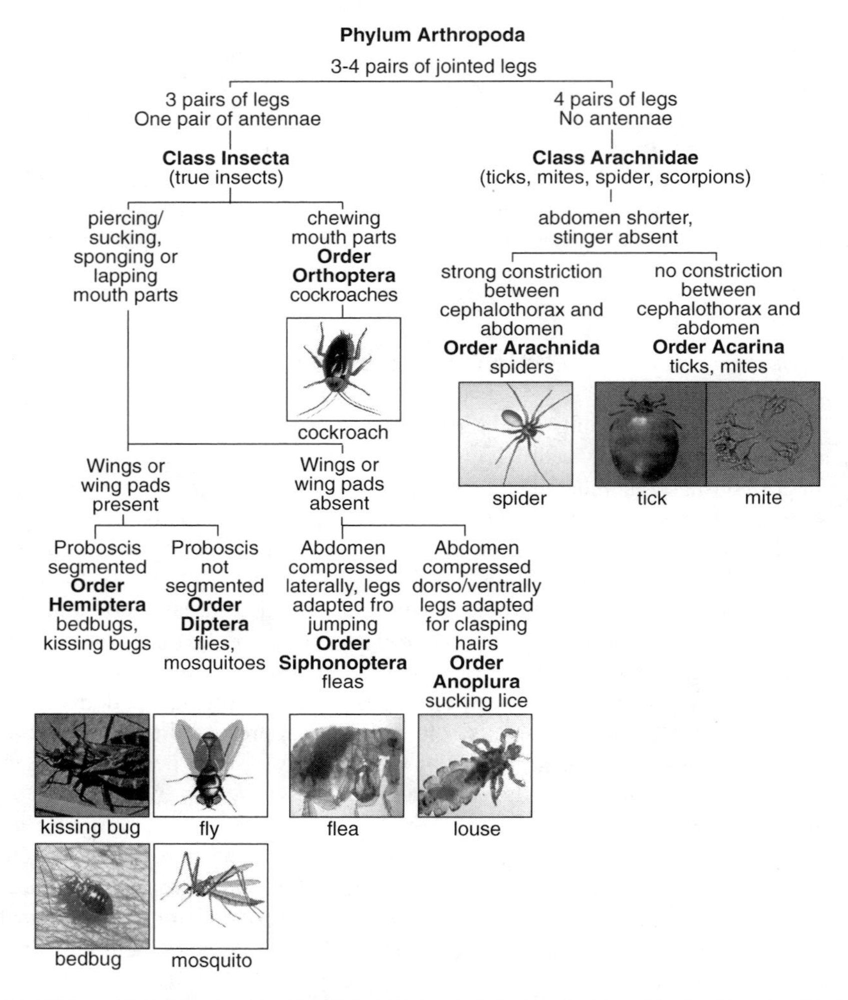

Figure 6.1 Pictorial key to selected orders of arthropods of public health significance. (Adapted from U.S. Dept of Health, Education and Welfare (HEW), *Pictorial Keys to Some Arthropods of Public Health Importance,* Washington, D.C., U.S. Department of HEW Public Health Services, 3, 1964.)

predators; and (6) climate changes, including warming trends, has promoted increases in some insect populations.[4] For these reasons, many insects are rebounding to large numbers. There have been significant increases in vector-borne diseases such as malaria, yellow fever, dengue fever, chagas disease, typhus, relapsing fever, the encephalitides, schistosomiasis, and plague. These diseases affect hundreds of millions of people worldwide and suppress the vitality and economies of many nations. Understanding the role of insects in this process is important to the development of public health policy and programs. In order to develop workable

WEB PICK **Go to http://www-unix.oit.umass.edu/~envhl565/index.html. Click on "Chapter Web Links" to find Insects on the World Wide Web under Chapter 6**	
Chapter 6	The Trouble with Pests
Whose Site?	This web site was developed by Jun Fan, a graduate student in the entomology department at Virginia Tech. The work was supported by Dr. Nicholas D. Stone and Dr. Timothy P. Mack.
URL	http://www.isis.vt.edu/~fanjun/text/Links.html
What's Here?	This site contains over 6000 URLs related to the study of insects and the field of entomology. There are sections on general interest, insect pests (field, forests, pets, poultry, vectors), common insects, (i.e, ants, fleas, flies, mosquitoes, ticks), and topics (anatomy, behavior, biology, ecology, forensic, medical, and taxonomy).

biological, physical, or chemical controls, it is critical to understand the biology and mechanisms for the spread of disease by these organisms.

INSECTS AND OTHER ARTHROPODS

General Structure and Development

Most insects start their life cycles with the fertilization of an egg, and they pass through either a complete or incomplete metamorphosis. Incomplete metamorphosis refers to insects such as roaches, body lice, and grasshoppers, that go through three developmental stages including egg, nymph, and adult stages. Mosquitoes, flies, and butterflies go through complete metamorphosis and pass through four developmental stages including egg, larvae, pupa, and adult stages. As insects grow and develop, they shed their protective exoskeleton while a new one develops underneath in a process called molting. The exoskeleton is a hard outer skin that protects internal organs and serves as an attachment site for muscles to provide movement. The hardened outer body wall of the exoskeleton may be covered with hairs, scales, or spines and is normally divided into segments joined by flexible intersegmental structures. In insects, these segments consist of a head, thorax, and abdomen as seen in Figure 6.2. Members of the order Arachnida have only one or two main segments as in ticks, mites, or spiders. The head usually contains mouth parts, antennae, large compound eyes and/or simple eyes. The numbers of eyes and antennae help to

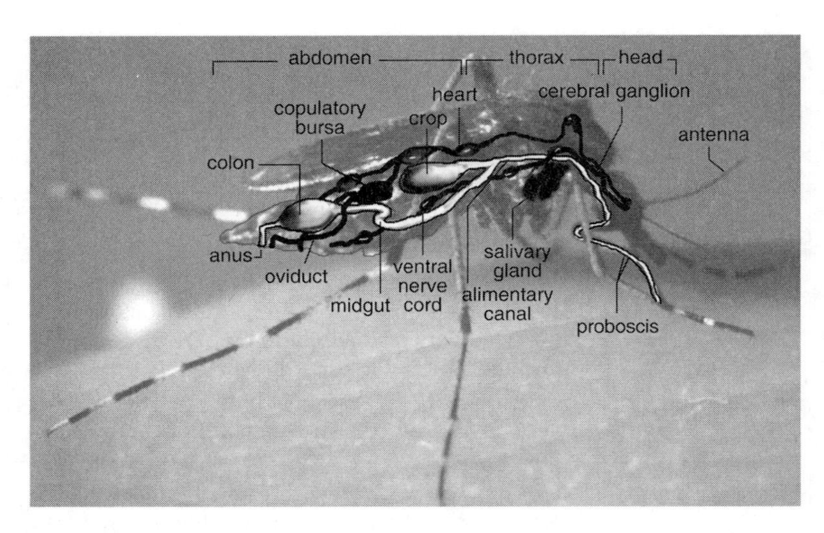

Figure 6.2 Internal and external structures of typical insect. (Adapted from Koren, H. and Bisesi, M., *Handbook of Environmental Health and Safety, Principle and Practices,* 3rd ed., CRC Press Inc./Lewis Publishers, Boca Raton, FL, chap. 4.)

distinguish among species. Mouth parts vary and are adapted to forms of feeding including piercing/sucking (mosquitoes and fleas), chewing and grinding (roaches), and lapping/sponging (houseflies, blowflies). Houseflies and similar sponging insects regurgitate fluids onto solid foods, liquefy them, and suck them up. They can carry disease organisms on body parts or in the digestive tract, depositing them on food or open wounds and so serving as mechanical vectors of disease. Mosquitoes and fleas can pierce the skin of animals and humans to reach blood. In many instances, the microbial disease agents go through a developmental stage in the vector involving sexual reproduction (i.e., malaria). In these instances, the vector is known as a biological vector. In the case of malarial transmission, the infective stage or sporozoites migrate to the salivary glands where they are inoculated into the bloodstream of the next human victim as the mosquito takes a blood meal (see Chapter 7, Figure 7.10). The thorax of insects bears three pairs of jointed legs and may also contain two sets of wings that are often reinforced with structures called veins. The abdomen, or third body part, contains external openings to the respiratory system known as spiracles and also contains external reproductive organs. Internally, the abdomen is composed mostly of reproductive and digestive organs (Figure 6.2).

Bedbugs and Kissing Bugs

The insects belonging to this category are of the order Hemiptera and have three pairs of jointed legs, sucking mouth parts, wings or wing pads, and a segmented proboscis. The kissing bug has well-defined wings, while the wings are absent in the dorso-ventrally flattened bedbug (Figure 6.3). True to their name, bedbugs hide in bedsprings, mattresses, and cracks in the wall, coming out at night to feed on humans and warm-blooded animals. People sensitive to the bites will develop large

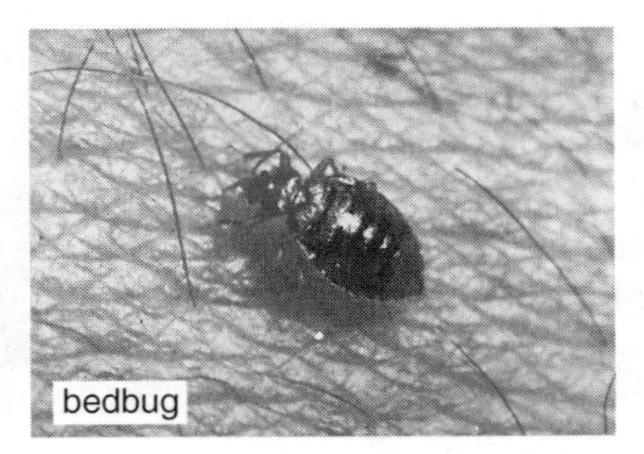

Figure 6.3 Examples of typical kissing bug and bedbug.

and very itchy welts. No disease transmission has been associated with bedbugs. The kissing bug is the arthropod vector for American trypanosomiasis, or Chagas' disease, caused by the flagellated protozoan parasite *Trypanosoma cruzi*. The disease occurs predominantly in Mexico, South America, and Central America, where reservoirs include rodents, armadillos, and opossums. The kissing bug lives in the crevices and cracks of stone or mud huts with thatched roofs. This reduviid bug appears at night while people are sleeping and bites them near the lips (hence "kissing bug"). The Trypanosome is deposited to the skin surface in the bugs' feces. The irritated skin is scratched by the person who inadvertently rubs the organism and feces into the open bite wound. The disease affects more than 50% of the population in some rural South American communities. Once inoculated, the Trypanosomes are carried by the blood to tissues throughout the body where they can multiply and produce severe inflammation. Fatalities occur in about 10% of those infected. These are usually children, and the heart appears to be the primary target organ disrupted in these fatalities. Damage to the central nervous system with paralysis of certain body functions is also common.[9]

Flies

Flies belong to the same order of insects (Diptera) as mosquitoes. Adult flies have one pair of functional wings and three distinct body parts including the head, thorax, and abdomen (Figure 6.4). The middle thoracic section is larger than the

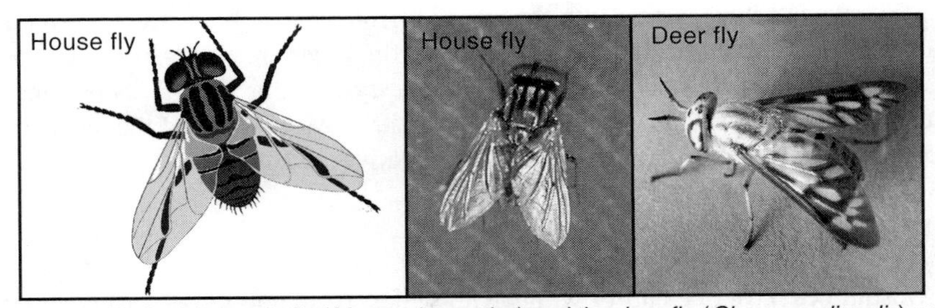

Figure 6.4 Typical domestic fly (*Musca domestica*) and the deer fly (*Chrysops discalis*), which is a biting fly that transmits deerfly fever, a form of tularemia.

others because it supports the highly developed wing muscles. Flies may have sponging mouth parts as does the house fly, while other fly species may have rasping or sucking mouth parts. All have compound eyes and a single pair of antennae. There are thousands of species of flies, and they are significant vectors of human disease and discomfort. Black flies, deer flies, sand flies, horse flies, and stable flies can all inflict painful bites or bleeding punctures. Humans have died from anaphylactic shock produced by the bulk of foreign protein injected into the skin from the simultaneous bites of hundreds of flies. Some species may lay eggs or larvae in wounds, resulting in invasion of the living tissue by fly larvae that produce a condition known as myiasis, where maggots grow in the tissue or in open cavities. The common house fly, *Musca domestica*, is an important vector of infectious disease organisms and foodborne illnesses. House flies can travel several miles in a single day, feeding upon human wastes, and transporting disease organisms on their bodies and in their feces to human food and beverages. Diseases that may be mechanically transmitted by flies include cholera, bacterial and amoebic dysentery, typhoid and paratyphoid fevers, salmonellosis, hookworm, pinworm, and whipworm. Biting flies have been responsible for the transmission of a number of diseases including sandfly fever, onchocerciasis (blinding filariasis), African sleeping sickness, deerfly fever (a form of tularemia), and loaisis (African eyeworm disease).

Flies breed rapidly, with the female laying several batches of 75 to 150 eggs onto garbage, manure, or any moist organic material within one to two weeks after reaching the adult stage. Flies undergo complete metamorphosis, with the eggs hatching in 24 hours to a larval stage that lasts 4 to 7 days. The larvae crawl into the soil or under debris, where they enter the pupa stage for 3 days to several weeks, depending on the temperature. The adult stage emerges from the pupa stage ready to start breeding. Several generations can be produced during the warm seasons.

The control of flies begins with the elimination or covering of breeding materials such as garbage, feces, or dead animals. These materials need to be buried, tightly covered, or incinerated to prevent access by adult flies. Tight-fitting screens of a 16-mesh size or smaller are an effective physical barrier. In heavily trafficked areas, air shields that blow down and away from the entrance have been successfully used to limit access by flies to a room or building. Biological controls have involved the radiation of male screworm flies in Florida and on the island of Curacao. The screworms mate once; and if mating takes place with a sterile male, no progeny develops. This process requires sterilization of large numbers of males by radiation or chemicals and their release into the environment. This is most effective when the area is somewhat isolated, making reinvasion from other areas less likely. Chemical attractants can be used to draw flies onto sticky flystrips, or DDVP resin strips that are useful in killing flies in areas of 1000 ft^3 per strip. As with any of the insecticides, precaution must be used around foods and sensitive persons.[8]

Mosquitoes

Mosquitoes have one pair of wings with scales, belong to the order Diptera (two wings), and the females are characterized by having piercing/sucking mouth parts

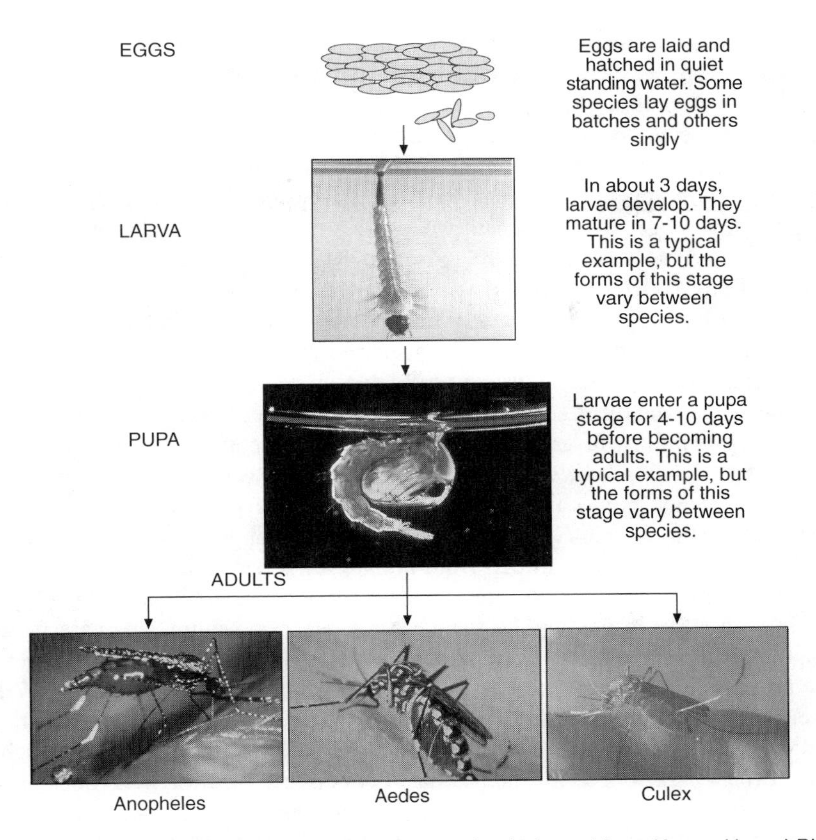

EGGS
Eggs are laid and hatched in quiet standing water. Some species lay eggs in batches and others singly
LARVA
In about 3 days, larvae develop. They mature in 7-10 days. This is a typical example, but the forms of this stage vary between species.
PUPA
Larvae enter a pupa stage for 4-10 days before becoming adults. This is a typical example, but the forms of this stage vary between species.
ADULTS
Anopheles
Aedes
Culex

Figure 6.5 Development stages of a typical mosquito. (Adapted from Koren, H. and Bisesi, M., *Handbook of Environmental Health and Safety, Principle and Practices,* 3rd ed., CRC Press Inc./Lewis Publishers, Boca Raton, FL, chap. 4.

with an elongated proboscis. Mosquitoes undergo complete metamorphosis. Some of the female species may lay several batches of 50 to 200 eggs, while others lay single eggs. Depending on the species, the eggs may be deposited in still water such as swamps, wet depressions, or in the collected water of old tires or containers. The eggs hatch in about three days to a larval stage that may molt three to four times over a 7- to 10-day period. The larvae enter a pupa stage, and in several more days become adult mosquitoes (Figure 6.5). The adult mosquitoes may travel from 100 yards to several miles per day, depending on the species, to feed on preferred hosts that may include humans, cattle, horses, other domestic animals, or a variety of wildlife. Once a host is found, the female mosquito takes a blood meal during which time it injects saliva into the skin. The saliva prevents clotting and also causes the itching which is experienced by most people. The time required for digestion of the blood meal, laying more eggs, and then seeking another blood meal may be as little as two days. Some disease agents require 10 to 14 days of development in the mosquito before they become infectious to humans or animals. A mosquito may therefore breed multiple times before passing on the disease. There are more than 2600 known species of mosquitoes. Many of the species of public health interest belong to the genera Anopheles, Culex, and Aedes. *Aedes aegypti* is an important vector for yellow fever. The human virus circulates in the blood only 3 to 4 days during the first stages of illness; and when the Aedes mosquito takes a blood meal during this time, the virus enters the mosquito. The disease can then be transmitted in 10 to 14 days. *Aedes albopictus* is also known as the Asian Tiger mosquito and is thought to have been inadvertently transported from Japan before it was found breeding in the pooled water of tires transported over Texas highways in 1985.[8,10] This mosquito bites severely, survives cold winters, and is capable of transmitting the agents of dengue fever and dengue hemorrhagic fever (the encephalitides group of viruses), and yellow fever. Although it has not yet become an important agent of disease in the U.S., *Aedes albopictus* has the potential to become a significant public health problem just as it is in Asia.

Species of the female Anophelene mosquito are important vectors of malaria. Malaria is biologically transmitted from human to human in a process that begins when male and female gametocytes of the malarial parasite *Plasmodium* sp. are taken in with a blood meal by the female Anophelene mosquito. The parasite goes through a sexual stage in the developmental cycle within the mosquito that results in the formation of slender sporozoites in a few weeks (See Chapter 7, Figure 7.10). These sporozoites migrate to the salivary glands of the mosquito and enter the host's blood during the mosquito's next blood meal. Here, the parasites pass through an asexual cycle characterized in some forms of the disease by the synchronous growth of the parasites in the red blood cells. The periodic destruction of the red blood cells with the release of parasites into the circulating blood often causes characteristically periodic episodes, including headaches in the prodromal stage followed by chills and then fever. The chills and fever often occur in regularly timed waves of 36 to 72 hours depending on the malarial species involved and results from inflammatory responses of the body to the explosive release of foreign bodies into the bloodstream.

Culex sp., or house mosquitoes, tend to bite fiercely and continuously on humans, birds, horses, and other animals. They can breed in most any standing water, especially when contaminated with organic matter such as waste food or sewage. *Culex* sp. is a night feeder and rests by day. They have the capacity to travel several miles though usually stay within a mile of the breeding area. They have been implicated primarily in the spread of viral encephalitis.

Control of mosquitoes can occur at the larval and adult stages. Larval stages may be controlled by chemicals, biological controls, or mechanical means. The most effective and environmentally friendly control involves the elimination of standing or stagnant water. Old tires, tin cans, plastic bottles, and land depressions where water can stagnate need to be buried or filled. Irrigation ditches should be smoothed and straightened to permit water to flow easily and drain quickly. In bodies of water too large to fill in, top-feeding minnows have been used to control the larvae. The addition of chemicals such as oil or pesticides to surface water has been effective in the past, but environmental concerns and regulations may restrict most applications. The adult mosquitoes can be temporarily or seasonally controlled by area spraying with compounds such as carbaryl or malathion. Most community spraying programs now occur in conjunction with a public health threat such as the potential spread of mosquito-borne viral encephalitis. Within the home, screens are an effective control when in good repair and having at least a 16 × 20 mesh. In many LDCs and in areas where screens are unavailable, mosquito netting is a useful deterrent. In LDCs, a contact poison such as DDT has been used on the netting to increase its effectiveness. Repellents such as DEET (diethyl toluamide) and Indalone offer some short-term protection against mosquito bites but will vary in effectiveness depending on the mosquito density, the species, and many other factors.

Fleas

Fleas are true insects with three pairs of legs and no wings. They are small and in the range of 1 to 8 mm (.04 to 0.3 inches) with a siphon or tube as a mouth part that is used to feed on the blood of warm-blooded animals (Figure 6.6). There are several species of fleas that have the capacity to bite humans, including the dog flea and cat flea (*Ctenocephalides* sp.), the human flea (*Pulex irritans*), and the rat flea (*Xenopsylla cheopis*). The female flea takes a blood meal from its host, mates, and then lays eggs. The eggs drop to the nearest surface such as carpet, concrete floors, or furniture where they undergo complete metamorphosis to an adult stage within 2 to 3 weeks. At the adult stage, the fleas look for a blood meal and will feed on animals or humans, multiplying in large numbers and spreading rapidly through living quarters, livestock pens, and yards. Their ability to take blood meals and move swiftly among animals and humans makes them important vectors of disease including plague (*Yersinia pestis*), murine typhus, tularemia, and even salmonellosis. The rat flea carries the plague bacillus in its intestinal tract, spreading it from rat to rat and from rat to human. The dark blue-black hemorrhages under the skin seen in bubonic plague have caused the disease to be called the "black death," when it ravaged Europe in the fourteenth century and killed more than 25% of the population. India continues to see periodic episodes of plague with nearly 10 million people dying from it in the early 1900s; many died in more recent outbreaks in 1994 in the

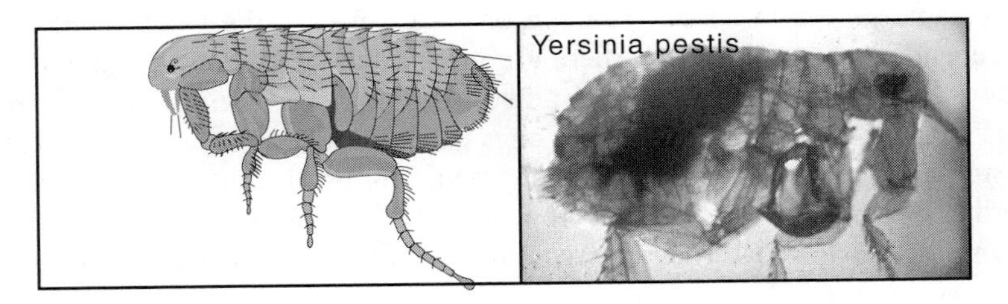

Figure 6.6 Oriental rat flea (*Xenopsylla cheopis*). The dark area in the abdominal area is where the plague bacillus, *Yersinia pestis*, has blocked the intestinal tract.

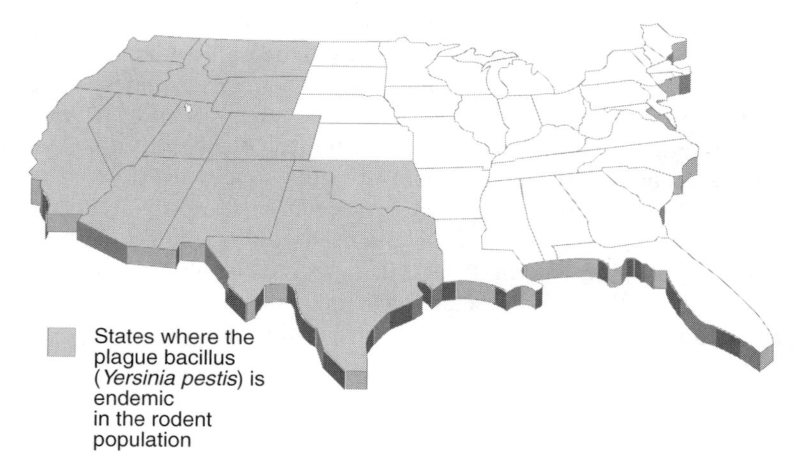
States where the plague bacillus (*Yersinia pestis*) is endemic in the rodent population

Figure 6.7 States in which plague is endemic in the wildlife rodent populations.

Indian city of Surat.[9] The disease is endemic in the wild rodent population in the far West and Southwestern U.S., where it is found in prairie dogs, chipmunks, and squirrels. The disease has continued to spread eastward since its introduction into the wildlife rodent population early in this century; it now has been identified in 13 western states with more than 360 cases of plague identified from these areas over the last 50 years (Figure 6.7).[9,10]

Flea control usually involves treating infested domestic animals and the domiciles in which they live with a pesticidal dust containing either carbaryl or methoxychlor. The mildly toxic pesticides have a prolonged residual time. Powder should be rubbed thoroughly into the animal fur and especially around the head and neck areas.

Lice

Lice that feed on humans or animals are called sucking lice. They have the capacity to bite severely and inject saliva into the bite, resulting in irritation and severe itching to the skin. Lice have no wings, a dorso-ventrally depressed abdomen (flat), and legs adapted for clasping hairs (Figure 6.8). Lice pass through an incomplete metamorphosis to adults. The egg or nit attaches to the head, pubic hair, or skin, depending on the species. There are three kinds of human lice including the head louse, *Pediculis humanis capitis*; the body louse, *Pediculis humanis corporis*; and the crab louse, *Pthirius pubis*. The eggs hatch in the presence of body heat, and the resulting nymph molts three times before becoming an adult within 1 to 4 weeks depending on how long the nymphs are in contact with the body. The lice are easily and rapidly transferred through person-to-person contact, bedding, brushes, and combs.

Louse-borne typhus is caused by *Rickettsia prowazekii* and carried by the human body louse. The organisms grow in the louse intestinal tract and are excreted by it. When the person scratches the itching bites, the feces and bacteria are rubbed into

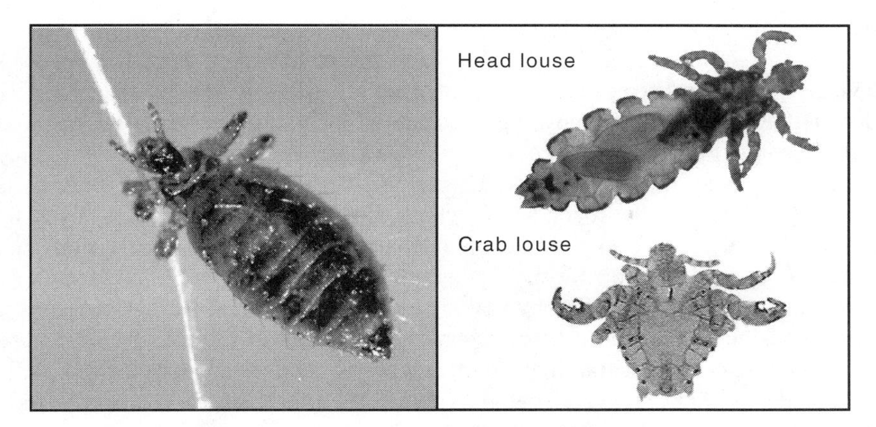

Figure 6.8 Typical louse shown clinging to hair shaft, a head louse (*Pediculis humanis capitis*), and a crab louse (*Pthirius pubis*).

the wound, causing an infection. Inflammation from the infection causes local blockage of small blood vessels and subcutaneous hemorrhaging. Untreated diseases can result in high mortality. The disease is normally seen in very crowded and unsanitary conditions. In addition to spreading epidemic typhus, lice also cause trench fever (*Rickettsia quintana*), and relapsing fever (*spirochetes* belonging to the genus *Borrelia*).

Roaches

Cockroaches are true insects having three pairs of legs, one pair of antennae, and chewing mouth parts (Figure 6.9). Roaches are oval and flattened dorso-ventrally with the head usually not apparent or hidden. There are several common species of roaches including: the American roach, *Periplaneta americana*; the german roach, *Blattella germanica*; and the oriental roach, *Blatta orientales*. The largest of these widely distributed roaches is the American variety which reaches lengths of 2 inches. The adults are dark brown to reddish brown in color. Roaches go through incomplete metamorphosis involving egg, nymph, and adult stages. The eggs are usually encased in a capsule, carried for a few days to a month depending on the species, and then attached to some object out of sight. The nymphs will hatch in 2 to 3 months and pass through many molts over several months before reaching the adult stage. Roaches prefer the dark and are usually concealed during daylight or when bright

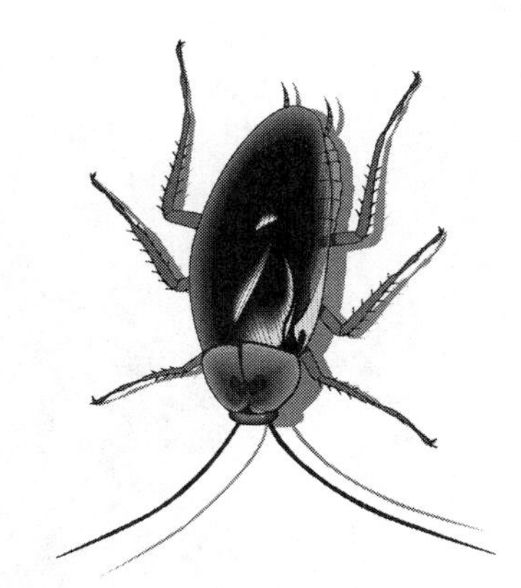

Figure 6.9 Typical American cockroach (*Periplaneta americana*).

lights are shining on an area. The roaches will conceal themselves behind cabinets, in drawers, behind moldings or furniture, and any place where there is a crack or crevice. The appearance of roaches in lighted areas usually means the area is heavily populated. Roaches feed on most organic material including fecal material, decayed food remnants, glue, and any accessible food. During the process of feeding, they regurgitate. Disease-producing microbes may be carried on the external body parts or in the intestines of the roach and mechanically transferred to food. No specific disease outbreaks have been attributed to roaches, but they may serve as vectors and create a level of discomfort for most people.

Roaches are best controlled by making certain no potential food sources exist. Waste foods should be immediately cleaned, and leftovers tightly covered and/or refrigerated. All areas, including the dark reaches of cabinets and closets, should be periodically cleaned to remove potential food stuffs from surfaces. Items from food stores including bags or cartons should be discarded as they may harbor roaches. Borax powder has proven effective in controlling roaches by dusting in areas where roaches are likely to hide. Organophosphate poisons such as malathion have also proven effective for roach control.

Mites and Ticks

Ticks and mites have four pairs of legs, no antennae or wings, and belong to the class Arachnida. A shorter abdomen and the absence of a constriction between the cephalothorax and abdomen place organisms such as ticks and mites in the order Acarina.

Mites

Mites do not have a clearly differentiated head, thorax, and abdomen (Figure 6.10). They are extremely small and may not be easily seen without the aid of magnification. Mites pass through several stages of development that include

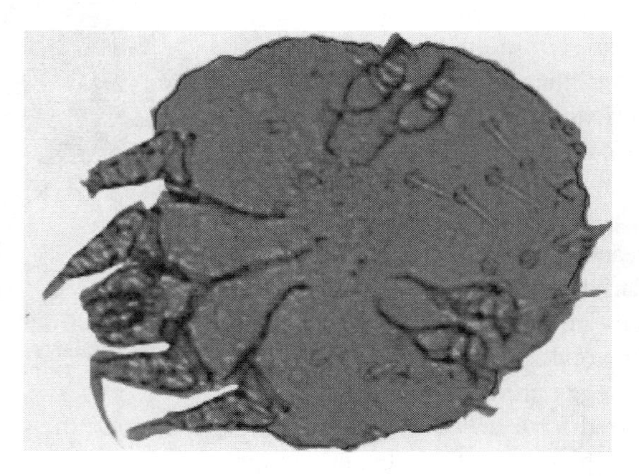

Figure 6.10 Typical scabies mite (*Sarcoptes scabiei*).

eggs, larvae, nymphs, and adult within 2 to 3 weeks, which is about their average length of life. Because mites may proliferate in areas where human body cells are shed and where body oils accumulate (mattresses, couches, beds), they can infest and irritate humans as well as transmit diseases. The mite *Sarcoptes scabiei* burrows under human skin to lay its eggs. This burrowing causes slightly elevated, serpentine lines, or a variety of inflammatory skin lesions resulting from secondary infections as a result of scratching. Intense itching, rashes, and redness often develop. The mites are transmitted through close personal contact with an infected person. Dogs and cats are often infested and develop a condition called mange. The mites frequently infest birds and rodents, which become a major source for infestations of homes and domestic animals. A type of mite called a chigger or redbug is the six-legged larvae of mites. Chiggers attach to the skin and inject a salivary secretion which dissolves the surrounding tissue and is then sucked up the tubular stylostome of the chigger. Scrub typhus is a rickettsial disease passed to rodents and humans by the chigger, causing death in up to half of those infected. The mite-infesting mice has been associated with the transmission of rickettsial pox caused by *Rickettsia akari*. Certain forms of hemorrhagic fever and encephalitis have also been associated with transmission by mites.

Preventing mite infestations begins with eliminating rodent and bird populations in the house or building and keeping everything clean and vacuumed while allowing plenty of sunlight into the living spaces. When mite infestations have occurred, these steps should be preceded by using a pyrethrum bomb, the area cleaned and vacuumed, and malathion applied around windows, doors, and baseboards for residual action.

Ticks

Like mites, ticks belong to the class Arachnida. Consequently, they have four pairs of legs, no antennae, and a fused head, thorax, and abdomen (Figure 6.11).

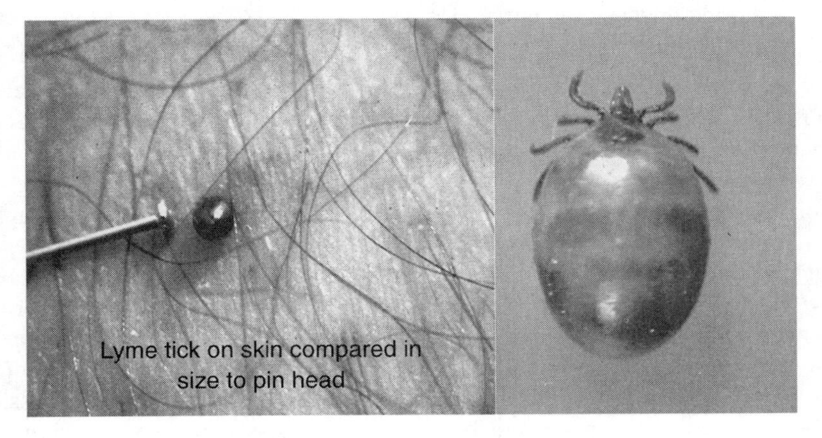

Lyme tick on skin compared in
size to pin head

Figure 6.11 An engorged tick associated with Lyme disease (*Ixodes scapularis*).

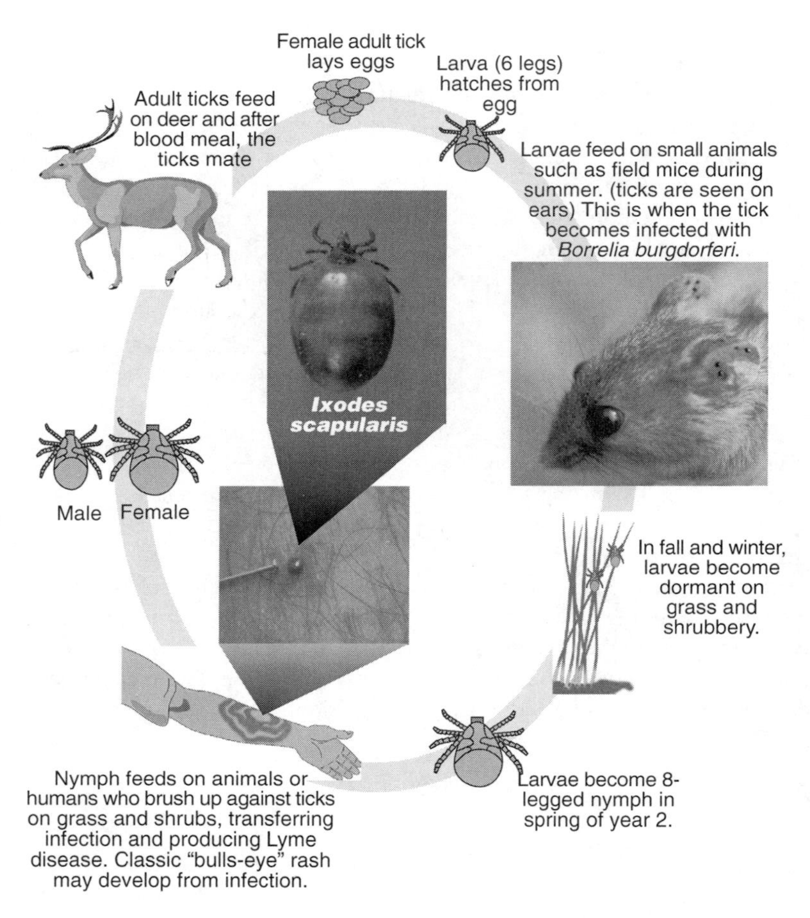

Figure 6.12 The life cycle of *Ixodes scapularis*, the tick vector of Lyme disease. (Adapted from Tortora, G.J., Funke, B.R. and Case, C.L., *Microbiology: An Introduction,* 5th ed., The Benjamin/Cummings Publishing Company, Inc., 1995, chap. 23.)

Ticks are particularly adapted to feed on blood by: (1) the use of barbed feeding organs that pierce the skin to reach the blood and also anchor the tick; (2) a flexible leathery body that easily distends when filled with blood; and (3) uniquely adapted pharyngeal muscles for sucking blood. There are two tick categories including soft (family Argasidae) and hard (family Ixodidae) ticks. Ticks undergo a complete metamorphosis, often involving a fairly complex cycle such as the deer tick, *Ixodes scapularis*, which is the vector of lyme disease caused by the spirochete *Borrelia burgdorferi* (Figure 6.12). The adult female Ixodes tick feeds once on a deer and then mates. She lays several thousand eggs in a batch. The eggs hatch to larvae in a few weeks to several months. The larvae of Ixodes fall to the ground where they come into contact with rodents and other small animals. The larvae (now possibly infected with *Borrelia* sp.) would be deposited on grass and leaves, where they become dormant over the winter. During spring of the second year, the larvae develop into eight-legged nymphs that appear on the tips of grass and brushes at the forest edge, lawns, and gardens, where they can cling to humans or animals that brush by them. The ticks will spread the disease to humans on which they feed.

Ticks have the capacity to transmit diseases mechanically since disease organisms may be carried on tick mouth parts or other body parts and easily infect other persons who are bitten by the tick. Many other organisms pass through developmental stages in the tick by a process known as *transstudial transmission*. In this process, organisms are passed from the tick to the egg and back through the adult tick developing from that egg. Thus, the disease agent may be carried over several generations. Among some of the more important diseases transmitted by ticks are: (1) the rickettsial diseases including tularemia, Q fever, and Rocky Mountain spotted fever; (2) the disease caused by *Borrelia* sp., including lyme disease and relapsing fever; (3) viral diseases such as Colorado tick fever,; and (4) bacillus diseases such as *Pasteurella tularensis*. Ticks also produce a disease known as tick paralysis.

Keeping ticks off of exposed skin is the best control method. Before hiking, biking, or working in high grass or brush, it is good to protect yourself by tucking pants in to the socks, tightly buttoning or taping long-sleeved shirts at the wrist, and wearing clothing that fits tightly to the neck. It is also good practice to wear light, uniformly colored clothing so that ticks can be easily spotted and removed. Spraying clothing and exposed skin with DEET insect repellent offers some protection. Some people have treated clothing with benzyl benzoate or indalone pesticides. Family dogs and cats should be frequently inspected for ticks using a fine-toothed tick comb, especially around the neck area. Tick and flea collars are helpful but do not guarantee against some infestation. When ticks are found on humans or animals, they can be removed by pulling directly out from the body while trying to keep the mouth parts intact to the tick. This process can sometimes be aided by applying nail polish or vaseline to the tick. Malathion dust and DDVP have proven effective in controlling ticks when applied over several months.

Rodents and Pests

Rodent Characteristics

Rats and mice are members of the order Rodentia, which include mammals with teeth and jaws adapted to gnawing. Rodent teeth can grow 5 to 6 inches a year and must be constantly used to keep them short. The constant gnawing by rodents is the source of much destruction. Other members of Rodentia include beaver, squirrels, porcupines, chinchillas, and lemmings. The two major rat species in the U.S. and many parts of the world include the roof rat, *Rattus rattus*, and the Norway rat, *Rattus norvegicus*. The house mouse, *Mus musculus* and deer mouse, *Peromyscus maniculatis*, are also pests in the U.S. The two rats mentioned, and the house mouse are commensals in that they live in or near humans, eat our food, and share diseases without providing any benefits. Because of their close association with humans, they are classified as murine rodents belonging to the subclass Murinae. The roof rat prefers to live in barns, attics, warehouses and other human structures. *Rattus rattus* is highly adapted to balancing along narrow beams and squeezing through cracks and crevices or tiny holes that it can create by gnawing. The roof rat is slender, has

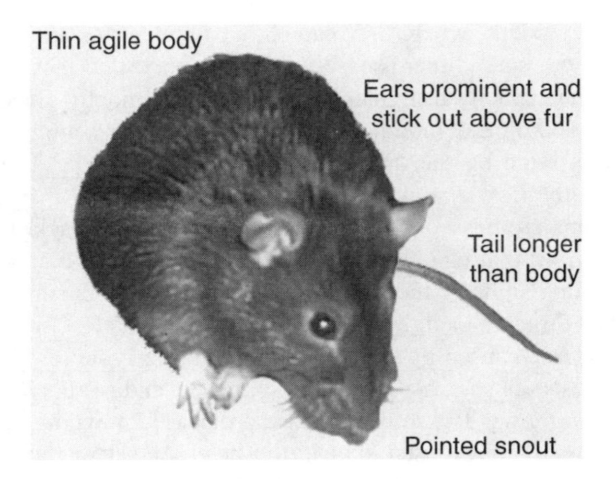

Figure 6.13 The roof rat (*Rattus rattus*).

Figure 6.14 The Norway rat (*Rattus norvegicus*).

a pointed snout, large prominent ears, and a tail longer than its body. It weighs between 8 to 12 ounces and reaches lengths of 8 to 17 inches (including tail) (Figure 6.13). The Norway rat tends to dig burrows under foundations, along banks of rivers, in fields, under solid waste, and under lumber piles. They live in sewers and between floors and walls. These are muscular rats with thick bodies, a blunt snout, and a tail shorter than its body. They reach lengths of 12 to 18 inches (including tail), and weigh up to 16 ounces (Figure 6.14). These features are adapted to burrowing. Rats have several abilities that help them to survive and adapt around humans. Rats can pass through a quarter-sized hole, which is about the size of its head, climb straight up a brick wall provided there are tiny holds for their toes, survive falls up to five stories high, jump over obstacles 3-feet high, cross over and balance on wires, and swim more than half a mile in open water while periodically diving to locate possible escape openings.[1,8]

Figure 6.15 House mouse, *Mus musculus*, and deer mouse, *Peromyscus maniculatis*.

The house mouse thrives everywhere people live. It weighs from ½ to ¾ ounces with a length of 6 to 7 inches (including a tail longer than its body). Its eyes and ears are large and prominent, and the nose is pointed (Figure 6.15a). They have a short gestation period of about 29 days, producing up to eight litters of five to six offspring per litter. They tend to build small nests of shredded paper, fabric, or grass anywhere, including in stored goods, cabinets, furniture, and drawers, while outside they live in weeds, rubbish, and tall grass.

The deer mouse, *Peromyscus maniculatis*, is a vector of the hanta virus (see Chapter 7). It is a thin agile creature weighing up to 1.25 ounces with a tail longer than its body. Its eyes and ears are large and prominent, and the nose is pointed (Figure 6.15b). It has white feet and a white underside with a brownish upperside. It tends to live apart from humans in the forest and at the forest edge or in meadows. It produces two to four litters per year with three to five offspring per litter. The deer mouse lives mostly on nuts, acorns, fruits, and insects.

Importance as Pests and Vectors of Disease

Rats and mice are considered to be pests by most societies because they harbor a variety of organisms that can produce diseases in humans, produce crop damage throughout the world, eat and contaminate food produce, cause structural damage, afflict harm directly by biting humans, and cause extreme discomfort in some people by their visible presence.

Carriers of Disease

Rodents harbor and transmit a variety of disease organisms including bubonic plague (*Yersinia pestis*), rickettsial pox (*Rickettsia akari*), murine typus (*Rickettsia typhi*), rat bite fever (*Streptobacillus moniliformis*), Weil's disease (*Leptospira incetrohaemorrhagiae*), salmonellosis (*Salmonella enteriditis* and other species), and trichinosis (*Trichinella spiralis*). Hantavirus is an emerging disease (Chapter 7) of severe consequences with a mortality in humans exceeding 50%. It is normally transmitted by inhaling virus particles in the dried urine and feces of the deer mouse.

Crop Damage

Rodents produce crop damage throughout the world. In some areas, rats destroy as much as one third of the entire harvest.[11] In the Phillipines, there have been a number of rat population explosions such as those occurring in 1952 and 1984, with rat populations reaching 200 to 2000 rats per hectare and destroying 90% of rice crops and 20 to 50% of corn.

Contaminate Food

Rodents eat and contaminate food and produce structural damage. Rats contaminate ten times more food and materials than they use, damaging from $1 to $10 worth of food for each of the estimated tens of millions of rats in this country alone. Rats and mice also gnaw their way through walls and doors and eat the coating off of wires, causing electrical shorts and fires. Rodents are thought to produce one billion dollars in damage every year in the U.S.[8]

Rat Bites

Rats bite an estimated 45,000 people each year in the U.S., producing disfiguring injuries, infections, and some fatalities. In some reported cases, the rats had to be beaten unconscious to release human victims.[8] Since there are no requirements for reporting rat bites, the actual number of bites may be much higher. The number of rat bites may likely increase since rats thrive in areas of poverty and urban decay, where garbage and trash, poor sanitation, and inadequate rat control measures abound along with increasing population density.

Discomfort

Rats and mice cause extreme discomfort in some people. Persons experiencing rat bites or who have a direct confrontation with rats may be traumatized psychologically by the encounter; some have lasting emotional consequences.[8] Others are simply terrified by the visible presence of rats or mice.

Rodent Control

Establishing Presence

The murine rodents tend to live close to humans and have a very limited range of 100 to 150 feet for rats and 10 to 30 feet for the house mouse. Deer mice range over one to four acres. Rodents tend to be most active at night when there is plentiful

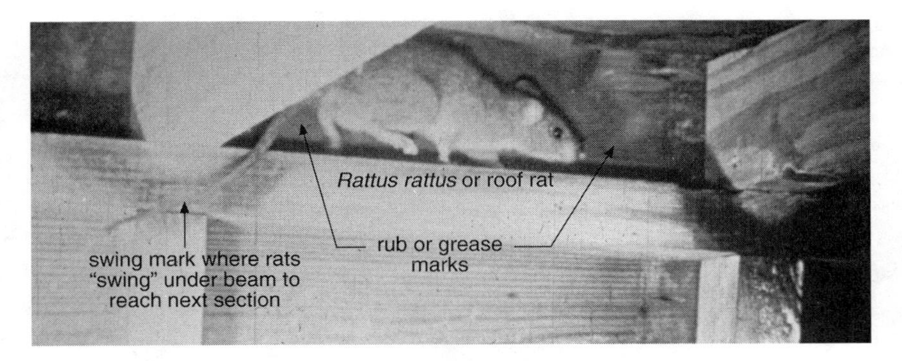

Figure 6.16 Rub marks or grease trails left by rats frequently traveling along the edge of a structure. (Source: Training materials from HEW, *Control of Domestic Rats and Mice, Public Health Service,* Communicable Disease Center, 22, 1964.)

food. If rats and mice are seen during the day, that is usually evidence of many more in the vicinity. Evidence of the rodents is often established by seeing the rodent or its droppings; seeing nesting materials in drawers, cabinets, near foundations, or other places around or in the structure; seeing holes gnawed into walls, doors, or other structures; or seeing grease trails or runways, tracts, swing, or rub marks along the wall edges (Figure 6.16). Rodents prefer to travel familiar and established routes along room edges and will leave their body oils along those edges.

Rodent Proofing

Rodents live in proximity to humans because they have easy access to food. Any effective control program must begin with the removal or securing of food and denial of harborage. Whether a private home, commercial establishment, or an entire community, every effort must be made to store food and food waste in areas or containers inaccessible to rodents. Refuse should be stored in 30-gallon metal containers with tight-fitting lids that are placed on platforms about 12 inches off the ground. The areas should be frequently cleaned and inspected for signs of rodent activity. Dried or canned foods should be stored in clean, dry areas on shelves above the floor or on palettes placed 24 inches from the wall so that all sides can be easily inspected and cleaned. Brush, weeds, and grass should be cut outward from the building to 100 feet, if possible. Any rubbish, debris, brush piles, or old lumber should be removed. The structure should be examined for any holes greater than ¼ inches in diameter. The building can be made resistant to rodent entry by installing metal doors or metal edges around doors; installing metal screens of 19 gauge or smaller; installing metal cones around pipes or wires entering the building; placing sheet metal plates where pipes or wires enter the building; and installing a curtain wall around the foundation made of sheet metal or concrete (Figure 6.17). An L-shaped curtain wall of concrete at least 2 feet deep may be poured along the foundation walls to prevent burrowing. Alternatively, sheet metal can also be used that is 12 inches above the ground and 6 inches below the ground.

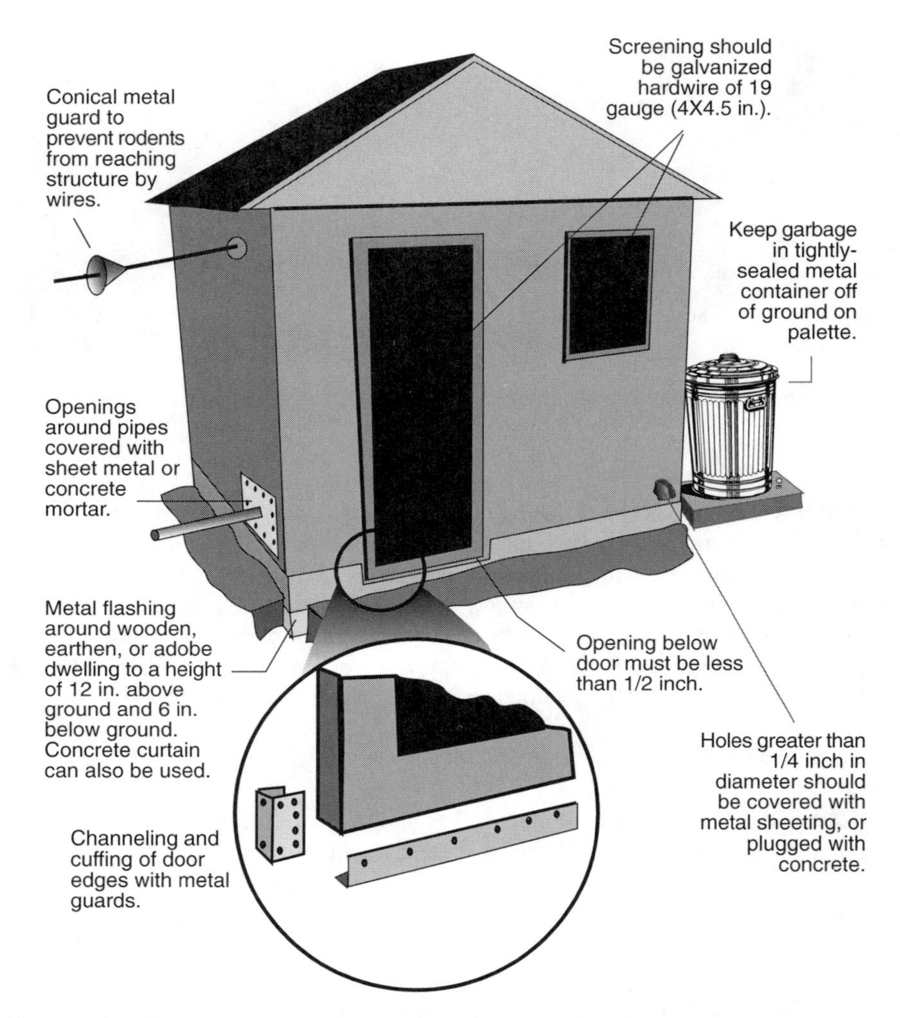

Figure 6.17 Methods of rodent proofing. (Adapted from HEW, *Control of Domestic Rats and Mice, Public Health Service,* Communicable Disease Center, 22, 1964.)

Poisoning or Trapping Rodents

Poisoning or trapping rodents is often necessary to complete the task of control. There are numerous varieties of traps available for the homeowner including some that are disposable plastic shelters where the bait may be inserted at the inside rear, the trap set by pulling a tab at the outside rear, and pushed into place where the rodents frequent. When sprung, the mouse and its container may be discarded without contacting the dead rodent. Traps for rats are less successful because the rats drag them or chew through them or themselves to escape (Figure 6.18). No rodents should be handled with unprotected skin.

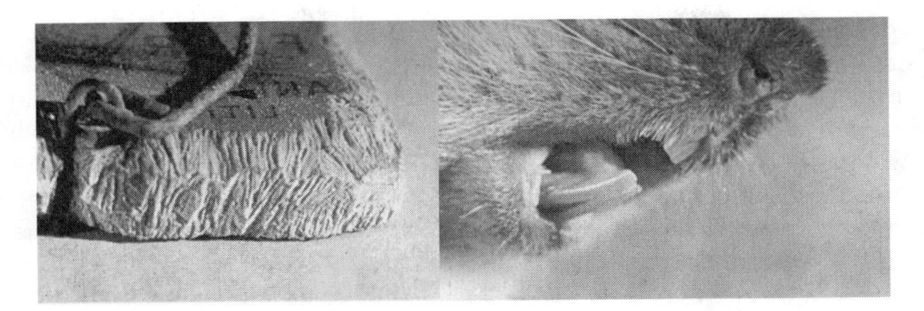

Figure 6.18 Evidence of rat gnawing itself free of trap. Rodents have teeth and jaws adapted to gnawing. (Source: Training materials from HEW, *Control of Domestic Rats and Mice, Public Health Service,* Communicable Disease Center, 22, 1964.)

Rat and mouse poisons available to consumers usually are composed of powerful anticoagulants such as warfarin, fumarin, chlorophacinone, PNP, and diphacinone. The anticoagulants cause internal hemorrhaging, resulting in the death of the rodent over a period of several days. Consequently, the rats or mice seldom associate the baited food with their illness and the colony continues to feed. The baits may be purchased as ready-to-distribute. Further discussion of rodent poisons is given in this chapter under the Pesticides section.

Once the areas of concern have undergone rodent proofing — the removal of food, denial of harborage, and poisoning or trapping — then continued vigilance is necessary. Cleanliness must be maintained and food made unavailable to the rodents. There should be periodic checks for signs of rodent activity.

PESTICIDES

History

Pests compete with humans for food, serve as vectors of disease, destroy crops or depress their market quality, cause structural damage to buildings and homes, and attack people directly, causing annoyance, injury, or even death. Prior to the 1940s, there was very little most people could do but suffer the consequences of pest invasions and cohabitation. In the middle and late 1800s, copper arsenate compounds (Paris Green) and lead arsenate (Bordeaux mixture) were introduced as fungicides and pesticides. These substances remain indefinitely in the environment and are toxic to insects and humans while creating potential harm to the environment through contamination of soils and water.[10] Compounds made from copper, chromium, and zinc were used to control insects, while hydrogen cyanide was used to control red scale on plants. All of these compounds are highly toxic, most are long-lived in the environment, and species have developed resistance against many of them. All arsenicals were banned following the introduction of better alternatives.[8] The advent of the chemical revolution in pesticides began in 1939 with the discovery of the usefulness of dichlorophenyltrichloroethane (DDT) by Paul Muller of the Geigy

Corporation, a chemical that was found to be highly toxic to most insects but not to humans. The use of DDT became critical to the effort of the allied troops in World War II who were combatting malaria, yellow fever, and a host of mosquito-borne, louse-borne, and other vector-borne diseases. Within the next 15 years, more than 25 new pesticides were synthesized, including many chlorinated organics (chlordane, heptachlor, dieldrin, endrin, and iodrin) and the toxic organophosphate, parathion. These compounds were credited with saving millions of lives by suppressing insect vectors of malaria, filariasis, dengue fever, and typhus.[13]

These pesticides were widely used in the U.S. since their introduction after World War II until the 1970s, when it became evident that many of the chlorinated organics produce significant effects on ecology and human health. Consequently, the use of most organochlorine insecticides were canceled for use in the U.S., although DDT and a few other of the organochlorines introduced about the same time remain in use in other parts of the world, such as LDCs where vector-borne diseases remain a major threat and the need for low-cost insecticides is great.[11] Nevertheless, the promise of a world free of vector-borne disease and unmarred crops free of insects and fungi has vanished in the avalanche of unforeseen events over the last 40 years. These problems include the resistance of vectors to pesticides, the adverse health and ecological effects of pesticides, and the proliferation of pesticides globally.

Insect Resistance

The previous 4 decades witnessed a dramatic shift between people and pests and raised our expectations that the battle over insects, rats, and fungi could be won. After all, there was a revolution in chemical warfare against these organisms that promised low toxicity to humans and high toxicity to insects. These chemicals were less toxic than arsenicals and heavy metals while being more effective against a broad array of insects. They were revolutionizing agriculture and freeing humanity of vector-borne diseases. However, almost immediately after their introduction, DDT and other organochlorine pesticides surfaced as potential villains to ecology and human health. Just as important was the realization that many key pests were developing resistance. U.S. production of chemical pesticides rose from 464,000 pounds in 1951 to levels approaching 1.5 billion pounds today without any decline in crops lost to pests.[15,16] From 1970 to 1980, the number of arthropods resistant to insecticides nearly doubled from 224 to 428, while the numbers of resistant species of rodents, bacteria, fungi, and weeds are increasing as well.[15] There are presently more than 500 species of resistant arthropods. Farmers use ten times the amount of pesticides, while crop losses are almost double that when DDT was first introduced.[10] Furthermore, many weed species are becoming resistant to herbicides; it is becoming an international problem since many of the genetically engineered plants associated with the Green Revolution may be easily overgrown by weed grasses. Such overgrowth could reduce crop yields by one third or more.[17]

The types of resistance most often encountered in arthropods are physiological or behavioral. Physiological resistance may occur from enzymatic detoxification of pesticides into a less harmful form; reduced permeability of exoskeleton to pesticides; and storage or excretion of pesticides without harm. Some insects survive pesticide exposure by landing less frequently, changing landing areas, or avoiding baits such as fly paper.[8] Some insects such as mosquitoes and flies have developed cross-resistance and are now resistant to both organochlorines and organophosphates.

A related issue with resistant species is the destruction of competitor organisms. Predators and competitors restrain pest populations. The application of pesticides upsets this ecological relationship by killing off predatory and competitive species that often permit the original target pests to resurge dramatically to even greater numbers when the pesticides are diminished. This process is known as *target pest resurgence*. Very often, the target pest is controlled by the pesticide, but once-harmless species grow numerous and become pests themselves to produce a secondary pest outbreak. Examples of this include eruption of the tobacco budworm on crops in Texas and Mexico, as well as the cottony cushion scale in citrus trees in California.[10,14]

The Health and Ecological Effect of Pesticides

Exposures

Humans are exposed to pesticides through contaminated drinking water, eating foods contaminated with pesticides, pesticidal use in the home, garden or lawn, exposure on transcontinental flights, use in agriculture, the production of pesticides, and other occupations.[18] Nearly 75% of pesticides are used in agriculture, accounting for 66% of the 7 billion dollars spent annually on pesticides.[19] There are four to 5 million Americans working in agriculture, with more than half of these migrant or seasonal workers; and many of these are children under the age of 16.[18] Poisonings from pesticides have been estimated to affect between 500,000 to 2.9 million people worldwide annually,[20] causing fatalities for 1% of the cases or more than 10,000 deaths per year worldwide.[21] There may be 300,000 cases of pesticide-related illness in the U.S., although less than 2% are reported, and most of these are neurotoxic pesticides.[18] There are reports that only 25 deaths occur in the U.S. each year from exposure to agricultural chemicals.[22] The major neurotoxic pesticides are organophosphate insecticides, which make up nearly 40% of all pesticide use in the U.S. and are the most commonly reported source of worker illness. Many organochlorine pesticides used prior to the 1970s are now banned in the U.S.[18] Most acute or one-time exposures include such symptoms as headaches, dizziness, muscular weakness, and fatigue. Poisoning from organophosphates may lead to diarrhea, vomiting, severe abdominal pain, muscular twitching, breathing difficulty, sweating, and, in severe cases, convulsions, coma, and death.[18] Four to nine percent of acutely poisoned persons experience delayed or persistent neurological or psychiatric effects including agitation, insomnia, confusion, and depression.[18] The consequences of chronic or low-level pesticide exposure are uncertain and unresolved. The concerns for the present tend to be directed at the possibilities of cancer and/or reproductive problems associated with organochlorine pesticides. There is concern that pesticides may be depressing the immune response in the general population especially in countries continuing to use pesticides banned in the U.S. Studies in the former Soviet Union found dose-dependent effects of pesticide exposure on the immune function and the development of chronic and infectious diseases, while a Canadian study demonstrated effects of pesticide-laden breast milk on babies' immunologic deficiencies.[23] The major groups of people at risk from the adverse risks of pesticides are children and persons in LDCs.

Children at Risk

Children are exposed to pesticides through prenatal maternal exposure; food and water sources; presence in agricultural fields with parents, or working the fields themselves; contact with residues on parents' clothing; or exposure to common household pesticides while in the home. Based on data from the American Association of Poison Control Centers, 74,000 children a year are involved in poisonings in the home from household pesticides.[24] Nearly half of all households with children under 5 had at least one pesticide stored in an unlocked cabinet less than 4 feet off the ground, with bathrooms and kitchens the most likely areas for improper storage. Common pesticides included roach sprays, bath disinfectants, insect sprays, flea and tick shampoos and dips, chlorine bleach, insect repellents, weed killers, and pool chemicals. Children may be protected from exposure to household pesticides according to the steps outlined in Table 6.1. Children are at greater risk to the toxic effects of these chemicals because they absorb more pesticides per pound of body weight, tend to receive higher doses than adults, have immature detoxification systems compared to adults, and may be more susceptible to neurotoxic effects because the central nervous system is still developing.[18] Food may be another potential source of pesticide exposure to children. The Natural Resources Defense Council (NRDC)

Table 6.1 Recommendations for Preventing Accidental Poisoning (Adapted from U.S. EPA, Office of Pesticide Protection (OPP), *Pesticides and Child Safety: Tips to Protect Children from Poisonings Around the Home, Prevention, Pesticides, and Toxic Substances,* EPA#735-F-93-050R, update Nov. 19, 1997, http://www.epa.gov/oppfead/ childsaf.htrn.

PREVENTING ACCIDENTAL POISONINGS IN CHILDREN
1. Store out of reach in a locked cabinet or shed, use child-proof safety latches
2. Read label and follow instructions for use including precautions and restrictions
3. Before applying pesticides, remove children, toys, and pets until pesticide dries, or according to recommendations on label
4. If use is interrupted (as by phone call), reclose the package and place out of reach of children
5. Never transfer pesticides to other containers
6. Teach children that pesticides are poisonous — something not to touch
7. Keep the number of the Poison Control Center near telephone
8. Alert caregivers to potential hazards of pesticides

released a report in February 1989 that attempted to analyze the extent of childrens' exposure to pesticides and the hazards based on information from the EPA, the Food and Drug Administration (FDA), and the Department of Agriculture (DOA). The report concluded that 17% of preschoolers, or 3 million children in that age group, are exposed to hazardous levels of pesticides in fruits and vegetables exceeding federal safe level standards.[25] The concern about cancer-causing pesticides in our food supply and the risk to children continues to be of public health interest,[26] despite the fact that the FDA found that 99% of 19,962 food samples contained no illegal residues.[27] Prompted by concerns of pesticides in foods, President Clinton signed the Food Quality Protection Act in August 1996. The new law requires special consideration of infants and children when setting standards, consideration of endocrine disruptors, consideration of pesticide benefits, and expanded consumer right-to-know efforts.[28]

Exposures in LDCs

Pesticides are widely misused in LDCs, and that has become a significant problem. It is estimated that less than 25% of the world's production of pesticides is used in these countries, but more than 50% of acute poisonings of pesticide applicators occur in these countries. Seventy-three to ninety-nine percent of worldwide deaths due to acute pesticide poisoning occur in LDCs.[18,29] The World Health Organization has estimated that every minute one person in an LDC is poisoned by pesticides.[30] The reasons for this are numerous, but may include failure to use protective clothing (sometimes workers are clad only in bathing suits) when applying pesticides; workers in the field during pesticide spraying; pesticides used in LDCs that are banned or severely restricted in developed countries; pesticide containers recycled to store food, milk, or cooking oil; and pesticide products labeled in languages not readable to the native populations, many of whom are also illiterate.[18] In an effort to control part of this problem, the United Nations Food and Agriculture Organization (FAO) adopted in 1985, and amended in 1987, a code that included a provision for prior informed consent (PIC). Under PIC, any pesticide restricted or banned in one country cannot be exported to another without its consent. In May 1995, countries agreed to establish this as a legally binding document and targeted 1997 as the year to make PIC a legally binding treaty.[31]

While DDT has not been used in the U.S. since it was banned in 1972, it is still in wide use in LDCs. In 1999, it was used in at least 27 countries. Its main purpose in these countries, however, is not to protect crops but to prevent the spread of illness and death from malaria. Malaria kills 2.7 million people each year, most of whom are children living in LDCs. Normally, environmental protection and efforts to protect the health of populations go hand in hand. In the case of DDT use in LDCs, however, environmentalists and many international public health advocates are at odds.

The United Nations has been attempting to draft a treaty banning the use of DDT (as well as a number of other pesticides) worldwide for several years. A number of prominent medical researchers, however, are fighting to ensure that the treaty will allow for DDT to be sprayed in small quantities on the interior walls of homes, where it acts as a repellent to potentially malaria-carrying mosquitoes. Advocates of the ban (including the World Wildlife Fund and Physicians for Social Responsibility) argue that DDT has been found in the breast milk of mothers living in homes sprayed in this fashion. Unfortunately, many of the malaria victims living in LDCs cannot afford alternatives such as drugs available to treat malaria, and drug resistance is increasing in any case. Other pesticides that are less harmful to the environment are available, such as pyrethroids, but there are two significant problems with their use. First, spraying a house with pyrethroids is three to four times more expensive than DDT, and only a few of the LDCs can afford the cost. Second, a reliance on only pyrethroids may ensure an increase in resistance to the pesticide among mosquitoes. Environmental groups such as the World Wildlife Fund argue that more creative solutions must be pursued, as the short-term benefits of using DDT are not worth the long-term consequences on the environment.[32]

The Counterarguments

The arguments have been advanced that pesticides are designed to be toxic and are made specifically to kill insects, fungi, and other pests. Pesticides in our foods and environment are claimed to cause a variety of toxic effects including reproductive damage, birth defects, interference with the immune and endocrine systems, nervous system damage, and cancer.[23,33] The counterview is that:

1. Trace exposure to pesticides shows no evidence of increasing the risk of cancer, birth defects, or other human ailments.[22]
2. No documented cases of deaths or illness from pesticide use in the U.S. has occurred when used in standard and acceptable ways.[22]
3. Many of the banned organochlorine pesticides have shown carcinogenicity only in mice.[22]
4. Pesticides help prevent grains and seeds from being contaminated with fungi that produce carcinogenic aflatoxins, while also providing an abundance of low-cost attractive fruits and vegetables, which are humanity's best defense against cancer and heart disease.[33,34]
5. Naturally occurring carcinogens may cause an estimated 38,000 of the nation's 500,000 annual cancer deaths, while pesticidal residues are thought to contribute to less than 40 cancer deaths per year.[33]

Ecological Concerns

Much more than 1.1 billion pounds of pesticides are used in the U.S. each year with more than five times that amount used globally. The greatest potential for the unwelcome effects on health and ecology comes from the contamination of surface and groundwater. Water is one of the major pathways by which pesticides are transported long distances to other parts of the environment, moving from streams to rivers to lakes and oceans.[16] Nearly 50% of the nation's drinking water is provided by surface supplies, while surface water also provides vital aquatic ecosystems for fish, birds, plants, and other wildlife. Pesticides enter the waterways from streams through agricultural runoff, from wastewater discharges, groundwater inflow, atmospheric deposition, and spills. Pesticide residues, especially of the persistent organochlorines, are now detected in the tissues of animals in virtually every location on Earth from the Antarctic to the Everglades, small New England streams, and the deepest ocean trenches. Concern is so great that the U.S. Geological Survey's National Water Quality Assessment (NAWQA) Program was designed to provide information about the extent and consequences of such pollution. The program, fully implemented in 1997, is designed to provide consistent and reliable information on water resources in 60 important river basins and aquifers across the nation. Early findings on pesticides in the atmosphere show "the pesticides were ubiquitous and generally detected wherever they were sought."[16] Further, a review of studies indicated that most water treatment plants are ineffective in removing the pesticides and that drinking water derived from some surface water sources contains pesticides above federal guidelines for at least part of the year.[16] Additionally, both the EPA and the National Academy of Sciences and Engineering (NAS/NAE) have developed or recommended water quality criteria for a total of 43 pesticides in order to protect aquatic life from short- and long-term exposure. One or more of these pesticides have exceeded aquatic life criteria in the river basins tested. There are few such criteria for herbicides and none for fungicides.[16] The significance of pesticides in surface waters has not been assessed because little is known about most pesticide toxicity at low level exposures, nor the possible cumulative effects of multiple pesticide exposures.[16]

The consequences of surface water contamination are usually evident in their effects on wildlife. At low concentrations, the persistent, bioaccumulative, organohalogen pesticides have disrupted the endocrine systems of birds, fish, turtles, and mammals, causing decreased fertility, metabolic and behavioral abnormalities, feminization of male species, and masculinization of female species. The effects have been manifested in the offspring because of exposures at critical times during embryonic development.[35]

Acute exposures to larger concentrations of pesticides can produce immediate and dramatic effects, including massive fish kills in streams where pesticide runoff or spills have occurred, deaths of geese browsing on pesticide-treated golf courses, or a variety of more than 2 million birds dying from carbamate pesticides contaminating water and plants.[36] Further, the destruction of certain insects or plants may have negative effects when they serve as a food source or habitat for other organisms. This is especially evident in birds that feed on insects killed by pesticides or use particular plants for nesting that have been eliminated by herbicides.

Groundwater can also be contaminated. Extracting water by pumping it from depths within the ground is no guarantee of chemical or biological purity. Thousands of private wells, and those serving entire communities, have been contaminated with pesticides and/or organic solvents. The EPA conducted the National Pesticide Survey in all 50 states between 1988 and 1990. The preliminary results revealed 10% of the nation's public well supplies had detectable levels of at least one pesticide.

Federal guidelines were exceeded in 1% of the community wells and 0.8% of the private wells.[8] In 1992, the EPA reported that more than 30% of private (rural) wells exhibited unacceptable levels of pesticide contamination resulting mostly from agricultural use. The complex and usually slow movement of groundwater plumes, combined with the highly variable application of pesticide amounts and types, makes it difficult to assess the full extent of the problem. The increasing use of pesticides in agriculture, garden, and lawn use may combine with the already contaminated plumes to cause more extensive contamination in the future. The Food Quality Protection Act of 1996 established a new safety standard for pesticide residue limits in food, and the EPA is required to periodically reassess these 9721 limits (tolerances). By August 3, 1999, as required, the EPA had completed 3290 of these reassessments, two thirds of which fell into the category of pesticides most threatening to humans. These reassessments included lower tolerance levels and restricted use for some of the pesticides that represent the greatest risk to humans, including two of the most toxic organophosphates, methyl parathion and azinphos methyl. Still, certain environmental advocacy groups have felt that the EPA's new regulations fall short of the EPA's mandate to protect the public safety.[37]

The Proliferation of Pesticides Globally

More than 300 tons of DDT were shipped to Peru in 1992, 20 years after it was banned in the U.S. The U.S. shipped nearly 12,000 tons of other restricted or banned pesticides in the same year. Since then, the export of pesticides banned for use in the U.S. have escalated by nearly 50%. According to studies performed by the Foundation for Advancement in Science and Education (FASE), about 250,000 tons of pesticides are exported from the U.S. annually, only a quarter of which are reported.[38] Nearly 5000 tons of the exported pesticides were never approved for use in this country, and more than 11,000 tons were severely restricted for use. Many of the LDCs still continue to import harmful pesticides because they are inexpensive and effective. In Kenya, small farms have increased output significantly in the last few years, making it the third largest source of income from foreign sources. The consequence has been increasing reliance on pesticides banned or restricted in the U.S.[31] The continued use of DDT in Mexico is evidenced in increased levels of related compounds (DDE) in breast milk and tissues, indicating to the authors of the research article that DDT use in Mexico is a public health problem.[39]

The banned insecticides exported to LDCs are thought to be evaporating or volatilized to the atmosphere in the warmer climates and 'hopping' in stages about the globe to cooler climates such as the arctic through the global distillation effect.[40] Many of the chemicals banned for decades in the U.S. are emitted from the hotter climates in LDCs, carried to colder climates in the atmosphere, and condensing where it shows up in the bark of shrub trees near the Arctic Circle. The more volatile the chemical, the faster it travels these distances to eventually condense in colder climates. It is thought to be the best explanation for how large quantities of volatile pesticides and other chemicals find their way into the flesh of polar bears and the breast milk of Inuit natives in the arctic.[6]

TYPES OF PESTICIDES

Insecticides

Insecticides generally produce their toxic effects on insects by poisoning the nervous system. Well-designed insecticides are harmless to humans, highly toxic to insects, easily dispensed and inexpensive, nonhazardous, and rapidly degradable to nontoxic forms. They often act as contact poisons by penetrating the foot pads or body wall, enter the insect breathing pores as a fumigant, act as a stomach poison after ingestion, or desiccate (dryout) the body wall, causing it to crack or break.[6] Insecticides are also capable of exerting their toxic effects on humans, and the major difference between its use as an insecticide or human poison is the dose. There are several types of insecticides that include organochlorines, organophosphates, carbamates and a variety of botanical and biologic insecticides.

Organochlorines

The chlorinated hydrocarbons attack the central nervous system, sending insects into hyperactivity, tremors, convulsions, and death. The organochlorines may be categorized into three subgroups including dichlorodiphenylethanes (DDT), hexachlorocyclohexanes, and chlorinated cyclodienes.

Dichlorophenylethanes

This subgroup of pesticides consists of DDT (Figure 6.19), methoxychlor, and dicofol. DDT is more acutely toxic and more slowly cleared from the body than methoxychlor or dicofol and is also more environmentally persistent. Consequently, less toxic and persistent methoxychlor has been used in many household insect sprays. DDT appears to disrupt sodium and potassium permeability in the nerve axon.[14] The storage of DDT in adipose (fat) tissue, biomagnification in the food chain, and disruption of calcium metabolism in fish-eating birds causing thinning egg shells and declines in bird populations, led to the banning of DDT for use in the U.S. in the 1970s. A closely-related pesticide, DDD, is also banned for use in the U.S.

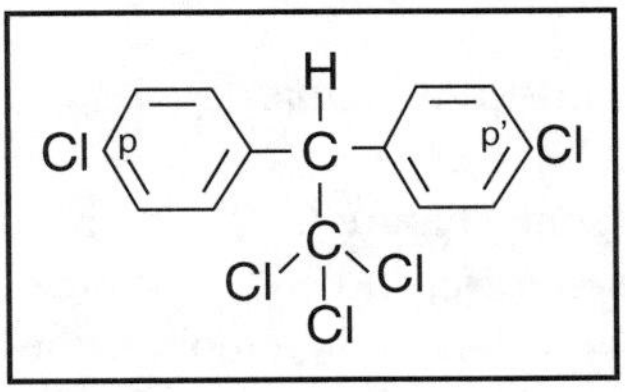

Figure 6.19 Chemical structure of p,p′-DDT.

Hexachlorocyclohexanes

There are many isomers for benzene hexachloride (HCB) with the g- isomer known as lindane the most toxic. Lindane (Figure 6.20) is used primarily for treating seeds in the U.S. and is effective in the treatment of ectoparasites such as fleas, lice, and ticks. As such, it is available in a powder, shampoo, lotion, or cream and is

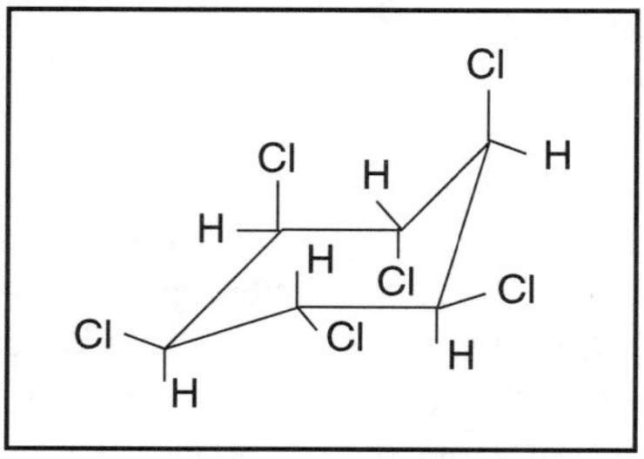

Figure 6.20 Chemical structure of lindane.

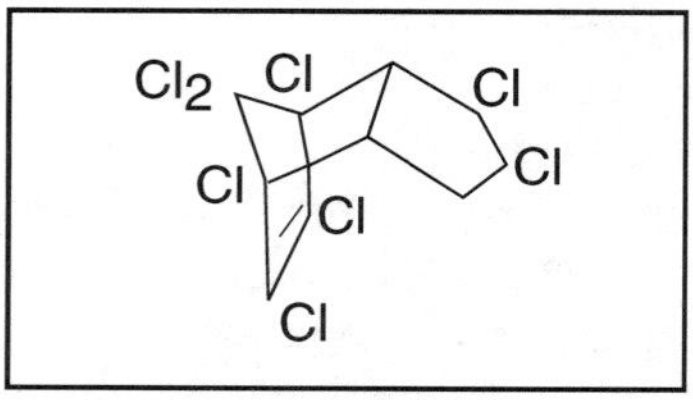

Figure 6.21 Chemical structure of chlordane.

often used in the treatment of scabies as well as the treatment of fleas on household pets. Part of the toxicity of lindane is based on its ability to inhibit g-aminobutyric acid (GABA)-dependent chloride flux into the neuron.[14]

Chlorinated Cyclodienes

Chlordane (Figure 6.21), aldrin, dieldrin, endosulfan, and heptachlor belong to this class of compounds that are characteristically more toxic than DDT and its related compounds. Because of their toxicity, rapid absorption through the skin, and animal carcinogenicity, the EPA suspended the use of most chlorinated cyclodienes in the U.S. in the mid-1970s. These substances are effective as stomach poisons, fumigants, and contact poisons and have been used effectively against roaches, silverfish, ants, and termites.[14,36]

Chlorinated Cyclodiene

Kepone (Figure 6.22) and mirex cause liver tumors in mice, and both pesticides were banned for use in the U.S. although they were never widely used in the nation. Kepone has caused severe neurological damage among production workers in Virginia and contamination of the nearby waters including the Chesapeake Bay.[41]

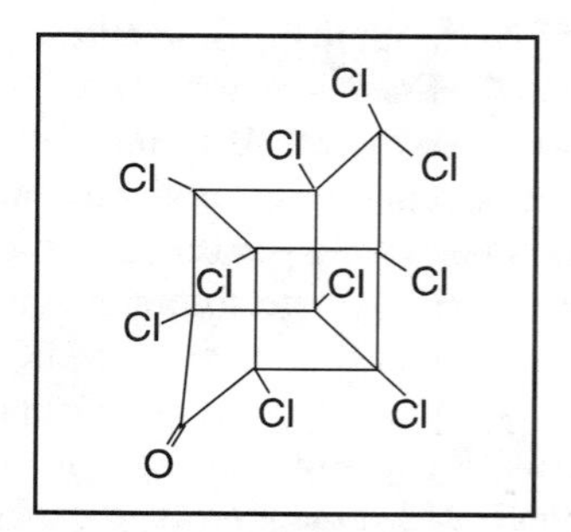

Figure 6.22 Chemical structure of

Organophosphates

Organophosphates are chemical derivatives of phosphoric acid developed in Germany during World War II as a poison for the control of aphids. Many of the organophosphates are short-lived in the environment because they are easily biodegraded and are effective against pests resistant to the organochlorine pesticides.[36] These compounds vary dramatically in their toxicity. Parathion is the pesticide most likely to be involved in human fatalities, although used extensively in agriculture (Figure 6.23). A single drop to the eye may be sufficient to kill a person, so only licensed pesticide operators should use the more toxic organophosphate pesticides. In an ongoing search to find organophosphates less toxic to humans, more than 100 such compounds have been developed.[42] These compounds vary in their toxicity, with the most toxic being phosdrin, parathion, dichlorvos, and TEPP. Moderately toxic organophosphates include DDVP, dimethoate, baytex, and diazinon. The least toxic of the organophosphates include ronnel, gardona, abate, and malathion.[36] Toxicity of these compounds is based on cholinesterase inhibition at the neuromuscular junction of nerve fibers. Acetylcholine is a neurotransmitter compound that accomplishes the transmission of a nerve impulse from the nerve ending to smooth muscle, the heart, and endocrine glands (Figure 6.24). The motor neuron stimulates muscle tissue by releasing the neurotransmitter acetylcholine from synaptic vesicles in the synaptic end bulbs at neuromuscular junctions occurring between the motor neuron and muscle fibers. A nerve impulse causes acetylcholine to be released that diffuses across the synaptic cleft and combines with receptor sites on the muscle fiber. This causes changes in the flow of sodium and potassium ions across the sarcolemma, resulting in a muscle action potential that travels down the sarcolemma resulting in muscle contraction. Under normal conditions, the acetylcholine is immediately hydrolyzed by the enzyme acetylcholinesterase. Organophosphates inhibit acetylcholinesterase, causing acetylcholine to accumulate at the neuromuscular junction and producing a variety of symptoms that include involuntary twitching, muscle weakness, paralysis of respiratory muscles, constriction of pupils, involuntary urination, vomiting, abdominal cramps, increased salivation and lacrimation, bronchoconstriction, increased sweating, and increased peristalsis. Symptoms that result from accumulation of acetylcholine in the central nervous system (CNS) may include anxiety, restlessness, headache, insomnia, nightmares, slurred speech, tremors, and coma. Fortunately, therapy is available for diagnosed cases of organophosphate poisoning that includes the administration of atropine plus 2-PAM. Parathion weakly inhibits cholinesterase but is quickly converted to paraoxon in tissues which is several thousand times more toxic. Malathion is much less toxic than parathion because humans produce a carboxyesterase enzyme that renders malathion nontoxic.[14] The highly volatile organophosphate DDVP (also known as vapona or dichlorvos) is often used as a spray or in pest strips. Diazinon is an effective insecticide for use in vegetables and fruits and is useful against roaches. Some of the more toxic organophosphates such as phosphorous have been banned by EPA, while others such as parathion are rarely used because of its toxicity. However, organophosphates continue to cause more deaths and poisonings of humans than any other category of pesticides.[14,36]

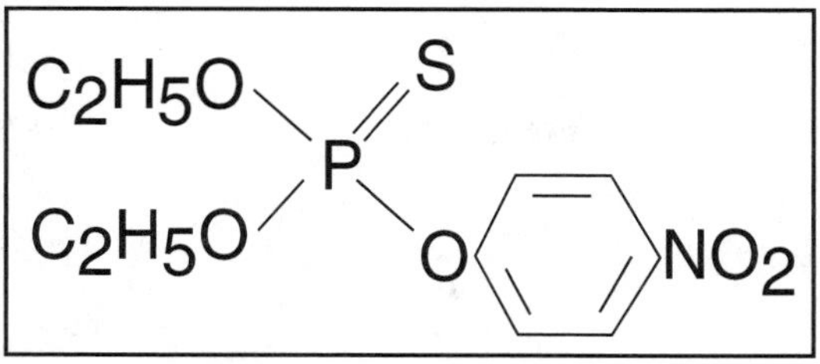

Figure 6.23 Chemical structure of parathion.

Carbamates

Carbamates are derivatives of carbonic acid and are contact poisons that inhibit cholinesterase in a manner similar to organophosphates. They vary widely in toxicity depending on the specific carbamate pesticides as do the organophosphates, but the

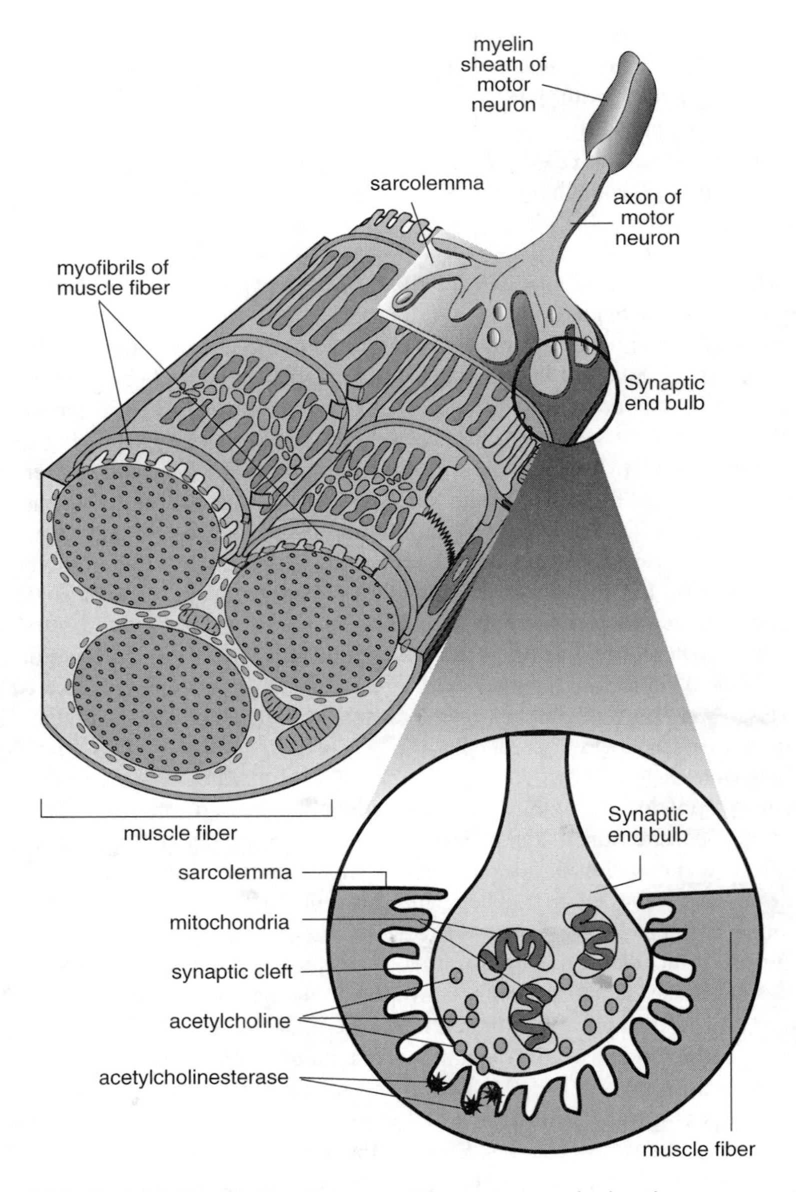

Figure 6.24 Acetylcholine, cholinesterase, and the neuromuscular junction.

treatment is not similar. Examples of carbamates include carbaryl, which is mildly toxic to mammals and is used to control fleas on pets. It is also effective in killing the rat flea. Another example of a carbamate is propoxur or baygon, which is both a contact and stomach poison that is effective against ticks, flies, sandflies, roaches, and mosquitoes.

Botanical and Biological Insecticides and Other Alternatives

The chrysanthemum plant produces a natural, highly insecticidal alkaloid known as pyrethrum. This is nontoxic to mammals. Because it is difficult to get enough chrysanthemums to supply the materials needed to produce the alkaloid, a number of pyrethrum compounds have been synthesized including permethrin, deltamethrin, and cypermethrin. Pyrethroids exhibit a toxicity to insects similar to the mechanism of DDT.

Another alkaloid naturally derived from the roots of plants is known as rotenone. It is grown in South America or Southeast Asia. This pesticide paralyzes insects by closing down chemical respiration in the nerve cell but is fairly harmless to humans. For this reason, rotenone is widely used in home insect sprays and other insecticide formulations for the household.

Methoprene is a growth regulator for insects that causes mosquitoes to die in the pupa stage by halting normal metamorphosis. It was first developed in the 1970s to control mosquito larvae, but several versions have been subsequently produced to control some beetles, flies, and fleas.[8,14]

Bacillus thurengensis produces a protein ("Bt" toxin) that is selectively toxic to insects. Once ingested, the toxin disrupts ion transport in the gut of the insects. Bt toxin has been used with great success against a wide variety of leaf-eating insects including gypsy moth larvae, although some insects appear to be developing resistance to the Bt toxin. It is also a bioengineered pesticide in that the gene required for its production has been inserted into a plant genome. This genome then manufactures the protein Bt toxin that selectively kills insects feeding on the plants. The Bt gene has been inserted into corn, tomatoes, and potatoes to target insects that feed on these plants.

A variation on bioengineered crops is the development of soybeans that are resistant to a powerful herbicide manufactured by Monsanto called Roundup™. Consequently, the crops can be sprayed with herbicide to kill weeds without effecting soybeans. Critics are concerned about the herbicidal resistance being spread to unwanted weeds. They are also concerned that pests will develop resistance to the Bt toxin engineered into the plants.[43] The critics may be justified since similarly bioengineered cotton crops were severely attacked by cotton bollworms that appeared unperturbed by the Bt toxin. This is especially distressing since several pests commonly migrate between the cotton and corn crops.[44]

The introduction of corn that has been genetically modified to include the Bt gene, however, reduced the need to spray crops with the Bt organism, but also raised concern about potential environmental effects. Use of Bt crops reduces use of pesticides and insecticides applied to corn and cotton crops to prevent destruction of the crops by the European corn borer. As mentioned above, the use of this technology could lead to an increase in the population of Bt-resistant corn borers. The *high dose/refuge strategy* is designed to delay the development of resistant pests. By surrounding Bt crops with non-Bt hybrids, the very small number of pests who are resistant to the toxin are likely to mate with non-Bt resistant pests which thrive in the adjacent non-Bt-treated corn. The offspring they produce should have only low or moderate resistance to the Bt toxin. This strategy, however, depends on certain assumptions about the mating patterns of Bt-resistant pests.[45]

Some concern has arisen over this seemingly harmless pesticide because of its close genetic relation to the anthrax virus, *B. anthracis*. Some scientists argue that the genetic similarity points to the potential of the Bt bacteria to mutate into strains that are more dangerous to humans.[46]

Naturally occurring disease agents such as fungal spores (Metarhizium flavoviride), pathogenic viruses, and protozoan parasites have been formulated as bait or sprayed on crops. In the case of protozoans and fungi, they grow inside the target pest and dissolve or digest internal organs. A fungal pathogen, *Lagenidium giganteum*, produces a motile biflagellate zoospore that infects mosquito larvae in their native water state. It has been shown recently to decrease populations of selected mosquitoes by 77% in trial assessments.[47]

A novel approach to control vector-borne diseases is being developed by a team of researchers at the Colorado State University in Fort Collins. Instead of developing methods to kill the mosquito vector for dengue fever, they are using bioengineering techniques to prevent the dengue virus from replicating in the mosquito. The researchers attached a small section of the dengue virus RNA to a harmless virus called Sindbis and inserted it into the mosquito, where it is replicated. The dengue RNA component halted any replication attempts by complete dengue viruses in the mosquito, thereby establishing a form of intracellular immunity. If the researchers can figure out how to get the mosquitoes to pass the disease resistance onto offspring, this may represent a worthwhile approach to the control of the disease.[48]

A significant alternative to the indiscriminate and runaway use of pesticides is the use of integrated pest management (IPM). IPM is a process that combines the best current knowledge about host–plant resistance, farming practices, and biological control with some limited chemical controls to effectively manage weeds and insect pests in an environmentally sound way. It is a process requiring flexibility, strategy, and knowledge. The agriculturist needs to know which pests to treat, when to treat, and how long to treat in order to keep crop losses from becoming economically unacceptable. Organic farming, which employs several different strategies to control pests, is increasing in popularity.[49] While conventional farmers use genetically-modified crops to control pests, organic farmers are using naturally occurring pesticides, predators, crop rotation, cover crops, and other nonlaboratory means of pest control.[50] Conventional farmers are also incorporating some of these methods using IPM. In part, the conversion to natural methods is due to numerous pests that have evolved resistance to pesticides. Increases in consumer awareness of the adverse effects of pesticides is also increasing market demand for agricultural products grown with reduced pesticide use.

Herbicides

Most herbicides are thought to have low toxicity to humans, although their roll as possible carcinogens has been an ongoing controversy.[51] There have been some reports of Hodgkins lymphomas and other soft tissue sarcomas associated with occupational exposure to phenoxyherbicides.[14] There are sufficient differences in plant species such that herbicides can be selective by killing weeds that are broad-leafed dicotyledenous plants while leaving members of the desirable grass family untouched. Atrazine, alachlor, 2,4-D and 2,4,5-T are examples of such selective herbicides (Figure 6.25). These are chlorophenoxy agents that destroy the plants by stimulating abnormal growth and interfering with the transport of nutrients. Here again, there is extraordinary controversy over the potential teratologic effects of these herbicides spurred on by the use of Agent Orange to defoliate areas of jungle in Vietnam.[52] Agent Orange contained 2,4-D, 2,4,5-T, and an unwanted but always present contaminant 2,3,7,8-tetrachlorodibenzo-p-dioxin (TCDD), also known as dioxin. Dioxin is an environmentally persistent substance that is highly toxic to humans and other mammals and is discussed in more detail in Chapter 5. Other selective herbicides include paraquat, diquat, and atrazine, which are used on crops such as sugar cane, pineapple, maize, cotton, and potatoes.

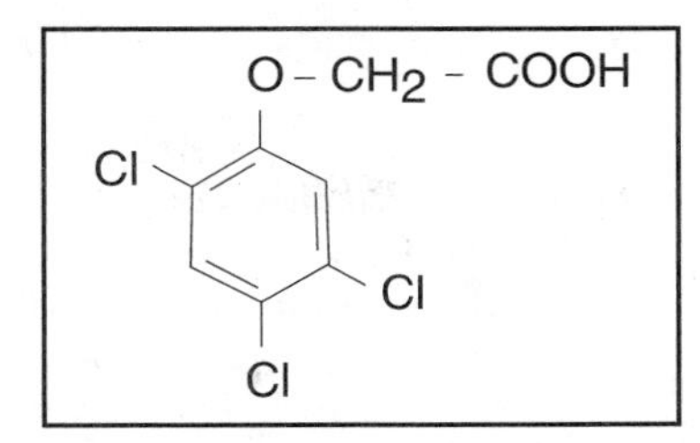

Figure 6.25 Chemical structure of 2, 4, 5-T herbicide.

Rodenticides

Although rodent control was discussed earlier in the chapter, a more detailed discussion of the rodenticide warfarin is offered here. Warfarin is among the most widely used rodenticides because it is effective in small quantities, reasonably nontoxic to humans, and does not promote bait shyness because it is a slow-acting poison. Warfarin is a coumadin compound and therefore a vitamin K antagonist that inhibits prothrombin synthesis. Repeated dosing over several days causes an overall reduction in the blood level prothrombin, which is necessary for coagulation. The consequence of such dosing is internal hemorrhaging, blood in the sputum and stools, and externally visible hematomas. The animals ultimately bleed to death. Some populations of rats have developed a resistance to warfarin. Consequently, more powerful versions known as superwarfarin, with a potency up to 200 times that of warfarin, have been developed that are more effective against these resistant strains.[53] Such compounds include brodifacoum, difethialone, and flupropadine. As might be expected, these compounds have much higher toxicity to humans than warfarin.[54] Other multiple dose poisons effective in controlling rats include fumarin, PMP, and diphacinone.

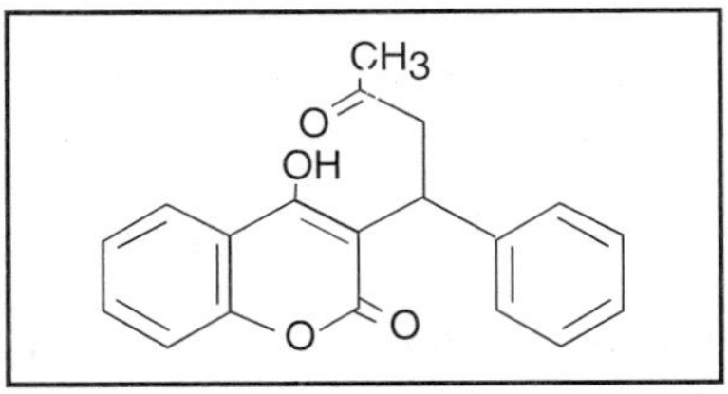

Figure 6.26 Chemical structure of warfarin.

REFERENCES

1. Vandermark, P., Oh, rats, *The Boston Globe Magazine,* 24, Nov. 5, 1995.
2. Blank, J., Bombay's rat patrol takes on the plague, *U.S. News and World Report,* 14, Oct. 14, 1994.
3. Mrotta, A., Rat meat, poor man's delicacy in Mexico, *Reuter's News Service,* Jul. 14, 1998, http://www.infoseek.com/Content?am=a2873LBY209reulbl 9980713&qt--rats&sv=IS&Ik=noframes&col=NX&kt=A&ak=newsl486.
4. Dispatch, Dengue hemorrhagic fever: The emergence of a global health problem, *EID,* 1:2, Apr/Jun, 1995.
5. World Health Organization, Control of Tropical Diseases (CTD): Malarial Control, WHO Office of Information, Geneva, Switzerland, 1995.
6. Koren, H., *Illustrated Dictionary of Environmental Health and Occupational Safety,* CRC Press Inc./Lewis Publishers, Boca Raton, FL, 407 pp.
7. U.S. Dept of Health, Education and Welfare (HEW), *Pictorial Keys to Some Arthropods of Public Health Importance,* Washington, D.C., U.S. Department of HEW Public Health Services, 3, 1964.
8. Koren, H. and Bisesi, M., *Handbook of Environmental Health and Safety, Principle and Practices,* 3rd ed., CRC Press Inc./Lewis Publishers, Boca Raton, FL, chap. 4.
9. Tortora, G.J., Funke, B.R. and Case, C.L., *Microbiology: An Introduction,* 5th ed., The Benjamin/Cummings Publishing Company, Inc., 1995, chap. 23.
10. Nadakavukaren, A., *Our Global Environment: A Health Perspective,* 4th ed., Waveland Press, Inc., Prospect Heights, IL, 1995, chap. 8.
11. Morgan, M., Gordon, L., Walker, B. et al., *Environmental Health,* 2nd ed., Morton Publishing Co., Englewood, CO, 1997, chap. 8.
12. U.S. Department of Health, Education and Welfare (HEW), *Control of Domestic Rats and Mice, Public Health Service,* Communicable Disease Center, 22, 1964.
13. Joy, R.M., Chlorinated Hydrocarbon Insecticides, in *Pesticides and Neurological Diseases,* edited by Ecobichon, D.J., and Joy, R.M., CRC Press, Inc., Boca Raton, FL, 1982, p. 91–150.
14. Saunders, D.S., and Harper, C., Pesticides, In *Principles and Methods of Toxicology,* 3rd ed., edited by Hayes, AW., Raven Press, Ltd., New York, 1996, chap. 11.
15. Dover, M., Getting off the pesticide treadmill, *Technology Review,* Nov/Dec, 53, 1985.
16. United States Geologic Survey (USGS), Pesticides in Surface Waters, U.S. Geological Survey Fact Sheet, FS-039-97, May 30, 1997, http://wwwrvares.er.usgs.gov/nawqa/NAWQA.ofr94.70.html.
17. Ganesse, L.P. and Puffer, L.A., Herbicide-resistant weeds may threaten wheat production in India, *Resources,* No. III (Spring), Resources for the Future, 1993.
18. U.S. Congress, Office of Technology Assessment, Neurotoxicity: Identifying and Controlling Poisons of the Nervous System, OTA-BA-426, Washington, D.C., U.S. Government Printing Office, April 1990, chap. 10.
19. U.S. Environmental Protection Agency (EPA), Office of Pesticide Programs, Pesticide Industry Sales and Usage:1987 Market Estimates, U.S. EPA, Washington, D.C., 1988.
20. Jeyaratnam, J., Health problems of pesticide usage in the third world, *Br. J. Indust. Med.,* 42, 505, 1985.
21. World Health Organization (WHO), Pesticide deaths, what's the toll?, *Ecoforum,* 6, 10, 1981.
22. Whelan, E.M., *Toxic Terror: The Truth Behind the Cancer Scares,* 2nd ed., Prometheus Books, Buffalo, NY, 1993, chap. 4.

23. Repetto, R. and Baligo, S.S., Pesticides and the immune system: the public health risks, *Cent. Eur. J. Public Health,* 4, 263, 1996.
24. U.S. EPA, Office of Pesticide Protection (OPP), *Pesticides and Child Safety: Tips to Protect Children from Poisonings Around the Home, Prevention, Pesticides, and Toxic Substances,* EPA#735-F-93-050R, update Nov. 19, 1997, http://www.epa.gov/oppfead/childsaf.htrn.
25. Siontral Resource Defense Council (NRDC), *Intolerable Risk: Pesticides in Our Children's Food,* Washington, D.C., 1989.
26. Vegej, J., Cancer causing pesticides abound in supermarkets' bounty — washing, peeling, buying organic can help, *Health News Review,* Jun. 12, 14, 1996.
27. Food and Drug Administration Pesticide Program, Residues in Food, 1990, *J. Assoc. Official Analyt. Chem.,* 74, 1991.
28. Hanson, D., Regulation after Delaney - new health-based standard replaces controversial clause dealing with cancer-causing pesticides in foods, *C&EN,* 38, Sept. 23, 1996.
29. Conservation Foundation, *Opportunities to Assist Developing Countries in the Proper Use of Agricultural and Industrial Chemicals,* Washington, D.C., Vol. 1, 1988.
30. U.S. Department of State, Proceedings of the U.S. Strategy Conference in Pesticide Management, U.S. Government printing Office, Washington, D.C., Jun. 1979.
31. Ngunjiti, P., Environment: Stemming the flow of dangerous chemicals. *Interpress Service English Newswire,* Sep. 25, 1996, -dnV://www2.elibrary.com/id/2525/getdoc.cgi?id=52564954xOy445&oids=OqOOI001 2&forW=Tl.
32. Stolberg, Sheryl DDT, target of global ban, finds defenders in experts on malaria, *The New York Times*, Saturday, Aug. 28, 1999.
33. Ames, B. and Wiles, R., Pesticide pros and cons: Are residues really bad for your health?, *Vegetarian Times,* 68, Nov. 1995,
34. Avery, D., How pesticides help prevent cancer, *Consumer Research,* 10, June 1995.
35. Colburn, T. and Clement, C., eds., *Advances in Modern Environmental Toxicology,* Vol. XXI, Chemically Induced Alterations in Sexual and Functional Development, The Wildlife Human Connection, Princeton Scientific Publishing, Princeton, NJ, 1992.
36. Koren, H. and Bisesi, M., *Handbook of Environmental Health and Safety, Principles and Practices,* 3rd ed., CRC Press Inc./Lewis Publishers, Boca Raton, FL, chap. 6.
37. Wald, M., Citing children, *The New York Times*, EPA is Banning Common Pesticide, New York, 1999. EPA, Tolerance Reassessment Progress EPA Attains Statutory Requirement by Completing Over 33% of All Required Tolerance Reassessments by August 3, 1999, http://www.nytimes.com/library/nat…ience/080399sci-epa-pesticide.html.
38. Raloff, J., The pesticide shuffle, *Science News,* 149, 174, Mar. 16, 1996.
39. Lopez-Carillo, L., Tores-Arreola, L., Torres-Sanches, L., et al., Is DDT use a public health problem in Mexico? *Env. Health Perspect.,* 104(6), 5, 84, June 1996.
40. Naj, A.K., Insecticides found to be traveling throughout the world, *Wall Street Journal, Science and Health,* 1, Sep. 29, 1995.
41. Reich, M.R., and Spong, J.K., Kepone: A chemical disaster in Hopewell, Virginia, *Int. J Health Ser.* 13, 227, 1983.
42. Meerdink, G.L., Organophosphorous and carbarnate insecticide poisoning in large animals, *Vet. Clin. North Am. Food Anim. Pract.,* 5, 374, 1989.
43. Tortora, G.J. and Anagnostakos, N.P., *Principles of Anatomy and Physiology,* 6th ed., Harper and Row Publishers, New York, 1990, chap. 10.
44. Greene, R., Hope, Concerns, sprout over biotech crops, *Los Angeles Times, Science, Medicine and the Env.,* Oct. 20, 1996.

45. Beardsley, T., Picking on cotton, *Scientific American, News & Analysis,* 43, Oct. 1996.
45. Mackensie, D. Friend or foe?, *New Scientist*, 22–23, Oct. 9, 1999.
46. Renner, R., Will Bt-based pest resistance management plans work?, *Environ. Sci. Technol.*, 33, 410, Oct. 1, 1999.
47. Cuda, P., Hornbey, J.A., Cotterill B. and Cattell, M., Evaluation of Lagenidium giganteum for biocontrol of *Mansonia* mosquitoes in Florida (Diptera: Culicidae). *Biol. Control,* 8, 124, 1997.
48. Adler, T., Keeping mosquitoes healthy for human's sake, *Sci. News,* 149, 295, May 11, 1996.
49. Bourne, J., The Organic revolution, *Audobon*, Vol. 101, i2 64(1), March 1999.
50. Bourne, J., Bugging Out, *Audobon*, Vol. 101, i2 71(1), March 1999.
51. Morrison, H.I., Wilkins, K., Semenciw, R. et al., Herbicides and cancer, *J. Natl/ Cancer Inst.,* 84, 1866, 1992.
52. Timoc, R., Parsonnet, J. and Halperin, D., Paraquat poisoning in southern Mexico: a report of 25 cases, *Arch. Environ. Health,* 48, 78, 1993.
53. Haug, B., Schjadtlversen, K. and Rygh, J., Poisoning with long-acting anticoagulant, *Tiasskr Nor Laegeforen,* 112, 195, 8, 1992.
54. Routh, C.R., Triplett, D.A., Murphy, M.J., et al., Superwarfarin ingestion and detection, *Am. J. Hematol.,* 36, 50, 1991.

CHAPTER 7

Emerging Diseases

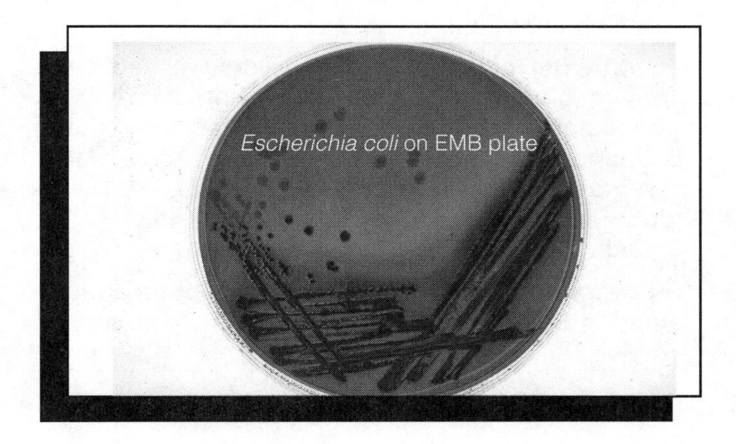

INTRODUCTION

Infectious diseases continue to be the foremost cause of death worldwide. The Centers for Disease Control and Prevention (CDC) reported a 58% rise in deaths from infectious diseases since 1980. Infections and parasitic diseases caused more than 17 million deaths in 1995, more than a quarter of all human deaths.[1] Although AIDS accounts for most of the increase, deaths from all other infectious diseases rose a dramatic 22%.[2] These figures are even more extraordinary because the International Code of Diseases lists many infectious diseases in categories other than infectious. For example, meningitis and cirrhosis are classified as diseases of the nervous system and liver, respectively, and only 17% of deaths attributable to infections are actually included in the codes for parasitic and infectious diseases.[3] About one quarter of the visits to physicians in the U.S. each year are attributable to infectious diseases, and the costs associated with these infections are estimated at more than $120 billion.[4]

OBJECTIVES FOR THIS CHAPTER

A student reading this chapter will be able to:

1. **Differentiate the emerging infectious diseases in the U.S. and those occurring worldwide**
2. **List and recognize the six major reasons associated with the emergence of infectious diseases**
3. **Explain the likely reasons for the emergence of specific infectious diseases**
4. **Identify, list, and explain the etiological agents, the epidemiology, and the disease characteristics of the major emerging infectious diseases including influenza, hanta virus, dengue fever, ebola, AIDs, Cryptosporidiosis, Malaria, Lyme disease, Tuberculosis, Streptococcal infections, and E. coli infections**
5. **Recognize and explain the practical approaches to limiting the emergence of infectious diseases**

Despite more than 100 years of scientific progress and the development of sanitary methods, infectious diseases still cause enormous human suffering, deplete scarce resources, impede social and economic development, and contribute to global instability. The potential for even greater human misery looms as an unrelenting danger. Recent outbreaks underscore the potential for the sudden appearance of infectious diseases in currently unaffected populations in the U.S. and the world.

Emerging Diseases in the U.S.

We are continuously challenged in the public news media with the terrifying prospect of new and emerging infections. The British tabloids coined the term "flesh-eating bacteria" to describe an invasive necrotizing infection caused by Group A Streptococci and suggested that we are about to experience epidemics from this organism, producing high mortality, amputations, and grotesque disfigurement.[5] New parasitic threats to water supplies have erupted into the nation's view. One of those parasites, Cryptosporidium, has caused more than five major outbreaks in the U.S., with the most severe in Milwaukee in 1993. This outbreak affected more than 400,000 people.[6] The population of patients with acquired immunodeficiencies is rapidly increasing among certain African American and heterosexual groups in the U.S.

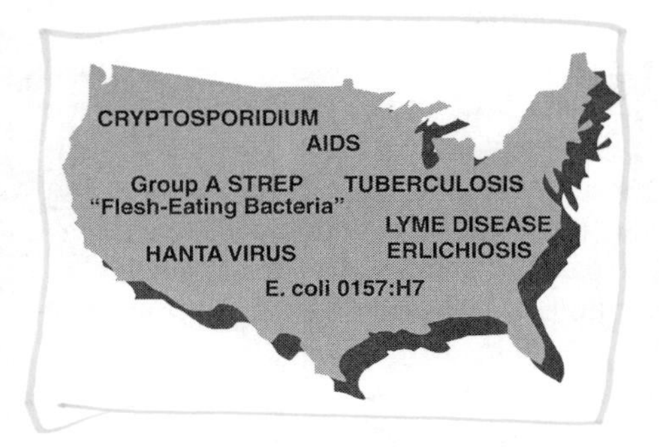

Nearly 225,000 persons over 13 years of age in the U.S. had AIDS in June 1996, representing a 65% increase over January 1993. AIDS remained one of the leading causes of death among persons 25 to 44 years of age in the U.S., accounting for 19% of deaths from all causes in this age group.[7] Infection with *Escherichia coli* serotype 0157:H7 was first described in 1982 and rapidly emerged as the major cause of bloody diarrhea and acute renal failure. The American public became aware of this disease when 700 persons became infected and four died after eating undercooked hamburger from a regional fast food chain in a 1993 outbreak.[8] There have been more than 30 outbreaks in the U.S. since, with many additional cases worldwide.[9] In 1993, a 6-year drought followed by heavy rains and flooding contributed to the increase of deer mice in the American Southwest, leading to an outbreak of hantavirus pulmonary syndrome. First appearing in the Navajo reservations, the disease has since spread to 20 additional states and has taken the lives of 45 people, nearly half of those infected.[10] Lyme disease is the second most prevalent emerging infectious disease in the U.S.; more than 65,000 cases have been reported to the Centers for Disease Control and Prevention since the disease was first described by Steere and colleagues in three Connecticut communities in 1977.[11]

Emerging Diseases Worldwide

These and other infectious diseases have potential worldwide significance. As an example, the possibility of Dengue fever reaching the U.S. is very real. According to the World Health Organization, there are almost 30 "new" recognized infectious diseases including HIV, Ebola, Hanatvirus, Hepatitis-C, the Legionella bacterium, and the Cyclospora parasite.[1] Until 1995, dengue fever had been isolated to Costa Rica's coastal mountain regions along the Pacific shore. But rising temperatures permitted the mosquitoes responsible for dengue fever to break the coastal mountain barriers and invade the rest of the country. By September of 1995, the epidemic had infected 140,000 people and claimed the lives of 4000. Dengue fever, a debilitating tropical disease transmitted by mosquitoes, is reemerging with such force that it has now reached epidemic proportions in Central America and is bearing down on the U.S.[12] Each year 35 to 60 million people worldwide contract dengue fever.[13]

The Ebola virus grabbed global attention with a flurry of news articles and films as one the deadliest viruses known, killing nearly 90% of those infected. There is no treatment or vaccine. Ebola and other hemorrhagic fever viruses typically start suddenly with malaise, fever, and flu-like symptoms, which are often followed by rashes and bleeding, along with kidney and liver failure as these organs and other body parts dissolve into mush.[14] According to the Pan American Health Organization, Chagas is the most important parasitic illness in Latin America, causing over 40,000 deaths each year. The protozoan disease is transmitted by a beetle known as the kissing bug (*Rhodnius prolixus*), which thrives in dark houses with cracked floors and walls and comes out at night to feed on blood sucked from the face of the human host. The inoculated Trypanosome parasite invades the stomach, heart, and muscles of the human host to produce disabling fatigue, energy loss, and often death.[15] In the 1990s, epidemic cholera reappeared in the Americas after being absent for nearly a century; from 1991 through June of 1994, more than one million cases and nearly 10,000 deaths were reported.[16]

In 1994, an extended monsoon season in Northern India followed by 3 months of extremely hot weather (100°F) drove rats into the cities, causing pneumonic plague from *Yersinia pestis* in Surat. The subsequent panic claimed the lives of many people and cost India $2 billion.[10]

A cluster of CJD (Creutzfeldt-Jakob disease) cases with a unique neuropathologic picture among unusually young patients in the U.K. was reported during a widespread epizootic of bovine spongiform encephalopathy (BSE).[17] This became notoriously referred to by the press as "mad cow disease" and prompted the uncommon action of destroying thousands of cows. This appearance of the disease alerted many countries, including the U.S., to update their surveillance for CJD and look for similar cases.[18] This increased attention, and the fact that CJD is rapidly and invariably fatal, raised public attention about this emerging disease to a heightened level.[19]

Malaria is the world's largest widespread mosquito-borne illness and is potentially the worst threat to humans in the future. Global warming provides an environment for mosquitoes to multiply, migrate, and increase their metabolism, forcing them to feed (bite) more often. Despite the fact that mosquitoes die when temperatures rise above 104°F, so do people and crops. Programs to eradicate malaria, along with socioeconomic development, have reduced endemic areas; but malaria has seen an increase in many parts of the world despite continued efforts to eradicate it. More than 100 million clinical cases occur annually; and more than one million people, mostly children, die each year in tropical Africa.[20] Mainly due to drug-resistant strains, malaria has reemerged as a major international infectious disease with nearly a half billion people worldwide developing the disease each year.[21] Malaria has also begun to reappear in the U.S. and is now reemerging at a geometric rate in South America.[22]

The World Health Organization (WHO) reported that efforts to control tuberculosis have failed and that drug-resistant strains of Mycobacterium tuberculosis are rising. Tuberculosis is considered one the world's leading infectious diseases and kills over 3 million worldwide each year, with many more becoming ill.[23]

What is an Emerging Infectious Disease?

New infections are emerging and old ones are expanding their geographical ranges faster than we can deploy resources to stop the organisms responsible. International travel and commerce are increasing, and technology is rapidly changing the conditions that increase the risk of exposure to infectious agents. The spectrum of infectious disease is changing quickly in conjunction with dramatic societal and environmental changes. Worldwide, explosive population growth in developing nations, expanding poverty, and urban migration are contributing to proliferation of both new and old infectious diseases. At the turn of the century only 5% of the population lived in cities, but by the year 2025 it is estimated that this will increase to 65%. Therefore, more and more microbes will migrate from rural areas to the urban areas.[24] Once expected to be eliminated as a public health problem, infectious diseases remain the leading cause of death and disability-adjusted life years (DALYs). Some of these diseases may be new to the human species. Others may simply have never been seen in North America before. The term *emerging infectious diseases* is now being applied to this phenomenon, and awareness of these emerging diseases has reached international levels. The term *emerging infectious diseases* refers to diseases of infectious origin whose incidence in humans has either increased within the past 2 decades or threatens to increase in the near future.[25]

Emerging infectious diseases, often with unknown long-term public health impact, continue to be identified. Table 7.1 lists major diseases or etiologic agents identified just within the last 20 years. New agents are regularly added to the list, particularly with the availability of nucleic acid amplification techniques for detecting and identifying otherwise noncultivable microorganisms.[26] In many cases, diseases became better recognized or defined (e.g., Legionnaires' disease, Lyme disease, human ehrlichiosis). Others are completely new. A previously unknown and deadly disease, acquired immunodeficiency syndrome (AIDS), originated from uncertain sources in one part of the world and spread globally at a speed that would have been unthinkable in earlier times.[27]

REASONS FOR THE EMERGENCE OF INFECTIOUS DISEASES

The emergence of so many new diseases and the reemergence of those thought to be controlled is difficult to understand or explain. However, rarely do emerging infections surface without reason. There are a number of specific explanations responsible for disease emergence that can be identified in most all cases.[28] Table 7.1 summarizes the known causes for a number of infections that have recently emerged. Most emerging infections, and even antibiotic-resistant strains of common bacterial pathogens, usually originate in one geographic location and then disseminate to new places. Morse suggested that infectious disease emergence can be viewed operationally as a two-step process: (1) the introduction of the agent into a new host population (whether the pathogen originated in the environment, possibly in another species, or as a variant of an existing human infection), followed by (2) the establishment and

WEB PICK **Go to http://www.unix.oit.umass.edu/~envhl565/index.html. Click on "Chapter Web Links" to find Emerging Infectious Diseases Home Page on the World Wide Web under Chapter 7**	
Chapter 7	Emerging Diseases
Whose Site?	Centers for Disease Control Health. *Emerging Infectious Diseases* is published six times a year by the National Center for Infectious Diseases, Centers for Disease Control and Prevention (CDC), 1600 Clifton Road, Mailstop D-61, Atlanta, GA 30333, USA. Telephone: 404-371-5329; fax: 404-371-5449.
URL	http://www.isis.vt.edu/~fanjun/text/Links.html
What's Here?	The Emerging Infectious Diseases (EID) journal is an electronic journal available in its entirety online for free and as individual articles in the Adobe™ Acrobat™ (.pdf) file format. This is part of CDC's plan for combating emerging infectious diseases. One of the main goals of CDC's plan is to enhance communication of public health information about emerging diseases so that prevention measures can be implemented without delay. Emerging Infectious Diseases is peer reviewed and provides information on emerging infections in several categories including *perspectives*, addressing factors that underlie disease emergence, including microbial adaptation and change, human demographics and behavior, technology and industry, economic development and land use, international travel and commerce, and breakdown of public health measures; *synopses*, summarizing existing information on specific emerging diseases or syndromes; *dispatches*, reporting preliminary but promising laboratory or epidemiologic data; *research studies*, reporting laboratory and epidemiologic results within a public health perspective; and *policy reviews*, reporting public health policies based on research and analysis of emerging disease issues.

Table 7.1 Major New and Reemerging Etiologic Agents of Infectious Diseases and the Reasons for Their Emergence (Adapted from Morse, S., Perspectives: Factors in the emergence of infectious diseases, *EID,* I([]), Jan. Mar., 1995.)

VIRAL

Viral diseases that have been identified since 1973*

1977 Ebola, Marburg
Origin undetermined (importation of monkeys associated with outbreaks in these primates in Europe and the U.S.)
1980 HTLV Influenza (pandemic)
Pig duck agriculture thought to contribute to reassortment of avian and mammalian influenza viruses
1983 HIV
Transmission by intimate contact as in sexual transmission, contaminated hypodermic needles, transfusions, organ transplants. Contributing conditions that spread the disease include war or civil conflict, urban decay, migration to cities, and travel
1989 Hepatitis C
Transmission in infected blood such as by transfusions, contaminated hypodermic needles, and sexual transmission
1993 Hantaviruses
Increased contact with rodent hosts because of ecological or environmental changes

Viral diseases that have reemerged

Argentine, Bolivian hemorrhagic fever
Agricultural changes that promote growth of rodents
Bovine spongiform encephalopathy (cattle)
Alterations in the rendering of meat products
Dengue, dengue hemorrhagic fever
Travel, transportation, urbanization, and migration
Lassa fever
Conditons such as urbanization that favor rodent host, increasing exposure (usually in homes)
Rift Valley fever
Irrigation dam building, agriculture: possibly change in virulence
Yellow fever
Conditions favoring mosquito vector (in "new" areas)

BACTERIAL

Bacterial diseases that have been identified since 1973*

1977 Legionella disease
Cooling and plumbing systems that allow the organism to grow in biofilms that form on water
1982 Hemolytic uremic syndrome *(Escherichia coli* – O157:H7)
Modern food processing on a large scale permitting contamination through meat storage tanks and unsterile plumbing
1982 Lyme borreliosis (*Borrelia burgdorferi*)
Close contact between homeowners encroaching on forested areas and the mice and deer (a secondary reservoir host) that maintain the tick vector for Borrelia
1983 Helicobacter pylori
Newly recognized as agent involved with gastric ulcers, probably widespread before recognition
1987 Toxic shock syndrome *(Staphylococcus aureus)*
Ultra-absorbency tampons
1992 Cholera (type 0139)
Likely introduced from Asia to South America by ship, with spread made possible by reduced water chlorination; Strain (type O139) from Asia newly spread by travel (similarly to past introductions of classic cholera)

(continued)

Table 7.1 (continued) Major New and Reemerging Etiologic Agents of Infectious Diseases and the Reasons for Their Emergence

Bacterial diseases that have reemerged
Tuberculosis Breakdown in public health measures such as reduction in prevention programs, inadequate sanitation, homelessness, AIDS
Streptococcus, group A (invasive necrotizing) Unknown, may be increased use of aspirin and related substances
PARASITES
Parasitic diseases that have been identified since 1973*
1976 Cryptosporidium, other waterborne pathogens Contaminated surface water, lack of proper filtration methods
Parasitic diseases that have reemerged
Malaria (in "new" areas) Spread of mosquito vectors, worldwide travel or migration, "Airport" malaria
Schistosomiasis Agriculture, dam building, deforestation, flood/drought, famine, climate changes

*Compiled by CDC staff. Dates of discovery are assigned on the basis of the year the isolation or identification of etiologic agents was reported.

further dissemination within the new host population.[25,27] The majority of emerging infections are thought to be caused by pathogens formerly existing in the environment and released from obscurity or presented with a selective advantage by altered conditions and given an opportunity to infect new host populations. Occasionally a variant may evolve and cause a new disease.[25,28] The term *microbial traffic* is given to the process by which infectious agents transfer from animals to humans or spread from isolated groups into new populations.[29,30] There are identifiable activities that increase microbial traffic and promote emergence and epidemics. In a number of instances, including many of the more unusual infections, the agents are zoonotic, crossing from their natural animal hosts into the human population. In separate instances, pathogens existing in geographically isolated populations are able to disseminate further. Very often, the emergence of a disease results from human actions, usually unintended. Changes in climate and other natural causes can also be responsible at times.[31]

Factors responsible for emerging infectious diseases include: (1) ecological changes such as those due to agricultural or economic development or to anomalies in climate; (2) human demographic changes (urbanization and behavior); (3) travel and commerce; (4) technology and industry (globalization); (5) microbial adaptation and change (resistance); and (6) breakdown of public health measures.[25]

Ecological Changes (Agriculture, Climate)

Agriculture

Changes in ecology, such as those resulting from agricultural or economic development, are among the most commonly identified reasons in the emergence of infectious diseases. Ecological changes are often present in outbreaks of formerly unrecognized diseases with high case–fatality rates. These diseases are often introduced from animal hosts because people are placed in close contact with a natural reservoir or host where the disease organism is present (often a zoonotic or arthropod-borne infection). People may expand into an area where the animal host thrives, the animal host may expand into human living areas, and/or the infectious agent may have increased in the animal host population.[25,29,30]

Deforestation is reported to be a major reason for the emergence of Lyme disease in the U.S. and Europe, because it increased the population of deer and the deer tick, the vector of Lyme disease. The movement of people into these areas placed a larger population in close proximity to the vector.[32] In Asia, the field mouse, *Apodemus agrarius*, is abundant in rice fields where people contract Korean hemorrhagic fever. This usually occurs during the rice harvest from contact with infected rodents. Argentine hemorrhagic fever has increased almost in direct relation to the expansion of maize agriculture and favoring the growth of a rodent that carries the disease organism.[33] Current researchers report that pandemic influenza arises when two influenza strains recombine, resulting in a new virus that is infectious to humans. Ducks are a major source of avian influenza. The avian flu virus may enter pigs that are in close proximity to the ducks. The gene segments of the viruses may then recombine (or reassort).[34] The agricultural practice of pig–duck farming as practiced in China places these animals in contact and creates an ideal setting for the creation of influenza recombinants capable of infecting humans.[35]

Water is also frequently associated with disease emergence. Infections transmitted by mosquitoes or other arthropods include some of the most serious and widespread diseases.[36] Mosquitoes are often stimulated by the expansion of standing water, simply because many of the mosquito vectors breed in water. There are many cases of diseases transmitted by water-breeding vectors, most involving dams, water for irrigation, or stored drinking water in cities.

Climate

Virtually all climate scientists believe that the increase in potentially fatal diseases is likely to be the single most important threat that climate changes pose to human health.[37] Extreme weather patterns as a result of natural fluctuations in the atmosphere or man-made changes (i.e., global warming) have routinely been followed by outbreaks of disease. "Dry spells in wet areas or torrential rains in normally dry spots tends to favor so-called opportunistic pests—rodents, insects, bacteria, protozoa, viruses—while making life more difficult for the predators that usually control them.[10] The same synergies that empower microbes also weaken our defenses against them. Heat, increased ultraviolet radiation resulting from ozone depletion, and pollutants like chlorinated hydrocarbons all suppress the disease-battling immune systems—both for humans and for other animals."[10]

The Earth's climate is becoming a critical factor in enhancing the spread of infectious diseases. WHO predicts the temperature of the planet will rise 2.7° to 8°F by the year 2030 and will profoundly affect human health. One of the predictable results will be the expansion of mosquito-borne disease from the tropics into temperate zones. The expansion may move northward as far as New York and could include malaria, dengue, and yellow fever (the latter has a 50% mortality). In 1990, WHO warned that global warming would enlarge the territories of disease-carrying insects and increase the at-risk population by 620 million people by the year 2050.[38]

The Pacific Ocean warming events influence weather patterns and jet streams across North and South America, western Europe, Asia, and Africa. El Niño events normally begin in the latter part of December and determine weather patterns for a year. The El Niño warming events have historically occurred about twice every 10 years.[39] These sea-warming events cause evaporation of water, producing extensive rainfall in some areas and drought in others. The power of the El Niño events has increased dramatically in recent years, producing +2° to 3°C temperature increases in the ocean well into the summer months, with the possibility of extending over much longer periods of time. It is predicted that El Niño events will be more frequent with greater intensity, onset, and duration consistent with a continued increase in greenhouse gases.[40] A warmer climate initiated by increased amounts of carbon in the atmosphere has caused El Niño events to last longer and to be more dramatic.[41] The sudden eruption of Hantavirus infection in the southwestern U.S. in 1993 to 1994, killing 48% of the 94 persons infected, has been attributed to the sudden population explosion of deer mice that harbor the organism. The probable events are that heavy rains fell following 6 years of drought, causing pine nuts and grasshoppers to flourish, which fed the population eruption of deer mice. This enlarged population was driven from the flooded burrows, increasing the possibility of contact with people.[42]

It has been suggested that certain organisms in marine environments are natural reservoirs for cholera vibrios and that large-scale effects on ocean currents may cause local increases in the reservoir organism with consequent flare-ups of cholera. Pollution and global climate changes induced by humans adversely affect the health of marine populations and have the potential to harm the humans because of the ocean's ability to act as incubator for diseases like cholera.[37] Changes in coastal ecology are generating "hot systems" in which random mutations of vibrio cholera may be selected and amplified and then transferred to human hosts through the consumption of contaminated fish and shellfish.[42] Aquatic plants, seaweeds, and free-floating phyto- and zooplankton are known to harbor vibrios. The phytoplankton or algae are growing in epidemic proportions, spurred on by climate-related warmer sea surfaces and the increase of nitrogen-rich wastewaters from coastal cities.[44] Additionally, the species of algae that are proliferating are less palatable to fish, and the numbers of fish have shown a decline in 14 of the world's 17 major fishing grounds.[42]

Human Demographic Changes (Urbanization) and Behavior

Changes in human demographics and behavior, or human population movements and upheavals caused by migration or war, are often significant factors in disease emergence. Increased population density in urban areas—migration in hopes of a better, more comfortable lifestyle—has surpassed basic services, including clean water supplies, sanitary conditions such as sewage disposal, and adequate housing.

Some of the most profound effects were among the poor of the inner-city slums. It was a frustrating paradox for many in public health that those most susceptible to infection had the worst facilities. Their socio-economic status led to a heightened risk of disease because of little or no access to clean water, running sewers, and pest control. In one example, until the mid-1950s, large areas around and within Washington, D.C. had no running water and were served by outhouses.[45] These areas not only infected the people who lived there but they also acted as a reservoir of disease that would periodically spread across entire regions of the country. The diseases were incubated by unhygienic practices as well as brought in by immigrants who could often not afford to live anywhere else.

Depressed economic conditions and the disappearance of viable agricultural land in rural areas in many parts of the world are prompting the mass movement of workers from rural areas to cities. It has been estimated by the United Nations that 65% of the world's population will be living in cities by the year 2025 as this migration continues.[25] Diseases such as HIV are thought to have originated in isolated rural areas and to have been brought to the densely-populated cities where the disease could spread locally and travel by infected humans along highways and air routes to cross state and national boundaries. Public health measures in overcrowded cities are often strained or unavailable to large groups of the urban impoverished living in inner city slums or in shanty towns on the periphery, thereby increasing the opportunity for emerging infectious diseases.[25]

Emerging infections such as HIV, cholera, and dengue have followed this pattern. An example of this is HIV-2, which was isolated from a man living in rural Liberia and thought to have originated from contact with animals where a virus similar to HIV-2 was isolated from the sooty mangabey monkey.[47,48] This monkey is often used for food. Such zoonotic infections may often occur and remain isolated, but the swelling migration of people from rural to urban environments increases the probability that the disease will spread to densely populated areas and be transported regionally along highways and airways to reach distant locations, permitting a pandemic as was the case for HIV-1. Simply transporting the virus with naturally low transmissibility would not be sufficient to produce a worldwide disease.

There were social behaviors that permitted intimate contact with the HIV organism through sexual transmission, intravenous drug use, or the use of technology in the form of blood transfusions and blood products. These factors associated with the very long duration of infectivity have combined to increase the opportunity for transmission, resulting in a pandemic that has affected much of the world and is now spreading through Asia.[25] Human activity influences a number of factors associated with the emergence of disease. The control of such diseases will require the motivation of appropriate individual behavior and constructive action regionally and even internationally. Unfortunately, the continued increase of infectious diseases on many fronts points to the need for additional scientific knowledge in human behavior.

Travel and Commerce

The main factor affecting the emergence and spread of new diseases is believed to be transportation.[1] Increased economic growth into national and international boundaries has led to increased travel, contributing to the notion of "diseases without boundaries." The incubation period for infectious diseases can be a period of hours to weeks. As a result, diseases have the potential to spread from one continent to another in a matter of hours before symptoms occur.

Travel

The spread of HIV through travel has been suggested.[47] Historically, infections introduced into a remote population area might occasionally be transported to a new place through commerce, travel, or war.[49] The bubonic plague, carried by the rat and transmitted to humans by the rat flea, is thought to have been introduced into Europe with the silk trade routes from Asia and continuing with the Crusades. Smallpox was introduced by the Spanish conquistadors into the Aztec nation in the 15th century and decimated more than 80% of the highly susceptible population. The 16th and 17th centuries saw ships transporting slaves from West Africa to the New World infected with yellow fever and carrying the mosquito vector, *Aedes aegypti*, to the new territories. Two centuries later, cholera traveled with its human hosts from its likely origin in the Ganges plain to the Middle East. From there cholera traveled to Europe and a substantial part of the remaining world.[25]

The Asian tiger mosquito, *Aedes albopictus*, entered the U.S., Brazil, and parts of Africa in the collected water within used tires shipped from Asia.[50] The mosquito's introduction is problematic because it bites aggressively, survives in both forest and suburban habitats, and is a vector for many human pathogens. It is a vector for epidemic dengue fever transmission in Asia and is a competent laboratory vector of La Crosse, yellow fever, and other viruses.[51] Fourteen strains of eastern equine encephalitis virus have been isolated from *A. albopictus* in Florida.[52] The mosquito has become established in at least 21 of the contiguous states in U.S. and Hawaii.[53] Malaria is another disease transmitted by a mosquito vector and is one of the most repeatedly imported diseases in nonendemic-disease areas. Cases of malaria contracted in airports are sometimes identified. Rats have carried hantaviruses virtually worldwide.[54] These infections were originally localized but were carried to exotic and foreign parts of the world. Similar histories are being repeated today with the added problem of greatly increased volumes of traffic moving at enormous speeds. Travel may involve short distances or the crossing of international borders. In 1998 an emigrant from Ukraine in a plane from Paris to New York infected 13 other passengers with drug-resistant TB.[24] The volume is huge. In the early 1990s, in excess of 500 million persons crossed international borders yearly on commercial airplane flights.[53] As many as 70 million persons are thought to work either legally or illegally in other countries, and the majority are from developing countries. Many are migrant workers following crops and moving seasonally.[55] Large populations are moved among countries as part of military maneuvers. The consequences of armed conflict and political unrest also displace millions. In the early 1990s, there were an estimated 20 million refugees and 30 million displaced persons worldwide.[53]

Commerce

Marine organisms may be transported in the hulls of ships and in their ballast water. There were 367 different species identified in the ballast water of ships traveling between Japan and Coos Bay, Oregon.[56] There is evidence that disease organisms are also transmitted by shipping such as the introduction of vibrio cholerae into South America.[57] Vibrio cholerae was isolated from the ballast, sewage, and bilge of 21% of cargo ships docked at ports in the Gulf of Mexico. The ships were previously in Brazil, Colombia, and Chile.[58] Vibrio cholerae O139 has traveled along waterways in Asia carried by people and boats generally engaged in commerce.[59] A strain of vibrio cholerae identical to an epidemic strain found in Latin America was detected in Mobile Bay, Alabama, within oysters and the fish that eat oysters.[60]

Researchers have regularly demonstrated that mosquitoes can survive international flights. When random searches were made of airplanes in London, mosquitoes were located on 12 of 67 airplanes from tropical countries.[61] Arthropods can survive even more extreme environments. In another study, mosquitoes, house flies, and beetles placed in wheel bays of Boeing 747B aircraft survived flights of 6 to 9 hours with external temperatures of –42°C.[59] Airplanes have also carried infective mosquitoes that caused human infection outside malaria-endemic areas (in Europe, for example).

More Than Transportation Required

Travel will permit a potentially pathogenic organism to be introduced into a new geographic area. But if it is to be established and cause disease, the organism must survive, reproduce, and enter a susceptible host. Fortunately, the majority of such introductions do not result in disease. Microbes that survive exclusively in the human host and are contagious through intimate contact and droplet nuclei can be readily transmitted worldwide. Many diseases such as tuberculosis, AIDS, pertussis, measles, diphtheria, and hepatitis are routinely transported around the globe. In those populations where a majority of the people have been vaccinated, the diseases find little opportunity to progress. Organisms are much more difficult to transplant when they have complicated life cycles that may involve animal hosts, vectors, or specific environmental requirements. Yellow fever, malaria, and dengue fever require specific mosquito vectors or they cannot proliferate in a particular geographic area. A snail intermediate host is required for the development of certain life stages in Schistosomiasis, so the disease organism cannot spread in the absence of this host. Even when all the conditions are present to permit the successful introduction of a parasite in a geographic area, the development of human disease is not an absolute consequence. The hog tapeworm, *Taenia solium*, is often not successfully transmitted in developed nations because there are adequate methods for confining and treating human sewage that prevent the eggs from infecting other persons. Briefly, the probability of transmission involves many biological, social, and environmental variables.[53]

Technology and Industry (Globalization)

High-volume rapid movement characterizes not only travel but also other industries in modern society. In operations that include food production, the use of products of biological origin or the use of modern production methods increases the chances of accidental contamination and amplifies the effects of such contamination.[62] The problem is complicated by globalization because agents are introduced from far away. Globalization has been defined as the process of denationalization of markets, laws, and politics in the sense of interlacing peoples and individuals for the sake of the common good.[63] Countries are finding it increasingly difficult to solve the problem of emerging infectious diseases, and public health policy is becoming denationalized to attack this problem.

Globalization is influencing public health in three ways.[62] First, the diseases are moving rapidly around the globe because of technology and economic interdependence, which has increased international travel and the international nature of food processing and handling.[53] As an example of technology increasing the risk of emerging diseases, the concentrating effects that occur with blood and tissue products have inadvertently disseminated infections unrecognized at the time, such as HIV and hepatitis B and C. Medical settings are also at the leading edge of exposure to new diseases, and many emerging infections have spread nosocomially in health care settings. Additionally, the common use and misuse of antibiotic treatments has permitted the development of drug-resistant pathogens.[64,65]

Food processing is also a contributory factor to emerging diseases. As an example, bovine spongiform encephalopathy (BSE), which emerged in Britain within the last few years, was probably an interspecies transfer of scrapie from sheep to cattle.[66] This probably occurred when changes in rendering processes resulted in incomplete inactivation of the scrapie agent in sheep byproducts fed to cattle.[67] Second, the funding of public health programs has been reduced because of increased competition in the global market and increased pressures to cut expenditures, thereby negatively affecting public health programs.[68] Third, public health programs have become international through WHO and health-related nongovernmental organizations. Worldwide health has improved through these international efforts with the elimination or reduction of many diseases. Unfortunately, these successes have contributed to a population crisis and produce overcrowding, inadequate sanitation, and overstretched public health infrastructures, leading to the expansion of disease.[65]

Microbial Adaptation and Change (Resistance)

Antibiotic Resistance

Different mechanisms of genetic change for microbes are routinely discovered. Spontaneous mutation is just the beginning. Microbes most often survive in enormous populations and are influenced by a myriad of physical and chemical mutagens. Haploid organisms are able to instantly express their genetic variations. They have numerous repair mechanisms that are subject to genetic control. Some strains that do not repair their DNA are therefore highly mutable. Others are relatively more stable. Some organisms are highly flexible in responding to environmental stresses. There are abundant processes that allow bacteria and viruses to readily exchange genetic material. Plasmids now spread throughout the microbial world. They can intersect the boundaries of yeast and bacteria. Lateral transfer is very substantial in the development of microorganisms, influencing the expressions of pathogenicity, toxicity, and antibiotic resistance. Microbes have a wide variety of mechanisms that allow them to exchange genetic information[69] and are, therefore, constantly evolving.

Antibiotic-resistant bacteria are emerging from the environment in response to the wide distribution of antimicrobials. This is an evolutionary lesson on microbial adaptation and a display of the power of natural selection. Selection for antibiotic-resistant bacteria[70] and drug-resistant parasites have become common, generated by the wide and often unsuitable use of antimicrobial drugs.[71,72] Pathogens can also attain antibiotic resistance genes from other, often nonpathogenic, species in the environment, selected or perhaps even driven by the ongoing pressure of antibiotics.[73] Many viruses demonstrate a high mutation rate and can rapidly evolve to yield new variants.[74] A classic example is influenza.[75] Yearly epidemics result from antigenic drifts in a previously circulating influenza strain. An alteration in an antigenic site of a surface protein, usually the hemagglutinin (H) protein, permits the current variant to reinfect formerly infected persons because the immune system no longer recognizes the altered antigen. Rarely, the evolution of a new variant may result in a new expression of disease.[29] It is possible that some recently reported characteristics of disease by Group A Streptococcus, such as rapidly invasive infection or necrotizing fasciitis, may also fall into this category.[25]

Only during the last 50 of 3000 years of recorded history have humans been able to cure an individual of a communicable disease. To underline the significance of this apparently mild statement, consider that in 1925 the number one killer in America was diarrhea and enteritis caused by bacterial infection; 109 of every 100,000 men, women and children were killed. This is roughly three times the current death rate from all causes combined. In the same year, the top ten causes of death included scarlet fever, pneumococcal pneumonia, whooping cough, diphtheria, and tuberculosis. Sixty years later, diarrhea and enteritis in this country were responsible for one in one million deaths.[76]

There is growing concern that bacterial pathogens are developing a resistance to antibiotics as a result of patients not completing the prescribed course of treatment or the inappropriate, and overprescribing, of common antibiotics by physicians. Most of the bacteria are destroyed in the early course of treatment, generally the weakest bacteria. However, the strongest survive and mutate (build resistance) so that the drug becomes ineffective. Thus, bacteria that develop a resistance to a specific antibiotic will not respond to the drug if it is prescribed later.

The use of unsupervised prophylactic tetracycline administration to 100,000 pilgrims enroute to Mecca from Indonesia is thought to have been responsible for

50% of the cholera strains in that country which are now tetracycline resistant. For a country as poor as Indonesia, with an average of $4.52 annual medical expenditure per capita, this is a devastating blow.[77] Tetracycline is one of the easiest and cheapest drugs to produce, which makes it safe and widely available in the third world. Other drugs often need to be imported and are so expensive as to be out of reach for the general populace.

A different example of resistance developed in shigella. This disease is the cause of dysentery throughout the developing world. After the development of multi-drug-resistant shigella in Africa, it appeared almost immediately in Bulgaria (resistant to four drugs), on the Navajo and Hopi reservations in the U.S. (resistant to two drugs), and in Vietnam, where 90% of the cases were resistant to three or more antibiotics.[78] Shigella was being carried to the four corners of the Earth and breaking out in areas with insufficient water treatment facilities. This demonstrates the disturbing properties of drug-resistant bacteria to travel vast distances.

While it took 15 years from the introduction of penicillin for widespread penicillin resistance to be reported to N. gonorrhoea in 1962, physicians who began using the flouroquinolone group of antibiotics (ciproflaxacin, ceftriaxone) in the late 1980s began reporting a large number of cases resistant to the drug by 1992.[16] This acceleration in conversion times was not limited to gonorrhea but is universally seen across most common bacterial infective organisms. It suggests something greater than a cause-and-effect when antibiotic misuse by physician "A" leads to a mutant strain in patient "B;" these drug-resistant strains are displaying an evolutionary trait. Over the last 50 years, we have not only bred-resistant organisms, but have selected for organisms who have the ability to develop resistance quickly.

By the last quarter of the 20th century, the threat of bacterial infections had begun to wane from the consciousness of the American mind. People who died from diseases such as pneumonia, like Jim Henson, creator of the Muppets, caused a stir because they were the exception to the rule that modern medicine could tackle bacterial infections. Part of the popular faith in medicine came from the perception of research as a constant fountain of new compounds. An array of new drugs seemed to appear every few years to be aggressively marketed by the pharmaceutical industry. In the early 1950s, there were three classes of antibiotics to choose from, including the beta-lactams (penicillin), tetracyclines, and aminoglycerides. In the 1990s, there are seven classes, although several are close cousins to each other.

Drug-resistant infections could be treated with other more powerful drugs like Vancomycin and Rocephin, which many hospitals consider their "big guns" in the disease war. The appearance of vancomycin-resistant strains in hospitals during the late 1980s has encouraged a view among many epidemiologist and bacteriologists that hospitals are the source of many forms of drug resistance. A recent report by the CDC found that Vancomycin resistance measured at 0.3% in 1986 and rose to 7.9% across several facilities in 1994.[78]

Antibiotics in Livestock Feed

Many studies from various sources, including the CDC, the FDA, and the Department of Agriculture, suggest a strong link between the use of antibiotics in livestock feed and the transmission of drug-resistant bacteria to humans. About 40% of antibiotics manufactured today are given to livestock, with 50 million pounds made each year.[24] Antibiotics have been added to livestock feed for many years to promote growth. Farmers have found that animals are able to absorb more nutrients from the feed and therefore grow more if the bacteria in the gut are killed. Unfortunately, organisms such as salmonella and camplylobacter, which are harmful diseases if contracted by humans, are beginning to appear increasingly more resistant to the antibiotics.[79] While people are not susceptible to all of these organisms, the bacteria are able to transfer their resistance to disease-causing bacteria through genetic material called plasmids.

In response to this discovery, the British Ministry of Health in 1989 adopted a White Paper on Livestock Feed and Antibiotics that regulated the administration of antibiotics to animals intended for human consumption. While the CDC has considered similar guidelines, they have never been adopted by congress due to pressure from livestock interest groups who fear a competitive burden if they are not able to use prophylactic levels of antibiotics in feed. While these practices do help to prevent some diseases in animals, over the long run they have been shown to select for resistant organisms, which can harm animals as well as humans.

Viruses

Antibiotics have no effect on viruses, and vaccines are often ineffective against bacterial infections. We have traditionally been able to combat such viral pathogens as rabies, polio, measles, and influenza, and have even eradicated smallpox through preventive vaccines. However, there are few antiviral drugs that exist, mainly because drugs that kill viruses also harm or kill its host cell (i.e., healthy cells). In addition, new viruses are continuously emerging, making it difficult for vaccines to be developed in sufficient time.

Breakdown of Public Health Measures

During the 18th and 19th centuries, advancements in public health vastly improved the overall health of the populace, particularly in urban settings. Improvements included reducing the exposure of drinking water and food to contaminated waste as well as improvements in the populations' natural mechanisms of resistance though better nutrition.[77] Vector control, chlorination of water, pasteurization of milk, immunization, and proper sewage disposal are classical public health and sanitation measures that have successfully minimized the spread of infectious diseases in humans. The pathogens, however, may exist in smaller numbers within reservoir hosts or the environment, and they represent a continuing threat for reemergence if these sanitary methods are not maintained. Well-understood and recognized diseases such as cholera are rapidly increasing because once-active public health measures have lapsed. This retrogression in sanitary methods unfortunately now appears frequently in both developing countries and the inner cities of the industrialized world. The reemergence of infectious diseases may, therefore, signal the disruption of public health measures and should demand vigilance against indifference in the war against infectious diseases.[25] For example, cholera has been fiercely erupting in South America and Africa for the first time in this century.[80] The swiftly expanding cholera outbreak in South America may have been aided by recent reductions in chlorine levels used to treat water supplies.[81] Cholera and other enteric diseases follow the fecal oral route of transmission, and so are frequently promoted by the lack of a reliable water supply. Problems such as these are aggravated in developing countries, but are not confined to these areas. The U.S. outbreak of waterborne Cryptosporidium infection in Milwaukee, Wisconsin, in the spring of 1993, with over 400,000 estimated cases, was partially the result of a nonfunctioning water filtration plant.[82] Water purification is a problem found in many other cities in the U.S.[83]

SPECIFIC EMERGING DISEASES

Viruses

According to Stephen S. Morse, the viruses with the greatest potential for emergence in the near future include hantaan (hantavirus), dengue, influenza, and HIV.[27] Therefore, these will be discussed in more detail. A discussion of Ebola is also included because it has become globally feared. The numbers of persons who have acquired the disease are exceptionally small compared with all other new and emerging diseases, but the gruesome manifestations of the disease, the high mortality, and the lack of treatment methods have spiraled this organism into the public eye with books, televisions shows, and movies. Therefore, it is discussed here to provide realistic contrast to the real worldwide killers.

Hantavirus

Background: the saga of Hantavirus outbreak in the U.S. began in 1994 when a physician was called to care for a man who collapsed at a funeral in New Mexico and died of acute respiratory failure. Soon after this, the physician found that several other persons in the Four Corners region (New Mexico, Arizona, Colorado, and Utah) of the Southwest had also died from similarly aggressive respiratory illness dating back to 1993. Initially, the victims were Native Americans located on a huge Navajo reservation. The media called it the "Navajo Flu." However, new cases were reported in the ensuing months from California, Rhode Island, and Florida. The Centers for Disease Control and Prevention (CDC) was contacted since the etiology of the disease could not be determined. CDC personnel descended on the area and and began a systematic effort to identify the disease. They collected blood and tissue specimens and victim histories, performed database searches, and tested for many different groups of potential pathogens. They then suspected a hantavirus, which is carried by rodents. CDC personnel then trapped and tested rodents from the area including the deer mouse, *Peromyscus maniculatis*. They found this to be the primary reservoir in New Mexico. The outbreak was brought under control when people in the area were educated on how to rid their homes of the mice and their droppings.[32] Although there have been isolated outbreaks in New York and Louisiana associated with different species of rodents, they have been controlled by practicing acceptable rodent control techniques and avoiding enclosed areas where the rodents may thrive.[84]

The Disease: the hantaviruses that emerged in the Four Corners region were determined to be the cause of an acute respiratory disease now termed hantavirus pulmonary syndrome (HPS). The identification was delayed because the symptoms varied remarkably from the characteristic Hantaviral diseases that had been reported in Europe and Asia. Therefore, it was an unexpected finding when sera from patients in this area reacted with hantaviral antigens, and genetic identification revealed a novel hantavirus in patients' tissues and in rodents trapped near patients' homes.[85,86] Initial symptoms of the American version (HPS) are flu-like and manifested as fever, chills, headache, muscular aches, and pains. Within a short time, however, the disease turns ugly and the lungs fill with fluids, causing severe respiratory distress for which there is no specific treatment. In HPS, death occurs from shock and cardiac complications, even with adequate tissue oxygenation.

Epidemiology: contraction of HPS in the U.S. is uncommon and fewer than 100 cases have been reported, but the mortality is high and there were 52 deaths among those reported to have contracted the disease. In contrast, people in China have been afflicted with a strain for thousands of years that annually causes 100,000 illnesses and perhaps as many 5000 to 10,000 deaths (5 to 10% mortality).[87] The hantavirus outbreaks in the Eastern Hemisphere (Asia) are identified with kidney failure and have been referred to as "hemorrhagic fever with renal syndrome" (HFRS). In the 1950s, Hantaan virus (isolated near the Hantaan River) caused Korean hemorrhagic fever with renal syndrome in thousands of United Nations troops.[88] Still others have been isolated in Scandinavia and called Puumala (PUU). About five to ten deaths per year (0.1% mortality) are caused by this strain.[89] Since the discovery of the HPS-causing hantaviruses, intense investigation of the ecology and epidemiology of hantaviruses has led to the discovery of many other novel hantaviruses. None of the states where this virus is located wanted it to carry the name of their state or community, so there was some bitter controversy over the naming of the virus. Scientists at the CDC finally decided on the term, Sin Nombre or "No Name," for the virus. Most of the 174 cases of HPS in the U.S. and Canada have been caused by Sin Nombre (SN) virus.[90] Cases of HPS in the southeastern U.S., as well as many in South America, are caused by a newly recognized clade (a group that shares a common ancestor) of viruses that includes Bayou (BAY), Black Creek Canal (BCC), and Andes viruses. These viruses are widespread and present a significant public health issue.[91]

The virus is spread to humans from contact with rodents. The pathway was first revealed in the early 1940s, when a viral etiology for HFRS was suggested by Russian and Japanese investigators who injected persons with filtered urine or serum from patients with naturally acquired disease.[88] These studies yielded the first evidence that the natural reservoir of hantaviruses was a rodent. The Japanese investigators were able to produce disease in humans by injecting bacteria-free filtrates of tissues from *Apodemus agrarius*, or mites, that fed on the Apodemus mice. In 1978, investigators demonstrated that patient sera reacted with antigen in lung sections of wild-caught *Apodemus agrarius*, confirming a rodent reservoir for HFRS-causing viruses. They also demonstrated that the virus could be passed from rodent to rodent.[92] In 1981, the successful propagation of Hantaan (HTN) virus in cell culture provided the first opportunity to study this pathogen systematically.[93] It is now established that hantaviruses can be carried by at least 16 various rodent species including rats, mice, and voles. Additionally, chickens and cats may be infected as well.[94] The viruses tend to be identified with a specific rodent host. In the Four Corners region the host is *Peromyscus maniculatis*, while in Florida the cotton rat (*Sigmodon hispidus*) is thought to be the primary carrier. Hantavirus infection does not appear to incapacitate its rodent reservoir host, which tends to produce a significant antibody response against the virion envelope and core proteins while maintaining chronic, probably lifelong infection.[91]

The emergence of the HPS virus in the U.S. is reported to be associated with environmental factors that promote the multiplication of a natural reservoir (field mouse) and so increases opportunities for human infection.[88] Researchers now report that this outbreak of HPS was probably initiated by climate irregularities associated with El Niño. The warming of the seas off the West Coast resulted in extensive rainfall in many areas including the Four Corners of the Southwest. This dramatically increased the natural foods (pine nuts and grasshoppers) of field mice and their numbers expanded ten-fold.[92] The HPS outbreak in the population of field mice subsequently invaded buildings to produce greater opportunities for indoor contact. The most common route of transmission to humans is by aerosolized mouse droppings containing the virus particles, although there is evidence that bites may also transmit the disease. Investigations have linked virus exposure to such activities as heavy farm work, threshing, sleeping on the ground, and military exercises. Exposure indoors was associated with invasion of homes by field mice during cold weather or to rodents nesting in or near dwellings. Persons of lower socioeconomic status appear to be at higher risk because housing conditions and agricultural activities permit more immediate contact with the mice. However, suburbanization, wilderness camping, and other outdoor recreational activities have spread infections to persons of middle and upper incomes.[88]

Dengue Fever

Background: dengue is considered as one of the most consequential mosquito-borne viral diseases to affect humans. This disease has a global distribution that rivals malaria and has the capacity to infect the more than 2.5 billion people on the planet at risk from epidemic transmission. There are in excess of ten million cases of dengue fever annually, with several hundred thousand cases of the more severe dengue hemorrhagic fever (DHF). DHF accounts for the majority of the 5% to 15% mortality that occurs worldwide, mostly in children.[96]

The Disease: there are four antigenically distinct viral serotypes that cause DHF. The serotypes are DEN-1 through DEN-4 and belong to the genus *Flavivirus*.[97] The expansion of dengue/DHF is aided by the fact that there is no cross-protective immunity with any of the viruses, so that it is possible for a person to acquire multiple dengue infections during a lifetime.[96] Dengue is primarily an urban tropical disease with severe flu-like symptoms that causes high fevers, frontal headache, severe body aches and pains, nausea, and vomiting. The "critical stage" for DHF occurs at the point at which people start to feel better. When the fever eases, patients start to develop "leaky capillary syndrome" in which the blood vessels leak and untreated patients will go into shock and die. Some cases show no outward symptoms, while others show bruising or pinpoint rashes that signal internal bleeding.[98]

Epidemiology: the greatest emerging health menace from dengue/DHF has been in Central and South America, where previous campaigns in the 1950s and 1960s to eliminate the mosquito vector, *Aedes aegypti*, to combat yellow fever met with success. However, complacency in mosquito-control programs has allowed *Aedes aegypti* to return with a vengeance, and it is a competent vector for dengue viruses. In the late 1980s, memories of mosquito-borne diseases faded, and other public health programs competed for the monies normally given to mosquito spraying. A rapid population increase throughout Latin America combined with rapid urbanization, poverty, and insufficient sewage and waste treatment methods resulted in an explosive growth in the mosquito vectors and a sixty-fold increase in dengue/DHF as seen in Figure 7.1.[96] Further, air travel has increased the probability of the virus spreading. Based on viral envelope gene sequencing, the DEN-3 virus strains in Latin America appear different than the DEN-3 strains previously isolated in these areas, and identical to the DEN-3 virus causing the major dengue epidemics in India

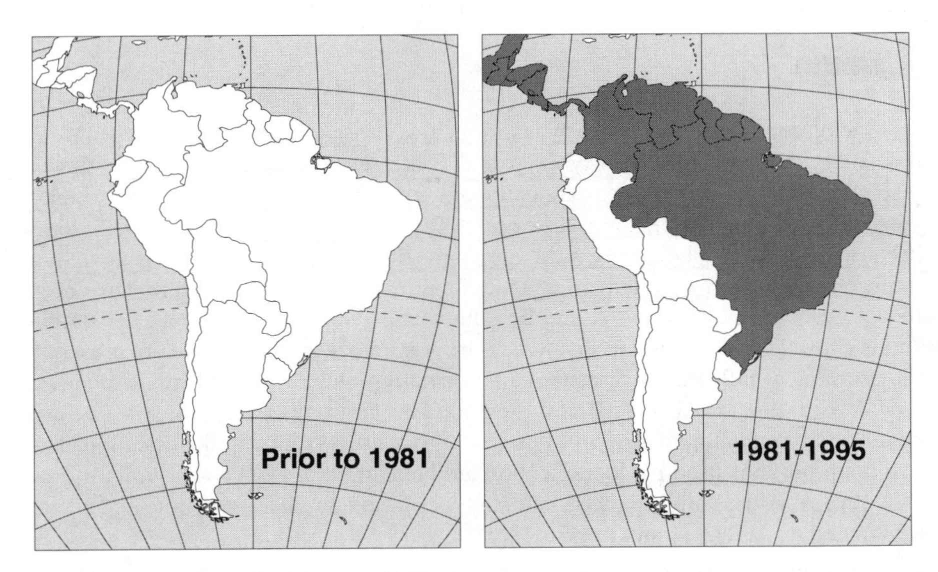

Figure 7.1 Distribution of *Aedes aegypti* (shaded areas) in the Americas in 1970, at the end of the mosquito eradication program, and in 1995. (Adapted from Dispatch, Dengue hemorrhagic fever: the emergence of a global health problem, *EID,* 1, 2, Apr.–Jun. 1995.)

and Sri Lanka.[96] Travel, therefore, appears to have been an important component in the spread of this disease.

Dengue fever and DHF has returned or emerged in many areas of the world (Figure 7.2). A dengue pandemic that began in Southeast Asia after World War II has increased dramatically since the 1980s, where it has emerged as the leading cause of hospitalization and mortality among children. Taiwan and China experienced a series of epidemics after being free of the scourge since the 1950s. When India and Sri Lanka experienced their first major epidemics, DHF once again reemerged in Asia. In 1994, Pakistan reported its first major epidemic with dengue. All four serotypes have been involved in the epidemics in China. In Asia, the Pacific Islands, Africa, and the Americas, the last 15 years have seen increasingly larger epidemics.[96,100] In 1998, 770,000 people in the Americas became infected.[101]

The likelihood of major outbreaks occurring in the U.S. is considered doubtful, but the possibility remains because the competent mosquito vectors are present, including *Aedes aegypti* and the Asian tiger mosquito, *Aedes albopictus*. These mosquitoes have the capacity to carry and transmit the dengue virus, and such transmission has been detected in south Texas in 1980 and 1986 and in northern Mexico.[102] More recently, dengue fever and DHF have been found within 10 miles of the U.S. border in Reynosa, Mexico, just across the line from McAllen, Texas.[98] Additionally, there have been more than 2000 suspected cases of imported dengue since 1977 in the U.S. with more than 20% of these cases confirmed as dengue.[103] The Unites States needs to maintain vigilance in controlling mosquito populations and maintaining the public health infrastructure that prevents uncontrolled urbanization, substandard housing, and inadequate water, sewage, and waste management.

Influenza

Background: there are three types influenza viruses designated as A, B and C, plus subtypes of A. Influenza types A and B are responsible for the epidemics of respiratory influenza that occur annually and are frequently connected with heightened rates of hospitalization and death.[104] Type C is quite different than A or B because it either produces very mild symptoms or none at all, and consequently has not been involved in any epidemics. Therefore, most public health efforts have been focused on controlling types A and B. Influenza is commonly referred to as the flu and is caused by these viruses that attack the respiratory system. There are presently three strains of influenza viruses that are circulating globally, and these include two type A viruses and one type B (Figure 7.3). The viruses contain proteins that can be detected immunologically, and this serves as a useful tool for identification. The two main proteins used for identification are hemagglutinin (H) and neuraminidase (N). Type A viruses are separated into subtypes based on these two proteins and are designated as A(H1N1) and A(H3N2).

The influenza type A viruses are also capable of undergoing gradual changes with a succession of mutations, or undergoing a sudden and significant change in the hemagglutinin (H) or neuraminidase (N) proteins. A gradual shift in the viral proteins is referred to as antigenic "drift," whereas abrupt changes in H or N proteins is called an antigen "shift." Changes in the type B virus appear limited to antigenic

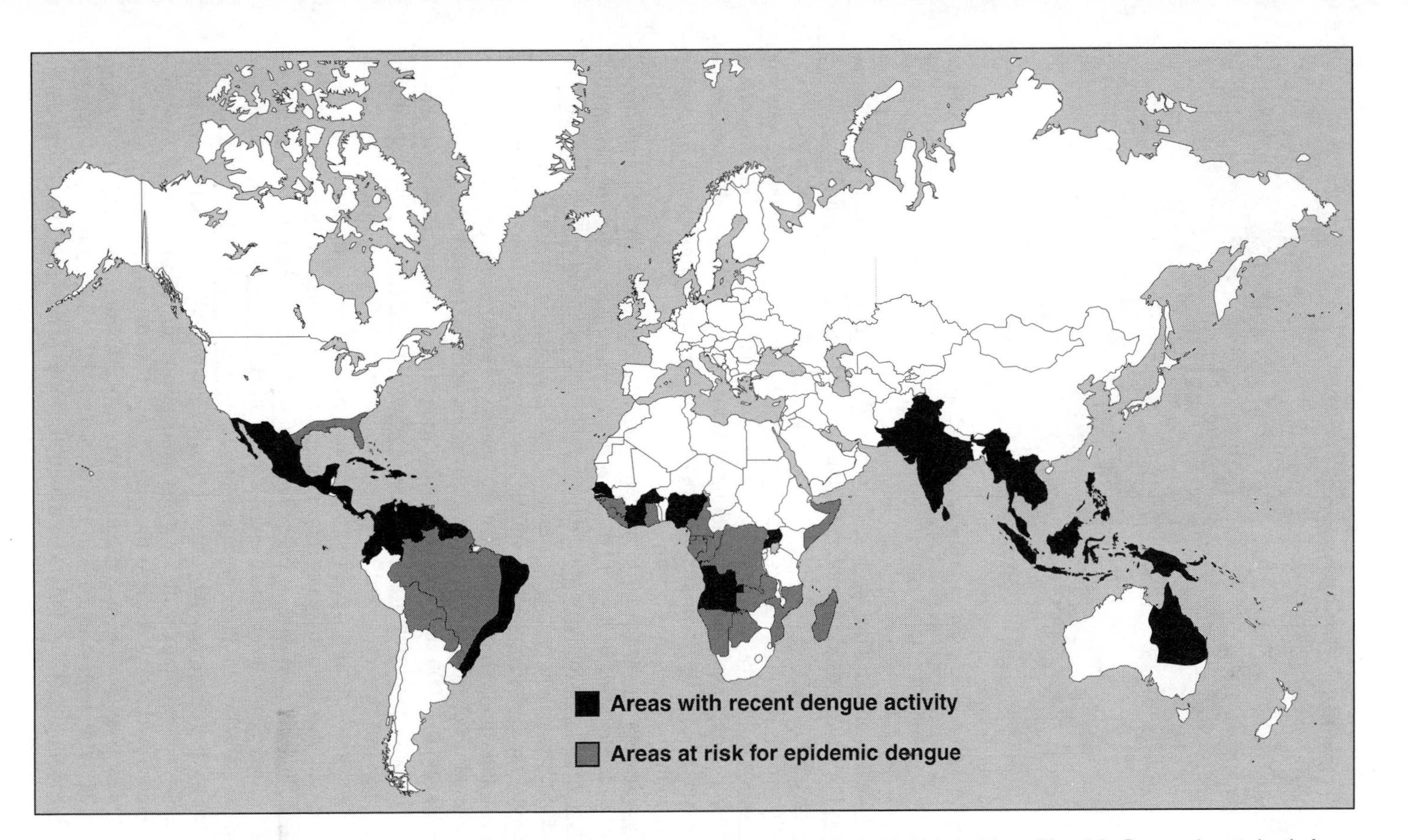

Figure 7.2 World distribution of dengue viruses and their mosquito vector, *Aedes aegypti*, in 1995. (Adapted from Dispatch, Dengue hemorrhagic fever: the emergence of a global health problem, *EID,* 1, 2, Apr.-Jun. 1995.)

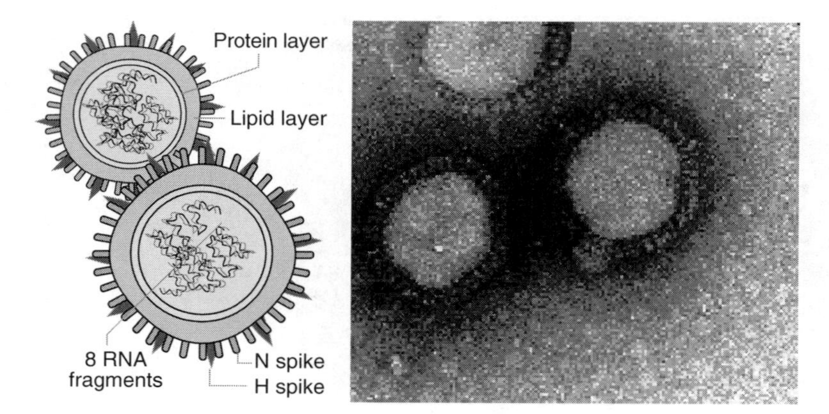

Figure 7.3 Photomicrograph of influenza virus. (Source: From World Health Organization (WHO), Emerging and other communicable diseases, Influenza A(H5NI) Fact Sheet 188, Jan. 9, 1999.)

drift. Changes in the viral protein through either of these mutation mechanisms permits the virus to evade immune reactions that may have been formed in response to contact with previous but slightly different influenza viruses. The altered viral proteins are no longer recognized by the immune cells and/or antibodies, resulting in a weakened response or no response at all. Influenza A and B viruses are continually expressing antigenic drift, resulting in seasonal flu outbreaks. When the more abrupt antigenic shift occurs, entire populations are left immunologically unprotected, and pandemics may occur.[104]

The Disease: the symptoms of influenza can vary from none to fatal. Influenza is normally characterized by a fever (100°F to 103°F); respiratory symptoms that include cough, sore throat, and stuffy nose; muscle aches and pain; and extreme fatigue. The course of the disease is usually self-limited and hospitalization is usually not required. Sometimes the disease produces vomiting and diarrhea, but this is normally in children and gastrointestinal symptoms are not the routine.[101] Most people recover within one to two weeks, but life-threatening complications such as pneumonia may develop. There are about 20,000 such deaths annually in the U.S., with the majority of serious illness and death occurring in the aged, very young, and debilitated. This is probably strain dependent, however, and recent infections with influenza A(H5N1) indicate severe symptoms in other age groups as well. However, the small number of cases involved with this strain may not reflect unreported and/or milder cases.

Epidemiology: influenza is an ancient disease, first described by Hippocrates in 412 BC. There have been to date more than 30 pandemics with three occurring within the last 80 years. There were epidemics in 1918, 1957, and 1968. The viruses involved were all from animals (pigs or birds). The mortality associated with these pandemics was enormous, with the worst pandemic occurring in 1918–1919 involving influenza A(H1N1) known as the Spanish flu. This disease caused an estimated 500,000 deaths in the U.S. and 20 million deaths worldwide. The 1957–1958 Asian flu or A(H2N2) resulted in 70,000 deaths in the U.S. followed by the 1968–1969 Hong Kong flu or A(H3N2) that killed 34,000 people in the U.S. The emergence of the Hong Kong flu in 1968 was relatively mild in terms of mortality, but it continues to circle the globe and to demonstrate antigenic drifting. Since the virus emerged in 1968, there have been an estimated 400,000 deaths worldwide with most of the mortality in elderly people. A(H3N2) remains the most destructive of the currently circulating influenza viruses.[104] A new influenza virus, A(H5N1), was isolated from a child in Hong Kong in May 1997 after the child died from Reye's Syndrome associated with taking aspirin in the presence of the flu virus. This is an avian virus that had a history of killing ducks and chickens with the latest avian outbreak in 1997 in Hong Kong. There were an additional 15 cases in Hong Kong, but no evidence of human-to-human transmission exists as of early 1998. There were four deaths recorded from A(H5N1) to the end of 1997, and most people infected had contact with chickens or ducks.

Following increased surveillance of A(H5N1) in Hong Kong and Guandong Province in China, there appears to be no risk of this virus being transmitted from raw, chilled, or frozen poultry foods. Investigations are now concentrating on the characteristics of the virus isolated from human cases, whether there are other persons who are becoming infected, and other natural reservoirs of the virus.

Additionally, efforts are under way to develop high growth reassortment H5 viruses for vaccine production, but no vaccine is being developed specifically for A(H5N1). This is made difficult because the viruses that have been isolated kill the fertilized eggs that are normally used to develop the vaccines. In order to reduce the risk of pandemic in the event the virus should become transmissible human to human, WHO has established a collaborative team that is providing global surveillance, identification of potential pandemic viruses, promotion of high growth seed virus for vaccine, and the facilitation of vaccine production and distribution.[106]

Ebola

Background: Ebola and Marburg viruses belong to a family of viruses called Filoviridae. These viruses have the appearance of long threads or filaments (Figure 7.4) and cause severe and often fatal hemorrhagic fevers in humans and nonhuman primates. Some strains have produced fatalities of nearly 90% in human outbreaks. Their extreme pathogenicity, combined with the lack of effective vaccines or antiviral drugs, classifies them as biosafety level four agents. These viruses are even more mystifying because their natural reservoir remains unfamiliar and the pathogenicity of the disease is unknown.[106] The Ebola virus was discovered in 1976 and was named for a river in Zaire, Africa, where it was first detected. The widespread public fear of the virus is probably due to the gruesome symptoms and high mortality, which has lead to a flurry of articles, books and movies. This fear is greatly exaggerated. There have been only 1400 deaths from Ebola in the last 20 years with nearly all of them being in Africa. This is contrasted with the millions of cases of HIV that occur in Africa and worldwide.[107]

The Disease: the Ebola virus is usually spread by close, personal contact with a person who is symptomatic with the disease. This most often occurs among hospital workers in Africa where sanitary conditions and infection control are limited and

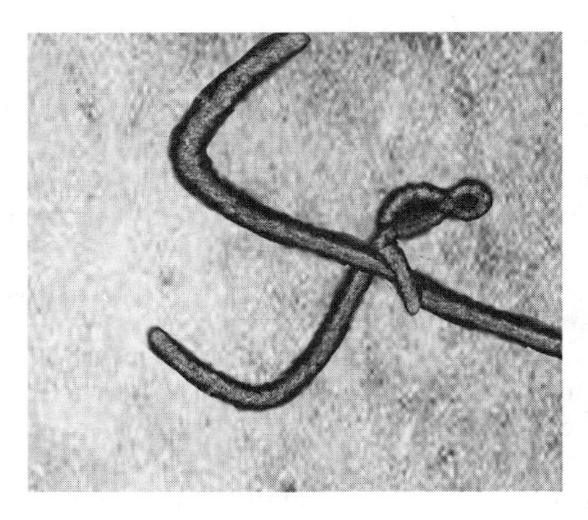

Figure 7.4 Ebola virus. (Simulated)

the reuse of hypodermic needles is common. Ebola fever typically starts suddenly 4 to 16 days after infection with malaise, fever, and flu-like symptoms, which can be followed by rashes, bleeding, and kidney and liver failure. Most patients appear overtly ill, dehydrated, apathetic, and disoriented. The fever associated with Ebola appears to begin with chills, headaches, muscle pain and aches, abdominal pain, sore throat, and chest pain. Generalized bleeding occurs with massive hemorrhaging of the internal organs, with bleeding into the gastrointestinal tract, from the skin, and even from injection sites as the clotting ability of the blood is diminished. The Ebola virus produces lesions in the liver, kidney, and spleen with areas of necrosis on the organs and the lymph nodes. The death of the patient usually occurs from shock within 7 to 16 days and is accompanied by extreme blood loss. The virus is spread through bodily fluids and secretions though not through casual contact.[106]

Epidemiology: infections from Ebola virus were first reported in 1976 when two outbreaks occurred at the same time, but in different locations and with different subtypes. One occurred in northern Zaire and the other in southern Sudan.[108] The total number of cases in these two outbreaks was 550 with 340 deaths, respectively. The case fatality rate from the Zaire subtype Ebola virus was 90% and the case fatality rate for the Sudan subtype was 50%. There was also an outbreak in Sudan in 1979 resulting in 22 deaths among the 34 cases identified, resulting in a case fatality rate of 60%, similar to the 1976 outbreaks. The Ebola Zaire strain emerged again on April 10, 1995, in an outbreak in Kikwit, Zaire. A patient admitted for suspected malaria infected the attending medical personnel with the Ebola virus, which initially spread rapidly but was brought quickly under control by a coalition of international health organizations, including CDC and WHO. The outbreak eventually resulted in 293 cases with 233 deaths. The U.S. escaped a major epidemic in 1989 when a new Ebola virus infected a colony of cynomolgus macaques at a Reston, Virginia, CDC facility that were imported from the Philippines.[109] This new virus was named Reston virus and was shown to be antigenically and genetically different than the African varieties. The Reston virus was highly pathogenic for the primates but produced no disease in the humans, although some people developed antibodies to the virus.[106] There was a cluster of chimpanzee deaths in the Tai forest on the Ivory Coast (Cote d'Ivoire) of western Africa in November 1995. A Swiss researcher contracted the disease and subsequently recovered. Researchers have been actively trapping small animals in a 20-square mile section of the canopied Tai National Forest in order to reveal the natural reservoir for the Ebola virus. Although there are plenty of suspects including bats, red colobus monkeys, and other small mammals, there has been no success in identifying the virus in any of the animals captured.[107]

The combination of impoverished populations, inadequate public health services, rapid transmission of the disease, and greater incursions into the African forests increases the possibility of more outbreaks. CDC and WHO have developed sensitive enzyme immunoassays and PCR (polymerase chain reaction) assays for the detection of filoviruses. The availability of rapid and reliable diagnostic tools will permit more effective disease surveillance programs that are necessary to minimize filovirus outbreaks.

AIDS/HIV

Background: AIDS is a virus that has been recently discovered to have originated from an endangered species of chimpanzee from Central Africa. It appears that the transmission of the virus from chimps to humans is related to rainforest destruction and the practice of human chimpanzee consumption.[110] The AIDS virus (Figure 7.5) belongs to a special group of viruses known as retroviruses and is referred to as human immunodeficiency virus (HIV). Its genetic information is not encoded as DNA, but instead as RNA (ribonucleic acid) and therefore has to be reverse transcripted into DNA. The tools for this are provided by the host cell itself, except for a little helper protein (reverse transcriptase) that the virus has brought with itself. When the virus enters the body, it prefers to attack certain types of white blood cells

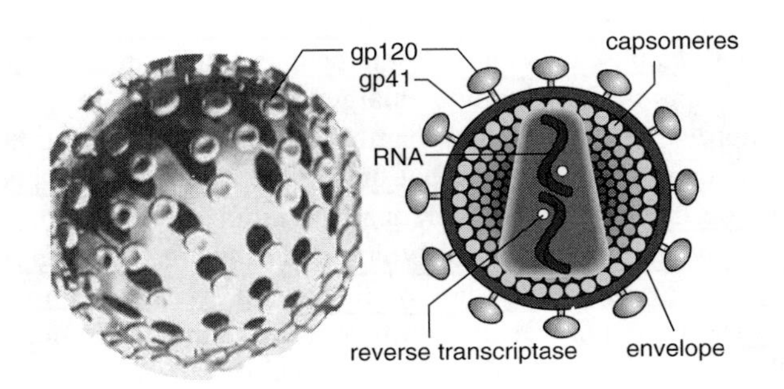

Figure 7.5 Human immunodeficiency virus (HIV).

that are a component of the cellular immune system. These cells are called helper T cells, which are an integral component of our immune system. The AIDS virus almost exclusively focuses on these white blood cells since these helper T cells have CD4 molecules on the surface to which the AIDS virus binds. The virus attaches itself to a special protein (CD4) on the surface of the helper T cell, resulting in a fusion of the viral membrane to the host T cell's membrane. The viral genetic information is then able to enter the cell and is transferred to the nucleus. This process may be accomplished in less than 12 hours after infection. The viral piece of DNA is then inserted randomly into the host DNA and ready for transcription. At the beginning of AIDS, the viral DNA is being transcribed to form many RNA molecules. The signal that causes this is yet unknown. The accumulating RNA is transported to the cytoplasm of the cell, where it can begin making proteins. The RNA, with the help of the host's resources, begins to make many copies of the different parts of the AIDS virus (the protective shell and the helper and anchor proteins). Once everything has been copied, thousands of partial virus particles migrate to the surface of the cell membrane and fuse with it. Finally, a copy of the RNA genetic information is added to the viral particle. Then this section of the cell membrane turns inside out and new viruses leave the cell.[111]

The Disease: HIV is transmitted most commonly by sexual contact with an infected partner and can enter the body through the vaginal lining, vulva, penis, rectum or mouth. Since the virus appears in the blood and many body fluids, it can be transmitted by infected blood and through contaminated needles. The risk of spreading HIV through transfusions has been significantly reduced since the introduction of heat-treating techniques in 1985. HIV has been transmitted to fetuses during pregnancy and birth. Fortunately, the administration of the drug AZT during pregnancy has dramatically reduced the risk of transmitting the virus to the fetus from its nontreatment level of 20% to 30% infection rates. HIV has also been transmitted to babies through the breast milk of HIV-infected mothers. There is no evidence that casual kissing or contact will result in infection. Simply touching objects such as telephones, towels, and the like that have previously been used by HIV-infected individuals will not spread the disease to uninfected individuals. There is also no known insect vector for HIV.[112]

Many people remain symptomless for months or years after acquiring the infection, while symptoms may develop in some people within one to two months after infection. The symptoms may include enlarged lymph nodes, fever, headache, and malaise. The more severe and lasting manifestations of HIV infection may surface from months to years after infection, but during the asymptomatic period the virus is infecting and destroying T4 cells. Gradually, a number of symptoms may emerge that include prolonged enlargement of lymph nodes, energy and weight loss, recurrent sweating and fevers, skin rashes or flaky skin, yeast infections, and pelvic inflammatory disease. A number of victims develop herpes infections with painful sores about the mouth and genitals, or a more painful disease known as shingles. During the early and intermediate periods of the disease, diagnosis depends on the detection of HIV antibodies in the blood. The antibodies may take as many as 6 months to appear in sufficient quantities in the blood to be detected by available tests such as ELISA and Western Blot.

The disease will often advance to a stage referred to as AIDS or acquired immunodeficiency syndrome. CDC has defined AIDS to include those people who have evidence of HIV and also fewer than 200 T4 cells. The definition also includes more than 25 opportunistic infections that have historically affected people with AIDS because of the immunocompromised state. One of the most common and frequently fatal opportunistic infections is Pneumocystis carinii pneumonia. The opportunistic infections produce a myriad of debilitating symptoms from respiratory distress, severe headaches, extreme fatigue, nausea, and vomiting, to wasting and coma. The continued decline in T4 cells and associated immunity increases the opportunity for the development of cancer. Many cancers may develop in AIDs patients, with some of the most common being Kaposi's sarcoma or lymphomas. Karposi's sarcoma is often evidenced by pigmented spots that develop on the skin.[112]

Although treatments continue to enter the war on AIDS, including reverse transcriptase inhibitors (AZT) and protease inhibitors, these drugs are likely to produce severe side effects, and HIV may become resistant to many of them. A number of drugs are also available to treat many of the opportunistic infections that are likely to arise when the immune system is compromised. No vaccine has been successfully introduced to date, and the best recommendation is to avoid high-risk behaviors that include unprotected sex or the sharing of needles.

Epidemiology: a joint surveillance effort by UNAIDS and WHO estimates that over 30 million people were living with HIV infection by the end of 1997. This translates to one in every 100 sexually active adults from 15 to 49 worldwide, with about 90% of the HIV-infected people living in the developing nations. This figure also includes over one million children under the age of 15. HIV continues to spread rapidly at the rate of 16,000 new cases daily and a total of over 5.8 million people in 1997. About 2.3 million people died of AIDS in 1997, and AIDS is responsible for about 20% of all deaths since the epidemic began in the 1970s. More than two thirds of the total number of people in the world living with HIV are from sub-Saharan Africa (Figure 7.6). This is contrasted against the 6% decrease in new AIDS

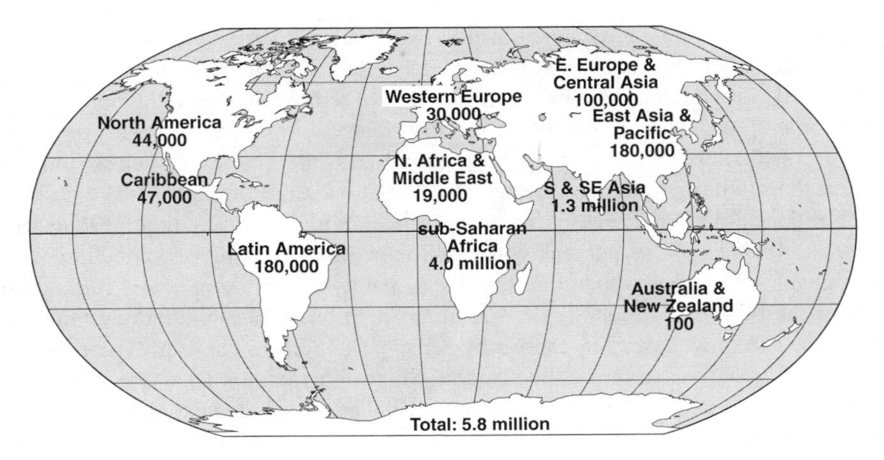

Figure 7.6 Distribution of HIV-infected pesons globally for 1997. (UNAIDS and WHO, Report on the Global HIV/AIDS epidemic, Dec. 1997, http://www.unaids.org/unaids/document/epideinio/report97.html.)

cases in the U.S. for the first time ever in 1996, which continued to decline in 1997. Unfortunately, AIDS cases rose in some sections of American society, including heterosexual African Americans and Hispanics.[113,114]

Bacteria

Escherichia Coli

Background: *Escherichia coli* are Gram-negative, facultatively anaerobic, short, straight rods that characteristically inhabit the intestines of humans and other animals and belong to the family Enterobacteriaceae. This family of enterics has a number of important pathogens including agents of bacillary dysentery (*Shigella dysenteriae*), gastroenteritis (Salmonella and Shigella), and the agents of typhoid (*Salmonella typhi*). Members of the enterics cause gastroenteritis, mostly, but have also been implicated in urinary tract infections, wound infections, pneumonia, septicemia, and meningitis. Strains of *Escherichia coli* have been implicated in many of these disorders. However, most strains of *E. coli* are harmless inhabitants of the intestines of warm-blooded animals and humans. Occasionally, however, a strain develops that is more pathogenic than others. *E. coli* 0157:57 is such a strain. The combination of letters and numbers in the name of this bacterium refers to the specific markers found on the surface of the organism and permits it to be distinguished by serological and other techniques from the hundreds of other strains. Infections with *E. coli* 0157:H7 were first described in 1982, and it has emerged along with a few other strains of *E. coli* as the predominant cause of hemorrhagic colitis in humans. The strains of *E. coli* capable of causing hemorrhagic colitis are referred to as Enterohemorrhagic *Escherichia coli* (EHEC).[115]

The Disease: *Escherichia coli* 0157:H7 is pathogenic for humans and has characteristically produced bloody diarrhea with abdominal cramps. Sometimes the infection causes non-bloody diarrhea with very few symptoms. There is usually little or no fever and the illness resolves in 5 to 10 days. In some persons, particularly children under 5 years of age and the elderly, the infection may progress into a more severe and life-threatening form of the disease known as hemolytic uremic syndrome (HUS). HUS is characterized by the destruction of red blood cells and kidney failure. HUS is a life-threatening condition usually treated in an intensive care unit with blood transfusions and kidney dialysis. Rapid and intensive care such as this can reduce the mortality rate to lower than 5%. About one third of the persons with HUS have abnormal kidney function many years later, and a few may require long-term dialysis. Another 8% of persons with HUS have other lifelong complications such as high blood pressure, seizures, blindness, paralysis, and sometimes removal of part of the bowel. In the U.S., HUS is the leading cause of acute kidney failure in children, and most cases are caused by serotype 0157:H7. This serotype produces a group of cytotoxins known collectively as Shiga-like toxins (SLTs) that are thought to be factors producing some of its virulence.[116] The toxins are very similar to toxins produced by a pathogenic Shigella, another member of the family Enterobacteriaceae.[117]

Epidemiology: the American public was startled into awareness with this disease in 1993 following a foodborne outbreak traced to the undercooked hamburgers eaten at a fast-food chain restaurant.[118] The outbreak involved the infection of 700 persons from four different states, with 51 of these persons developing HUS and four people dying from the syndrome. Serotype 0157:H7 was implicated in this outbreak, but it was first isolated 11 years earlier in 1982. There were more than 30 outbreaks to follow in the 10 years before the 1993 tragedy.[116] Since the 1993 outbreak, there have been increased reports of 0157:H7 infections giving rise to more than 60 outbreaks. Part of this increase may be due to elevated awareness among doctors, microbiologists, and consumers. As there is little evidence that this organism caused infections before the 1970s, it is an example of the emergence or evolution of a foodborne pathogen.

The majority of infections with serotype 0157:H7 have come from eating undercooked beef products, but many other sources of infection have been identified. Outbreaks have been associated with acidic foods such as fresh-pressed cider in which 27 persons were infected with serotype 0157:H7.[119] The apples used to make the unpasteurized cider without preservatives were thought to have been contaminated with cow manure.[120] In another outbreak involving several restaurants and 48 infected persons, the implicated food was mayonnaise which is also acidic to a pH of 3.6.[121] The serotype 0157:H7 organism was thought to have been transferred to the mayonnaise by cross-contamination with meat products, and this bacteria has been shown capable of surviving up to 55 days in this acidic environment.[121] Serotype 0157:H7 infections have also been traced to both drinking water and recreational water. An outbreak in Missouri involving more than 240 people with four deaths occurred in 1989 and was associated with contaminated drinking water.[122] Fecal contamination of lake water in Portland, Oregon was thought to be responsible for an outbreak of serotype 0157:H7 among swimmers where 21 children were infected. Based on these findings, infection with serotype 0157:H7 is most often associated with eating undercooked beef products, but contamination can come from drinking contaminated raw milk or other unprocessed raw beverages that might come into contact with human or bovine feces. Swimming in or drinking water contaminated with sewage is also a risk. Finally, the organism has been found capable of surviving in highly acidic foods once thought to be low risk for the transmission of bacterial foodborne infection.

Lyme Disease

Background: Lyme disease is caused by the spirochete *Borrelia burgdorferi*, a Gram-negative, slender, flexible bacteria that is helically coiled (Figure 7.7). The organism is anaerobic and fermentative in its energy metabolism and it is spread to humans by the bite of ticks of the genus *Ixodes* (Figure 7.8). The tick responsible for the spread of Lyme disease in the northeastern and North Central U.S. is known as the deer tick and characteristically feeds on the white-tailed deer, the white-footed mouse, other mammals, birds, and even reptiles.[123,124] The black-legged tick is responsible for the transmission of the disease along the Pacific Coast and in the

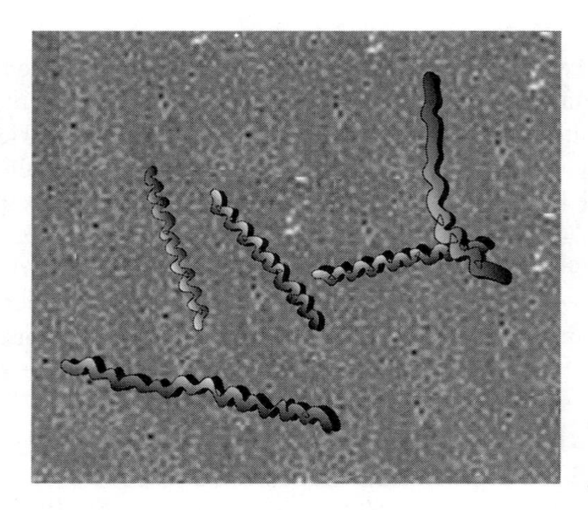

Figure 7.7 The sphirochete, *Borrella burgdorferi.* (Simulated)

Figure 7.8 Female *Ixodes scapularis.* This tick species is capable of transmitting Lyme disease to humans.

southeastern U.S. The Ixodes ticks are very small, about the size of a pinhead in their lymph and larval stages. The adult stages are just slightly larger.

The lymph stage is most likely to be the infective stage of the vector since it actively feeds, and its smaller size allows it to remain undetected for the 24 to 48 hours needed to complete a blood meal and transmit the infective spirochete. The tick is usually transmitted from the tips of grass and brushes to persons who brush against them. The ticks will search out any unprotected body parts but often attach to hidden and hairy areas such as the scalp, groin or armpits. Here they will insert their mouth parts into the skin and begin a several-day process of feeding on the blood that enlarges their bodies. The life cycle of these ticks is complex and requires 2 years to complete. Typically, the adult ticks will feed on large animals such as the deer in the fall and early spring. The females will eventually drop to the ground where they will lay their eggs, which hatch into larvae by the summer. The larvae will feed on mice and other small animals and birds throughout the summer and fall. They will become inactive through the winter and will molt to nymphs in the spring. This stage will feed on small rodents, other small mammals, birds, and the unprepared human that comes into contact with them on the tips of grass and brush, usually at the forest edge, but also lawns and gardens. During the normal cycle, the nymphs will develop into adults, completing the 2-year cycle.

The Disease and Epidemiology: Lyme disease was first reported in 1975 near Lyme, Connecticut, following a mysterious outbreak of arthritis. Reports of Lyme disease escalated since then, placing Lyme disease in the category of an emerging disease. First reported in 1975, the Centers for Disease Control and Prevention reported 226 cases in 1980 and more than 500 cases in 1983. The disease occurs primarily in the coastal regions of the northeastern U.S. including Connecticut, Rhode Island, New York, Maryland, and Massachusetts. It is also reported in the western states of California, Nevada, Oregon, and in many midwestern states. The early stages of Lyme disease are characterized by headache, fever, chills, swollen lymph glands, muscle and joint pain, and a characteristic skin rash (erythema migrans). This red rash is usually circular and appears 3 to 30 days after the tick bite. The rash often appears on the thigh, trunk, groin, or armpits. This rash (or rashes) will expand to include a large area and will often show clearing in the center to give a bulls-eye appearance. Diagnosis of Lyme disease is made difficult in the absence of the typical rash, and serologic tests may be required. Untreated, Lyme disease may progress to arthritis with pain and swelling in the large joints, paralysis of the facial muscles, and a meningitis characterized by stiffness in the neck, fever, and severe headache. Lyme disease rarely results in death, but chronic Lyme disease can lead to permanent damage to joints or the nervous system.

A newly discovered bacteria carried by the same tick that carries Lyme disease is also emerging as a significant agent of disease. The organism causes Human Granuloctyic Ehrlichiosis (HGE). It produces high fever, headache, muscle aches, chills, and sometimes death. Most cases have been identified in Wisconsin and Minnesota, but the disease can be found everywhere there is Lyme disease, and about 10% of patients with Lyme disease may also have HGE. This is important because HGE responds only to tetracycline and doxycycline.[125]

Prevention is best accomplished by avoiding tick-infested areas in the summer periods. Wearing light-colored clothing that covers arms and legs and is tucked in at the sock and belt line (with tape at the sock line) will help to prevent the ticks from crawling underneath clothing and make them easier to spot against a light background. Spraying with DEET on exposed skin surfaces other than the face will also be helpful. A vaccine has been approved by the FDA called LYMErix.[126] It must be administered in three injections spread out over the course of several months. It is limited to people 15 to 70 years of age and has only an 80% rate of effectiveness. Booster shots may also be necessary to keep the vaccine effective. The FDA recommends that all other preventative measures be taken whether or not one has received the vaccine. One out of five people who have had the vaccine will still be at risk for acquiring the disease.

Streptococcus

Background: Streptococci are Gram-positive cocci (spheres) arranged in chains or in pairs, and include many species that are pathogenic. There are several groups of Streptococci based on serological reactions to surface antigens, according to the original work of Lancefield in 1933. These groups include A-H, J, and K. The group A Streptococci (GAS) are commonly found in the throat and on the skin. They may cause no symptoms, but may also cause infections that range from mild to severe and even life threatening. The major pathogens are included in groups A and B, and their pathogenicity is associated with certain enzymes and surface proteins including hemolysins, erythrogenic toxins, and M-protein.[127] Hemolysins are enzymes capable of breaking or lysing blood cells. When Streptococci are grown on sheep blood agar, the types of hemolysis produced place these organisms into three groups. There are gamma hemolytic strep that produce no hemolysis, alpha hemolytic strep that produce a zone of green discoloration around the colony, and beta hemolytic strep that produce a zone of clear hemolysis around the colony. Beta hemolytic, group A strep are most often associated with strep throat infections. The ability to hemolyse blood cells is associated with powerful hemolysins such as oxygen-sensitive streptolysin O and oxygen stable streptolysin S enzymes. Streptococci may also produce erythrogenic toxins (A, B, and C) that harm blood vessels beneath the skin and appear to be produced in strep that carry a provirus with the genetic materials capable of coding for the production of these toxins.[127] The streptococci may produce a broad array of enzymes including neuraminidases, hyaluronidases, streptokinases, ATPases, DNAses, and many others that participate in the destruction and invasion of human tissue.

The Disease: the group A Streptococci produce a variety of diseases that include strep throat, impetigo, and scarlet fever. Occasionally, the strep invade deeper tissues such as blood, lungs, or even fat and muscles. The more severe of these invasions results in necrotizing fasciitis and/or streptococcal toxic shock syndrome. The popular media characterized the strep that caused necrotizing fasciitis as "flesh-eating bacteria" because of their ability to produce a progressing destructive infection of the underlying tissues. Streptococcal toxic shock syndrome is caused by a highly invasive strep that produces shock and injury to internal organs such as the kidneys, liver, and lungs. Necrotizing fasciitis usually begins with a trivial or even unnoticed trauma that appears in 24 hours as a lesion with swelling and redness. This is a deep-seated infection of the subcutaneous tissue that progressively destroys the underlying connective tissue and fat while sometimes sparing the skin and muscle. The early signs and symptoms of necrotizing fasciitis include fever, severe pain and swelling, and erythema (redness) at the site of the wound. The wound deeply colors to blue with blisters that contain a clear yellow fluid. Within 4 to 5 days, the purple areas become gangrenous, and at 7 to 20 days, dead skin separates at the margins of infection, revealing marked necrosis of the underlying subcutaneous tissue. In a few cases, this process quickly escalates until several large areas of gangrenous skin are apparent and the patient may become dull and unresponsive due to the toxins in his or her system. Many of the present GAS cases of necrotizing fasciitis inevitably lead to severe systemic illness with high morbidity even in otherwise healthy patients who receive extensive support in the form of antibiotics, dialysis, and surgical techniques.[127] These recent developments, along with the increased mortality from GAS, suggest the emergence of a more virulent streptococci.[128]

Streptococcal toxic shock syndrome (strep TSS) is defined as any group A streptococcal infection associated with the early onset of shock and organ failure. The initial symptoms of strep TSS are sudden and severe pain in an extremity or may mimic peritonitis, pneumonia, or heart attack. About one fifth of the patients develop flu-like symptoms with fever, chill, nausea, vomiting, and diarrhea. Fever and evidence of soft tissue infection are present in most patients, with a majority of these progressing to necrotizing fasciitis.[127] Patients are often admitted already in shock; and despite treatment, this shock continues combined with renal dysfunction and acute respiratory distress in about 55% of the patients. Mortality rates of 30 to 70% have been reported.[129]

Epidemiology: beginning in the 1980s there was a sudden elevation in the reporting of a highly invasive group A streptococci infection with or without necrotizing fasciitis associated with shock and organ failure. The number of these cases appeared to increase from almost none to over 2000 per year.[130] There are about 10,000 to 15,000 cases of invasive GAS in the U.S. annually, resulting in 2000 deaths. The mortality rate for streptococcal TSS is about 60% of the 2000 to 3000 cases reported per year. Annually, about 20% of the 500 to 1500 patients who acquire streptococcal fasciitis have died. These figures should be contrasted to the millions of people who acquire strep infections annually. In fact, GAS is an organism with historic significance that has produced many outbreaks in the past, and the number of people involved were far greater than the several thousand cases of invasive GAS that are presently seen on a yearly basis. The argument has been made that GAS may have become more virulent and so produced these severe cases. This, however, may not be the only reason for the more intense invasions. Stevens argues that the emergence of a more virulent strain of GAS would have produced many more cases in the form of an epidemic.[128] It has been suggested that people are becoming more susceptible to GAS infections because of the use of nonsteroidal anti-inflammatory drugs (NSAIDs) such as ibuprofen. These observations have been anecdotal but reported from more than one medical group.[130] According to Stevens, NSAIDs block the production of prostoglandins, prevent fever, and encourage overproduction of certain chemicals (TNF or tumor necrosis factor) in macrophages that mask symptoms associated with strep TSS, and contribute to shock and organ failure.[131]

Tuberculosis

Background and Disease: tuberculosis (TB) is a chronic infectious disease of the lower respiratory tract caused by Mycobacterium tuberculosis, a slender, acid-fast rod with cell walls containing high lipid levels. The slow-growing bacilli are transmitted by aerosols from persons with active disease and is an obligate aerobe and an intracellular parasite that survives inside alveolar macrophages in the lungs. After multiplying in the macrophages, they enter the lymph nodes through the lymph channels where they continue to multiply intracellularly. They can spread from here to other organs, including the lungs. Symptoms normally begin to develop at this

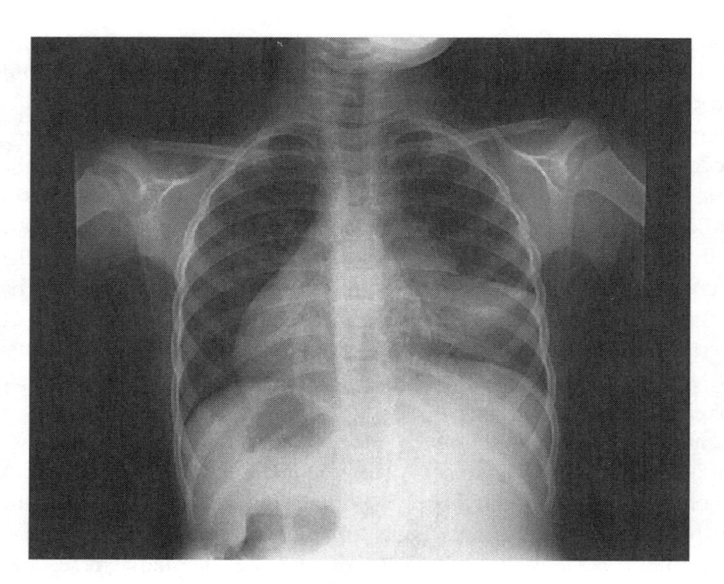

Figure 7.9 X-radiograph of patient with tuberculosis.

stage from a cell-mediated immunity that walls off the pathogen within multinucleated giant cells surrounded by lymphocytes and macrophages. These structures can be seen on x-ray and are known as tubercles (Figure 7.9). Such tubercles may become scarred or calcified over time while the bacilli remain dormant but alive within. The disease is clinically manifested when there is an immune lapse or reinfection that allows the organisms to multiply. This stimulates the immune response, causing the tubercles to break down and erode through the bronchi wall and spilling mycobacterium-laden liquid caseum into the bronchi. The characteristic cough of active tuberculosis derives from this process and is also the reason why mycobacterium appears in the aerosols created from the cough of an infected person.[127]

Epidemiology: in 1999 there were 18,000 cases of TB in the U.S., with 1% having been resistant to antibiotics.[132] After the introduction of antibiotics, TB was thought to have been under control, but the disease persisted in the poorer reaches of society (inner-city dwellers, the homeless), and it is reemerging dramatically. The contributing factors are many, but may include the emergence of AIDS, which has increased the susceptibility of infected persons to TB and diminished resources allocated by governments for the control of TB. Tuberculosis was declared a U.S. public health emergency in 1992, and CDC is attempting to help control the spread of the world's top infectious disease killer. Tuberculosis kills over 3 million people worldwide each year, and many more become ill from it.[133] Unfortunately, a WHO report released in March 1995 reveals that efforts to control tuberculosis have been ineffective and that drug-resistant strains of M. tuberculosis are on the rise.[134] Even though chemotherapy and effective measures of prevention exist, TB is increasing in the developing world as well as in the industrialized world.[135] WHO estimates

8 million new cases and 3 million deaths per year from TB with 450,000 of those deaths in children under 15 years of age.[136] WHO estimates that the 1990s will see 90 million new cases and 30 million deaths with annual rates of infection in developing countries exceeding 2%.[137] In 1995, there were 22,860 cases of TB in the U.S. with about 8% of the cases appearing in people born in Mexico and living near border cities in California, New Mexico, Texas, and Arizona. Antibiotic-resistant strains of TB are also much higher in this group than in the general U.S. population. South Africa currently has the highest rate of TB in the world, and little is being done to curb the outbreak. There are 350 cases of TB per 100,000 population in South Africa with a rising incidence of "multi-drug-resistant" TB and a death rate of 54% of those who develop the disease. In Russia, there were 100,000 cases in 1999, 30% having been resistant to antibiotics.[132] Certain environmental factors appear to accelerate the disease. These factors include crowded hospital rooms shared with TB patients, lack of sanitary conditions, unfiltered air, closed windows at night, and crowded living and working conditions.[138]

Patients often feel better after a few weeks of medicine and stop taking it. The consequence is a reinfection with TB that has a greater probability of being resistant to medication. WHO predicts that by the year 2005, TB will kill 4 million people a year, up from the 3 million dying annually now.[134] To combat the outbreaks and reemergence of TB, the WHO report titled *Stop TB at the Source* recommends international adoption of a control strategy called Directly Observed Treatment–Short Course (DOTS), which is intended to assure that all patients complete their entire course of treatment. This program requires that patients take their medicine under the eyes of a guardian every day for 6 months and take regular sputum tests to see if the pills are working. The program has been effective in New York, where TB has been reduced by over 20% to about 3000 new cases per year. Unfortunately, results of recent studies conducted by the World Health Organization have questioned the effectiveness of DOTS.[139] There is increasing support for a new method of treatment called DOTSPlus, which requires taking up to seven drugs daily for 18 to 24 months.[139] Despite these efforts, there have been a small number of patients resistant to all anti-TB drugs. Starting in 1993, the New York health department has placed patients who resist taking the complete regimen of medicine under "detention until cured."[134] Multiple-resistant strains of TB will become increasingly less responsive to even more vigorous treatment programs such as DOTSPlus.[139] Seventeen million new cases and 60 million deaths due to multi-resistant TB are expected between 1998 and 2030.

Parasites

Cryptosporidium

Background: Cryptosporidium is a single-celled microscopic protozoan parasite that belongs to the class Sporozoa. These organisms are not normally motile and asexually reproduce by multiple fission and sexually by gamete fusion. The resistant form of the parasite is called an oocyst, which is characterized by an outer protective shell that protects the organism against environmental extremes such as heat, cold, dryness, and chemical insult. There are several species of Cryptosporidium, but there is only one pathogenic species, *Cryptosporidium parvum*. This organism is widely distributed in the intestines of animals and humans and normally causes infections in humans by the fecal oral route. It is estimated that as few as 30 or even one oocyst(s) may cause infection when swallowed.[140] Cryptosporidium is therefore a highly infectious enteric parasite, widely distributed, that is resistant to chlorine and difficult to filter, making it a serious threat to water supplies.

Disease: persons who swallow Cryptosporidium cysts may develop symptoms within 2 to 10 days. The oocysts reach the upper small bowel where they excyst and produce four infectious sporozoites that attach to the surface epithelium of the digestive tract and reproduce, forming more oocysts and sporozoites. Many of the organisms are excreted while others maintain the infection. The symptoms include watery diarrhea, stomach cramps, nausea, and a slight fever. These symptoms are self-limiting and will normally last 2 weeks or less. The organism may be shed from the intestines for up to 2 months after the initial illness and spread this disease to others through fecal–oral transmission. Persons who have a weakened immune system, such as those with HIV/AIDS, cancer, and transplant patients or those on immunosuppresive drugs, are at increased risk from infection and may develop serious and life-threatening illness from this organism. There is no pharmaceutical treatment for Cryptospordiosis and the more serious illness is managed by fluid replacement.[141]

Epidemiology: contamination of drinking water by Cryptosporidium is a growing concern, especially after the outbreak in Milwaukee in 1993 that affected some 400,000 people. There have been five major outbreaks associated with public water supplies of Cryptosporidium gastroenteritis in the U.S. and seven in the U.K. since 1983.[142] There have also been numerous smaller outbreaks. Most of these outbreaks have been drinking and recreational waterborne in nature. Cryptosporidium has also been spread by unpasteurized cider and by person to person in households, day care centers, and hospitals.[143] Cryptosporidium is found in animal droppings and human feces, soil, drinking water and recreational water, food, hands, and surfaces contaminated by such wastes. Many persons have become infected even swimming in chlorinated and filtered pools because the organism has the capacity to survive chlorination and to pass through most pool filters. Cryptosporidium was first reported as a human pathogen in 1976 and came to prominence in the U.S. in 1982, with increased infections in the highly immunocompromised population of people living with AIDS.[144] Cryptosporidium is not a new organism but is thought to have emerged because it was newly spread among immunocompromised patients and became symptomatic.[145] In many areas of the world, the organisms have been shown to have infected a large portion of the population. In some areas of Brazil, as many as 90% of the children under 5 have been infected, and over 50% of the children in certain rural areas of China show serological evidence of infection.[146,147] Similar seroepidemiologic studies suggest that 17% to 32% of persons in Virginia, Texas, and Wisconsin with healthy immune systems showed evidence of Cryptosporidium infection by the time they reached young adulthood.[148]

Malaria

Background: malarial diseases are caused by protozoan parasites belonging to the genus *Plasmodium*. There are four species known to infest humans — *P. falciparum*, *P. vivax*, *P. ovale*, and *P. malariae*. These parasites are transmitted from human to human by the bite of a female anophelene mosquito in which the parasite has gone through a complex development, taking several days to 3 weeks depending on the ambient temperature. Since the sexual reproduction of the parasite occurs in the mosquito, it is the definitive host, and humans are the intermediate host and reservoir. The female anophelines become infected when they feed on blood from a person infected that contains male and female gametocytes of the parasite. A complex cycle takes place that involves the union of the gametocytes in the stomach of the mosquito and results in the development of slender, microscopic sporozoites that appear in the salivary glands and are infective for humans (Figure 7.10). This maturation, or sporogenic cycle, must be completed within the mosquito before transmission to humans can take place. The night-feeding female anopheline takes a blood meal and injects the malarial sporozoites into the human bloodstream. The sporozoites enter the liver where they infect liver cells and begin the process of development. Depending on the species of malaria, the stages in the human show some variation. In general, once in the liver cells, the parasites begin the process of schizogeny (splitting stage) and they divide in the liver cells to produce merozoites. These merozoites are infective for red blood cells. The merozoites enter the bloodstream where they attach to red blood cells and invade them. Inside the cell, the parasite forms the classical signet ring stage (Figure 7.11) and feeds on the cells' contents as it grows through the stages of trophozoite and schizont. Female and male gametocytes are also occasionally produced, and these are the sexual forms of the parasite that continue the cycle of the organism when taken into the anopheline mosquito through a blood meal. While developing within the red blood cell, the parasite converts hemoglobin to hemozoin, thereby depleting the hemoglobin content of the blood and giving rise to many of the symptoms of malaria. The parasite proliferates in the red blood cells until it can no longer be contained, and the red blood cells are destroyed as they break free into the blood stream where they can reinvade more red blood cells. This cycle often becomes synchronous, causing regularly timed waves of parasitemia from 36 to 72 hours, depending on the particular malarial parasite.[149]

Disease: the symptomology and pathogenesis of malaria infection is related to the parasite's stage of growth and the host's parasitemia. High parasitemias result in decreased hemoglobin and a lower oxygen-carrying capacity. This quickly leads to fatigue and repository distress. The detritus of lysed red blood cells as well as the explosion of foreign bodies in the bloodstream from logarithmic growth of malaria results in massive immune responses, primarily driven by mononuclear phagocyte-released cytokines. Since the explosions of parasitemia occur in regularly timed waves (from as little as 36 hours in *P. falciparum* to 72 hours in *P. malariae*), the characteristic periodic fever develops. The fever is a response to the massive interferon and interleukin release that subsides as the body clears the infection and

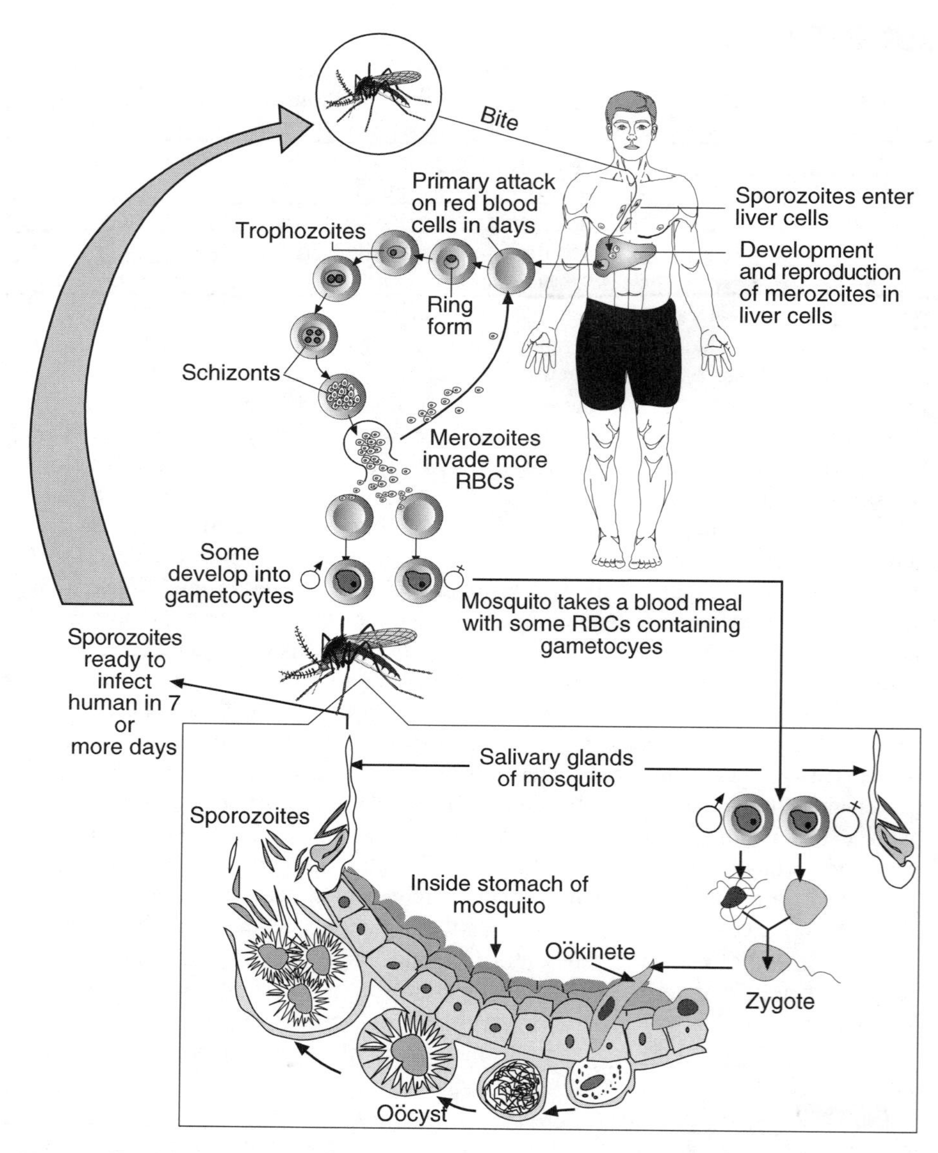

Figure 7.10 Life cycle of *Plasmodium* species in female *Anopheline* mosquito.

the merozoites invade new cells to begin a new cycle. The systemic damage of malaria is not only from septicemic reactions to cytokines but also the proliferation of knobs on the red blood cell surface, sometimes called pigmotic cells. These projections can result in agglutination of parasites at capillary pathways and ischemia to many organ areas. Untreated infections lead to splenomegaly (enlarged spleen) and, particularly in *falciparum*, to cerebral malaria and death. Children are particularly susceptible to cerebral malaria, and the bush doctor cure for urgent cases is an

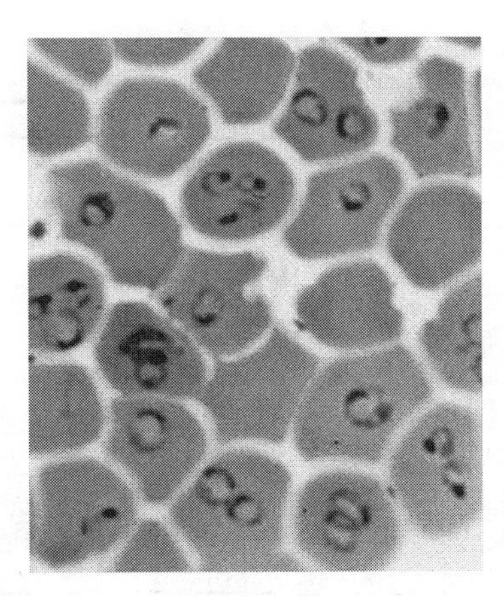

Figure 7.11 Ring stage of malarial parasite in red blood cells.

immediate blood transfusion from a close member of the family. This has resulted in a high rate of pediatric AIDS in Africa.

Epidemiology: WHO estimates that there are 300 to 500 million people worldwide infected with malaria.[150] More than 805,000 deaths occur in sub-Saharan Africa alone on an annual basis. The impact of malaria on the human condition is even more startling when measured in terms of disability-adjusted life years (DALYs). This is an indicator developed to estimate the disease burden so that cost-effective prevention strategies can be developed.[151] On a worldwide basis, malaria has caused the loss of over 35 million DALYs, which is slightly behind tuberculosis and slightly less than half of all cancers globally. The contribution of malaria to global disease and disability is of major public health significance.

The majority of malarial transmission occurs in tropical and subtropical countries (Figure 7.12). The reemergence of malaria as a major threat in these areas is due to

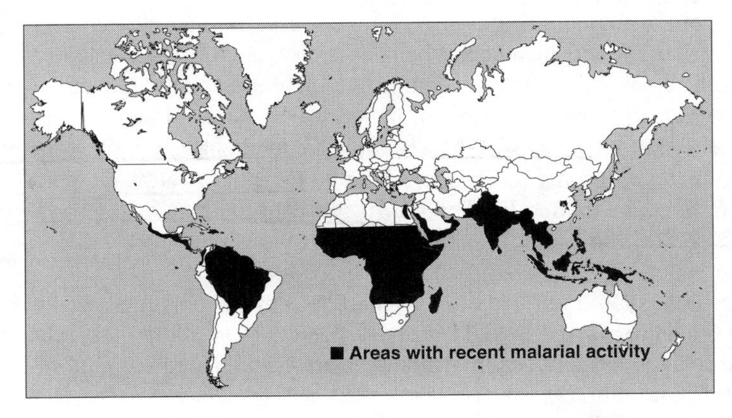

Figure 7.12 Global distribution of malaria. (Adapted from Zucker, J.R., Changing patterns of Autochthonous malaria transmission in the United States: a review of recent outbreaks, *EID,* 2, 1, Jan.-Mar., 1996.

a number of reasons. Mosquito control programs have historically been the primary mechanism of attack. However, such programs alone have not been effective in many areas such as sub-Saharan Africa, where the applications of spraying were neither complete nor periodic at effective intervals.

Decreased spraying of homes with DDT is also blamed for the dramatic reemergence of malaria in South America, where there has been a geometric growth in malaria incidence since the 1970s.[152] A major reason for this decreased use is the concern for ecological consequences of DDT spraying. Some authors reason that the use of DDT in homes or on nets used over the sleeping areas at night represents barely a fraction of the DDT used in agriculture. In fact, the amount used to treat 1000 acres of cotton could be used to treat the homes for the entire country of Guyana. However, most countries have looked to alternative treatments and have found them too expensive. Furthermore, there has been a steady decline in international and U.S. support for malarial control in these countries, and similar reductions in the budgets of the endemic countries themselves.[153] The problem is aggravated by the development of pesticide-resistant mosquitoes in the endemic areas. There is little probability that more money will be allocated for more expensive insecticides.

There are other forces at work to power the reemergence of malaria in the world. These include the appearance of drug-resistant malaria and the predicted expansion of malaria associated with global warming. The malarial parasite has a family of genes called var genes (for variability) that permit it to change its signature proteins just enough to stay one step ahead of the host antibodies aimed against it in a process called antigenic variation. These var genes account for nearly 6% of the parasite's genome and permit it to mix and match genes to create millions of antigenic proteins. The malarial parasite developed resistance to antimalarial drugs as well. In particular, it developed resistance to the inexpensive and highly effective drug chloroquine, and resistant strains began appearing in the 1960s. Unfortunately, the alternative drugs are too expensive for most people in the malaria-endemic areas, so the disease continues to kill nearly 2 million people a year.[154]

Global warming spurred by greenhouse gases is an internationally debated concern. One of the potential consequences of such a warming trend may be the increase in the distribution and incidence of malaria.[155] A team of four scientists from the Netherlands created a complex mathematical model that can potentially predict the impact of global climate changes on malarial risk. A simulation experiment using this model showed a widespread increase in the potential for malarial transmission.[156] A global increase of only a few degrees could expand malaria's domain from 42% to over 60% of the planet. Further, the increasing temperatures would increase the metabolism of the mosquito, making it a more voracious biter while speeding up the development of the malarial parasite inside the mosquito.[10]

Americans should not be complacent about this reemergence. Malaria has also begun to reappear in the U.S. There have been at least three outbreaks of locally acquired malaria in densely-populated areas of the country that point to the danger of a significant outbreak. Malaria was a once a major disease in the U.S. with over 600,000 cases occurring in 1914, and more than 125,000 cases in 1934, when systematic reporting was in effect.[157] The processes of urbanization, increased drainage,

DDT spraying, screening of houses, and introduction of quinine all contributed to the substantial control of malaria. Malaria was eradicated as a problem disease in the U.S. by the 1950s. Nearly all cases of malaria detected in the U.S. since then have been imported. There have been 76 cases of malaria reported from 1957 through 1994, including the three outbreaks occurring in the densely populated areas of New Jersey (1991), New York (1993), and Texas (1994). In these cases, the weather was more humid than usual, there were larger mosquito populations, and the patients lived in areas with large migrant populations from countries with endemic malaria.[149] The likelihood of widespread malarial outbreaks in the U.S. is limited by adequate measures of sanitation, health care, and targeted mosquito control. However, a continued warming trend predicted by global climactic changes could expand the range of malaria by permitting the parasite to complete its cycle in the mosquito. Greater pressures would be exerted on the public health support systems within the U.S.

PRACTICAL APPROACHES TO LIMITING THE EMERGENCE OF INFECTIOUS DISEASES

The emergence of 29 new infectious diseases and reemergence of many others is creating national and international crises. The control of these diseases is made difficult because of the large number of disease-causing organisms, the enormous diversity of geographic areas from which they can emerge, the potential for rapid global dissemination, and the tremendous number of ecological and social factors that influence emergence.[26,30] Primary responsibility for addressing new and re-emerging infectious diseases rests squarely with the custodians of public health. Indeed, the fundamental maxim of public health must guide current prevention programs: the health of the individual is best ensured by maintaining or improving the health of the entire community.

D.A. Henderson, the driving force in the successful international smallpox eradication program, proposes the establishment of an international network consisting of disease centers strategically located in tropical areas, especially near metropolitan areas. Each center would have a clinical facility, diagnostic and research laboratory, an epidemiological unit to investigate disease and local response capabilities, and a professional training unit. This would also require an academic and government laboratory necessary for coordinating data collection and analysis, conducting research, providing feedback and support, and when necessary, activating an "international rapid response system."[26] The World Health Organization has created a Division on Emerging Diseases in 1995 to try to deal with a global problem of new and emerging infectious diseases. WHO offices, including the Pan American Health Organization, are developing a regional plan and setting goals that include strengthening regional and national laboratory and surveillance networks, establishing national and regional infrastructures for early warning and rapid response to infectious diseases, promoting and developing applied research for rapid diagnosis, epidemiology and prevention, and strengthening regional capacity for effective implementation of prevention and control strategies.[158]

The Centers for Disease Control and Prevention began four Emerging Infections Programs in Oregon, California, Connecticut, and Minnesota, to deal with some of the nation's underfunded health threats. Each of these states has different projects that range from foodborne illnesses and meningitis to Valley Fever, Ehrlichia, and Cryptosporidium. The programs collaborate with federal agencies, local health departments, academic institutions, professional societies, and international organizations. A Prevention Strategy Plan for the U.S. was developed with four major goals: (1) promptly investigate and monitor emerging pathogens, the diseases they cause, and factors of emergence; (2) integrate laboratory science and epidemiology to optimize public health practice; (3) enhance communication of public health information about emerging diseases and ensure prompt implementation prevention strategies; and (4) strengthen local, state, and federal public health infrastructures to support surveillance and implement prevention and control programs.[4]

Ewald reports that control could be made more manageable if it were possible to focus attention on the pathogens most likely to become dangerous. Although there is disagreement about the ability to predict and prevent the emergence of these new pathogens,[159] Ewald emphasizes that the identification of these pathogens rests upon two major characteristics that include an ability to spread easily from human to human through vectors, and transmission features that select for high levels of virulence. The transmission features that promote virulence include one or more of the following: (a) being vector-borne with humans as part of the life cycle; (b) durability in the external environment if directly transmitted; (c) being needle-borne or borne by attendants in medical care; and (d) being sexually transmitted and mutation prone with an attraction to critical cell types.[158] Knowledge of these evolutionary characteristics should help to distinguish the long-term threat of the emerging disease organism and adjust the amount of resources invested in keeping pace with the anticipated threat.

Unfortunately, we cannot accurately predict where, when, or how the next disease will emerge or reemerge and what impact it will have on the individual, society, nation, or the world. Environmental factors such as unusual weather patterns, agricultural practices, deforestation, and other human behavioral factors such as increased population density, compliance, and overprescribing antibiotics, and mere chance will continue to play a significant role in the emergence or reemergence of disease. Increased vigilance is necessary, and the application of more resources is the ultimate answer — contradicting the current erosion of resources to public health nationally and internationally.

REFERENCES

1. Mulvaney, K., The scourge of mankind: infectious diseases, once near eradication, show remarkable persistence, *Environ. Mag.,* 42–44, 42, 1999.
2. Walsham, C., Our planet is environmentally sick, *PlanetENN,* http://www.enn.com/planetenn/I 11 196/feature3.htm, Nov. 11, 1996.
3. Bennett, I.V., Holmberg, S.D., Rogers M.F., and Solomon S.L., Infectious and parasitic diseases, in: *Closing the Gap: The Burden of Unnecessary Illness.* Antler, R.W., Dull, H.B., eds., Oxford University Press, New York, 1987.

4. Centers for Disease Control and Prevention, Addressing emerging infectious disease threats: a prevention strategy for the United States, Atlanta, Georgia, U.S. Department of Health and Human Services, Public Health Service, 1994.
5. Stevens, D.L., Streptococcal toxic-shock syndrome: Spectrum of disease, pathogenesis, and new concepts in treatment, *Emerging Infectious Diseases* (EID), 1, 3, Jul/Sep 1995.
6. Newman, A., EPA considering quick action on Cryptosporidium, *Environ. Sci. Technol.,* 29:1, 17a, 1995.
7. Centers for Disease Control and Prevention, Update: trends in AIDS incidence, deaths and prevalence, United States, 1996, *MMWR,* 46, 165, 1997.
8. Centers for Disease Control and Prevention, Update: multistate outbreak of *Escherichia coli* 0157:117 infections from hamburgers-westem United States, 1992–1993, *MMWR,* 42, 258, 1993.
9. Feng, P., Synopsis: Escherichia coli Serotype 0157:H7: novel vehicles of infection and emergence of phenotypic variants, *EID,* 1, 2, Apr.-Jun., 1995.
10. Linden, E., Global fever, *Time,* Jul., 8, 56, 1996.
11. Steere, A.C., Maliwista, S.E., Snydman, D.R., Shope, R.E., Andirnan, W.A., Ross, M.R., and Steele, F.M., Lyme arthritis: an epidemic of oligoarticular arthritis in children and adults in three Connecticut communities, *Arthritis Rheum.,* 20, 7, 1977.
12. Rohter, L., U.S. is now threatened by epidemic of dengue, *New York Times Int.,* Sept. 23, 5, 1995.
13. Gubler, D.L., *Vector-Borne Diseases, an Encyclopedia of the Environment,* Houghton Mifflin Co., New York, 1994.
14. Altman, L.K., 56 die in a month in Zaire outbreak of mystery disease, *New York Times Int.,* Al, May 10, 1995.
15. Mint, C., Insect-borne disease: curing the carrier, *Sci. News,* Apr. 12, 151, 223, 1997.
16. Satcher, D., Perspectives: emerging infections: getting ahead of the curve, *EID,* 1 (1), Jan.-Mar., 1995.
17. Willq, R.G., Ironside, J.W., Zeidler, M., Cousens, S.N., Estibeiro, K., Alperovitch, A., et al., A new variant of Creutzfeldt-Jakob disease in the U.K., *Lancet,* 347, 921, 1996.
18. Centers for Disease Control and Prevention, World Health Organization consultation on public health issues related to bovine spongiforin encephalopathy and the emergence of a new variant of Creutzfeldt-Jakob disease. *MMWR;* 45:303, 295, 1996.
19. Brown, P. and Cathala, F., Creutzfeldt-Jacob disease in France, in *Slow Transmissible Diseases of the Nervous System,* Hadlow, W.J., Prusiner, S.B., eds., Academic Press, New York, 1979, Vol. 1, 213.
20. Martens, W.J., Potential Impact of global climate change on malaria risk, *Env. Health Perspect.,* 103(5), 458, May, 1995.
21. Angier, N., Malaria's genetic game of cloak and dagger, *New York Times,* C I, Aug. 22, 1995.
22. Roberts, D.R., et al., Global strategies, and a malaria control crisis in South America, *EID,* 3(3), Jul-Sep., 1997.
23. Nowak, R., WHO calls for action against TB, *Science,* 267, 1763, Mar. 1995.
24. Drexler, M., Welch, M.B., Public health-in our battle to stay healthy, we are up against a daunting new array of … clever germs, *The Boston Sunday Globe*, Sun., Nov. 21, 1999.
25. Morse, S., Perspectives: Factors in the emergence of infectious diseases, *EID,* I(1), Jan. Mar., 1995.
26. Satcher, D., Perspectives: emerging infections, getting ahead of the curve, *EID,* I(1) Jan./Mar., 1995.

27. Morse, S.S., Examining the origins of emerging viruses, *Emerging Viruses,* Morse, S.S., ed., Oxford University Press, New York, p. 10, 1993.
28. Morse, S.S., Regulating viral traffic, *Issues Sci. Technol.,* 7, 81, 1990.
29. Morse, S.S., Toward an evolutionary biology of viruses, *The Evolutionary Biology of Viruses,* Morse, S.S., Ed., Raven Press, New York, p. 1, 1994.
30. Morse, S.S., Emerging viruses: Defining the rules for viral traffic, *Perspect. Biol. Med.,* 34, 387, 1991.
31. Rogers, D.J. and Packer, M.J., Vector-borne diseases, models, and global change, *Lancet,* 342, 1282, 1993.
32. Barbour, A.G. and Fish, D., The biological and social phenomenon of Lyme disease, *Science,* 260, 1610, 1993.
33. Johnson, C.M., Emerging viruses in context: An overview of viral hemorrhagic fevers, in *Emerging Viruses,* Morse, S.S., Ed., Oxford University Press, New York, p. 46, 1993.
34. Webster, R.G., Bean W.J., Gorman, O.T., Chambers, T.M., and Kawaoka, Y., Evolution and ecology of influenza A viruses, *Microbiol. Rev.,* 56, 152, 1992.
35. Scholtissek, C. and Naylor, E., Fish farming and influenza pandernics, *Nature,* 331, 215, 1988.
36. World Health Organization, Geographical distribution of arthropod-borne diseases and their principal vectors, Geneva: World Health Organization (WHONBC/89.967), 138, 1989.
37. Kingsnorth, P., Human health on the line, *The Ecologist*, 29, 92, 1999.
38. Washam, C., Our planet is environmentally sick, *PlanetENN,* Nov. 11, 1996.
39. Galntz, M.H., Katz, R.W., Nicholls, N., Eds. *Teleconnections Linking Worldwide Climate Anomalies,* Cambridge University Press, New York, 1991.
40. Manabe, S. and Stouffer, R.J., Century-scale effects of increased atmospheric C02 on the ocean-atmosphere system, *Nature,* 364, 215, 1993.
41. Harvell, C.D., et al., Emerging marine–diseases–climate links and anthropogenic factors, *Science*, 285, 1505, 1999.
42. Epstein, P.R., Ford, T.E., and Colwell, R.R., Marine ecosystems, *Lancet,* 342, 1216, 1993.
44. Anderson, D.M., Fifth International Conference on Toxic Marine Phytoplankton: A personal perspective. Hairnful Algae News, *Int. Marine Sci. (UNESCO),* Paris, 62 (suppl:6-7), 1993.
45. The Newshour with Jim Lehrer, An essay in the later half of the show on the troubled history of Washington, D.C. in the 20th century, Feb. 17, 1995.
46. United Nations, World Urbanization Prospects, 1990, United Nations, New York, 1991.
47. Han, J.S., Short, M., Taylor, M.E., et al., Species-specific diversity among simian immunodeficiency viruses from African greea monkeys, *J. Virol.,* 65, 2816, 1991.
48. Gan, F., Yne, L., White, A.T., et al., Human infection by genetically diverse SIVSM-related HIV-2 in West Africa, *Nature,* 358, 495, 1992.
49. McNeill, W.H., *Plagues and Peoples,* Anchor Press/Doubleday, New York, 1976.
50. Centers for Disease Control and Prevention, *Aedes albopictus* introduction into continental Africa, 1991. *MMWR,* 40, 836, 1991.
51. Moore, C.G., Francy, D.B., Eliason, D.A., and Monath, T.P., Aedes albopictus in the United States: rapid spread of a potential disease vector, *J. Am. Mosq. Control Assoc.,* 4, 356, 1991.
52. Mitchell, C.J., Niebylski, M.L., Smith, G.C., et al., Isolation of eastern equine encephalitis virus from *Aedes albopictus* in Florida, *Science,* 257, 526, 1991.

53. Wilson, M., Perspectives: Travel and the emergence of infectious disease, *EID,* 1:2, Apr./Jun., 1995.
54. LeDuc, J.W., Childs, J.E., and Glass, G.E., The hantaviruses, etiologic agents of hemorrhagic fever with renal syndrome: a possible cause of hypertension and chronic renal disease in the United States, *Annu. Rev. Public Health,* 13, 79, 1992.
55. Siem, H. and Bollini, P., Migration and health in the 1990s, *Intl. Migration,* 30, 1992.
56. Carlton, J.T. and Geller, J.B., Ecological roulette: The global transport of non-indigenous marine organisms, *Science,* 261, 78, 1992.
57. World Health Organization, Cholera in the Americas, *Wkly. Epidemiol. Rec.,* 67, 33, 1992.
58. McCarthy, S.A., McPhearson, R.M., and Guarino, A.M., Toxigenic *Vibrio cholerae* 01 and cargo ships entering Gulf of Mexico, *Lancet,* 339, 624, 1992.
59. Rarnamurthy, T., Garg, S., Sharma, R., et al., Emergence of novel strain of *Vibrio cholerae* with epidemic potential in southern and eastern India, *Lancet,* 341, 703, 1992.
60. DePaola, A., Capers, G.M., Moters, M.L., et al., Isolation of Latin American epidemic strain of *Vibrio cholerae* 01 from U.S. Gulf Coast, *Lancet,* 339, 624, 1992.
61. Russel, R.C., Survival of insects in the wheel bays of a Boeing 747B aircraft on flights between tropical and temperate airports, *Bull. WHO,* 65, 659, 1992.
62. Fidler, D., Perspectives; Globalization, international law, and emerging infectious diseases, *EID,* 2, 2, Apr.-Jun., 1996.
63. Delbruck, J., Globalization of law, politics, and markets/implications for domestic law; a European perspective, *Ind. J.Global Leg. Stud.,* 1, 9, 1992.
64. Centers for Disease Control and Prevention, Addressing emerging infectious disease threats: a prevention strategy for the United States, Atlanta, U.S. Department of Health and Human Services, Public Health Service, 1994.
65. World Health Organization, Communicable disease prevention and control: new, emerging, and re-emerging infectious diseases, *WHO* Doc. A48115, Feb. 22, 1995.
66. Morse, S.S., Looking for a link, *Nature,* 344, 297, 1992.
67. Wilesmith, J.W., Ryan, J.B.M., and Atkinson, M.J., Bovine spongiform encephalopathy: epidemiological studies on the origin, *Vet. Rec.,* 128, 199, 1992.
68. Berkelman, R.L., Bryan, R.T., Osterholm, M.T., Leduc, J.W., and Hughes, J.M., Infectious disease surveillance: a crumbling foundation, *Science,* 264, 368, 1992.
69. Lederberg, J., Infectious disease as an evolutionary paradigm, *Eff,* 3(4), Dec. 1997.
70. Davies, J., Inactivation of antibiotics and the dissernination of resistance genes, *Science,* 264, 375, 1994.
71. Bloona, B.R. and Murray, C.J.L., Tuberculosis: commentary on a re-emergent killer, *Science,* 257, 1055, 1994.
72. Neu, H.C., The crisis in antibiotic resistance, *Science,* 257, 1064, 1994.
73. Davies, J., Inactivation of antibiotics and the dissemination of resistance genes, *Science,* 264, 375, 1994.
74. Domingo, E. and Holland, J.J., Mutation rates and rapid evolution of RNA viruses, in *The Evolutionary Biology of Viruses,* Morse, S.S., Ed., Raven Press, New York, 1994, 161–84.
75. Kilbourne, E.D., The molecular epiderniology of influenza, *J. Infect. Dis.,* 127, 478, 1978.
76. Bureau of the Census, Mortality statistics, 1927, Part 1, Washington, D.C., 1927 U.S. Government Printing Office; National Center for Health Statistics: Vital statistics of the United States, 1986, Col. 11 - Mortality, Part A, U.S. Government Printing Office, Washington, D.C., 1988.

77. Garret, L., *The Coming Plague,* Harper Collins Canada Ltd. New York, 1994, p. 445.
78. Jungkind, D.L., *Antimicrobial Resistance: A Crisis in Health Care, Advances in Experimental Medicine and Biology,* Plenum Press, New York, 1995, p. 33.
79. Torassa, U., Antibiotic overuse a growing problem, Sep. 28, 1999, http://www.Sfgate.com/cgi-bin/article.cgi?file=/examiner/hotnews/stories/29/animal.dtl.
80. Glass, R.I., Libel, M., and Brandling-Bennett, A.D., Epidemic cholera in the Americas, *Science,* 265, 1524, 1992.
81. Moore, P.S., Meningococcal meningitis in sub-Saharan Africa: A model for the epidemic process, *Clin. Infect. Dis.,* 14, 515, 1992
82. MacKenzie, W.R., Hoxie, N.J., Proctor, M.E., et al., A massive outbreak in Milwaukee of Cryptosporidium infection transmitted through the water supply, *N. Engl. J. Med.,* 33 1, 161, 1994.
83. Centers for Disease Control and Prevention, Assessment of inadequately filtered public drinking water, Washington, D.C., Dec. 1993, *MMWR,* 43, 661, 1994.
84. Ryan, M., They track the deadliest viruses, *Parade,* Apr. 23, 10, 1995.
85. Nichol, S.T., Spiropoulou, C.F., Morzunov, S., et al., Genetic identification of a hantavirus associated with an outbreak of acute respiratory illness, *Science,* 262, 914, 1995.
86. Schmaljohn, A., Li D., Negley D.L., Bressler D.S., et al., Isolation and initial characterization of a newfound hantavirus from California, *Virology,* 206, 963, 1995.
87. Hurt, H., A deadly new virus, *Self Mag.,* 88, Apr. 1993.
88. Gajdusek, D., Virus hemorrhagic fevers, *J. Pediatr.,* 60, 841, 1962.
89. Lee, H.W., Epidemiology and pathogenesis of hemorrhagic fever with renal syndrome, *The Bunyaviridae,* Elliott, R.M., Ed., Plenum Press, New York, 1996, p. 253–67.
90. World Health Organization, Hemorrhagic fever with renal syndrome memorandum from a WHO meeting, *Bull. World Health Organization,* 61, 269, 1983.
91. Schmaljohn, C. and Hjelle, B., Synopses Hantavirusus: a global disease problem, *EID,* 3, 2, 95, Apr.-Jun. 1997.
92. Lee, H., Lee, P., and Johnson, K., Isolation of the etiologic agent of Korean hemorthagic fever, *J. Infect. Dis.,* 137, 298, 1978.
93. French, G., Foulke, R., Brand, O., and Eddy, G., Korean hemorrhagic fever: propagation of the etiologic agent in a cell line of human origin, *Science,* 21, 1046, 1981.
94. Epstein, P.R., Emerging diseases and ecosystem instability: new threats to public health, *Am. J. Public Health,* 80/2, 16, 1995.
95. Wenzel, R.P., A new hantavirus infection in North America, *N. Engl. J. Med.,* 330, 1004, 1994.
96. Dispatch, Dengue hemorrhagic fever: the emergence of a global health problem, *EID,* 1, 2, Apr.-Jun. 1995.
97. Gubler, D.J., Dengue, in: *Epidemiology of Arthropod-Borne Viral Disease,* Monath, T.P.M., Ed., CRC Press, Boca Raton, FL, 1988, p. 223-60.
98. Manning, A., Painful, flu-like tropical illness can be deadly, *USA Today,* Sep. 18, 1, 1995.
99. Henig, R.M., The new mosquito menace, *New York Times,* Sep. 13, A21, 1995.
100. Gubler, D.J., Dengue haemorrhagic fever: a global update [Editorial], *Virus, Information Exchange News,* 8, 2, 1995.
101. Associated Press, TB, other diseases reemerge to threaten millions in the Americas, Sep. 27, 1999, http://www.my…./Go?template.html.
102. Gubler, D.J., Dengue and dengue haemorrhagic fever in the Americas, WHO, Regional Office for SEAsia, New Delhi, Monograph, SEARO, 22, 9, 1995.
103. Rigau-Plkrez, J.G., Gubler, D.J., Vomdarn, A.V., and Clark, G.G., Dengue surveillance-United States, 1986-1992, *MMWR,* 43, 7, 1994.

104. World Health Organization (WHO), Emerging and other communicable diseases, Influenza A(H5NI) Fact Sheet 188, Jan. 9, 1999.
105. Centers for Disease Control (CDC), 1998, http://www.edc.gov/ncidod/diseases/flu/fluinfo.htm.
106. Commentary, Remergence of Ebola virus in Africa, *EID,* 1, 2, Jul.-Sep., 1995.
107. Troy, K., The plague that wasn't, *Newsweek,* Dec. 9, 46, 1996.
108. Bowen, E.T.W., Platt, G.S., Lloyd, G., et al., Viral haemorrhagic fever in southern Sudan and northern Zaire: Preliminary studies on the aetiologic agent, *Lancet,* 1, 571, 1977.
109. Jahrling, P.B., Geisbert, T.W., Dalgard, D.W., et al., Preliminary report: isolation of Ebola virus from monkeys imported to USA, *Lancet,* 335, 502, 1977.
110. Clover, C., Highfield, R., AIDS started by humans eating chimps ..., *The Daily Telegraph (London),* Mon., Feb. 1, 1999.
111. Jaeger, P., 1998, http://jaeger.mnith-giessen.de:6423/aids.html.
112. National Institute of Allergy and Infectious Diseases (NIAID), HIV infection and AIDS, U.S. Dept. of Health & Human Services, Public Health Service, May 1997, http://www.aegis.com/topics/whataidsis.htnil.
113. UNAIDS and WHO, Report on the Global HIV/AIDS epidemic, Dec. 1997, http://www.unaids.org/unaids/document/epideinio/report97.html.
114. Centers for Disease Control and Prevention, Update: trends in AIDS incidence, deaths, and prevalence, United States, 1996, *MMWR Morb. Mortal Wkly. Rep.,* 46, 165, 1997.
115. Li'eng, P., Synopsis *Escherichia coli* serotype 0157:H7: novel vehicles of infection and emergence of phenotypic variants, *EID,* 1(2), Apr.-Jun., 1995.
116. Griffin, P.M. and Tauxe, P.V., The epidemiology of infections caused by *Escherichia coli* 0157:H7, other enterohemorrhagic *E. coli*, and the associated hemolytic uremic syndrome, *Epidentiol. Rev.,* 13, 60, 1991.
117. Karmali, M.A., Infection by Verocytotoxin-producing *Escherichia coli*, *Clin. Microbiol. Rev.,* 2, 15, 1989.
118. Centers for Disease Control and Prevention, Update: multistate outbreak of *Escherichia coli* 0157:H7 infections from hamburgers-westem United States, 1992–1993, *MMWR,* 42, 258, 1993.
119. Besser, R.E., Lett, S.M., Weber, J.T., et al., An outbreak of diarrhea and hemolytic uremic syndrome from *Escherichia coli* 0157:H7 in fresh-pressed apple cider. *JAMA,* 269, 2217, 1993.
120. Zhao, T., Doyle, M.P., and Besser, R.E., Fate of enterohemorrhagic *Escherichia coli* 0157:H7 in apple cider with and without preservatives, *Appl. Environ. Microbiol.,* 59, 2526, 1993.
121. Weagant, S.D., Bryant, J.L., and Bark, D.H., Survival of *Escherichia coli* 015TH7 in mayonnaise and mayonnaise-based sauces at room and refrigerated temperatures, *J. Food Prot.,* 57, 629, 1994.
122. Swerdlow, D.L., Woodruff, B.A., Brady, R.C., et al., A waterbome outbreak in Missouri of *Escherichia coli* 0157:H7 associated with bloody diarrhea and death, *Ann. Intern. Med.,* 117, 812, 1992.
123. Peisman, J., Schreifer, M.E., and Burkot, T.R., Ability of experimentally infected chickens to infect ticks with the Lyme disease sprichete, *Borrelia burgdorferi*, *Am. J. Trop. Hyg.,* 54, 294, 1996.
124. Levin, M., Levine, J.F., Howard, P., and Apperson, C.S., Reservoir competence of the southeastern five-lined skink *(Eumeces inexpectatus)* and the green anole (*Anolis carolensis*) for *Borrelia burgdorteri*, *Am. J. Trop. Med. Hyg.,* 54, 92, 1996.
125. Manning, A., Lyme-causing tick carries worse disease, *USA Today,* Sep. 18, 1, 1995.

126. Colburn, D., The clock is TICK-ing, New Lyme vaccine effectively sterilizes the blood-sucking tick, *Daily Hampshire Gazette,* Tues., July 6, 1999.
127. Cano, R.J. and Colonic, J.S., Disease of the Respiratory Tract, in *Microbiology,* West Publishing Co. MN, 1986, chap. 24.
128. Stevens, D.L., Invasive group A streptococcus infections, *Clin. Infect. Dis.,* 14, 2, 1992.
129. Demers, B., Shno, A.E., Vellend, H., et al., Severe invasive group A streptococcal infections in Ontario, Canada: 1987–1991, *Clin. Infect. Dis.,* 16, 792, 1993.
130. Seachrist, L., The Once and future scourge, *Sci. News,* 148, 234, Oct. 1996.
131. Stevens, D.L., Invasive group A streptococcal infections: the past, present and future, *Pediatr. Infect. Dis. J.,* 13, 561, 1994.
132. Hayden, T., Tuberculosis is making a comeback, *Newsweek,* Nov. 8, 1999.
133. Meyer, T., U.S., Mexico must fight disease that knows no boundaries, *The Boston Globe,* 01, 15, Nov. 30, 1996.
134. Nowak, R., WHO calls for action against TB, *Science,* 267, 1763, Mar. 24, 1995.
135. Starke, J.R. and Correa, A.G., Management of mycobacterial infection and disease in children, *Pediatr. Infect. Dis. J.,* 14, 455, 1995.
136. Kochi, A., The global tuberculosis situation and the new control strategy of the World Health Organization, *Tuber. Lung. Dis.,* 72, 1, 1991.
137. Raviglione, M.C., Snider, D.E., and Kochi, A., Global epidemiology of tuberculosis. Morbidity and mortality of a worldwide epidemic, *JAMA,* 273, 220, 1995.
138. McNeil, D., South Africa slow to abate spread of tuberculosis, *The New York Times,* Oct. 13, 1996.
139. Miller, J., Study says new TB strains need an intensive strategy, *The New York Times,* A6, Oct. 28, 1999.
140. DuPont, H.L., Chappell, C.L., Sterling, C.R., et al., The infectivity of *Cryptosporidium parvum* in healthy volunteers, *N. Engl. J. Med.,* 332, 855, 1995.
141. Centers for Disease Control and Prevention (CDC), *Cryptosporidium and Water: A Public Health Handbook,* NCID, Mailstop F-22 4770 Buford Highway N.E., Atlanta, GA 30341-3724, 1997, 134 pp.
142. Newman, A., EPA considering quick action on Cryptosporidium, *Env. Sci. Technol.,* 29(l), 17A, Jan. 1995.
143. Newman, R.D., Zu, S.-X., Wuhib, T., et al., Household epidemiology of *Cryptosporidium parvum* infection, *Ann. Intern. Med.,* 120, 500, 1994.
144. Nime, F.A., Burek, J.D., Page, D.L., et al., Acute enterocalitis in a human being infected with the protozoan Cryptosporidium, *Gastroenterology,* 70, 592, 1994.
145. Guerrant, R., Synopses Cryptosporidiosis: an emerging highly infectious threat, *EID,* 3, 1, Jan.-Mar., 1997.
146. Zu, S.-X., Li, J-F., Barrett, L.J., et al., Seroepiderniologic study of Cryptosporidiurn infection in children from rural communities of Anhui, China, *Am. J. Trop. Med. Hyg.,* 51, 1, 1994.
147. Ungar, B.L., Gilmn, P.H., Lanata, C.F., and Peres-Schael, L., Seroepiderniology of Cryptosporidiurn infection in two Latin American populations, *J. Infect. Dis.,* 157, 551, 1988.
148. Ungar, B.L.P., Soave, R., Fayer R., and Nash T.E., Enzyme immunoassay detection of inummoglobulin M and G antibodies to Cryptosporidium in immunocompetent and immunocompromised persons, *J. Infect. Dis.,* 153, 570, 1986.
149. Zucker, J.R., Changing patterns of Autochthonous malaria transmission in the United States: a review of recent outbreaks, *EID,* 2, 1, Jan.-Mar., 1996.
150. World Health Organization, Control of tropical Diseases (CTD): Malarial Control, Geneva, Switzerland: *WHO* Office of Information, 1995.

151. Murray, C.J.L., Quantifying the burden of disease; the technical basis for disability adjusted life years, *Bull. World Health Organ.,* 72, 429, 1994.
152. Roberts, D., Laughlin, L., Hsheih, P., and Legters, J., DDT, global strategies and a malaria control crisis in South America. *EID,* 3, 3, Jul.-Sep., 1997.
153. Pan American Health Organization, Status of malaria programs in the Americas. *XLIV Report,* Washington, D.C., PAHO, 1996.
154. Angier, N., Malaria's genetic game of cloak and dagger, *The New York Times,* Cl, Aug. 22, 1995.
155. Doll, R., Health and the environment in the 1990s, *Am. J. Public Health,* 82, 933, 1992.
156. Martins, W., Niessen, L., Rotmans, J., Jetten, T., and McMichael, A., Potential impact of global climate change on malaria risk, *Env. Health Persp.,* 103:5, 458, 1995.
157. Pan American Health Organization, Report for registration of malaria eradication from United States of America, Washington, D.C., Dec., 1969.
158. WHO, WHO reports on new, re-emerging diseases threatening world health, *The Nation's Health,* Vol 24, Nov. 1995.
159. Ewald, P.W., Guarding against the most dangerous emerging pathogens: insights from evolutionary biology, *EID,* 2, 4, Oct.-Dec., 1996.

CHAPTER 8

Foodborne Illness

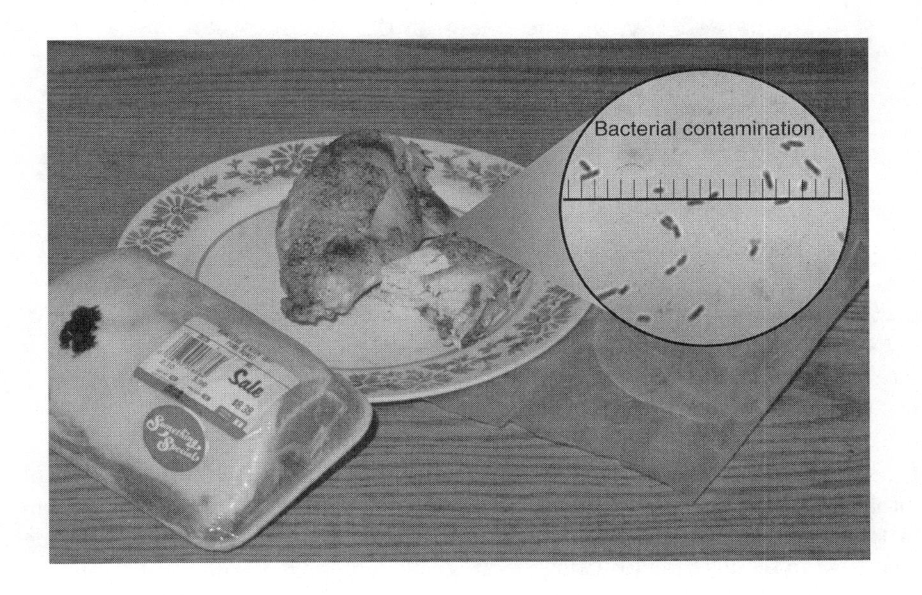

INTRODUCTION

Martin Molina is a health promoter in the small village of El Achiotal located on the Pacific Coast of El Salvador. Like most evenings, all lights were off by 8 pm and most had settled quietly into their hammocks — a custom left over from the civil war that had just ended. That night, Martin was awakened by a knock at the door. He remained silent and listened as the knock grew frantic and more rapid. Finally, a familiar voice called his name. It was Mrs. Cruz, an elderly mother of six. Martin was relieved. Since his childhood he had heard stories of those who had answered the door in the night — they were usually recruited immediately into the military or disappeared at the hand of the infamous death squads.

OBJECTIVES FOR THIS CHAPTER

A student reading this chapter will be able to:

1. **Recognize, list, and explain the major reasons for food protection programs**
2. **List and describe the major categories and subcategories of agents causing foodborne illness**
3. **Describe the major foodborne pathogens including parasitic, viral, and bacterial diseases**
4. **Explain the mechanisms by which these pathogens cause foodborne illness, and describe how the life cycles of these organisms are important in this transmission of disease**
5. **List and describe the major disease symptoms in humans for these foodborne pathogens**
6. **Describe and explain the HACCP system in protecting against foodborne disease**
7. **Discuss recent regulatory efforts in the area of food potation**

An upset Mrs. Cruz explained to Martin that her son had been vomiting with profuse diarrhea for several hours, a scenario that was all too familiar to Martin. Martin grabbed his small bag (which he constantly kept packed for emergencies) and went to see little Santos Cruz, the fifth patient of the day with these symptoms.

Martin had fled to the U.S. when the civil war broke out. Like many of his countrymen, he had managed to return to El Salvador with enough money to buy some property, build a house, and even finance a high school education. Martin did not know why he had returned to his small village with no potable water, no human waste facilities, and an electrical current so weak it could power little more than a light bulb. Martin did enjoy the prestige of being the most educated person in the village and found working as a health worker extremely rewarding.

When Martin arrived at Santos's side, the child was naked, dry heaving into an old bucket, and complaining of stomach cramps. Martin started to prepare a rehydration solution using a powder concentrate and water from a nearby well. Mrs. Cruz explained to Martin that she had been giving Santos the same solution that she had received at the community meeting concerning the sudden increase of sickness in the village, but Santos always vomited it up. Martin insisted that Santos try to drink the mix because he was severely dehydrated and was sure to only get worse. As soon as Santos swallowed a cup full, he vomited on Martin's feet, who looked down only to realize he was standing in what appeared to be diarrhea from little Santos.

Mrs. Cruz saw the frustration in Martin's face and began to cry fearing that Santos would not recover, like so many other children of the village had failed to do. Martin looked up at the rising moon, aware his village was experiencing an outbreak of some sort. He was aware, like so many times before, that the origin of the outbreak would remain unknown and that he was ill equipped to remediate an outbreak of this magnitude. Was it the food, the water he was using to mix the rehydrating solution, or something else about which he was unaware? Martin returned his glance back down to his feet. He thought about the nice apartment in New York that he had shared with four illegal aliens and wondered what life would be like if he had not returned.

—Christopher Landry, Peace Corps Volunteer in El Salvador, 1994–1996.

Worldwide Distribution of Foodborne Pathogens

Worldwide, 1.5 billion children under the age of 5 suffer from diarrhea, and tragically, over 3 million die as a consequence. A large proportion of these cases is due to foodborne illness, which puts in perspective the severity of this issue.[1] The CAST report, issued by the Council for Agricultural Science and Technology (CAST) in 1994, reported that in the U.S. there are as many as 33 million cases of foodborne illness that are responsible for an estimated 9000 deaths annually, and figures from the FDA indicate the number of cases per year could be as high as 80 million.

Foodborne illness is a major health and economic problem in the U.S. and abroad. In the U.S., surveillance systems have been put in place to monitor the incidence and impact of important foodborne pathogens such as *Campylobacter jejuni*, *Clostridium perfringens*, *Escherichia coli* O157:H7, *Salmonella* spp., and *Staphylococcus aureus*. Internationally, it is very difficult to accurately assess the prevalence of foodborne disease because of the differences in existing surveillance systems, if they exist at all. On an international level, cholera is the only foodborne disease monitored by international health regulations. When comparing statistics among countries, differences in disease classification often make it difficult to find data on specific diseases. For example, amoebiasis may be classified according to disease "amoebiasis," or as a symptom under "dysentery." The list of notifiable diseases also varies among countries. While some countries report all identified cases, others report only cases associated with outbreaks, leaving the sporadic cases unaccounted for. Lesser developed countries (LDCs) lack the infrastructure and resources necessary to monitor many of the causative agents of foodborne disease, and it may be in the best interest of some governments to minimize reports on foodborne illness to protect the food export and tourism industries. Unfortunately, it is these LDCs that suffer most from foodborne illness.[2]

In spite of these difficulties, organizations such as the World Health Organization (WHO) have attempted to work around the barriers of estimating the incidence of foodborne disease to provide us with global estimates. When making such estimation, certain principles apply. When severe diseases such as botulism are reported, one can assume that the degree of underreporting is small and reported numbers are considered close to the actual. When symptoms of disease are mild or short in duration, such as those caused by *Staphylococcus aureus*, investigators can expect a higher rate of underreporting, which is factored in during the estimation of the disease. When there is little or no acceptable data available, researchers base their estimates on the incidence of disease in infants, or by extrapolating data from similar regions where such information may be available.[1,2]

Differences in the prevalence of foodborne illness that may appear among the various regions of the world are due to the following:

- Climate influences lifecycles and growth patterns of foodborne pathogens. A colder climate may impede the growth of certain pathogens. A warmer climate will promote the spread of some pathogens.
- Populations with amplified numbers of elderly or infants, which are more susceptible to disease, will have higher rates of illness.
- The nutritional status of a culture plays an integral part in determining the health status of a population. A population with a healthy diet will most likely possess a competent immune system and will be capable of fending off potential health threats.
- Cultural aspects like the knowledge, attitudes, stress (such as illness, loss of a significant other, or poor living conditions), and practices of individuals may be the most important factor in determining the risk factors for disease. For example, African indigenous populations retain the custom of eating raw meat or pork which puts them at high risk of consuming viable *Trichinella spiralis* (roundworm) or *Taenia* spp. (tapeworm) organisms.[3,4]

Reasons for Food Protection Programs

The implementation of programs to minimize foodborne diseases is important because of the problems associated with morbidity, mortality, and economic loss. Foodborne diseases escalate health care costs, as those who are ill access the health system for treatment while imposing an enormous economic burden on related industries. This in turn impacts the communities dependent on these industries.

Morbidity and Mortality Due to Foodborne Disease

As mentioned earlier, the morbidity (sickness) and mortality (death) rates from foodborne disease are nearly impossible to estimate due to the inadequacies in surveillance and reporting systems. Even though the U.S. has a relatively safe food supply, the CDC shows there are still 76 million cases of foodborne illness each year (compiled using death certificates, hospital surveys, and academic studies).[5] Although in the U.S. the accepted number of deaths due to foodborne illness is 9000 annually, it is difficult to determine how accurate this number is because often foodborne illnesses are not reported. People affected do not believe their illness is caused by food, and states do not often have the resources to look into all foodborne illness cases.[6] Of the food-related outbreaks in the U.S., the origin of half remain unknown. The causative agents and modes of transmission (means through which an causative agent is spread) are known in less than 1% of the severe gastroenteritis cases, which leaves the true impact on society questionable.[4] Regardless of the exact numbers, symptoms from foodborne illness range from nausea and mild diarrhea to severe dysentery and death. Sickness can be self-limiting, as is illness from Cryptosporidium, or life threatening from complications as seen with *E. coli* O157:H7. The onset (time lapse between exposure and the expression of symptoms) of illness can be as quick as half an hour, or take years, as heavy metals accumulate in our body until they reach toxic levels.

Economic Consequences of Foodborne Illness

An outbreak of foodborne illness is almost always accompanied by a cost that reflects the amplitude of the outbreak. Medical costs incurred during the treatment of the illness, and the loss of wages for those who must stay home from work, are examples of expenses encountered by individuals inflicted with a foodborne illness. Upon identifying a food product as the origin of illness, the industries responsible for producing, processing, or distributing the foodstuff suffer losses reaching into the millions as they halt production, recall the contaminated product for disposal, lose sales, and increase marketing efforts to compensate for losses. Other costs result from investigations searching for the source of contamination, increased quality assurance measures to produce safer products, litigation fees from law suits, and the loss of trade (Figure 8.1).[5]

CAUSATIVE AGENTS OF FOODBORNE DISEASE

Foodborne illness is defined as any illness incurred from the consumption of contaminated food. Food contaminants come from a wide variety of sources. Certain plants and animals produce substances that are toxic to humans. Chemicals such as arsenic and lead may contaminate food through agricultural or industrial processes. Radionuclides from nuclear fallout are present in food products cultivated at or near testing sites. Parasites, viruses, fungi, and bacteria are all examples of living organisms that have the potential to contaminate a food supply. These agents are termed *foodborne pathogens* (Table 8.1).[4]

Radionuclides

Radiation is introduced into the food chain naturally from mineral deposits beneath the Earth's surface or from the atmosphere in the form of ultraviolet and cosmic rays. Radiation from these sources is emitted at low levels tolerable by plant and animal life. Radionuclides, deposited in the environment accidentally or intentionally as a direct result of human activity, are of much greater concern. The levels that result from these activities are tens to hundreds of times higher than those normally encountered by plant and animal life. Exposure to radiation at high levels

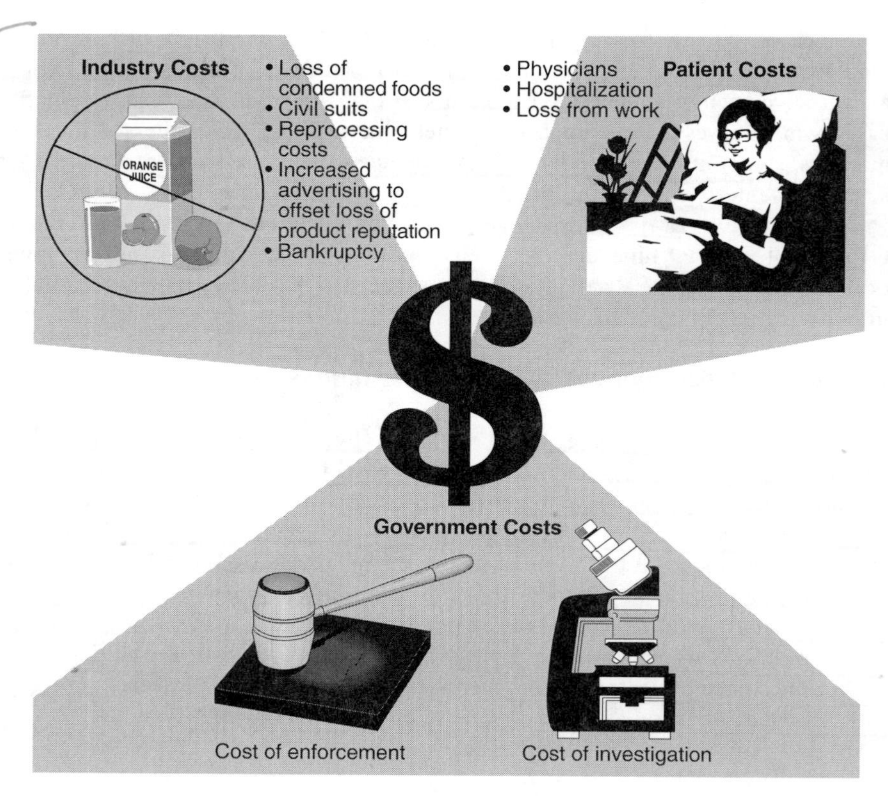

Figure 8.1 Economic losses associated with foodborne outbreaks.

has the potential to cause irreversible damage to whole organ systems, genetic mutations, or deformities in offspring. If the central nervous system is damaged, symptoms of disorientation, convulsions, and lack of muscle coordination are experienced.

The fallout from the Chernobyl accident of 1986 deposited Cesium 137 (137Cs) and Iodine 131 (131I) throughout the northern hemisphere. The isotopes 137Cs and 131I are absorbed by plants through the roots and foliage, which are in turn used as feed for farm animals. Humans ingest these radionuclides directly from the consumption of fruits and vegetables, or indirectly by consuming meat and dairy products.[5] There is recent concern that the concrete vault meant to contain the Chernobyl radioactivity is crumbling and that the danger of additional radioactive leakage is imminent. A similar scenario is put forth from nuclear weapons testing; and while we are aware of the dangers that arise from fallout, nuclear weapons testing continues to be a concern to the present day. Recently, India and Pakistan conducted a series of nuclear weapons tests in an attempt to achieve the upper hand in an arms race.[6] Although the nuclear testing sites are underground and usually in remote areas where there is little or no agricultural activity, radioactive plumes may travel across national boundaries where fallout could still pose a significant threat to human populations.

Table 8.1 Some Causative Agents of Food-Borne Disease

SOME CAUSATIVE AGENTS OF FOODBORNE ILLNESS
Chemicals
Antimony
Cadmium
Lead
Mercury
Polychlorinated biphenyls
Packaging materials
Pesticides
Industrial processes
Food Additives
Gras (Generally Recognized as Safe)
Saccharin
Monosodium glutamate
Nitrates and nitrites
Color additives
Poisonous Plants and Animals
Plant Sources
Alkaloids
Lectins
Saponins
Glucosinolates
Mushroom poisoning
Animal Sources
Paralytic shellfish poisoning
Ciguatera poisoning
Pufferfish poisoning
Foodborne Pathogens
Parasites
Nematodes
Trichinella spiralis, Tapeworms
Protozoans
E. histolyitca, G. lamblia, Cryptosporidium
Viruses
Hepatitis, Norwalk-type
Fungi
Aspergillus spp., Penicillium spp., Mucor spp.
Bacteria
Toxin producing
C. botulinum
Staphylococcus aureus
Vibrio cholerae
Infectious
Salmonella, Campylobacter, E. coli

Chemicals

Ironically, man is responsible for many chemical contaminants presently found in food. Between 80 to 90% of our exposure to potentially harmful chemicals is from food consumption.[4] Chemicals enter the food from packaging materials, agricultural applications of pesticides and fertilizers, by adding preservatives or colorings

Table 8.2 Some of the Many Chemicals Originating from Human Processes that May be Found in Food

UNINTENDED CHEMICALS FOUND IN FOOD
Insecticides DDT, parathion, pyrethrum, arsenicals, others
Fungicides Dithiocarbamates, mercurials, others
Herbicides Carbamates, chlorphenoxy cpds (2, 4-D), Bipyridyls
Fertilizers Nitrogen, others
Treatments and supplements Food additives and veterinary drugs
Accidental and inadvertent Mercury, PCBs, lead, dioxin, aluminum and cadmium from kitchenware
Migration from packaging Plasticizers, stabilizers such as alkylphenols, printing inks, tin, and lead
Chemicals resulting from processing or preparation Polycyclic aromatic hydrocarbons (PAHs), nitrosamines, mutagens

to foods, or by the release of industrial chemicals into the environment. Many of these chemicals have become ubiquitous. Table 8.2 lists some of the many chemicals found in foods. These chemicals largely originate from human processes.

Packaging Materials

Chemical poisoning from packaging materials can cause illness and death within 30 minutes. Substances such as plasticizers, stabilizers, and inks can migrate from the packaging to the food, causing illness.[4] Exposure to these substances is usually exceptionally low and the chemicals tend to be nontoxic. In the U.S., commercial packaging materials are tested for chemical migration and can be considered safe. There are a few chemical contaminants of this nature that are worth mentioning.

Antimony

Antimony leaches into foods stored in chipped enamel containers. Acidic foods such as lemon juice will cause antimony to leach out into the product. The ingestion of antimony can cause complications of the gastrointestinal, cardiovascular, and hepatic systems.

Cadmium

Cadmium, which is used as a plating material for trays and containers, is also leached into food from containers by acidic conditions. Chronic exposure to cadmium can lead to kidney damage. Cadmium accumulates in both the liver and the kidneys. However, in the liver, cadmium is tied up by metallothionein, whereas in the kidney, metallothionein is broken down by lysosomes that allow the unbound cadmium to accumulate.

Lead

Aside from contaminating food through pesticide application, lead is present in the solder of metal containers and cans. Again, acidic foods will cause exposed lead to leach into the food. But in canning there is a coating that prevents the food from coming in contact with the metal. Lead consumption causes neurological complications, kidney failure, and weakens bone integrity. For this reason, lead is being phased out of use in the industry. Other options are being explored.

Industrial Processes

Traces of chemicals from industrial processes are often found in foods. Levels are usually extremely low and they therefore go unnoticed. Occasionally, hazardous quantities are dumped into the environment by accident, carelessness, or with malicious intent, causing foodborne illness.

Mercury

Mercury is a byproduct of many industrial processes. Chronic exposure to mercury presents an emotional or psychological disorder. Methyl mercury, however, is an acute toxin that causes tremors, neurological complications, kidney failure, and birth defects. Methyl mercury is the primary form of mercury found in foods.

In Minamata Bay, Japan, industries were dumping waste contaminated with inorganic mercury into the bay. Bacteria in the bay converted the inorganic mercury to methyl mercury, which resulted in over 1200 cases of food poisoning from the consumption of contaminated fish.

Mercury is also a component of fungicides used on animal feed. Animals consume this feed and are then slaughtered for human consumption. Consuming meats and fish containing high levels of mercury can cause blindness, paralysis, or death.[4]

Polychlorinated Biphenyls

Polychlorinated biphenyls (PCBs) are widely used in industry. They are extremely stable compounds that do not degrade easily, they are resistant to heat, and are also highly toxic. In Japan in 1968, 1000 people suffered food poisoning from PCB contamination in rice oil. During the rice oil process, a heating pipe with a small hole allowed PCBs to contaminate the rice oil. Symptoms included swelling of the eyes, rash, and gastrointestinal illness, which resulted in five deaths. Alternatives to PCBs are being researched, but they are so widespread today that we can expect to see traces of them for many years to come.

Dioxin

Another chemical turning up in foods is dioxin. In early 1999, high levels of dioxin were found in Belgian cattle. The government admitted that it does not inspect cattle that are brought up near industrial sites (a common source for dioxin along with incinerators), nor does it have requirements to measure dioxin levels in beef. Following this discovery, new efforts are being made in Belgium and other industrial nations to keep dioxin-contaminated beef out of the global markey.[9]

Pesticides

Pesticides are introduced into the environment through the air, water, and soil. Although they are used to control insects, fungi, and rodents, they may reduce crop yield. Organochlorine compounds (such as DDT and chlordane), organophosphates (such as parathion and malathion), and inorganic compounds (such as arsenics) have been applied to food in the form of a pesticide. DDT was found to interfere with the reproductive cycle of some birds and was banned in 1972. Since then, the EPA has banned aldrin, dieldrin, hepaclor, and kepone. Yet traces of these compounds and their metabolites continue to be found in our food. Many Americans are eating imported foods that still contain these potentially dangerous residues. Many of these chemicals banned from use in the U.S. are sold to developing nations that use them extensively in producing crops for export to the American market.

The characteristics of some pesticides make them especially dangerous. DDT and other chemicals of its class accumulate in the environment. This is why we still find trace amounts of residue in foods produced in the U.S., especially shellfish. In response to this, pesticides were formulated to break down more readily. Because these pesticides spend less time in the environment, they tend to be more toxic. For example, parathion is a threat to farm workers because it can be absorbed through the skin. Carbamate compounds inhibit cholinesterase, which is an essential enzyme in transmitting nerve impulses at the synapses. The EPA has proposed tolerance levels for pesticides applied to food products. Tolerance levels are the maximum permitted amounts of pesticides that can be found on a raw product. Although these efforts are well intentioned, only a small portion of imported products entering the U.S. is tested for pesticides.[9,10]

Food Additives

Food additives are used to alter taste, color, texture, nutritive value, appearance, and resistance to deterioration. There are thousands of food additives ranging from simple coumpounds such as salt to more complex compounds such as monosodium glutamate. Food additives are also considered to be the least hazardous source of foodborne illness, ranking behind pesticides, environmental contaminants, natural toxins, and microbial toxins.

In the early part of the 20th century, increased knowledge of the toxic effects from long-term chemical exposure prompted the passage of the Food and Drug Act of 1906, to which several amendments have been added. In 1958, the Food Additive Amendment to the Food Drug and Cosmetic Act required the manufacturer of any new potential food additive to conduct testing and achieve FDA approval before marketing. Not included in this amendment are "generally recognized as safe" (GRAS) substances, and color additives which are covered under the Color Additive Amendment of 1950.[9]

Saccharin

Saccharin is a low-calorie sweetener that has been approved by the FDA but still remains highly scrutinized. In laboratory animals, saccharin promotes bladder cancer after exposure to a bladder carcinogen. In 1977, the FDA proposed a ban on saccharin as a food additive. However, the ban was postponed by Congress pending additional studies to support the results. Saccharin was not removed under the Delaney Clause because various studies have reported conflicting results. A considerably high amount of saccharin is needed to produce the observed effects. It has been speculated that sodium, not saccharin, may be responsible for the bladder cancer.[9]

Monosodim Glutamate

Monosodium glutamate (MSG) is used to enhance flavor in Chinese foods. It is also thought to be the cause of Chinese Restaurant Syndrome (headaches and possible nausea) and lesions of the retina. An allowable daily intake (ADI) of 120 mg/kg has been established for individuals over one year of age, and manufacturers in the U.S. have voluntarily halted the use of MSG in infant foods. The role of MSG in Chinese Restaurant Syndrome remains unclear, but recent studies indicate that MSG might not be reponsible for the development of the syndrome.[9]

Nitrates and Nitrites

Nitrates and nitrites are used to cure ham, bacon, and other meats. They give meats their characteristic taste and color while preventing the growth of *Clostridium botulinum* spores. Nitrate and nitrites can also be found in vegetables, water, saliva, and tobacco. In the body, nitrates can be reduced to nitrites, which in turn oxidize hemoglobin into methemoglobin. If methemoglobin accumulates in the blood, oxygen cannot be transported. Anoxia will result.

In food, nitrites react with amines to form nitrosamines. This is of concern because nitrosamines have caused cancer of the liver, kidney, bladder, stomach, and pancreas of laboratory animals. Because nitrosamines are not added to foods, they are not excluded by the Delaney Clause. The FDA has lowered the amount of nitrites added to meats, and has suggested the addition of ascorbate or sodium erythorbate to prevent the formation of nitrosamines.

GRAS (Generally Recognized As Safe)

GRAS substances are chemicals that had a history of safe use before the 1958 Food Additive Amendment passed. There are appoximately 700 GRAS substances. Currently, the FDA is reviewing the safety of these substances and reclassifying them if deemed necessary. One other piece of legislation worth mentioning is the Delaney Clause, which is included in the Food Additive and Color Additive Amendments of the Food Drug and Cosmetic Act. The Delaney Clause prohibits the use of any food additive found to induce cancer in laboratory animals under appropriate conditions. This clause has been challenged with much resistance from manufacturers because of the controversy surrounding the extrapolation of results from lab animals to humans.[9]

Color Additives

Colors are added to foods such as candy, cakes, cereal, ice cream and soft drinks to give them a color that is more appealing to the consumer. Natural colors such as grape skin extract, caramel, and fruit juices did not have to undergo certification after the Color Additive Amendment of 1960 passed. All synthetic colorings, however, required certification if there was any question as to their safety, even if the substance had been previously certified. There are seven synthetic colorings that are certified for unlimited use, according to good manufacturing practices (GMPs). They are FD&C Blue No. 1 and 7, Red No. 3 and 40, Yellow No. 5 and 6, and Green No. 3. Other colors such as Green No. 1 and Red No. 2 were removed from use because of concerns over their carcinogenicity and other effects from chronic exposure.[9]

Poisonous Plants and Animals

Although there are numerous species of plants and animals that are poisonous, the majority are not part of our diet. By the process of trial and error, humans have identified plants that were either harmful to man or possesed little nutritional value and excluded them from our diet. Some plants and animals known to be harmful have a significant nutritional value and are still part of our diet. Plants such as castor bean, alfalfa, radish, and some mushrooms produce substances that could be harmful if consumed. Currently, there are 1200 species of marine organisms that produce toxins. Fortunately, only a few are important in foodborne illness. Toxic plankton are consumed by shellfish and other marine animals, which in turn are harvested for human consumption. Likewise, a species of the puffer fish, which is important in China and Japan, also produces a toxin. In some instances, the toxic substance is inactivated by properly cooking the food (puffer fish, lectins). However, many plants and animals are served raw, leaving the consumer defenseless against any toxins that may be present in the food.

Plant Sources

Alkaloids

Alkaloids are part of a plant's natural defense. These nitrogen-containing organic compounds ward off predators such as insects and other herbivores that might consume a portion of the plant. Three major groups of alkaloids important to humans are:

1. The pyrrolizidine group of alkaloids are found in herbs such as coltsfoot, confrey, and petasite. These are used in home remedies. Humans consume these herbs in the form of teas and other herbal concoctions that are a common source of these alkaloids. Pyrrolizidine alkaloids cause extensive damage to the liver and lungs and have a high mortality rate among children.
2. Solanum alkaloids cause diarrhea, headaches, vomiting, and neurological problems in humans who have consumed green, sprouted, stressed, or storage-abused potatoes. Solanum can withstand normal preparatory procedures such as baking, boiling, or microwaving. It was once thought that solanum played a role in anencephaly-spina bifida (ASB), but research has refuted any such hypothesis.
3. Xanthine alkaloids are found in coffees, teas, and many other beverages in the form of caffeine. The caffeine found in ten cups of coffee (0.5–1.0 g) is enough to produce adverse effects in adults, and child mortality is possible after the ingestion of 5g of caffeine.

Lectins

Lectins are plant proteins that agglutinate red blood cells. More than 800 species of edible plants contain lectins, over 600 of which belong to the Leguminosae family. Ricin, found in the castor bean, is a highly toxic lectin. Effects in humans vary from stunted growth to fatality, according to the source and dose of the lectin. Lectins are destroyed by cooking in wet heat.

Saponins

Saponins are glycosides that hemolyze red blood cells. When alfalfa, which has a high concentration of saponins, is fed to lab animals, they suffer from depressed growth rates and anemia. Saponins are of significant concern in the U.S. As we are experiencing a dietary shift to healthier foods such as alfalfa and soy-based products, we can also expect an increase of saponin intoxications.

Glucosinolates

Glucosinolates are found in broccoli, cabbage, radish, horseradish, and mustard greens, and are thought to cause hypothyroidism (enlargement of the thyroid gland). Hence they are also termed goitrogens.

Mushroom Poisoning

There are approximately 100 species of mushrooms thought to be toxic and 12 species known to contain lethal toxins. Mushroom toxicity can be classified into two types of syndromes based on the latency time of the symptoms. Rapid onset syndrome is rarely severe and produces symptoms such as gastrointestinal illness, parathesia, and/or CNS syndrome within 3 hours of consumption. Delayed onset syndrome presents symptoms as headaches, diarrhea, hepatic incompetence, muscle pains, and chills 6 hours after consumption. Symptoms of mushroom poisoning vary with mushroom species. Mushroom-induced headaches are common to European species, while profuse diarrhea and hepatic complications are more common to North America.

Animal Sources

Paralytic Shellfish Poisoning

Paralytic shellfish poisoning results when humans consume shellfish that have fed on toxic plankton. Shellfish become toxic to humans when they feed on dinoflagellates such as *Gonyaulax catenella* in numbers greater than 200 ml of water. When theses toxic organisms are present in numbers greater than 20,000 ml they are visible to the naked eye. This phenomena has been termed *red tide* for its crimson appearance. Saxitoxin, which is just one of the various toxins involved in paralytic shellfish poisoning, can be lethal after consuming only four milligrams. Symptoms include a tingling or burning sensation of the lips and gums, ataxia, and paralysis of the diaphragm. If treatment is sought early enough, mortality can be avoided.

Ciguatera Poisoning

Species of barracuda, eels, jacks, groupers, red snapper, and sea bass specific to the Caribbean and South Pacific have been implicated in ciguatera poisoning. The ciguatera toxin, like the saxitoxin, is found in a dinoflagellate. The species of marine animals mentioned above do not feed directly on the dinoflagellate, *Gamblerdiscus toxicus*, which contains the ciguatera toxin, but the small feed fish they thrive on consume the plankton as a primary component of their diet. Signs of ciguatera poisoning are a tingling or numbing sensation of the lips, tongue, and throat, which gives way to nausea, vomiting, and diarrhea. In extreme cases, muscle and joint pains arise, making it difficult for the patient to walk.

Puffer Fish Poisoning

The puffer fish, also known as the blowfish, is a common delicacy in China and Japan. While not all puffers are poisonous, those belonging to the genus *Fugu* (fugu fish) contain the lethal tetrodotoxin, and are also a most sought-after delicacy. When the fish is prepared in acidic or alkaline conditions, the toxin is inactivated. If the fish are prepared incorrectly, victims experience symptoms similar to intoxication

Table 8.3 Foodborne Pathogens Included in the CAST Report and Fungi (Adapted from Motarjemi, Y. and Kaferstein, F.K., Global estimation of foodborne disease, *World Health Org. Stat. Q.*, 50, 8, 1997.)

FOODBORNE PATHOGENS	
Viruses	**Parasites**
Hepatitis A	*Anisakid nematodes*
Norwalk	*Cryptosporidium parvum*
Norwalk-like	*Diphyllobothrium latum*
	Entamoeba histolytica
	Giardia lamblia
	Taenia saginata
	Taenia solium
	Toxoplasma gondii
	Trichinella spiralis
Bacteria	**Toxins**
Aeromonas hydrophila	Ciguatera toxin
Bacilus cereus	Diarrheic shellfish poisons
Brucella abortus	Domoic acid
Campylobacter jejuni	Histamine
Clostridium botulinum	Scromboid
Clostridium perfringens	Paralytic shellfish poisons
Coxiella burnetti	Tetrodotoxin
Escherichia coli	
Listeria monocytogenes	**Fungi**
Mycobacterium bovis	*Aspergillus* spp.
Salmonella spp.	*Penicillium* spp.
Shigella spp.	*Mucor* spp.
Staphylococcus aureus	*Rhizopus* spp.
Vibrio cholerae	*Candida* spp.
Vibrio parahemolyticus	*Cryptococcus neoformans*
Vibrio vulnificus	
Yersinia enterocolitica	

from saxitoxin, and death from respiratory paralysis can occur from the consumption of only 1 to 2 mg of the toxin.

FOODBORNE PATHOGENS

Until recently, when one thought of a foodborne outbreak, Salmonella was assumed to be the causative agent unless investigation proved otherwise. Since 1975, 25 new foodborne pathogens have been identified. More than 40 potential foodborne pathogens have been listed by CAST, and a number of these are discussed in the following sections along with fungi, an area not considered by CAST (Table 8.3).[2]

Coupled with these new foodborne pathogens are changes in dietary habits and food processing techniques, which are partly responsible for the surfacing of new pathogens and an increase in the number of illnesses due to traditional pathogens. Some of these changes are mentioned below.

- A decrease in lactic acid-producing bacteria as a result of processing milk increases the chances of survival of Salmonella or Campylobacter, as their competition has been removed.
- Water polluted with raw sewage or manure containing harmful numbers of microbes comes in contact with food, causing a variety of illnesses.

- Due to the abuse of antibiotics and sulfur-containing drugs in farm animals, some bacteria such as Salmonella have acquired resistance to antibiotics.
- A change in dietary preference. In an attempt to eat healthier, we are consuming foods that are lower in fat but also more susceptible to microbial growth.
- The increase in demand for packaged food with a longer shelf life prevents aerobic growth but supports anaerobic pathogens like Clostridia, which can be deadly.
- The growing popularity of ready-to-eat or take-out foods that require only minimal heating before consumption raises the temperature of the food within the ideal growth range of many bacteria without destroying the pathogen.[10]

The presence of these pathogens in foods or in the body can result in illness due to infection or intoxication. For example, infections such as that caused by *Giardia lamblia* arise when the organism colonizes on the intestine wall, reducing the absorption surface, and therefore decreasing the intake of nutrients. Infections caused by *Salmonella typhi* or Hepatitis A have invasive properties and can spread beyond the gastrointestinal tract, entering the bloodstream or liver and causing typhoid fever and liver disease, respectively. Intoxication results when an organism releases a toxin while growing on a food or while inhabiting the gastrointestinal tract of humans. Illness from toxigenic pathogens has a rapid onset and is usually more severe than that from an infection. In many cases, food products that contain a toxin have no altered appearance, taste, or smell, making them especially dangerous. In the following sections, we will examine parasitic, viral, fungal, and bacterial pathogens and the illnesses they cause.[11]

Parasitic Infections

Parasites are organisms that require another animal (host) to complete its life cycle. Organisms such as *Entamoeba histolytica* or *Cryptosporidium parvum* require only one host, they are shed in the feces, and infection occurs by consuming food or water contaminated with these feces. *Trichinella spiralis*, on the other hand, causes infection in those who eat meat infested with the cyst of the organism. In the case of Trichinella, the organisms requires two hosts, the *intermediate* host (pig), which houses the cyst, and the *definitive* host (human), where the organism develops into an adult and is shed in the feces.

The Nematodes

Trichinella spiralis

Species of *Trichinella spiralis* are found in warm-blooded animals such as bear, hogs, walrus, wild boar, and rats. This organism causes trichinosis when meat from warm-blooded animals containing viable larvae are consumed by humans. In the 1940s, there were an average of 400 cases of trichinosis per year. This number has decreased markedly to just 32 cases in 1994 as a result of improved cooking habits, greater access to home freezers, and legislation that regulates the quality of feed intended for commercial hogs. Trichinella organisms can be destroyed by cooking meat until it turns a gray color or by freezing. In Idaho in 1995, an outbreak occurred from the ingestion of cougar jerky. The outbreak was contained to ten people who displayed symptoms of eosinophilia, fever, rash, and overall weakness.[12]

Tapeworms

The beef tapeworm *Taenia saginata* and the pork tapeworm *Taenia solium* infest humans through the ingestion of undercooked beef or pork that contains viable larvae. *Taenia solium* occurs from the ingestion of undercooked pork products. The pig is an intermediate host for *T. solium*, and the pig becomes infected when it consumes food that is contaminated with human fecal material containing the tapeworm eggs (Figure 8.3). The eggs enter the pig stomach and excyst where they penetrate through the mucosa into the lymphatics and bloodstream. Then they are carried to the striated muscles, the brain, and possibly other organs. They become embedded in the muscle of the animal where they develop into cysterici (encysted larvae), which are thin-walled, fluid-filled bladders containing an inverted scolex or head of the tapeworm (Figure 8.3). If undercooked pork containing viable larvae are ingested by humans, the larvae excyst in the intestinal tract. The scolex of *Taenia solium*, containing four suckers and two rows of hooks, protrudes from the cystericus and attaches itself to the intestinal wall, where it develops into an adult with the formation of proglottids. The proglottids mature to develop seven to 13 lateral uterine branches that produce tens of thousands of embryonated eggs. The most distal (and most mature) proglottids detach from the chain of proglottids and are either discharged from the intestine or migrate out. The adult tapeworm can grow up to seven meters in length and inhabit the intestine for 30 years, during which time it sheds gravid proglottids (tapeworm segments containing eggs) in the feces of the primary host (human). If careless hygiene is practiced, person-to-person transmission is possible via direct contact or by food contaminated with fecal matter. There is also a cycle that may occur in humans whereby the *T. solium* eggs may be ingested from external sources, from self-contamination with fecal matter, or from reverse peristalsis when the contents of the intestines are backflushed into the stomach from the process of vomiting. In this case, the eggs excyst and the larvae migrate to the human muscles, the brain, and possibly other organs such as the eye where they develop into cysticerci.

The cycle of the beef tapeworm, *Taenia saginata*, follows a similar route with the exception that reinfection by the egg stage of the tapeworm does not occur in humans (Figure 8.4). The consumption of undercooked or raw beef containing the cysticerci leads to the development of an adult beef tapeworm in the human intestinal tract, which has the capacity of reaching lengths of 25 meters, but is usually less than half that size.

Although tapeworm infections are rare in the U.S., they are more prominent in areas where beef or pork are consumed raw. In healthy individuals symptoms often go unnoticed. In malnourished individuals multiple infections can occur, leading to blockage of the intestine. If the tapeworm leaves the intestine and infects organs such as the heart or brain, effects can be severe regardless of the health status of the individual. Tapeworms can be prevented by appropriately disposing of contaminated feces, cooking meat to an internal temperature of 57°C, or freezing it at –10°C for 5 days.[13]

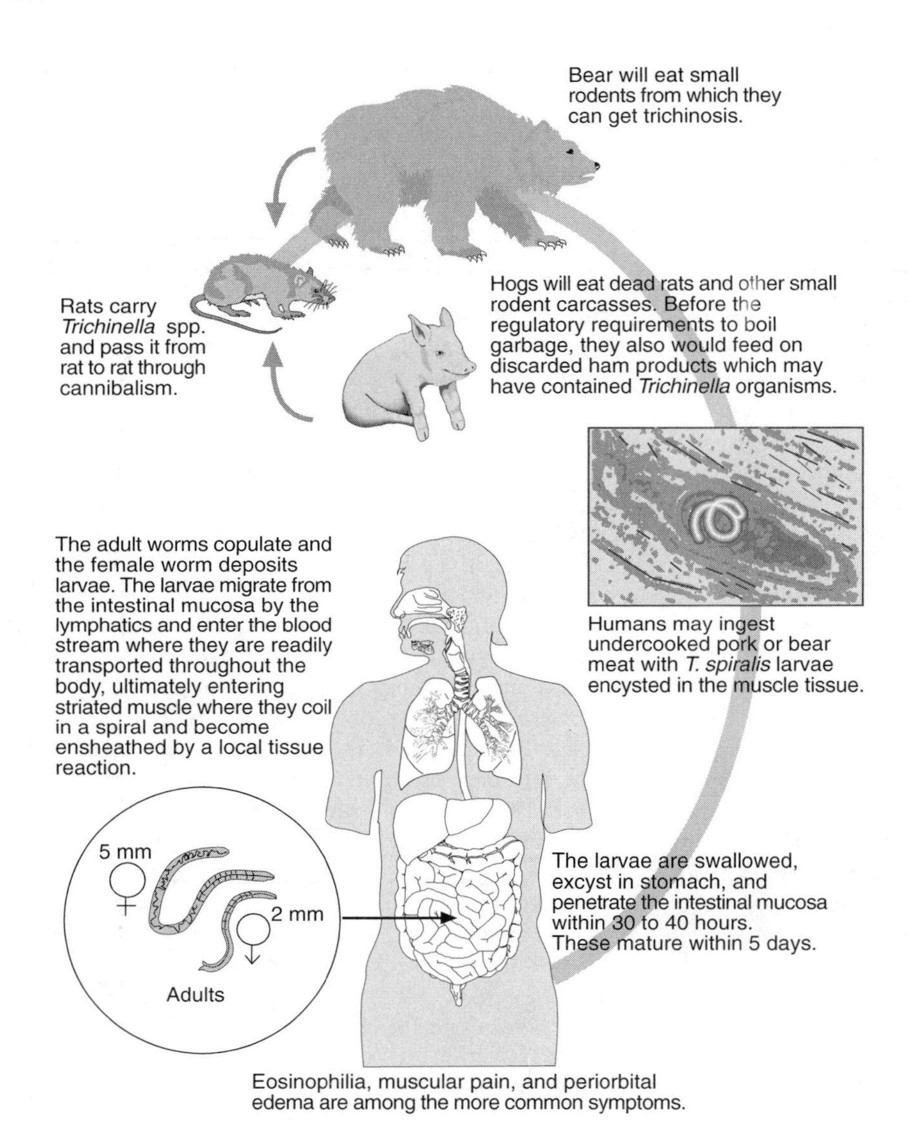

Bear will eat small rodents from which they can get trichinosis.
Rats carry *Trichinella* spp. and pass it from rat to rat through cannibalism.
Hogs will eat dead rats and other small rodent carcasses. Before the regulatory requirements to boil garbage, they also would feed on discarded ham products which may have contained *Trichinella* organisms.
Humans may ingest undercooked pork or bear meat with *T. spiralis* larvae encysted in the muscle tissue.
The adult worms copulate and the female worm deposits larvae. The larvae migrate from the intestinal mucosa by the lymphatics and enter the blood stream where they are readily transported throughout the body, ultimately entering striated muscle where they coil in a spiral and become ensheathed by a local tissue reaction.
5 mm
♀
2 mm
♂
Adults
The larvae are swallowed, excyst in stomach, and penetrate the intestinal mucosa within 30 to 40 hours. These mature within 5 days.
Eosinophilia, muscular pain, and periorbital edema are among the more common symptoms.

Figure 8.2 Life cycle of *Trichinella spiralis*.

The Protozoans

Entamoeba histolytica

Amebiasis, which is caused by *Entamoeba histolytica*, affects about 10% of the world's population. Outbreaks occur where sanitation is poor, where risky sexual habits are practiced, and in institutional facilities. In such settings, fecal material comes in contact with water and food products, increasing the likelihood of the organism

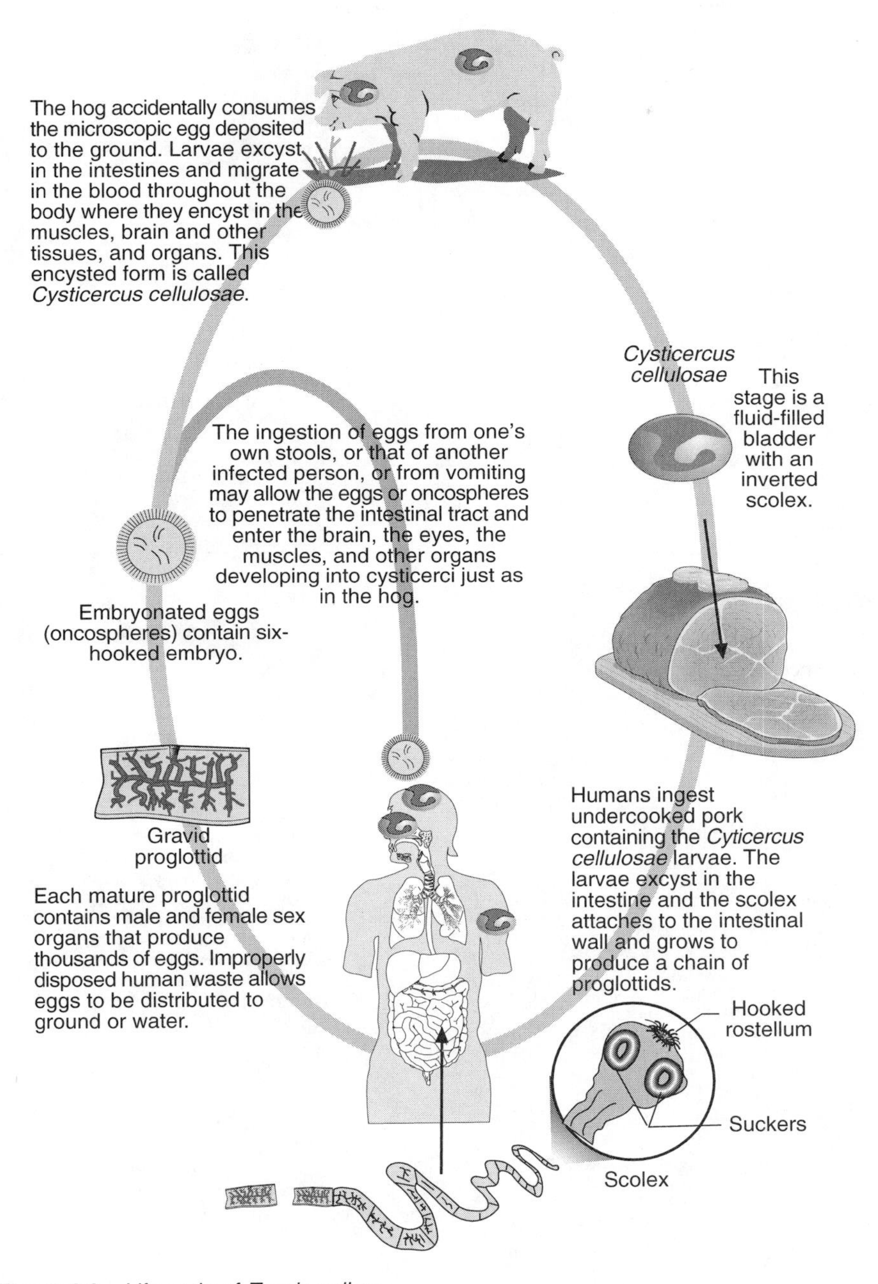

The hog accidentally consumes the microscopic egg deposited to the ground. Larvae excyst in the intestines and migrate in the blood throughout the body where they encyst in the muscles, brain and other tissues, and organs. This encysted form is called *Cysticercus cellulosae*.
Cysticercus cellulosae
This stage is a fluid-filled bladder with an inverted scolex.
The ingestion of eggs from one's own stools, or that of another infected person, or from vomiting may allow the eggs or oncospheres to penetrate the intestinal tract and enter the brain, the eyes, the muscles, and other organs developing into cysticerci just as in the hog.
Embryonated eggs (oncospheres) contain six-hooked embryo.
Gravid proglottid
Each mature proglottid contains male and female sex organs that produce thousands of eggs. Improperly disposed human waste allows eggs to be distributed to ground or water.
Humans ingest undercooked pork containing the *Cyticercus cellulosae* larvae. The larvae excyst in the intestine and the scolex attaches to the intestinal wall and grows to produce a chain of proglottids.
Hooked rostellum
Suckers
Scolex

Figure 8.3 Life cycle of *Taenia solium*.

being ingested. While in the cyst stage of its life cycle, the organisms can survive in the external environment where it is transmitted by water or the fecal–oral route (Figure 8.5). once ingested, the cysts are broken down by normal digestive action. Motile trophozoites are then released. Trophozoites multiply and invade the large

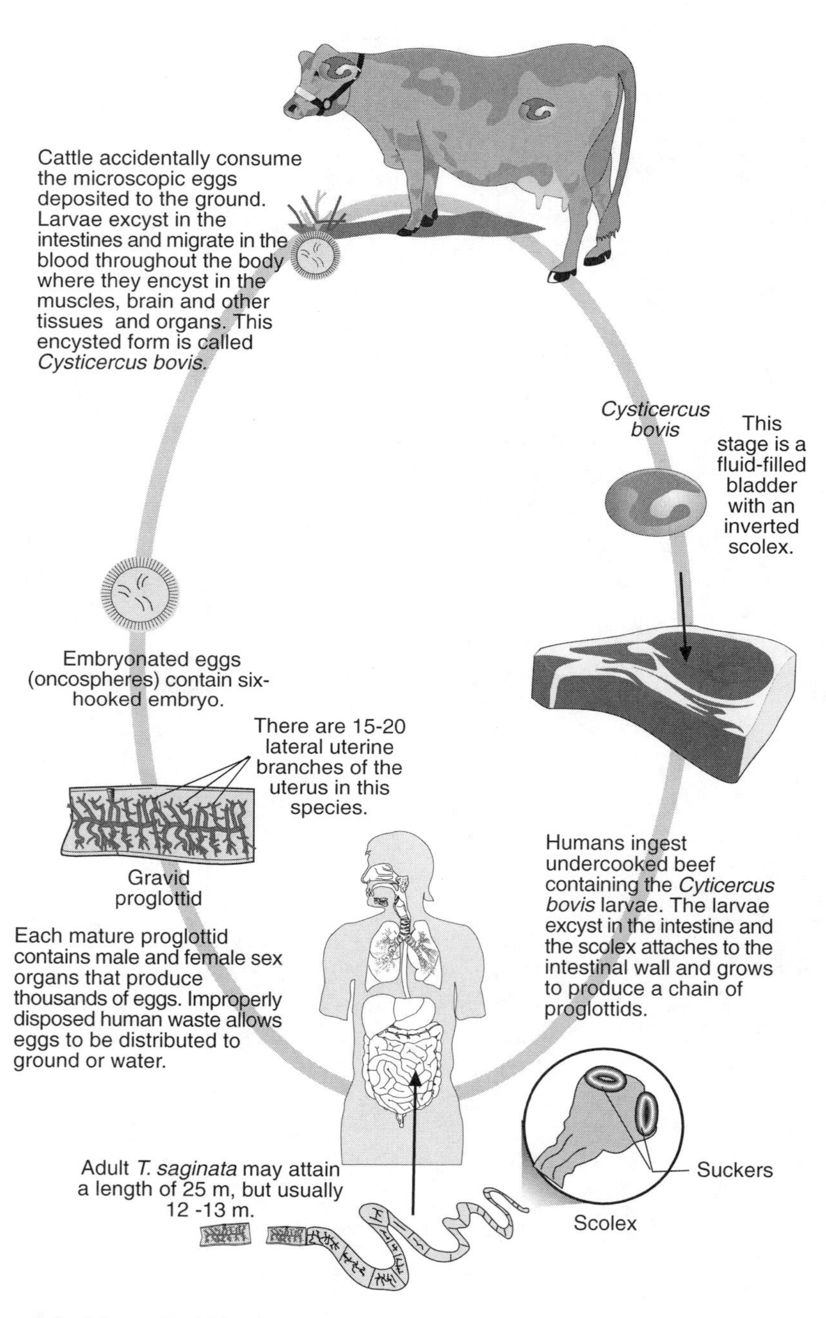
Cattle accidentally consume the microscopic eggs deposited to the ground. Larvae excyst in the intestines and migrate in the blood throughout the body where they encyst in the muscles, brain and other tissues and organs. This encysted form is called *Cysticercus bovis.*
Cysticercus bovis
This stage is a fluid-filled bladder with an inverted scolex.
Embryonated eggs (oncospheres) contain six-hooked embryo.
There are 15-20 lateral uterine branches of the uterus in this species.
Gravid proglottid
Each mature proglottid contains male and female sex organs that produce thousands of eggs. Improperly disposed human waste allows eggs to be distributed to ground or water.
Humans ingest undercooked beef containing the *Cyticercus bovis* larvae. The larvae excyst in the intestine and the scolex attaches to the intestinal wall and grows to produce a chain of proglottids.
Adult *T. saginata* may attain a length of 25 m, but usually 12 -13 m.
Suckers
Scolex

Figure 8.4 Life cycle of *Taenia saginata*.

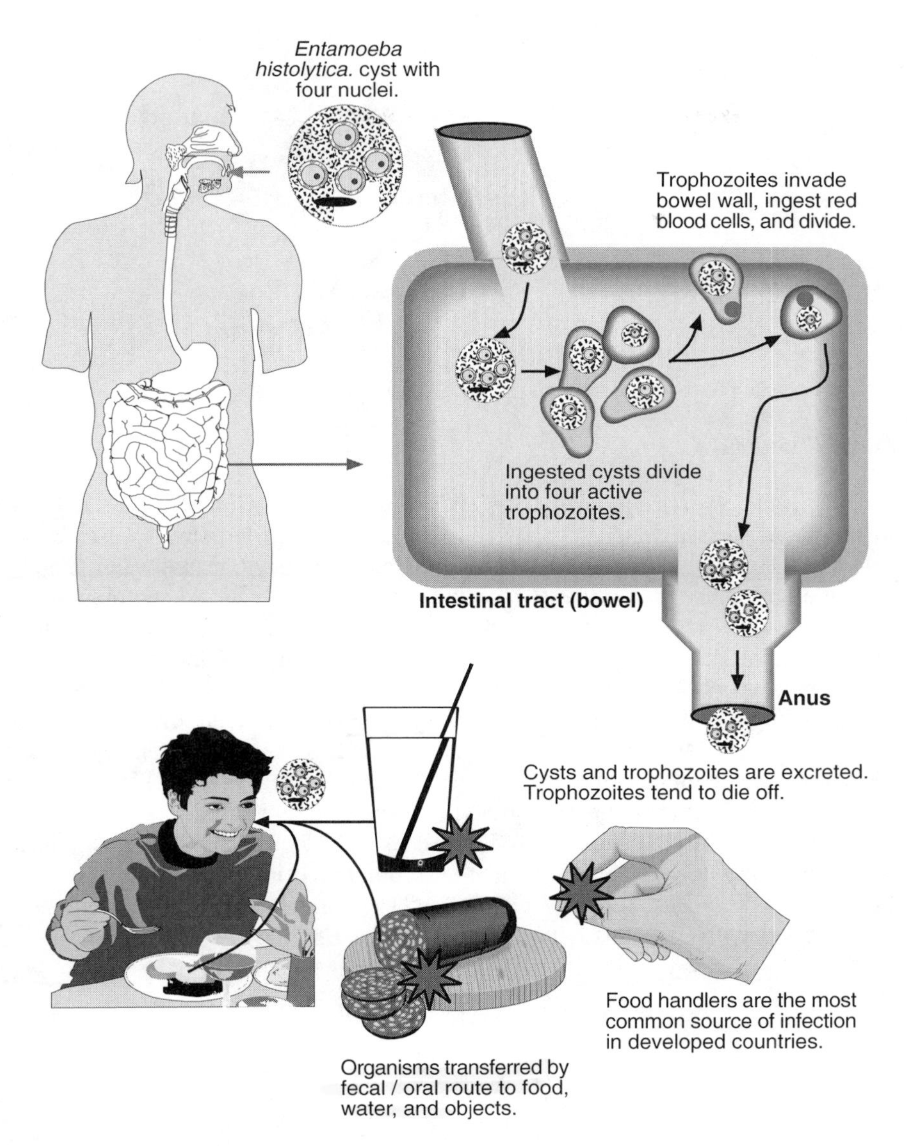

Figure 8.5 Life cycle of *Entamoeba histolytica.*

intestine, causing symptoms that vary from mild to bloody diarrhea with a dozen explosive liquid stools a day. This latter case is known as amoebic dysentery. The incubation time of the disease varies from 3 days to a month. In the case of chronic infections, the patient can be asymptomatic while continuously shedding viable cysts, or symptomatic, suffering from bouts of diarrhea and fever until treatment is sought.

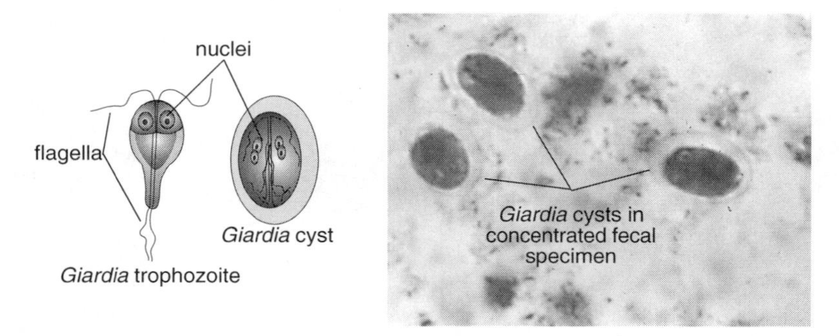

Figure 8.6 Trophozoite and cyst stages of *Giardia lamblia*.

Giardia lamblia

Giardia lamblia is a protozoan flagellate found in areas with poor sanitation and in unfiltered surface water supplies (Figure 8.6). *Giardia* is considered to be the primary agent for diarrheal outbreaks associated with contaminated water supplies having a prevalence of 1.5 to 20%.[14] In the U.S., giardiasis is most common among those who travel to endemic areas, in homosexuals, and in child day care settings where constant diaper changing facilitates the spread from one child to another. The organism is also found in surface waters used for drinking and recreational purposes. The cyst stage of the organism resists chlorination and can, therefore, pass through municipal supplies that do not also have adequate filtration systems. The cysts reach the surface water supplies through the fecal deposits of beaver and muskrats. When ingested cysts reach the small intestine, trophozoites are released. The trophozoites attach themselves to the intestinal wall by adhesive disks, causing inflammation and irritation of the bowel. This inflammation prompts the influx of fluids, causing a loose stool. As a result, food is malabsorbed. The patient suffers from dehydration. Symptoms usually manifest within one week of eating the contaminated food and consist of nausea, explosive diarrhea (up to ten movements per day), and fatigue. Without treatment, symptoms linger and reinfection is possible.[15]

Cryptosporidium

The first reported case of cryptosporidiosis, which is caused by the parasite *Cryptosporidium*, occurred in 1976. Cases were prevalent primarily among immunocompromised individuals whose immune systems were unable to protect the body from infection. In 1993, the city of Milwaukee experienced the most comprehensive outbreak to date, which involved 400,000 inhabitants of the city who drank from the contaminated public water supply. Primarily a waterborne pathogen, *Cryptosporidium* is transmitted via water contaminated with feces from human and agricultural origins. In 1994, the first documented case of foodborne cryptosporidiosis stemmed from fresh-pressed apple cider. Foodborne transmission of *Cryptosporidium* occurs via the fecal–oral route, usually from careless food handlers shedding the hardy oocysts (see "life cycle," Figure 8.7) of the organism. In healthy individuals, symptoms

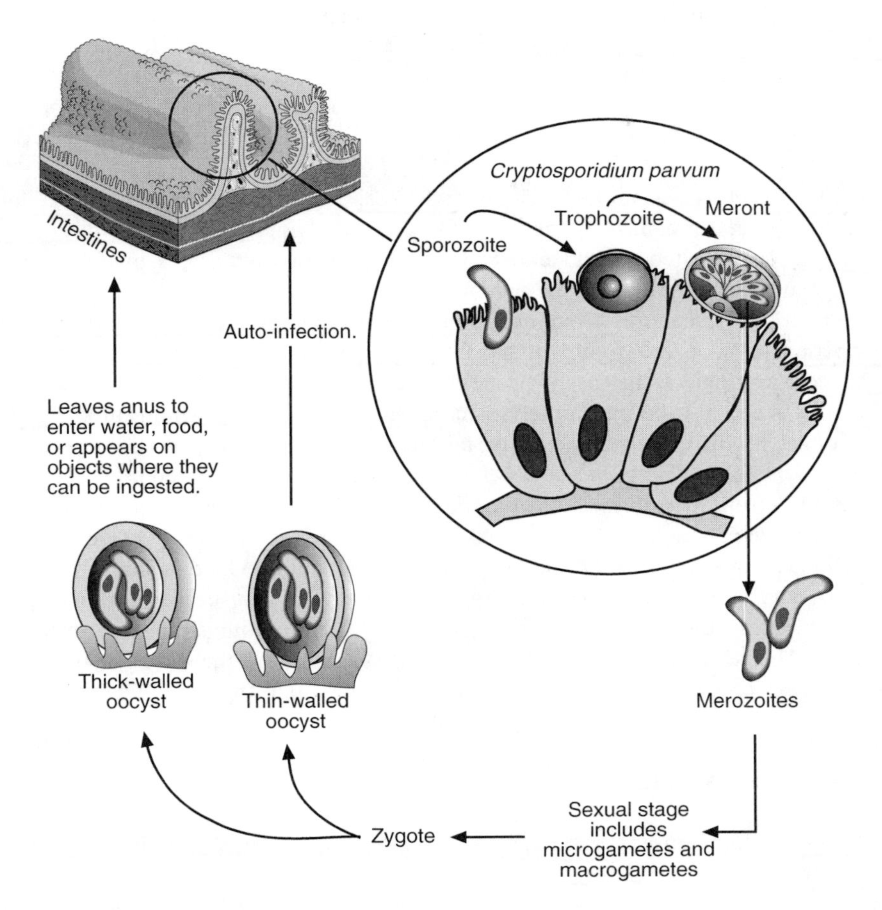

Figure 8.7 Life cycle of *Cryptosporiudium parvum.*

appear as mild diarrhea, nausea, cramps, and a low-grade fever. Immunocompromised patients such as infants, the elderly, or those with AIDS experience high volume diarrhea, weight loss, and severe abdominal cramps. *Cryptosporidium* is of particular concern because the oocysts remain viable in the chlorinated water of swimming pools and can survive the water treatment processes through which our drinking water passes. We can expect to see more outbreaks of cryptosporidiosis until researches and regulators devise more adequate means to destroy the organism in our water supplies.[17]

Viruses

Viruses are microscopic particles that usually contain a single strand of RNA. Although viral particles can exist outside a host cell, as during transmission, they require a host cell for replication to occur. After invading and replicating within a cell, death or damage to the host cell is common, but this is not always the case.

The two most prominent foodborne viruses of present day are Hepatitis A and Norwalk-like virus.

Hepatitis A

Hepatitis A is transmitted via the fecal–oral route and causes liver infection occasionally accompanied by jaundice. On average, symptoms occur 28 days after infection, but infected individuals shed the virus before symptoms manifest, producing an exceptionally risky situation as the infected person might handle food without added precaution. Contamination can occur by infected food workers handling foodstuffs, or from food products that have come in contact with water polluted with fecal matter, as is the case with shellfish harvested from sewage-infested waters. Hepatitis A can also be transmitted from patient contact or intravenous drug use. In each of the last three decades there have been sharp peaks in the number of Hepatitis A cases in the U.S. There was a 5% increase in the number of reported cases of Hepatitis A from 1992 to 1993. We will have to wait for further data, but it is possible that we are experiencing the beginnings of another upward trend.[18,19] There are new techniques, however, to determine the sources of outbreaks. Genetic analysis can be used by getting samples of Hepatitis A from patients in a suspected outbreak area and comparing them to determine if the strains came from similar or different sources. This technology was used most recently to determine the source of Hepatitis A outbreaks in Michigan, Maine, and other states in which frozen strawberries used in the schools were responsible for the outbreaks.[20]

Norwalk-Like Virus

Norwalk-like viruses are small, circular organisms thought to contain RNA. Little is known about these viruses, which do not multiply in laboratory conditions. In 1982, Norwalk-like viruses were the leading cause of reported foodborne illness in the U.S., responsible for 5000 cases from two different outbreaks. Both outbreaks had occurred in Minnesota, and the source of contamination in each outbreak was found to be infected food handlers. The food merely served as a vehicle for the spread of the virus. Since 1982, Norwalk-like viruses have not caused any outbreaks of these proportions, but they are recognized as a major cause of foodborne illness in Minnesota, New York, and the U.K.

Food products such as creams, cream fillings, and salads are efficient vehicles for viruses because they do not undergo any extensive heating before being served. Hence the virus is not destroyed. In New York, shellfish are the major vehicle of transmission. Norwalk-like viruses display an attack rate near 50% following the consumption of contaminated shellfish or other foods. The onset of illness occurs between 18 and 36 hours after consuming the virus, and typical symptoms include diarrhea and nausea, lasting 24 to 36 hours. Although little is known about the Norwalk-like viruses, it is believed that foods that serve as a vehicle for Hepatitis A can also transmit the Norwalk-like type of virus.[18]

A Case History — Almost a Tragedy

Colleen Tobin was a normal schoolgirl growing up in Northampton, Massachusetts. When Colleen was 13, she started to express symptoms of stomach aches, diarrhea, fatigue, and loss of weight. She had stopped growing and her weight was decreasing at an alarming rate. Naturally, her parents had brought her to meet with several specialists in the state, but the initial examinations turned up no answers or leads. Doctors had not considered a parasitic infection for two primary reasons. One, she had not traveled to any country where unsanitary conditions might facilitate the spread of parasitic disease, and two, she had not been on any backpacking excursions where she might have consumed untreated water. Unable to find any explanations for Colleen's illness, doctors suspected she was deliberately starving herself and suggested she be admitted to a psychiatric hospital. Appalled by this suggestion, Colleen's parents continued to seek out new doctors, as she had now lost 12 pounds and had a body temperature of 92°F. The gastrointestinal specialists from Springfield, Massachusetts, ran several tests, all of which turned up negative. Again, little Colleen was faulted, accused of taking laxatives. In her defense, Colleen mentioned that she did not know what a laxative was, and furthermore, there were no stores where Colleen could have purchased laxatives. Finally, Colleen was tested for parasites and the results returned negative, a phenomena Colleen's parents learned was not unusual, even among infected individuals.

Nearly one year after the onset of her initial symptoms, Colleen spent ten days at Massachusetts General Hospital. She now weighed only 90 pounds, 30 pounds less than when she expressed her first symptoms. She had a feeding tube down her throat, her fingernails and mouth were turning blue, and her hair fell off in clumps. After having been seen by the hospital psychiatrist, again, it was recommended that she be admitted to a psychiatric hospital.

Finally, after more than 2 years of illness, Colleen's family had received word of a similar case in Northampton were the illness had gone undiagnosed until they sought the help of a well-known parasitologist in New York City. Having exhausted every other resource, the Tobins went to New York City in hopes of gaining some insight into Colleen's condition.

Using a special rectal smear that the parasitologist had developed himself, three parasites were identified in Colleen's specimen within 2 hours. *Giardia lamblia*, *Entamoeba histolytica*, and *Enterobius vermicularis* parasites had stripped the lining of Colleen's intestine. So although she was eating, her body was unable to absorb the necessary nutrients. Colleen was put on a regimen of 11 different antibiotics which initially turned her skin green and yellow. Twice weekly she had to receive an intravenous injection of vitamins, while taking pills to enhance the absorption in her intestine. Since her treatment, Colleen has gained back the weight she lost, is now growing again, and leads the life of a normal teenage girl; but the origin of her illness still remains unknown.[16]

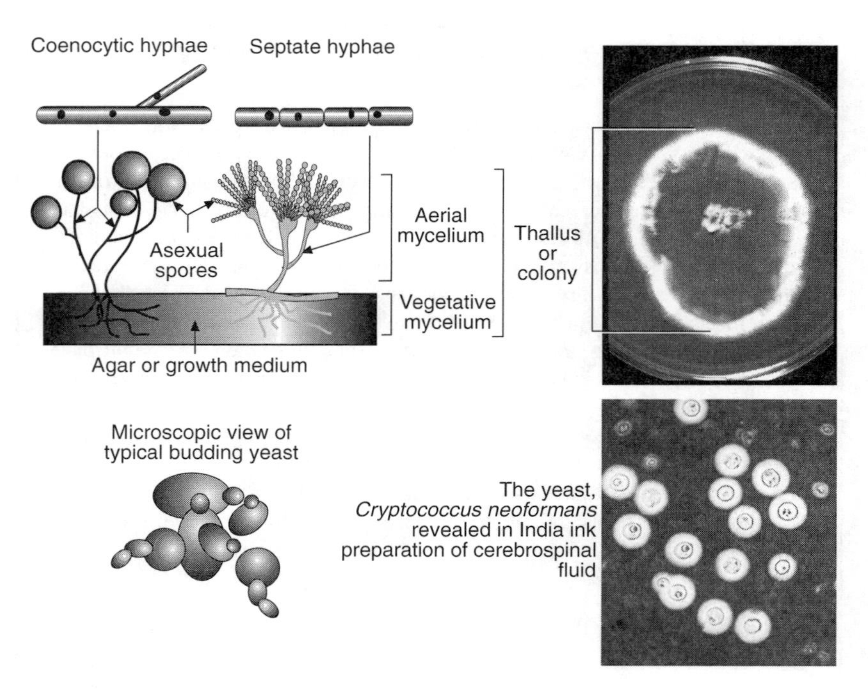

Figure 8.8 Morphology of molds and yeasts.

Fungi

Fungi such as molds and yeasts are single- and multi-celled, plant-like organisms that grow on cereals, breads, fruits, vegetables, and cheeses. For years man has used fungi for their beneficial properties. Antibiotics such as penicillin and vitamins such as vitamin C, are derived from molds. Yeasts are used in making wine and bread. They also serve as a protein source in foods. Aside from these beneficial effects, certain fungi possess properties that are harmful to humans. The potato blight in Ireland during the 1800s was caused by the fungus *Phytothora infestans*. Certain molds, such as *Penicillium* and *Aspergillus*, are toxigenic and can produce mycotoxins such as aflatoxin. Mycotoxins are mold metabolites produced on food that cause illness or death when ingested by man or animals. Other molds such as *Mucor* and *Rhizopus* species are common spoilage organisms and can also be pathogenic to man. Fungi can grow in conditions in which most bacteria could not survive. Molds can tolerate a pH as low as 2.0 and can thrive in low moisture environments. The majority of molds are aerobes and are therefore found on the surfaces of substrates where oxygen is present. In contrast to this, yeasts are facultative anaerobes which may be found within a substrate, require more moisture than molds, and prefer slightly acidic environments.

The body of a mold, called a colony or thallus, (Figure 8.8) gives rise to long filaments known as hyphae. Hyphae that are divided by septae into cell-like compartments containing nuclei are called septate hyphae. If septa are not present, the

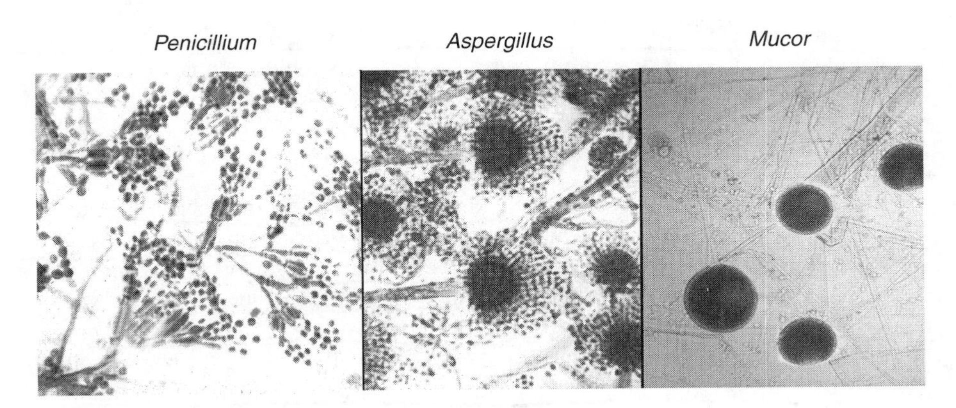

Figure 8.9 Microscopic views of slide cultures of *Pencillium, Asperigillus,* and *Mucor.*

hyphae is said to be coenocytic, with continuous cells and nuclei throughout the appendage. As hyphae grow they form a tangled mesh called a mycelium, which can be divided into two parts. The vegetative mycelium is responsible for the uptake of nutrients, and the aerial mycelium are where the reproductive spores can be found. The spores on the mycelium give color and texture to the molds, producing various shades of green, blue, brown, and even black, with a cotton, velvet, or waxy texture. When the molds are grown using a technique called *slide culture*, the spores may be viewed microscopically in an undamaged state in relation to the hyphae. These structures are very useful in identifying the molds. Examples of three common molds, *Aspergillus*, *Penicillium*, and *Mucor* are shown in Figure 8.9. Yeasts, on the other hand, are not made up of unicellular filaments, but usually come in the form of a sphere. Yeasts reproduce by mitosis, forming a colony of unicellular organisms that normally form a glistening, smooth, pasty colony on agar. Yeast growth on foods commonly has a white powder-like appearance.[21]

Aspergillus spp.

In England during the 1960s, turkey X disease killed thousands of turkeys, ducklings, and baby chicks, which resulted in a huge economic loss to the area. An investigation revealed that the peanut meal feed was contaminated with *Aspergillus flavus* and a toxin that appeared to be produced by the mold. Since this incident, much research has been conducted on molds and the mycotoxins they produce. Aflatoxin, produced by strains of *Aspergillus flavus*, causes hemorrhaging, anemia, ataxia, hematosis, cirrhosis of the liver, and is a very potent carcinogen. There are four primary aflatoxins — B1, B2, G1, and G2 — which are found in peanuts, corn, and cotton seed. In the U.S., aflatoxin is of concern during the storage of corn, and in milk from cows that have consumed the aflatoxin. Techniques such as treating milk with ultraviolet light or soaking corn in bisulfite during processing are ways that the food industries eliminate the toxin.

Another toxin produced by *Aspergillus* species is the ochratoxin, produced by *Aspergillus ochraceus*. It is found in peanuts, Brazil nuts, specific grains, red and black pepper, and citrus fruit. The ochratoxin causes swelling of the liver and necrosis of the kidney tubules, leading to blockage. The ochratoxin has been isolated in low-quality corn in the U.S.

Penicillium spp.

Rubratoxin, patulin, and yellow rice toxins are produced by members of the genus *Penicillium. Penicillium rubrum*, which produces the rubratoxin, can be found on plants such as peanuts, corn, legumes, and sunflower seeds. Patulin, which is produced primarily by *Penicillium expansum*, is found in rotting fruit due to the growth of organisms. Yellow rice toxins are produced by various species of *Penicillium* growing on rice during storage; the toxin causes the rice to turn yellow, hence the name "yellow rice toxins." Citrinin and Citreoviridin are examples of toxins that produce symptoms such as vomiting, difficulty breathing, low blood pressure, and respiratory arrest.[21]

Mucor and Rhizopus spp.

Mucormycosis is the disease caused by fungi in the order Mucorales. Species of the genera *Mucor* and *Rhizopus* are the common species responsible for this disease. Found in the soil, *Mucor* and *Rhizopus* species are common spoilage organisms of bread and fruit. Pathogenic strains such as *Rhizopus oryzae* are opportunistic organisms, infecting diabetics, leukemics, and those with suppressed immune systems. The onset of the disease is rapid and includes the invasion of blood vessels, causing embolisms and tissue necrosis. The disease can spread to whole organ systems, causing meningitis and death within 2 to 10 days.

Yeasts

Yeast are sometimes difficult to distinguish from molds. Yeasts are used in industry for the manufacture of beer, ale, bread, and glycerin. Yeasts also produce infections and disease as a result of food contamination. True yeasts, such as *Saccharomyces cerevisiae*, which is used in making beer, produce ascospores after sexual mating. In the laboratory, many yeasts do not display this sexual stage and have been termed "false yeasts," or yeast-like fungi. *Candida albicans* and *Cryptococcus neoformans* are two examples of yeast-like fungi. They are the most common causes of these infections. Small numbers of *Candida albicans* may be found in healthy individuals, but infection can occur in infants — when taking broad spectrum antibiotics during pregnancy — or when the body is burdened with a malignant disease. Infections vary considerably in severity. Infection can occur between fingers and toes, causing a rash, in the mouth, vagina, and lungs, or systemic involvement of mutliple organs. *Cryptococcus neoformans* is normally a ubiquitous saprophytic organism that is recovered in large numbers from the filth and debris associated with pigeon roosts. The organisms gain entry into the host through the lungs, normally producing a rapidly resolved pulmonary infection. In a few cases, it produces a serious and often fatal central nervous system infection, where the organisms appear in India ink preparations of the cerebrospinal fluid (Figure 8.8).[22]

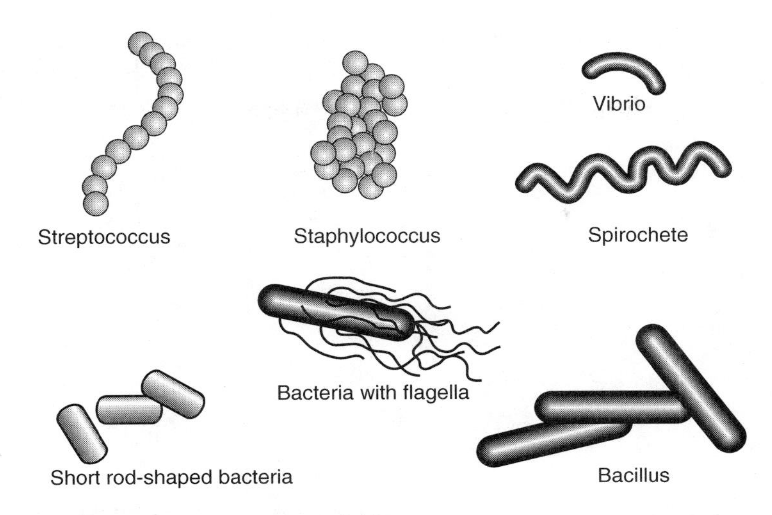

Figure 8.10 Examples of various bacterial forms.

Bacteria

Bacteria are the single-celled organisms that are responsible for more than 80% of foodborne illness.[4] Bacteria are much larger than viruses and can be viewed with a conventional microscope. Bacteria are categorized according to the physical and chemical properties they possess. Two broad groups of bacteria classification are Gram-positive and Gram-negative, which are based on the amount of peptidoglycan (layer of sugar derivatives that gives strength to the cell wall) in the cell wall of the organism. The cell wall of Gram-negative bacteria such as Salmonella and *Escherichia coli* contains only 10% peptidoglycan. In Gram-positives such as Staphylococci and Clostridia organisms, the cell wall contains up to 90% peptidoglycan, which is why they retain the violet crystal stain. After staining, examination under a microscope will reveal the different shapes of bacteria in the form of coccus, rods, spirilium, spirochete, and appendaged bacteria (Figure 8.10). Rod- and coccus-shaped bacteria are most commonly associated with foodborne illness. Another important characteristic useful in identifying bacteria is the ability to grow in the presence or the absence of oxygen. Aerobic bacteria require oxygen for growth, while the growth of anaerobes such as Clostridium is inhibited by the presence of oxygen (Figure 8.11). Other organisms such as Campylobacter are microaerophiles, which require low concentrations of oxygen for growth. If the anaerobe *Clostridium botulinum* is suspected, the investigator might search for endospores, which are structures produced during the life cycle of certain bacteria (Figure 8.12). Endospores can withstand extreme heat, harsh chemicals, and are very difficult to destroy. The dehydrated spore consists of bacterial DNA, a little RNA, some enzymes, and several layers around the spore, including a thick protein spore coat that confers resistance to harsh chemicals. The spore cytoplasm also contains dipicolinic acid that forms a complex with calcium. The dipicolinic acid is a major factor

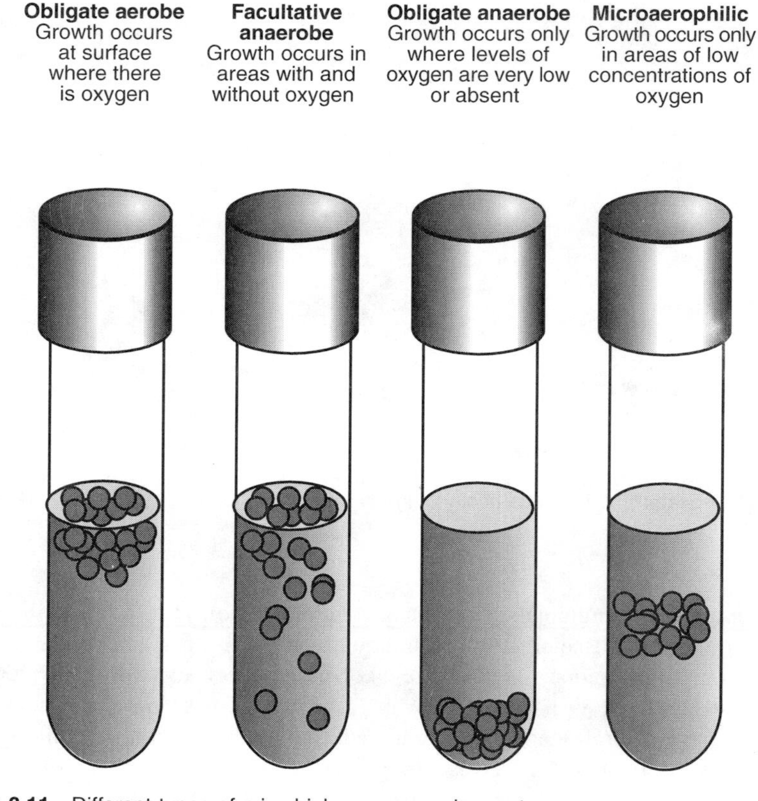

Figure 8.11 Different types of microbial oxygen requirements.

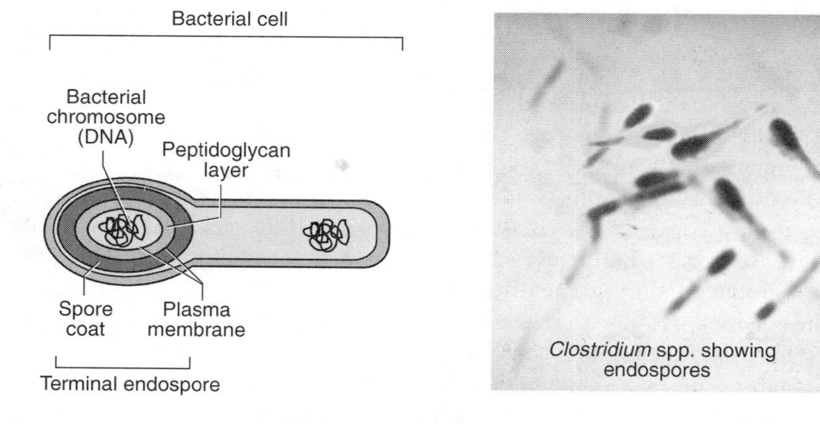

Figure 8.12 Endospores of *Clostridium* spp.

in the heat resistance of spores and is essential for later resuming metabolic functions. Aside from the characteristics mentioned above, several other factors such as motility, lactose fermentation, or growth on a specific medium are properties that aid in the identification of bacteria.

Salmonella spp.

Salmonellosis is the disease caused by the Salmonella organisms, Gram-negative facultative anaerobes. Although there are approximately 40,000 confirmed cases each year, investigators estimate that the actual number of cases is between 2 and 4 million.[23] Salmonella species are part of the normal intestinal flora of humans, cattle, poultry, pigs, and many other animals. The disease is transmitted via food and water, putting every person at risk. There are three syndromes caused by Salmonella species: typhoid fever, enteric fever, and gastroenteritis.

Typhoid fever, caused by *Salmonella typhi*, accounts for less than 2.5% of all salmonellosis in the U.S. Humans are the only reservoir for *S. typhi*, and transmission occurs through fecal contamination of food and water by an infected food worker or sewage contamination. Infection occurs when ingested organisms invade the epithelial cells of the small intestine. After penetrating the intestine wall, they invade the immune system, where they multiply and then are carried in the blood throughout the body, causing systemic infection. Symptoms include septicemia, high fever, constipation, diarrhea, vomiting, spotting on the chest and trunk, and bleeding from the bowels and nose. Not all who harbor the organism show symptoms, even though they may shed the organism and infect others. These people are called carriers, and the most notorious of these is "Typhoid Mary." The term "Typhoid Mary" originated with Mary Mallon, an Irish immigrant who infected over 1000 people through her unsanitary cooking habits until she was placed in exile on an island off of Manhattan.[24]

Enteric fever is the second type of syndrome, caused by *Salmonella paratyphi*. This illness comprises only 0.5% of the salmonellosis cases in the U.S., and is similar to typhoid fever, but less severe.

The third and most prominent syndrome of salmonellosis is gastroenteritis, caused by some 2300 serotypes of Salmonella. Only 150 of this number are actually known to cause human sickness, but all are assumed to be pathogenic. Gastroenteritis from Salmonella species does not include the invasive properties or systemic infection as seen with the typhi and paratyphi species. These organisms colonize in the small intestine, causing intestinal inflammation, resulting in diarrhea, abdominal cramps, chills, fever, and vomiting, which last one to four days. The severity of the infection varies with the involved Salmonella species, the health status of the individual before infection, and the implicated food. Many healthy individuals who become infected do not seek medical treatment, as symptoms are often mild, lasting one to three days. This is the primary reason why only 40,000 of the estimated 2 to 4 million cases are reported each year.

Salmonellosis stems from a wide variety of sources, including milk, eggs, ice cream, poultry, beef, and salads infected with such species as *S. typhirium*,

S. enteriditis, and *S. montevideo*, just to mention a few. Investigations of poultry products in the U.S. revealed that Salmonella species are present in 30% of the chicken available on the market; the numbers of bacteria are usually low, which is why we do not experience a greater number of outbreaks. Considerable efforts are being put forth to eradicate Salmonella in poultry, as it is implicated in one third of the foodborne outbreaks in the U.S.

In 1997, the FDA approved PREEMPT, the first competitive exclusion product for use in poultry. The concept of competitive exclusion is rather simple. When chicks are hatched they are especially vulnerable to Salmonella infection because their natural intestinal flora, which compete with Salmonella and other bacteria for nutrients, have not yet been established. By spraying newly hatched chicks with a mixture of live bacteria, researchers were able to eliminate 99.9% of the Salmonella in poultry. Although this technique may eliminate Salmonella in poultry, there are other means by which an individual may become infected.[25] Salmonella is also spread by infected people who are shedding the organisms in their feces. Unsanitary habits, such as not properly washing after using the rest room, facilitate the spread through a hand shake or other contact. This mode of transmission is known as the person-to-person mode.

Domestic animals such as turtles and iguanas should be considered potential carriers of Salmonella and handled accordingly. Before 1975, domestic turtles accounted for 280,000 cases of Salmonella annually. These numbers prompted the FDA to put a ban on the interstate shipment of turtles, which resulted in an 18% reduction of salmonellosis in children ages one to nine.[26]

Salmonella species are commonly found in iguanas (*S. marina*), but the species involved rarely causes human disease. Even though iguanas are imported into the U.S. from many foreign countries, there are no quarantines or health inspections upon entering this country.

Treatment of gastroenteritis from Salmonella is limited to hydrating fluids and rest. Antibiotic treatment is offered only to severe cases for two reasons. First, non-selective antibiotics will eliminate the natural flora of the intestine, leaving the body prone to other infections. Second, antibiotic treatment will not kill organisms that have acquired resistance to these drugs, leaving only them to flourish and infect. This second reason is a primary argument of advocates lobbying against the use of antibiotics in meat and poultry.

Staphylococcus spp.

Staphylococcus food poisoning, caused by the Gram-positive cocci *Staphylococcus aureus*, is one of the most common foodborne illnesses in the U.S. Staphylococci organisms can be found present in the nasal passages, pimples, throat, and wounds of humans. The organism can grow between 7°C and 48°C, and produces an enterotoxin between 20°C and 37°C. Sickness is due to the consumption of the enterotoxin produced, therefore Staphylococcus food poisoning is considered an intoxication as opposed to an infection. The intoxication is characterized by nausea, vomiting, and diarrhea, occurring within 30 minutes to 8 hours after consuming the enterotoxin. Sickness is usually short in duration and self-limiting, but death has occurred from intoxication. The enterotoxins can survive prolonged boiling and are considered to be heat stable. Therefore, once the enterotoxins are produced in food, they have the capacity to produce illness even if the food is reheated and the Staphylococcus organisms are killed.

Contamination occurs through the preparation of foods by infected food handlers. Foods such as creams, cream pies, potato salad, and ham have all been implicated in outbreaks of Staphylococci food poisoning, but prevalence of infection has decreased as a result of proper refrigeration. Currently, ham is the most common vehicle of transmission for the Staphylococci organisms.

Staphylococci growth is inhibited by the presence of competitors, but it can survive in high salt and sugar concentrations that destroy other organisms. Precooked ham contains salt in concentrations up to 3.5%, which kills off other organisms, leaving a perfect environment for Staphylococci to thrive.

In September 1997, in Northeast Florida, 18 individuals attending a retirement party suffered from Staphylococci food poisoning. Samples of leftover ham and rice pilaf were found to contain the enterotoxin produced by Staphylococcus aureus. Although investigators never found the source of contamination, it was likely due to the whole ham being placed in the freezer after cooking, which prolonged the cooling time, allowing Staphylococcus organisms to multiply.[27]

Although the foods mentioned above have been implicated in staphylococcal intoxications, virtually any food can serve as a vehicle if handled carelessly by infected food workers and left out at temperatures ideal for growth.

Clostridium spp.

Clostridium perfringens is an anaerobic spore-forming bacteria found in soil and the intestinal tracts of humans and other animals. *C. perfringens* grows at temperatures between 20° and 50°C. In optimum conditions (45°C), with the right media, *C. perfringens* can double its numbers in as little as 8 to 10 minutes, making it the fastest-growing bacteria known to man.[28]

In order for food poisoning to occur, numbers in excess of 108 bacteria must be ingested. When the organism senses harsh conditions, such as the lack of nutrients or water, it sporulates and releases the enterotoxin responsible for the intoxication. Two types of spores classified by their heat sensitivity are produced by various strains of *C. perfringens*. Heat resistance is measured by the D value (the amount of time and heat required to kill 90% of the population). Heat resistant spores have a D90°C value of 15 to 45 minutes, which implies that 90% of the organism population is eliminated by heating to 90°C for 15 to 45 minutes. Heat-sensitive spores have a D90°C value of 3 to 5 minutes. The enterotoxin produced by *C. perfringens* is destroyed by heating to 60°C for 10 minutes.

C. perfringens food poisoning usually occurs when cooked foods such as beef, turkey, or chicken are held for several hours at room temperature before serving, or when dense foods are refrigerated after cooking. These conditions allow the vegetative cells to grow to dangerous levels before being consumed.

In March of 1993, 156 persons suffered from food poisoning at a St. Patrick's day meal in Ohio. Investigation of the outbreak revealed that the delicatessen serving the meal had prepared corned beef sandwiches early in the day and held them at room temperature until they were purchased. As mentioned earlier, holding foods at room temperature for extended periods of time is a dangerous practice that gives many pathogenic organisms the chance to proliferate.[29]

Food poisoning by *C. perfringens* is mild, usually limited to diarrhea, abdominal pain, and vomiting. Symptoms can manifest as soon as two hours after ingesting the contaminated food and last about 24 hours. Death from *C. perfringens* is rare, but does occur among the elderly and those with an underlying condition.[29]

Clostridium botulinum

Clostridium botulinum, like *Clostridium perfringens,* is a spore-forming anaerobic bacteria found in soils throughout the world. Botulism is the illness that results when *C. botulinum* spores germinate and produce a toxin in the food to be ingested. Outbreaks of botulism are rare in the U.S. (34 cases in 1994); and although commercial foods have been implicated, home-canned foods are the most common source. *C. botulinum* organisms are not a threat in food unless they can germinate into the toxin-producing vegetative cells. By destroying the spores in foods before canning or storing products, risk of botulism can be eliminated. The spores, however, are resistant to heat (D90°C is 30 minutes) and chemicals, making it difficult to destroy. The neurotoxins produced by the organism are fortunately heat labile and can be easily destroyed by boiling for 3 minutes or by heating to 80°C for 30 minutes.

C. botulinum organisms can grow in temperatures between 3.3°C and 45°C, but this varies with the type of organism. Aside from heat and chemicals, pH and oxygen concentration also affect growth. A pH of 4.6 inhibits the germination of spores. Foods such as tomatoes may naturally have a pH as low as 4.6, while food of a higher pH may have acids added to them during processing to prevent spore germination. The oxygen concentration of the growth medium is a critical factor to *C. botulinum* growth. While the conditions on the surface of a food product may be aerobic, anaerobic conditions may exist within the food mass. In products that are smoked, such as fish and sausage, anaerobic conditions can develop next to the cell, producing a particularly dangerous situation because the organism may be present in lethal numbers without having affected the appearance of the food. Fresh mushrooms were previously packaged in containers sealed with a cellophane wrap. If aerobic bacteria consumed all the oxygen present in the package, they would die off, leaving anaerobic conditions without competitors, an ideal setting for an outbreak of botulism. The cellophane wrap now used is punched with holes to avoid this scenario.

Currently there are seven types of *C. botulinum,* A to G, which are identified by the toxin they produce. Only types A, B, E, and F have been found to cause foodborne illness in humans. Types C and D cause botulism in cattle and other animals, while G has been reported to cause infant botulism not associated with food. Botulism caused by the A toxin is the most common in the U.S., followed by B, E, and F. The A and B toxins have been isolated in fruits, vegetables, fish, condiments, beef, pork, and poultry. The E toxin has been isolated primarily in fish, and the F toxin only in meat products.

Food poisoning in adults occurs within 12 to 36 hours of consuming the toxin. At the onset, symptoms such as nausea, vomiting, and diarrhea are present, then as the condition develops, fatigue, blurred vision, and difficulty speaking and swallowing are experienced. If the condition goes untreated, death can occur from respiratory paralysis. Infant botulism occurs in children less than one year old. Ingested honey has been the source of most food-associated cases, but 50% of infant botulism occurs in children who were strictly breastfed, leaving air, dust, and soil as the likely sources of contamination.[30]

Campylobacter

Campylobacter is one of the leading causes of acute gastroenteritis in the U.S., responsible for approximately 2 million cases annually. *Campylobacter* species are part of the normal flora of the gastrointestinal tract of warm-blooded animals. Of primary concern are cattle and poultry. Contamination occurs primarily during food processing when the intestinal tract is lacerated, allowing feces to contaminate the food product. Infected food handlers and raw milk are also significant sources of infection. *Campylobacter jejuni* is the organism most commonly implicated in *Campylobacter* infection.

Campylobacter jejuni is a microaerophile (requiring little oxygen), capable of growing in 3% to 15% oxygen. It has a narrow temperature range for growth (between 30°C to 47°C), but more importantly, *Campylobacter* can survive for weeks in refrigeration at 4°C. Although no growth may occur during refrigeration, *Campylobacter* is still a threat because infection can occur from as little as a few hundred cells. Symptoms in the infected are usually mild including nausea, vomiting, and bloody diarrhea; but in severe infections, Gullian–Barre Syndrome develops, which causes neuromuscular paralysis.[31]

Escherichia coli

E. coli organisms can be found among the normal bacterial flora in the intestine of humans and other warm-blooded animals. The O157:H7 strain of *E. coli* has caused several outbreaks of severe sickness and death, which has given the public the impression that all *E. coli* strains are harmful, when in fact the majority are harmless and some strains can be found in the intestinal tracts of healthy, asymptomatic individuals.

E. coli organisms, which are important to foodborne illness, can be divided into four groups: enteroinvasive, enterotoxigenic, enteropathogenic, and enterohemorrhagic. The latter the one to which the infamous *E. coli* O157:H7 strain belongs. Enteroinvasive *E. coli*, as the name implies, invade the epithelial cells of the intestine, resulting in fever, chills, and bloody diarrhea. In contrast, enterotoxigenic *E. coli* are responsible for traveler's diarrhea, exhibiting cholera-like symptoms. Cells colonize on the wall of the intestine and then produce a toxin that prompts watery diarrhea, vomiting, abdominal cramps, and can result in dehydration. Enteropathogenic *E. coli* are most commonly found among infant nurseries in developing countries. During the 1940s and 1950s, enteropathogenic *E. coli* were a problem in the U.S.; but the development of more sanitary practices has diminished their importance in recent decades.

Enterohemorrhagic *E. coli*, also known as *E. coli* O157:H7, has become one of the more problematic foodborne pathogens during the 1990s. Cattle are the major reservoir for the organism, and human infection is the result of consuming improperly cooked ground beef, raw milk, or unpasteurized apple cider. Symptoms generally include abdominal cramps, watery to bloody diarrhea, vomiting, and possibly a fever. *E. coli* O157:H7 is especially dangerous because it can be accompanied by hemolytic uremic syndrome (HUS), which is the primary cause of renal failure in children.[32] In January 1993, Washington state experienced an outbreak of *E. coli* O157:H7 due to improperly cooked hamburgers from a fast-food chain. The outbreak consisted of 501 cases, 45 cases of HUS, and three deaths.[33] Outbreaks such as the one mentioned above can be avoided by properly cooking meat to an internal temperature of 68°C. Another outbreak of *E. coli* O157:H7 occurred in 1996 in Japan due to consumption of raw alfalfa sprouts. There were 9000 illnesses and 17 deaths due to the outbreak. This outbreak and several smaller outbreaks in the U.S. have prompted the FDA to put out a warning against consumption of raw alfalfa sprouts until new measures are put in place to ensure sprout safety.[34] There are also new techniques being developed to fight off *E. coli* O157:H7 in milk and other products. Cold pasteurization is a process by which an electron beam is directed towards the product, reacts with *E. coli* DNA, and inactivates it. Thus far, this new technology has not been found to alter the foods in any significant way.[35]

Vibrio Cholerae

The disease cholera, caused by the organism *Vibrio cholerae*, is most prominent in lesser developed countries where unsanitary conditions prevail. In the U.S., most cholera cases are associated with international travelers who either become infected while abroad or who bring contaminated food items back into the U.S. where the infection may occur. In 1992, 102 cases of cholera were identified in the U.S., the most reported to the CDC in a single year. Ninety-five of the cases were associated with travel, 75 of which occurred during an outbreak on an Argentinean airline.[36]

In Peru in 1991, an outbreak of cholera spread to 322,562 Peruvians. Following this initial outbreak, cholera spread form Peru throughout most of South America, north through Central America, Mexico, even reaching the U.S. Statistics from the Pan American Health Organization indicate that between 1991 and 1992 there were 645,030 cases with 5,694 deaths related to this epidemic.[35] A member of the genus *Vibrio*, this Gram-negative rod can proliferate in temperatures between 15 and 42°C, in a pH range of 6 to 10, and in a solution containing up to 6% sodium chloride. Hence it can be found in coastal waters contaminated with fecal matter. Crabs, shrimp, mussels, and clams can harbor the organism for long periods of time, and have all been implicated as sources of infection.[36]

Vibrio cholerae does not invade the intestinal lining. Instead it colonizes on the lining and produces the toxin choleragen, which causes the massive influx of fluids into the intestine. Symptoms precipitate 2 to 3 days after consuming the contaminated food and generally appear as abdominal pains, dehydration, and a characteristic diarrhea which has been termed "rice water stool" for its similar appearance to the water left behind after rice has been cooked. Complications associated with cholera include hypoglycemia, severe dehydration, and even renal failure. Cholera can have a case fatality rate of up to 50% in untreated patients, who usually die of complications surrounding the extreme loss of fluid from the body (up to one liter per hour in adults).

Although a vaccine exists, it is not practical for the inhabitants of endemic areas because it is effective for only 3 to 6 months. In those endemic areas, an increase in sanitation practices would be more practical. However, the vaccine is recommended for travelers spending short periods of time in endemic areas. Antibiotics can be given for treatment, although, as seen with Salmonella, many strains have acquired resistance to the traditional medicines. Rehydration solution can be used to treat mild cases which usually pass without any other treatment. In severe cases, rehydrating fluids are introduced intravenously to counter the rapid loss of fluid.[15]

Factors Frequently Cited in Foodborne Illness

During the food processing chain, many controls are put in place to prevent foodborne disease. To produce a food that is 100% free of contamination is not feasible due to the economic costs as well as the fact that such a rigorous sanitation process would alter the food so much that it would no longer appeal to the consumer. Instead, acceptable limits have been set which allow low numbers of certain bacteria to present in a food. If the food is treated properly, the presence of such low numbers of bacteria should not pose a threat. In many cases, however, food is abused. The result is often a foodborne outbreak.

Factors that contribute to foodborne illness can be broken down into three categories: (1) time and temperature, (2) poor hygiene practices by food handlers, and (3) cross contamination. Cited below are several factors within these categories that are commonly found to be the cause of foodborne outbreaks, which are largely preventable.[37]

1. Improperly refrigerated food
2. Improperly heated or cooked food
3. Food handlers who practice poor hygiene
4. Lapse of a dya or more between preparing and serving food
5. Introducing raw or contaminated materials to a food that will not undergo further cooking
6. Improper storage of foods at temperatures ideal for bacterial growth
7. Failure to properly heat previously cooked foods to temperatures that will kill bacteria
8. Cross contamination of ready-to-serve foods with raw foods, contaminated utensils or machinery, or through the mishandling of foods

Some suggestions for reducing the hazards associated with the contamination of food are shown in Figure 8.13.

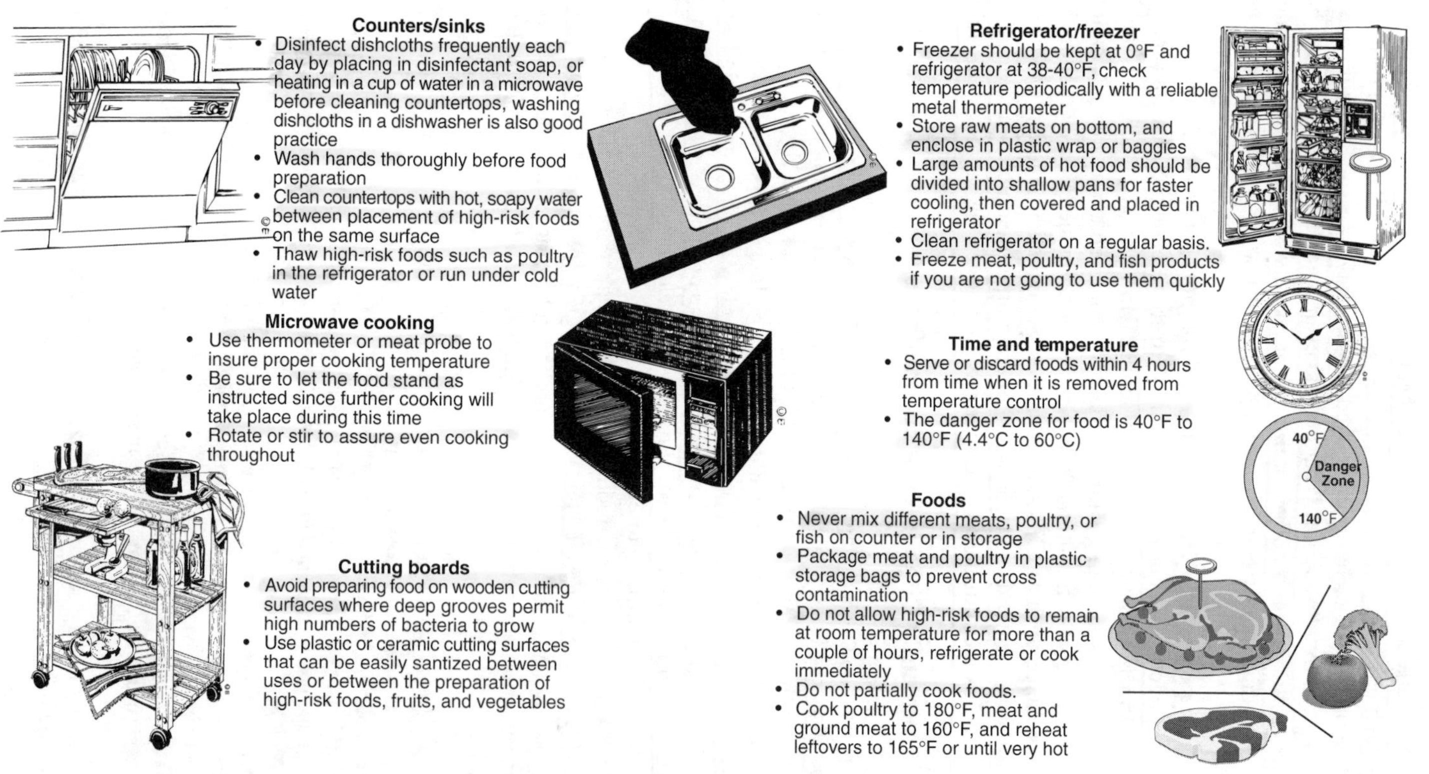

Counters/sinks
• Disinfect dishcloths frequently each day by placing in disinfectant soap, or heating in a cup of water in a microwave before cleaning countertops, washing dishcloths in a dishwasher is also good practice
• Wash hands thoroughly before food preparation
• Clean countertops with hot, soapy water between placement of high-risk foods on the same surface
• Thaw high-risk foods such as poultry in the refrigerator or run under cold water
Microwave cooking
• Use thermometer or meat probe to insure proper cooking temperature
• Be sure to let the food stand as instructed since further cooking will take place during this time
• Rotate or stir to assure even cooking throughout
Cutting boards
• Avoid preparing food on wooden cutting surfaces where deep grooves permit high numbers of bacteria to grow
• Use plastic or ceramic cutting surfaces that can be easily santized between uses or between the preparation of high-risk foods, fruits, and vegetables
Refrigerator/freezer
• Freezer should be kept at 0°F and refrigerator at 38-40°F, check temperature periodically with a reliable metal thermometer
• Store raw meats on bottom, and enclose in plastic wrap or baggies
• Large amounts of hot food should be divided into shallow pans for faster cooling, then covered and placed in refrigerator
• Clean refrigerator on a regular basis.
• Freeze meat, poultry, and fish products if you are not going to use them quickly
Time and temperature
• Serve or discard foods within 4 hours from time when it is removed from temperature control
• The danger zone for food is 40°F to 140°F (4.4°C to 60°C)
40°F
Danger Zone
140°F
Foods
• Never mix different meats, poultry, or fish on counter or in storage
• Package meat and poultry in plastic storage bags to prevent cross contamination
• Do not allow high-risk foods to remain at room temperature for more than a couple of hours, refrigerate or cook immediately
• Do not partially cook foods.
• Cook poultry to 180°F, meat and ground meat to 160°F, and reheat leftovers to 165°F or until very hot

Figure 8.13 Recommendations for reducing hazards associated with the contamination of food.

Hazard Analysis Critical Control Points

Although there are several measures in place to prevent foodborne disease, pathogens such as *E. coli* and *Salmonella* still continue to pose a serious health threat. In response to this threat, the federal government has mandated the implementation of hazard analysis critical control points (HACCP) strategies in the seafood, poultry, and meat industries.

In contrast to the traditional end product testing methods, HACCP intervenes at various stages in the food processing/preparation chain to ensure a safer product. The objective of a HACCP plan is to identify hazard points within a certain industry that may lead to foodborne illness, implement strategies to minimize the identified hazards, and monitor these practices to ensure that all implemented plans are adhered to and are effective. There are seven key principles to the HACCP system: (1) assessing the hazard; (2) identifying critical control points; (3) establishing standard procedures and critical limits; (4) monitoring procedures; (5) corrective actions, (6) record keeping; and (7) verification the system is working properly. By adhering to these principles, food establishments can reduce the risk of a foodborne illness (Table 8.4).

Assessing the Hazard

Adequately assessing the hazards in a food process is essential to an effective HACCP plan. The person or persons conducting the hazard assessment must review the food on site to determine which are potentially hazardous; employee activities should be supervised for dangerous practices; and the design of the facility should be examined for potential flaws.

Identifying Critical Control Points

Once the hazards have been assessed, critical control points (CCPs) are identified to control the potential hazards. For example, if raw chicken is being handled, Salmonella would be an identified hazard and the cooking process would be the critical control point, because it is there that Salmonella can be eliminated. Depending on the hazard, other critical control points could be refrigeration to prevent microbial growth or fluctuation of the pH to inactivate a toxin.

Establishing Standard Procedures and Setting the Critical Limits

Procedures such as washing hands, sanitizing machinery, and food preparation should be established and well defined to ensure the production of a safe product. Critical limits such as time, temperature, pH, and water activity are parameters that ensure the production of a safe product. Operating outside the critical limits could result in foodborne illness. In the case of raw chicken, the critical limits during cooking would be heating to an internal temperature of at least 73.9°C for 15 seconds or more. By not meeting these critical limits, Salmonella organisms can remain viable in numbers high enough to cause infection or possibly contaminate another food item.

Table 8.4 The Seven Key Principles of HACCP

HAZARD CRITICAL CONTROL POINT (HACCP) SYSTEM

Assess the Hazards
- Identify potentially hazardous foods
- Follow the flow of food to assess hazards at receiving, storing, preparing, cooking, holding, serving, cooling, and reheating
- Estimate risks

Identify Critical Control Points (CCPs)
- Develop procedure and flowcharts showing the flow of food and all of the CCPs

Set Up Procedures and Standards for CCPs
- Standards must be met at each CCP and should be: measurable, based on fact, correct for the recipe, clear directions with specific actions

Monitor CCPs
- Check to see if standards are met; employees should be involved in process; standards must be met

Take Corrective Actions
- If standard not met, correct it
- Have specific steps for correction

Set Up Record Keeping System
- Blank forms near equipment where they are to be used
- Notebooks to write down actions
- Flowcharts and recipes near work areas

Verify That the System Works
- Identify and assess all hazards
- CCPs selected
- Standards set with monitoring and schedules
- Corrective actions in place
- Monitoring being done
- Flaws or omissions corrected
- Monitoring equipment calibrated

Monitoring Procedures

The periodic monitoring of procedures is necessary to track the operation of the facility, indicate when critical values have not been followed, and provide documentation in the event of an inspection or other situation which may require verification. For instance, while cooking chicken, conduct periodic checks on the oven temperature. Inspection reveals that the oven is operating at 74°C, the time, date, and temperature are recorded, and the supervisor might request that the temperature be

WEB PICK Go to http://www.unix.oit.umass.edu/~envhl565/index.html Click on "Chapter Web Links" to find FDA/CFSAN Bad Bug Book: Introduction to Foodborne Pathogenic Microorganisms and Natural Toxins on the World Wide Web under Chapter 8	
Chapter 8	Foodborne Illness
Whose Site?	U.S. Department of Health and Human Services, Food and Drug Administration, Center for Food Safety & Applied Nutrition
URL	http://vm.cfsan.fda.gov/~mow/intro.html
What's Here?	The Center for Food Safety and Applied Nutrition (CFSAN) has prepared this handbook and Web site on foodborne pathogenic microorganisms (bacteria, viruses and parasites) and natural toxins. Each section focuses on either one foodborne pathogenic microorganism or natural toxin. In some chapters, a closely related group of organisms or natural toxins is covered. Each chapter provides information on their characteristics, habitat or source, associated foods, infective dose, characteristic disease symptoms, complications, recent and/or major outbreaks, and any susceptible populations. The chapters contain minimal information on the analytical methods used to detect, isolate, and/or identify the pathogens or natural toxins.

raised slightly. By monitoring the CCPs, improperly cooking the chicken can be avoided, preventing a possible outbreak of Salmonella.

Corrective Actions

Corrective actions must be taken when a CCP has operated outside of the critical limits. In some cases it may be necessary to discard a product, while other situations may simply require bringing the CCP back within the critical limits. In either case, the deviation from the critical limit should be documented and any corrective action should be approved by the proper regulatory bodies before implementing HACCP.

Record Keeping

If a facility is required to have an HACCP plan, a written copy of the plan must be kept on site. In the event government inspectors visit a facility, documentation regarding CCPs, critical limits, and deviations from those limits should be available.

Verification the HACCP System is Working Correctly

The purpose of verification is to check how effective the plan is in reducing the identified hazards, and possibly reveal any hazards that might not have been previously identified. Verification may include reviewing current documentation for patterns in errors or corrective actions, which may reveal false reporting, ineffective CCPs, or other flaws in the HACCP plan. Detection methods such as those discussed previously confirm the presence or absence of identified microbes and should reveal if the HACCP plan has resulted in decreased numbers of those pathogens.[45]

U.S. Regulatory Efforts with Regard to Food Protection

On December 18, 1997, the FDA required that all seafood processors — domestic and those importing to the U.S. — carry out a hazard analysis of their products and processes. In the event hazards are recognized, the facility is required to have a written HACCP plan to address the problem. The ruling also identified eight sanitation concerns that must be monitored and controlled. A formal program of sanitation standard operating procedures (SOPs) is not required by the FDA; however, the FDA does require that records on sanitation SOPs be filed.

The U.S. imports more than 50% of the seafood that Americans eat, and there are an estimated 6000 firms in 172 different countries that are involved in the seafood industry. Any importers of seafood have to verify that their product was produced in accordance with the HACCP ruling. HACCP plans undergo yearly evaluations to compensate for any changes in manufacturing practices and to add more controls if additional hazards are identified. As can be expected, this ruling was met with resistance from the seafood industry. Many importers have to find new producers who can comply with the ruling, while domestic producers may have to change their current practices in order to meet the new requirements. The meat and poultry industries are currently experiencing similar rulings.

On January 27, 1997, the USDA required meat and poultry slaughterers and processing facilities to have sanitation SOPs in place and to also conduct generic *E. coli* testing. As of January 26, 1998, large facilities employing 500 or more workers had to adhere to reduction-performance standards for Salmonella organisms to ensure the HACCP plan is reducing the risk of foodborne illness. On January 25, 1999, facilities harboring 10 to 499 workers were subject to this regulation, and by January 25, 2000, over 6200 meat and poultry plants were operating under this ruling. Pending the success of the mentioned legislation, we may expect similar rulings toward fruit and vegetable industries.[46]

Surveillance Efforts

Outside of the food industry, considerable efforts are also being put forth to control and better understand foodborne pathogens. The Foodborne Diseases Active Surveillance Network (FoodNet), which is part of the CDC Emerging Infections program, is the latest surveillance strategy aimed at detecting foodborne illness, which would normally go unnoticed under the traditional system of passive surveillance.

Previously, foodborne disease was monitored by a passive approach of reporting cases that had been confirmed in a clinical lab. This method only accounted for a fraction of the actual cases, hence, the actual importance of foodborne disease was underestimated. The FoodNet is an active approach to tracking disease, where health officials contact laboratories to request data on new cases of foodborne illness. The health official, in turn, reports the information to the CDC via a computer network. Surveillance sites have been established in several states where surveys are conducted among clinical laboratories, physicians, and the general population. The purpose of these surveys is to more accurately assess the incidence of foodborne illness and to monitor the procedures of physicians and laboratory personnel, which might influence the number of pathogens that are isolated and reported.

Case control studies are another component of the FoodNet, providing information about which food items are most commonly associated with these diseases. Isolates from these studies are sent to the CDC, where they are phagetyped and tested for antibiotic resistance. This information alerts health officials to the presence of new pathogens that have not been previously isolated, or to strains that have become resistant to antibiotics used to reduce their numbers in animals or to combat infections in the human body. The FoodNet has been up and running since January 1, 1996. Since then it has identified outbreaks of *Campylobacter* in California, Salmonella in Oregon, and two outbreaks of *E. coli* O157:H7 in Connecticut.[47]

Another agency recently set up in the U.K. is the Food Standards Agency. This agency will be used for creating policies regarding food safety and reporting of food poisoning. There is some controversy over this agency, however, because it is looking into issues that are normally covered by the Ministry of Agriculture, Fisheries and Food and local governments. Many in England felt that these agencies did not do a good job with the recent outbreaks of both *E. coli* and Salmonella and pushed for a new agency that will actually investigate the cases and make new regulations, something the other agencies have only advised on.[48]

Along with sources like FoodNet and agencies like the Food Standards Agency, there are temperature indicators that consumers can buy to test their foods. These indicators are adhesive strips that signal when foods have changed temperature and are therefore dangerous. The strips can also be used in restaurants and grocery stores to determine freshness.[49]

REFERENCES

1. Kaferstein F.K., Food safety: A commonly underestimated public health issue, *World Health Org. Stat. Q.,* 50, 3, 1997.
2. Motarjemi, Y. and Kaferstein, F.K., Global estimation of foodborne disease, *World Health Org. Stat. Q.,* 50, 8, 1997.
3. Buzby, J.C. and Roberts, T., Economic costs and trade impacts of microbial foodborric illness, *World Health Org. Stat. Q.,* 50, 57, 1997.
4. Theis, M. Case problem: Balancing the risk of foodborne illness with patient rights and quality of life. *J. Am. Dietetic Assoc.,* 99, 9, 1999.
5. Koren, H. and Bisesi, M., *Handbook of Environmental Health and Safety,* Lewis Publishers, Boca Raton, FL, 1996, chap. 2.

6. Stuart, D., Study puts U.S. food poisoning at 76 million yearly, *The New York Times,* Wed., Sep. 17, 1, 1999.
7. Anonymous, Food poisoning: The real extent of foodborne disease, *Consultant,* 39, 3, 1999.
8. Cliver, D.O., *Transmission of Disease Via Foods, Foodborne Diseases,* Academic Press, Inc., San Diego, 1990, chap. 1.
9. MCGeary, J., Nukes ... They're back, *Time,* May 25, 1998, p. 34.
10. Ennever, F.K., *Metals, Principles and Methods of Toxicology,* Raven Press, New York, 1994, chap. 12.
11. Taylor, S.L., *Chemical Intoxications, Foodbome Diseases,* Academic Press, Inc., San Diego, 1990, chap. 10.
12. Castle, S. Dioxin in meat is Europe's new food scare, *The Independent* (London), Wednesday, Sept. 22, 2, 1999.
13. Reddy, C.S., Hayes, A.W., *Food-Borne Intoxicants, Principles and Methods of Toxicology,* Raven Press, New York, 1994, chap. 9.
14. Hunter, B., Overlooked threats of foodbourne illness, *Consumer's Res. Mag.,* 78, Oct. 1, 1995.
15. Massachusetts Department of Public Health, *Foodborne Illness Investigation and Control Manual,* 1997.
16. CDC, Outbreak of trichinellosis associated with eating cougar jerky-Idaho, *Morbidity Mortality Wkly. Rep.,* 45:10, 205, 1995.
17. Cliver, D.O., *Parasites, Foodborne Diseases,* Academic Press, Inc., San Diego, 1990, chap. 21.
18. Markel, E., Voge, M. and John, D., *Medical Parasitology,* 7th ed., W.B. Saunders Co., Philadelphia, 1992, chap. 3.
19. Eckman, M., and Priff, N., *Diseases,* 2nd ed., Springhouse Corp., Springhouse, PA, 1997.
20. Cameron, J., Parasites, a 'select epidemic,' *Hampshire Daily Gazette,* Sep. 12, 23, 1995.
21. Juranek, D.D., *Cryptosporidiosis*: sources of infection and guidelines for prevention, *Clin. Infect. Dis.,* 21, Suppl. 1, S57, 1996.
22. Hutin, Y.J.F, et al., A multistate foodborne outbreak of Hepatitis A. *New Engl. J. Med.,* 340, 8, 1999.
23. Cliver, D.O., *Viruses, Foodborne Diseases,* Academic Press, Inc., San Diego, 1990, chap. 20.
24. CDC, 1998, http://www.cdc.gov/ncidod/diseases/hepatitis/h96trend.htm.
25. Tortora, G.J., Funke, B.R., and Case, C.L., *Microbiology, An Introduction,* 5th ed., The Benjamin/Cummings Publishing Co., Redwood City, CA, 1995, chap. 12.
26. Marth, E.H., *Mycotoxins, Foodborne Diseases,* Academic Press, Inc., San Diego, 1990, chap. 8.
27. Moore, G.S. and Jaciow, D.M., *Mycology for the Clinical Laboratory,* Reston Publishing, Reston, VA, 1979.
28. CDC, Salmonella serotype Montevideo infections associated with chicks–Idaho, Washington, and Oregon, Spring 1995 and 1996, *Morbidity Mortality Wkly. Rep.,* 46:11, 237, 1997.
29. Huckstep, R.L., *Typhoid Fever and other Salmonella Infections,* E. & S. Livingstone LTD., London, 1962, p. 227.
30. Murphy, J., FDA approves new anti-Salmonella spray developed by ARS, *Food Chem. News,* March 23, 26, 1998
31. Doyle, M.P. and Cliver, D.O., *Salmonella, Foodborne Diseases,* Academic Press, Inc., San Diego, 1990, chap. 11.

32. CDC, Outbreak of Staphylococcal food poisoning associated with precooked ham–Florida, 1997, *Morbidity Mortality Wkly. Rep.,* 46:50, 1190, 1997.
33. Johnson, E., *Clostridium Perfringens Food Poisoning, Foodborne Diseases,* Academic Press, Inc., San Diego, 1990, chap. 16.
34. CDC, *Clostridium perfringens* gastroenteritis associated with corned beef served at a St. Patrick's Day meals–Ohio and Virginia, 1993, *Morbidity Mortality Wkly. Rep.,* 43, 8, 137, 1993.
35. Sugiyama, H., *Clostridium Botulinum, Foodborne Diseases,* Academic Press, Inc., San Diego, 1990, chap. 6.
36. CDC, Outbreak of *Campylobacter* enteritis associated with cross-contamination of food–Oklahoma, 1996, *Stat. Q,*
37. Doyle, M.P, and Cliver D.O., *Escherichia Coli, Foodborne Diseases,* Academic Press, Inc., San Diego, 1990, chap. 13.
38. Kurtzweil, P. Questions keep sprouting about sprouts. *FDA Consumer,* 33, 18, 1999.
39. Bell, B.P., and Goldoft, M., Griffin, P.M., et al. A multistate outbreak of *Escherichia coli* O157:H7, *JAMA,* 272, 1349, 1994.
40. Earth Vision Reports, New Pasteurization Technology Makes Foods Safer. Missouri, 1999, http://www.earthvision.net/ColdFusion/News_Page1.cfm?NewsID=8790.
41. CDC, Cholera associated with international travel, *Morbidity Mortality Wkly. Rep.,* 41:36, 664, 1992.
42. CDC, Update: Cholera-Western Hemisphere, 1992, *Morbidity Mortality Wkly Rep.,* 41:36, 667, 1992.
43. Doyle, M.P., and Cliver, D.O., *Vibrio, Foodborne Diseases,* Academic Press, Inc. San Diego, 1990, chap. 17.
44. The National Restaurant Association, *Applied Foodservice Sanitation,* 4th ed., Educational Foundation of the Restaurant Association, 1992.
45. The National Restaurant Association, *Serving Safe Food,* The Educational Foundation of the National Restaurant Association, Kendall/Hunt Publishing Company, Dubuque, IA, 1995.
46. Reisman, M., Regulated HACCP moves in, *Food Qual.,* 5:1, 21, 1998.
47. CDC, CDC's Emerging Infectious Program, CDC/USDA/FDA Foodborne Diseases Active Surveillance Network (FoodNet), 11/04/97, http://www.cdc.gov.ncidod/foodnet/foodnet.htm.
48. Concentrated drive to end food poisoning. *BBC Online,* Jan. 28, 1999, http://new2this.bbc.co.uk/hi/end...Fsafety/newsid%5f263000/263873.stm.
49. Carter, J., New weapon aids food safety, *Springfield Sunday Republican,* Sun., Feb. 14, 11, 1999.

CHAPTER 9

Water and Wastewater

INTRODUCTION

Imagine a world without water. It is impossible. Water plays too important a role in our lives to conceive of life without it. We use water for drinking and cooking, showering, washing dishes, cleaning clothes, flushing the toilet, watering our lawns, and washing our cars. Water is cheap, accessible, plentiful, and relatively safe to drink. If tap water is not safe for drinking, it is usually simple enough to buy clean water at a store. We are rarely faced with any day-to-day difficulty in obtaining clean water. However, not everyone in the world has this luxury. In many areas of the world, fresh water is scarce; and tremendous time and resources are devoted to obtain fresh water. All too often, that water is contaminated with waterborne disease, due to lack of proper purification or water treatment systems. This situation results in thousands of cases of waterborne illness every year in countries with inadequate clean water.

OBJECTIVES FOR THIS CHAPTER

A student reading this chapter will be able to:

1. **List and describe the different stages of the hydrological cycle and the relative amounts of freshwater on the planet**
2. **List the three main consumers of water, noting each consumer's major use of water**
3. **Discuss water scarcity, pointing out the world areas at risk of water shortage and potential associated conflict; describe some methods of water management**
4. **List and describe several sources of freshwater, and describe the types of wells used to pump groundwater**
5. **Describe groundwater formation, including a discussion of contamination, recharge, water mining, fossil water, and some problems associated with overuse**
6. **List the various sources and types of pollution threatening water supplies, noting the difference between point and nonpoint sources**
7. **Briefly outline the Clean Water Act and the Safe Drinking Water Act, explaining the purpose of each**
8. **Describe the process and purpose of wastewater treatment; list the components of a septic system and the components of a typical wastewater treatment plant; define and describe BOD and sag curve**

Despite what many people think, waterborne disease and water contamination are not limited to developing nations. More and more frequently in the U.S., the news media report about contaminated water and resulting illness. Water is not always as pure or safe as we think. In countries throughout the world, lakes, rivers and oceans are used as dumping grounds for all kinds of waste. In the U.S., laws have been passed to protect our water. But these laws might not be enough to stem the tide of pollution discharged from factories, running off of fields, and dumped illegally. An awareness of the problem and a strong sense of responsibility are essential for proper management and preservation of a safe, plentiful water supply.

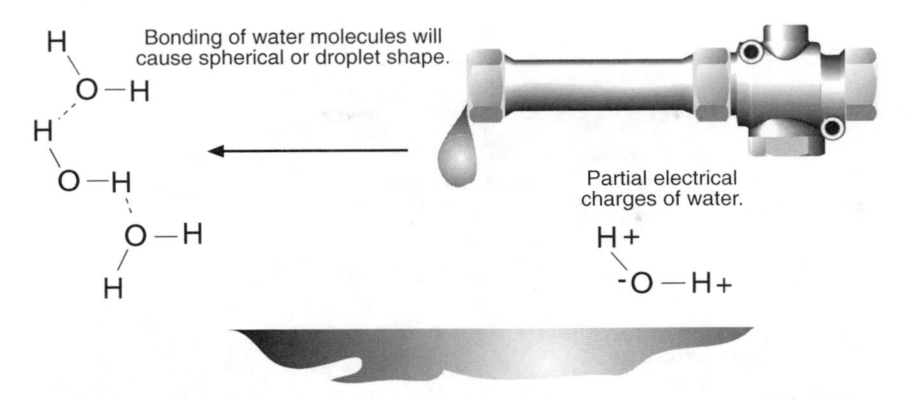

Figure 9.1 Configuration of water. (Adapted from Turk, J. and Mirk A., *Environmental Science,* 4th ed., Saunders College Publishing, Philadelphia, PA, 1988, chap. 17.)

However, different cultures and countries treat water according to their customs and traditions. Not everyone shares the same level of concern about water supply and water pollution. These differences in beliefs about the importance of water quality, water scarcity, and water use can cause conflict and hinder a joint effort to protect the world's water supplies.

The Properties of Water

Water is a unique compound. It is composed of one hydrogen molecule and two oxygen molecules and has a bent configuration. The oxygen carries a partial-negative charge, and the two hydrogens carry a partial-positive charge (Figure 9.1). Because of this shape and charge configuration, water exerts a major influence on the Earth's environments, dissolving polar molecules easily. Due to water's high boiling point and a high melting point, large amounts of energy are required to vaporize water or to melt ice. Clouds, rain, and snowmelt in winter all occur from the absorption of heat from the sun to transform water into different states. This heat absorption leads to the hydrological cycle.[1]

Hydrological Cycle

Consider all the different forms of water. Water exists in the atmosphere as clouds, rain, snow, sleet, and fog; on the ground as rivers, lakes, and oceans, and underneath the Earth's surface as groundwater. These different states of water comprise the hydrological cycle. The hydrological cycle is a process involving the sun, the atmosphere, the Earth, and water (Figure 9.2).[2] This cycle consists of evaporation, condensation, transportation, transpiration, precipitation, and runoff. When surface water absorbs heat from the sun, it evaporates or vaporizes into the atmosphere. Clouds or water vapor transport water through the atmosphere. Plants also add water to the atmosphere in a process called transpiration or evapotranspiration. When water condenses to rain, sleet, or snow, it falls to the Earth as precipitation. The precipitation that is not absorbed by the soil becomes runoff, flowing over the land.[3]

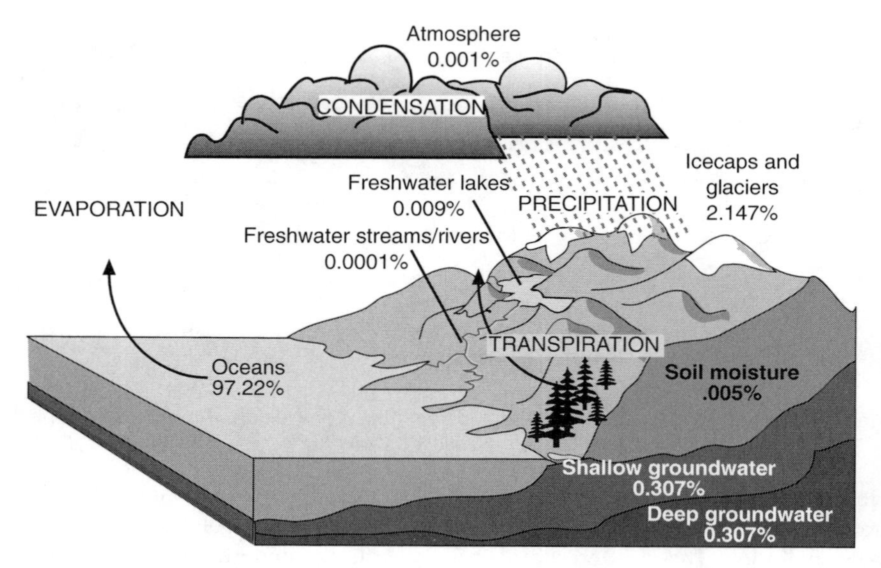

Figure 9.2 The hydrological cycle, and percentages of water in various Earth compartments. (Adapted from Grigg, N.S., *Water Resources Management,* McGraw-Hill, New York, 1996, chap. 1.)

Water Resources

A view of the ocean would convince even a desert dweller that vast amounts of water cover large regions of this planet. Water covers 71% of the Earth's surface. The approximate amount of water on the Earth is 1.3 billion cubic kilometers. Ninety-seven percent of this volume is salt water contained within the oceans and seas. However, humans and animals require fresh water for consumption, which makes up only 3% of the total amount of water. Much of this 3% of fresh water exists as icebergs or deep groundwater. Less than 1% of the Earth's water exists as accessible fresh water, available for human use.[4] Thus, despite the appearance of vast amounts of available water, freshwater sources are limited. In addition, most of the easily accessible sources of freshwater are currently in use. To obtain new sources would require significant expense and effort.[5] An increasing world demand for water, coupled with the threat of water pollution, is creating a growing crisis of water supply. As the demand for water continues to grow, new sources or new water management strategies will be required to meet people's needs.

Water and Health

Water comprises approximately 67% of the human body and is a major component of blood and lymph flowing through our bodies. Humans must drink water to survive; without water, a human will live only 3 days. Water also plays a major role in hygiene, as simple hand washing can prevent the spread of disease. Access to clean water is critical to human well-being and survival.

Many people in the world lack access to pure, fresh water, and their health often suffers from this lack. In fact, over 1.7 billion people in the world lack access to clean drinking water.[6] In an effort to provide all people with clean water and sanitation facilities, the United Nations declared the 1980s the "International Drinking Water Supply and Sanitation Decade." Unfortunately, much of the world's population remains in need of pure fresh water.[2] In some developing countries, poor people in the urban outskirts must purchase water from vendors at rates forty times the municipal rate.[6] Some people must use polluted water for drinking and bathing, as it is the only water source available. The Ganges River in India demonstrates this problem. Revered as a goddess by Hindus, the Ganges serves as a source of water for drinking and bathing, religious rituals, cremation, and sewage outflow. All of this human activity occurring in and along the river leads to rampant waterborne disease in this region, claiming the lives of 2 million Indian children a year. Yet people still drink, cook, swim, and immerse themselves in this holy but polluted river.[7] While modern technology may help clean up polluted waters and improve public health, it is not always welcome. A water supply project in Quiescapa, Bolivia, initially stalled because of villager resistance to abandoning unsanitary traditions, such as allowing animals free access to the water source. Teaching the villagers about water purity helped elicit local cooperation toward completion of the project.[8]

Drought, military conflicts, inadequate government support, and insufficient funds all contribute to the problem of supplying clean water to the world's population. Rural areas are especially hard hit. Worldwide, estimates of investments in water and sanitation projects indicate that 80% is allocated to urban needs and only 20% to rural facilities. Unfortunately, in developing countries, water facilities in rural areas are often in disrepair. The lack of simple spare parts can render a water supply such as a hand pump useless.[9]

Water Shortage and Scarcity

Water on this planet is not evenly distributed. Some areas of the world have plentiful water supplies; others have little or none. Nearly a billion people in 50 countries live with severe water shortages every day of their lives.[10] A 1995 World Bank report indicated that 40% of the world's population lives in countries facing water shortages. These shortages threaten public health by causing dehydration, use of polluted water, and lack of proper sanitation facilities.[11] One out of every five people on this planet lacks a clean water supply. People living in regions of water scarcity often must devote large amounts of time and energy to obtain water, sometimes walking several miles to reach a source of drinking water. These water shortages occur in areas with high rates of population growth, such as Central Asia and Sub-Saharan Africa. Since increased population means an increasing demand for water, the problems will only escalate. Parts of China, India, and South America are also experiencing water shortages.[12]

Globally, the demand for water has been increasing at about 2.3% annually, with a doubling of demand occurring every 21 years. Such a demand for water cannot be met by our current water supplies. In addition to the pressure of increased demand for water, water pollution also threatens the available water supply. A startling 95% of untreated sewage throughout the world and an increasing amount of industrial waste is discharged into the world's rivers.[12] At some point in the future, worldwide water use will be limited by physical, economic, and environmental limitations.[5]

Water Rights and Conflicts

With clean, fresh water in demand, conflicts over water rights and resources are inevitable. Many major water sources cross national boundaries, ensuring disputes between countries staking their claim to this valuable resource. Most of the undeveloped sources of freshwater traverse two or more countries. These international waters are actual and potential sources of contention.[13] Asia and the Middle East suffer from water-related disagreements. Such turmoil is not new. In 1950, a quarrel over water occurred between Pakistan and India, bordering on war.[14] More recent disputes have involved other countries. Bangladesh and India have conflicted over use of the Ganges and other shared rivers.[15] Turkey's water projects have met opposition from Syria and Iraq. Egypt opposed Ethiopia's plans to use additional water from the Blue Nile for irrigation.[16] The Jordan River basin has increased tension between Israel and Palestine. In the U.S., opposing factions also bristle over water rights. New York City's attempts to control the upstate New York watersheds that supply the city's water angered upstate residents.[17] Water disputes have also occurred in the western U.S., where water is scarce and demand high. The Colorado River serves many cities and agricultural areas, all of which want their fair share of the river.[18] In such cases, choices must sometimes be made regarding water allocation. Clearly, water scarcity and unclear water rights can cause or exacerbate national and international conflict.

These examples of water conflict highlight the importance of water as a pressing global issue. Interestingly, water scarcity has not yet garnered the level of attention global warming or ozone depletion have received, even though it poses a more immediate health threat to billions of people. The countries most affected by water scarcity are developing countries, where a lack of fresh water could seriously limit or hinder their growth. The amount and level of water conflict will grow with the level of water scarcity. Proper water management and international cooperation could halt some of the disagreements over water allocation and even prolong the use of scarce water resources.[19]

Water Consumption and Management

Water is an integral, necessary part of our daily lives. We use water for food production, power generation, dishwashing, even recreation. While most people agree that water is a valuable resource, it does not appear to be properly valued. Water abuse and overuse demonstrate that it is not. Many factors impact the amount of consumption and the way a country uses water, including the economy, available technology, level of industry, agriculture, culture, and climate. For example, as standards of living increase in developing countries throughout the world, water use will increase as more people use flush toilets, dishwashers, and washing machines.[5]

Effective water management would improve water efficiency, decrease consumption, and help preserve remaining resources.

Water management encompasses a number of techniques to preserve water resources, such as adequate water pricing, community education, and leak prevention. Part of the problem with excess consumption and water abuse through pollution has to do with inadequate water pricing. The consumer's price of water rarely reflects the actual cost of storage, treatment, and delivery. In the U.S., water is generally inexpensive, which contributes to an attitude of excess. A higher water price that includes the true cost would stimulate conservation, as consumers would bear the cost of overconsumption. If water has little or no value, then it will be wasted without a second thought. As the value of water increases, so does the attention to its effective use. For example, Germany has the highest price for water in all of Western Europe and North America. It also has the lowest overall average water use per person. On the other hand, the U.S., with the least expensive water, has the highest per capita use.[20] One drawback to accurate water pricing is the unfair burden it places on the poor. Urban poor often already pay more for water than nonurban consumers. Increasing water prices to truly reflect its costs while ensuring that everyone's basic water needs are met would stimulate water conservation to no one's detriment.[21] Education is another key component to a successful water management plan. Informing consumers about the impact of their water use will engender a greater sense of responsibility in the user and hopefully aid in conservation efforts. In addition, large quantities of water are lost through leakage. Stopping this loss would save tremendous amounts of water. All of these techniques encourage sustainable water use. Sustainable water use implies that current needs are met without diminishing the resource for future generations' use and at no expense to environmental need.[21] As the world's water demands increase, water management might be an increasingly valuable tool to combat scarcity and pollution.

WATER USE

Overview

Water use changes over time as countries and economies develop, standards of living rise, populations grow, and technology expands the realms of water use. Water use even varies with time of day and time of year, with the greatest uses occurring during the evening and the summer, respectively. The three major water consumers in the world are agriculture, industry, and households or individual use. In the U.S., the average daily per capita water use from public supplies is approximately 180 gallons per day. This includes domestic, industrial, commercial, and public use.[22]

Agriculture

Agriculture consumes the largest portion of the freshwater supply, with over two thirds of the world's water demand used for irrigation. Sixty percent of this water is lost to evaporation or runoff. Such an inefficient system cannot continue as water scarcity escalates in scope. As water resources become more scarce, agriculture will need to develop a more efficient system of water usage, using less water to grow more food.[22] In addition, agricultural water use tends to be heavily subsidized, both in the U.S. and worldwide. Water subsidies provide little incentive to economize water use and may in fact promote wasteful practices. Droughts can wreak havoc on agriculture, since they rely so heavily on water withdrawals. In 1998, Texas and Oklahoma experienced severe droughts that devastated the agricultural sector through crop loss, costing the state and farmers billions of dollars.[23] Globally, agriculture's share of world water use has been declining.[5]

Industry

The industrial sector also uses large quantities of water for numerous purposes, including manufacturing, cooling, and condensation by power plants, and waste disposal. Approximate industrial water use in the U.S. is over 200 billion gallons per day.[25,26] Worldwide, industrial demand for water is expected to increase approximately three times the current rates in the next 20 years.

Domestic

U.S. domestic water use includes cooking, showering, flushing toilets, washing dishes and laundry, and watering the lawn. Water is also used extensively for recreation. In the U.S., household water use is extensive. Technology allows for large volumes of water to be used in dishwashers, washing machines, and flush toilets. Per capita domestic use in the U.S. ranges between 75 and 135 gallons per capita per day (gpcd) (Figure 9.3).[27] The bulk of domestic water use serves for

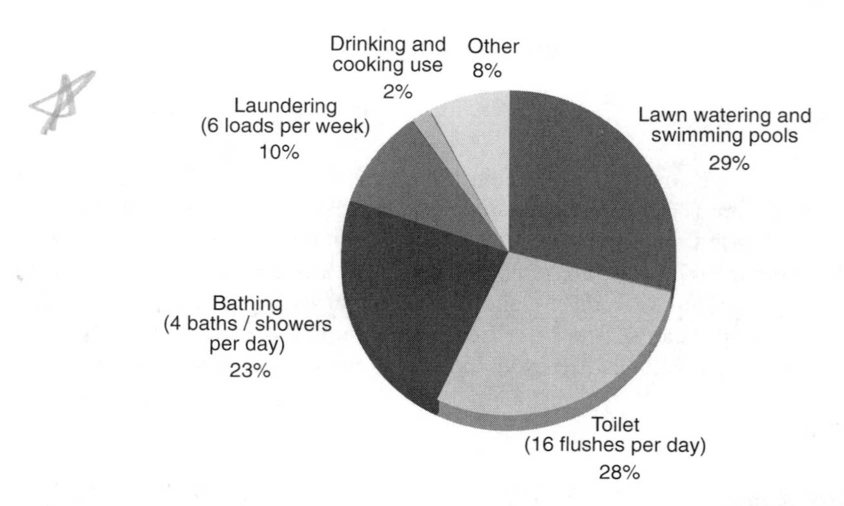

Figure 9.3 Water usage in the U.S. based on four-person family. (Adapted from Grigg, N.S., *Water Resources Management,* McGraw-Hill, New York, 1996, chap. 1.)

flushing toilets, showering, and watering lawns. These domestic uses do not occur in many developing countries where water is not so available. In those countries, water serves to meet basic domestic needs such as cooking and drinking.

SOURCES OF DRINKING WATER

Surface Water

The amount of surface water on the planet is approximately 230,000 km3, or about .017% of the total volume of water.[2] The U.S. has vast amounts of surface water, with 3.5 million miles of rivers and streams and 41 million acres of lakes. This surface water serves as drinking water to a large number of Americans. Sources of surface water develop through rainfall or snowmelt. Precipitation that is not adsorbed into the soil flows along the ground as runoff, entering streams, rivers and lakes. These waterways are sources from which water may be drawn off for human use. The areas surrounding rivers and lakes from which water flows into the waterways are called watersheds. Surface water is accessed through intakes, concrete structures on or below the surface of the water that divert or draw water from the source. Due to its exposed state, surface water is prone to contamination from a number of sources, including diffuse pollution such as agricultural runoff. It is estimated that each year about 450 cubic kilometers of wastewater are discharged into rivers, streams, and lakes.[28] Surface water sources and watersheds require protection and management to limit or prevent contamination. Watershed fencing, limited recreational access, and public education can help protect surface waters against pollution.[29]

Groundwater

Not all drinking water sources exist aboveground. In fact, much of the available fresh water exists as groundwater, below the surface of the Earth. Groundwater volume worldwide is estimated to be 8.5 million km3 or 0.62% of the total water volume.[2] Groundwater sources supply drinking water to 50% of the people living in the U.S. and to 90% of people living in U.S. rural areas. Groundwater develops in the following manner. When rain falls on the Earth, some of the water percolates downward through the spaces in the soil, pulled by gravity. At a certain point, the water reaches an impermeable layer of rock. At this layer, the water stops moving (Figure 9.4). As additional water flows downward, it fills all of the spaces between the rocks and sand that make up the soil above the impervious layer. This area of saturated soil is the zone of saturation. The top of the zone of saturation is the water table. The water table can rise and fall depending on the amount of rainfall and outflow of water from the zone of saturation into a river or a lake. Groundwater sources able to economically deliver water are referred to as aquifers. Some groundwater sources have a layer of impermeable rock above and below the zone of saturation. This type of water formation, called an Artesian aquifer, is under pressure greater than atmospheric pressure due to the weight of the water.[1]

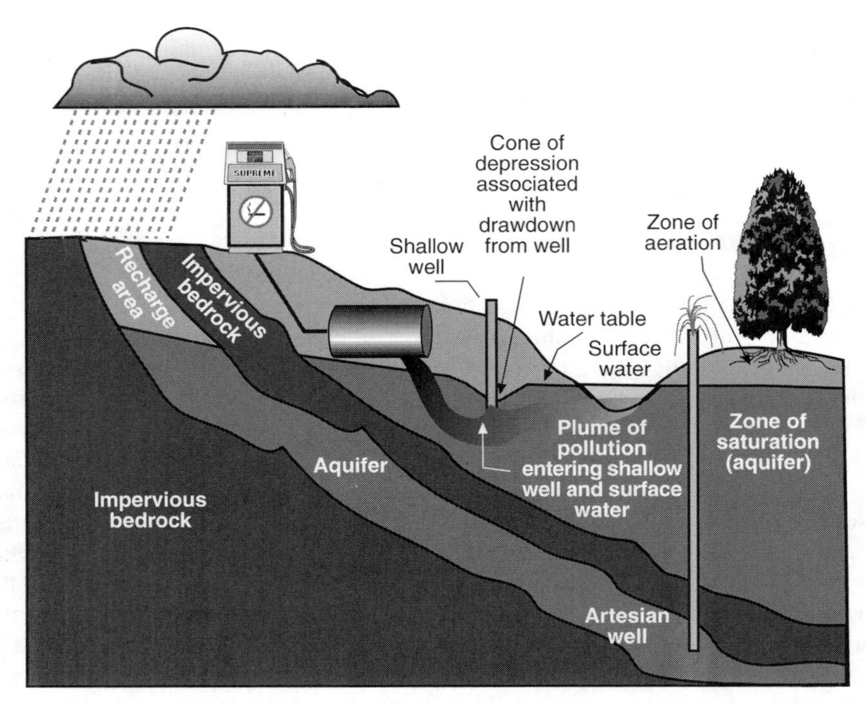

Figure 9.4 Groundwater configuration. (Adapted from Koren Koren, H. and Bisesi, M., *Handbook of Environmental Health and Safety,* 3rd ed., Lewis Publishers, Boca Raton, 1996, p. 273.)

Wells

Wells provide drinking water for millions of people throughout the world. In some areas of the U.S., no access to a public water supply exists, so wells serve as the central water supply. Wells may be private, serving a single user, or municipal, serving many people. A well site requires careful consideration of several factors, such as the source of the well water, its purity, the location of the well, and well water protection, both above and below the ground. Several types of wells exist, ranging from the crude to the sophisticated. Some wells are dug (Figure 9.5) or bored; others are driven or drilled (Figure 9.6). Wells vary in their susceptibility to contamination. Open hand-dug wells are frequently contaminated by animal feces, exposing the user to potential waterborne disease. Such wells tend to be shallow and less than 25 feet in depth. The shallow nature of these wells makes them prone to drought and contamination. Covering a dug well decreases the likelihood of contamination. However, if the groundwater itself is contaminated, the well water should be treated before it is consumed. More than 90% of residential wells are drilled. These wells are much deeper, reaching depths of up to 350 feet. They resist contamination and yield 2 to 5 gallons per minute more than 98% of the time. Septic tanks and farm runoff pose a significant pollution threat to well water.[33,34]

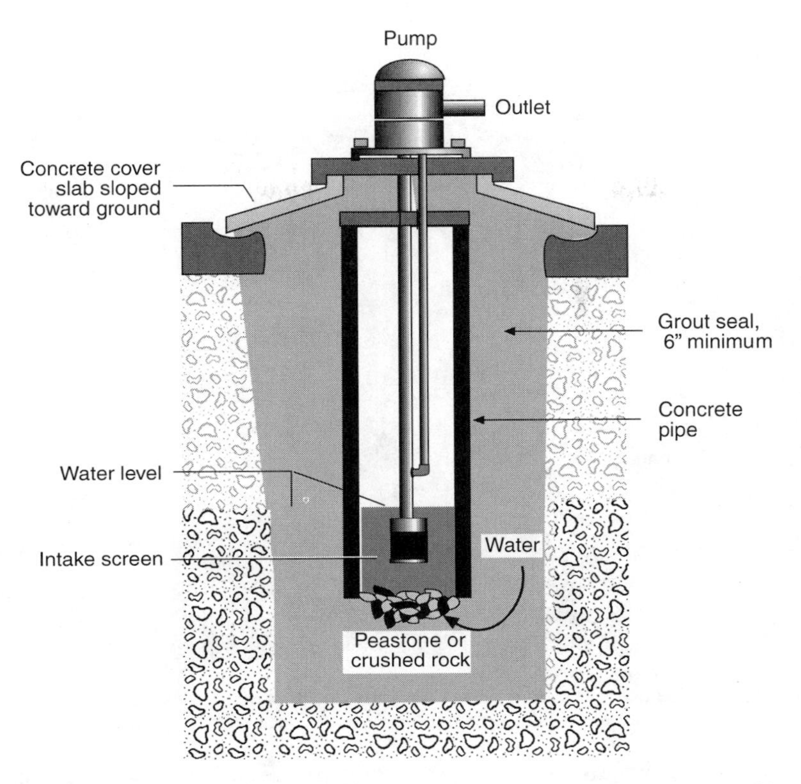

Figure 9.5 Dug well with two-pipe jet pump installation. (Adapted from USEPA, *Manual of Individual Water Supply Systems,* U.S. Government Printing Office, Washington, D.C., 1973, p. 34.)

Groundwater Contamination

Although groundwater may appear protected, it is not immune from contamination. As far back as the late 1960s, synthetic organic chemicals were found in groundwater sources. Additional instances of groundwater contamination occurred in the 1970s and 1980s. Despite its seemingly inaccessible location underground, groundwater is vulnerable to several means of pollution. Rainfall may transport surface pollutants deep into the soil, down to the water table. Polluted surface water may infiltrate a groundwater source. Improperly disposed hazardous waste can seep into groundwater. A staggering number of potential pollution sources exist in the U.S. including over 23 million septic systems; between 5 and 6 million underground storage tanks; millions of tons of pesticides and fertilizers; and municipal landfills and abandoned hazardous waste sites. Agricultural runoff, accidental spills, stormwater, and wastewater are all potential sources of groundwater contamination. Sources of groundwater pollution may be divided into planned and unplanned activities, involving surface and subsurface pollution from point and non-point sources (Table 9.1). A point source refers to pollutants entering the environment

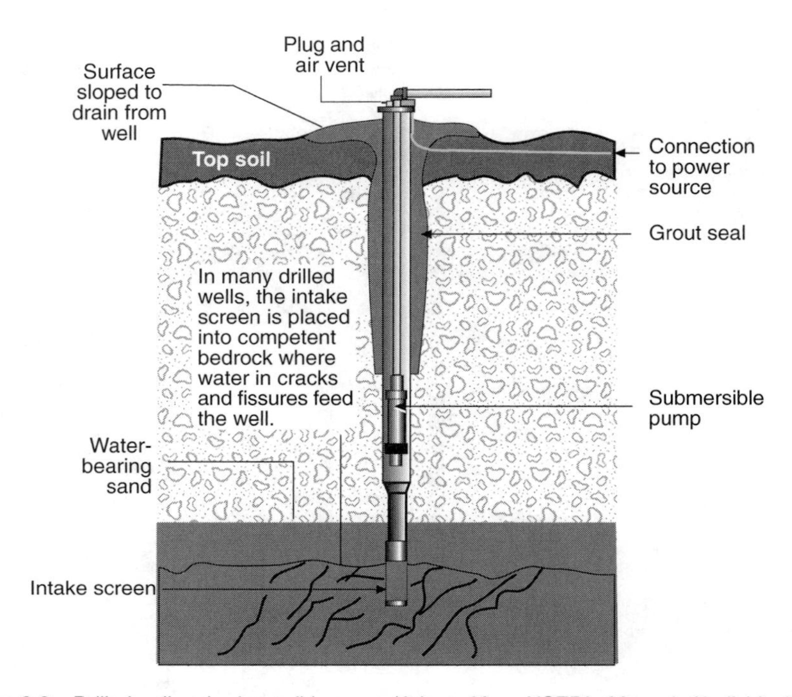

Figure 9.6 Drilled well and submersible pump. (Adapted from USEPA, *Manual of Individual Water Supply Systems,* U.S. Government Printing Office, Washington, D.C., 1973, p. 34.)

from a specific point such as a pipe or a specific source such as a factory or treatment plant. Nonpoint source refers to pollutants entering the environment from a broad area and may include scattered sources. The types of pollutants found in groundwater include nitrates, inorganic ions, synthetic organic chemicals, and pathogens. Many of these contaminants pose significant health risks.[35–37]

Recharge and Water Mining

Recharge is the replacement of groundwater by natural processes. Recharge of aquifers can take hundreds and possibly thousands of years. The rate of groundwater use in the U.S. presently exceeds the rate of recharge of these aquifers, which may lead to water shortages in the future. Overuse of some water systems in the U.S. has caused severe water stress. Water stress is an actual term defined as "the ratio of water withdrawal to water availability." Water stress falls into four categories: low, moderate, medium-high, and high. It occurs when more than 10% of the available freshwater is being used.[27] Currently, the Ogallala Aquifer that underlies eight states in the western and midwestern U.S. and water systems in Florida are stressed from the large quantities of water being pumped out of these underground sources (Figure 9.7). In addition, certain sources of groundwater are considered nonrenewable resources, like coal or oil, because of the length of time they took to develop. The overuse of these "fossil" waters is called water mining, because the

Table 9.1 Sources of Groundwater Pollution

SOURCES OF GROUNDWATER POLLUTION				
	PLANNED ACTIVITIES			
	Waste Disposal	**Storage or transport**	**Agricultural**	**Other**
SURFACE	Landfills NP Open dumps NP Surface impoundments P Waste tailings P Land application of sewage sludge P	Material stockpiles P	Pesticides NP Fertilizers NP Irrigation practices NP Animal feeding P	Road de-icing NP
	Waste Disposal	**Storage or transport**		
SUBSURFACE	Injection wells P Septic tanks P Cesspools P Other wells P Sewage pipelines NP	Undergound storage tanks P Pipelines NP		
	UNPLANNED ACTIVITIES			
SURFACE	Urban runoff NP Acid deposition NP Accidental spills P Natural leachates NP			
SUBSURFACE	Saltwater intrusion NP	Groundwater / surface water interaction NP		

NP = nonpoint source
P = point source

resource is being permanently depleted. An example of fossil water is the Ogallala Aquifer. Water from this aquifer is used for irrigation by farmers in Kansas, Nebraska, Oklahoma, and Texas. The large quantities of water used in irrigation greatly exceed the rate of replenishment. Unfortunately, rainfall in this region is low compared to other parts of the country, only 25 to 100 cm/year, compared to the eastern U.S., which receives 100 to 150 cm/yr.[1,36]

Subsidence and Salination

Aquifer depletion can result in subsidence and saltwater intrusion. The first, subsidence, involves a settling of the soil as the water is pumped out. As the rocks and dirt settle more tightly together, the volume of the soil decreases. This decreased volume of Earth cannot hold as much water as it did previously. The second problem with aquifer depletion is saltwater intrusion, which occurs in most coastal areas. As water is pumped out of the aquifer, the zone of saturation decreases at both the upper and lower levels. Saltwater can seep into the aquifer at the lower level, polluting the fresh water.[1]

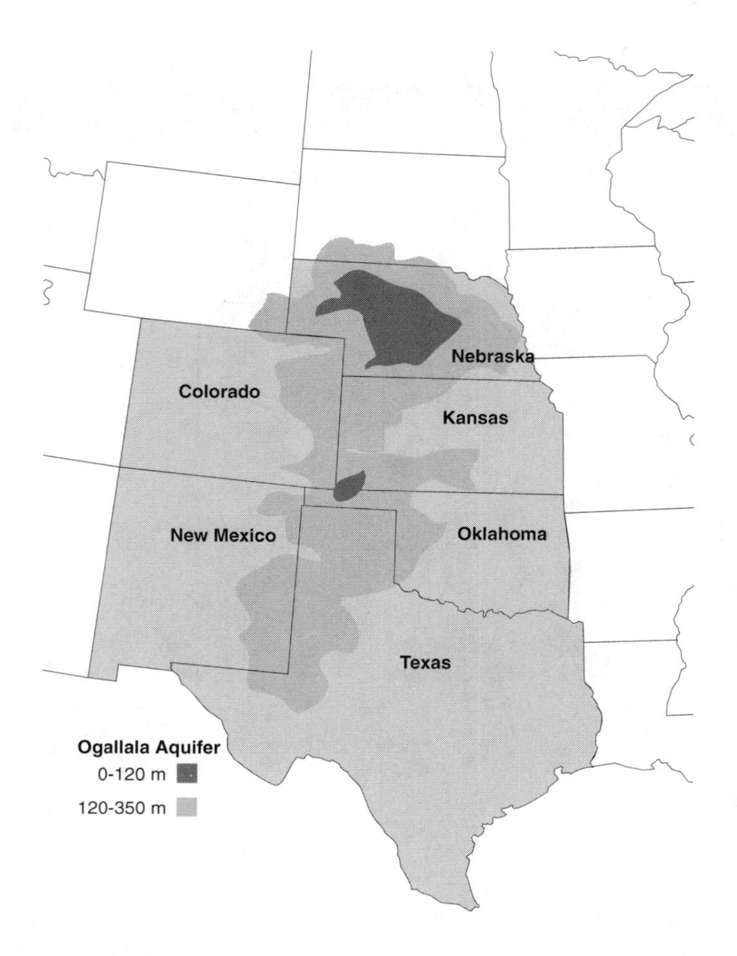

Figure 9.7 Geographic location of the Ogallala Aquifer. (Adapted from Turk, J. and Mirk A., *Environmental Science,* 4th ed., Saunders College Publishing, Philadelphia, PA, 1988, chap. 17.)

Groundwater Protection

In the U.S., groundwater is protected directly and indirectly by various laws including the Resource Conservation and Recovery Act (RCRA), the Comprehensive Environmental Response, Compensation and Liability Act (CERCLA), the Safe Drinking Water Act, (SDWA), the Federal Insecticide, Fungicide, and Rodenticide Act (FIFRA), the Toxic Substances and Control Act (TSCA), and the Pollution Prevention Act of 1990. By regulating the use, transportation, and disposal of thousands upon thousands of chemicals, these laws are an attempt to preserve the quality and purity of U.S. groundwater. However, a national comprehensive groundwater protection plan has not yet been established by the EPA.[37]

<table>
<tr><th colspan="2">WEB PICK
Go to http://www.unix.oit.umass.edu/~envhl565/index.html
Click on "Chapter Web Links" to find Water on the World Wide Web under Chapter 9</th></tr>
<tr><td>Chapter 9</td><td>Water and Wastewater</td></tr>
<tr><td>Whose Site?</td><td>U.S. Environmental Protection Agency</td></tr>
<tr><td>URL</td><td>http://www.epa.gov/ebtpages/water.html</td></tr>
<tr><td>What's Here?</td><td>This is a collection of water topics including: drinking water, groundwater, storm water, surface water, wastewater, water pollutants, water pollution, water pollution control, water quality monitoring, and water pollution legal aspects. Each topic has several subtopics that facilitate your search on information about water. Click on the topic "drinking water" and find several subtopics including:

• Providing Safe Drinking Water in America: Drinking Water Compliance Reports — The 1996 Safe Drinking Water Act Amendments require that each state submit an annual report on public water system violations to EPA.
• Public Drinking Water Systems Programs — The public drinking water systems that EPA and delegated states and tribes regulate provide drinking water to 90% of Americans.
• Who Is Responsible for Drinking Water Quality? — Local governments and private water suppliers have direct responsibility for the quality of the water that flows to your tap.

Locate your watershed at this site and retrieve an assessment of your watershed health, restoration efforts, an interactive mapping tool for environmental quality, and toxic releases into the watershed.</td></tr>
</table>

Other Sources

Desalinization

Desalinization is a technique of removing the salt from ocean water to provide drinking water. First, filters remove sediments, suspended solids, and salt from the feed water. With the use of pressure, this water is then pushed through a semi-permeable reverse osmosis filter, resulting in water that is 99% free of salts, minerals, and other ions. The product water then passes through a charcoal filter, removing unwanted tastes and odors. In the last step, the filtration process is completed by an ultraviolet sterilizer where 99.8% of all microorganisms, including viruses and bacteria, are destroyed.[34] While this process currently carries a high price tag, it does offer a means to access plentiful ocean water. This practice is currently in use in some areas of water scarcity, such as the Middle East.

Bottled Water

Even though Americans have easy access to tap water, as a country we consume large quantities of bottled water. In 1994 Americans consumed 2.43 billion gallons of bottled water at a cost of $2.1 billion. Many people believe that bottled water is safer and cleaner than tap water. However, bottled water is sometimes no better and can be worse than regular tap water. Tests by government agencies have discovered arsenic, bacteria, and concentrated chemicals in bottled water. In addition, many of the claims on bottled water such as "pure" and "natural" are meaningless, as these terms are not regulated by any government agency. While many people visualize crystal clear streams filling their water bottles, the source could be from anywhere, including a kitchen faucet. Despite these drawbacks, bottled water can serve as an excellent substitute for tap water, if tap water is unsafe for drinking. More responsible bottlers take additional precautions in purifying their water products. Specific wording on the product reveals information about the level of purity. A water labeled "Artesian" is well water from a contained underground body of water. Such water is less likely to be contaminated by chemicals from the surface. A bottled water that has been "purified" has been treated by one of several means of purification, such as distillation or disinfection with ozone or filtration through carbon or by reverse osmosis. Both Artesian and purified water are lead free.[38]

Dams

Dams have long been used as a means of controlling and storing water for human purposes. Reservoirs created by dams allow access to large quantities of water that would otherwise have flowed downstream. In addition to providing means to meet water demand, dams provide a method to manage river basins. Despite these positives, dams are not popular, as they may disrupt the natural ecosystem. Despite the immediate benefits of increased power production and better navigation, the Egyptian-built Aswan Dam was one such unpopular national project. The Aswan Dam succeeded in the dislocation of 60,000 indigenous people, the flooding of archeological sites, a decline in overall fish populations, and the destuction of many natural bank ecosystems.[39] By the same token, large-scale dams such as the proposed Three Gorges Dam in China have generated tremendous international environmental controversy and protest. Nevertheless, dams continue to provide water access to areas around the world.[40]

Water Reuse

The possibility of reusing water has become a reality in some countries as freshwater resources become depleted. Reuse typically involves the use of treated wastewater for nonconsumptive purposes, such as agricultural irrigation, landscaping, and industrial uses. Due to the high purity of treated effluents, water reuse does not appear to pose a health threat when used for nonpotable purposes. An excellent means of water conservation, water reuse will continue to expand into areas of the world with increasingly limited water supplies.[41]

Reusing wastewater for human consumption is a more controversial, perhaps less appealing option. However, it is a reality in some U.S. cities such as Los Angeles and El Paso. In these cities, treated wastewater is injected directly into aquifers in a process called artificial groundwater recharge. This technology has attracted some criticism, as the safety of this practice is still in question.[42] Interestingly, some areas of the world already use wastewater for drinking and bathing, albeit indirectly. Large cities at the downstream end of major rivers, such as London and New Orleans, draw drinking water from the river. By the time the river reaches these cities, the water has been used several times over for various purposes upstream, including many polluting uses.[43]

WATER POLLUTION

Overview

Water in its natural state is not pure. Water naturally picks up dirt, minerals, and particles as it flows over the ground. Much of the pollution that threatens our water supply today is anthropogenic in that it is generated by humans and not part of a natural process. Industry, agriculture, and overpopulation have all contributed to pollution of the world's water supply. In this country, media reports abound about beaches closed due to dangerously high bacterial counts or about wells contaminated with carcinogenic chemicals. Many different sources of contamination and pollution exist, from microscopic organisms to organic chemicals. In the U.S., laws and regulations exist to try to maintain pure water supplies. Worldwide, water protection and water purity vary from country to country. Some strategies to protect water supplies include protection of areas near sources of drinking water, limitations on pollutant discharges into our waterways, and the processes of chlorination and filtration. However, these strategies are not always effective. Sometimes a disinfection method, such as chlorination, might not kill all the microorganisms that might lead to waterborne illness. Other times, a contaminant might go undetected in a public well for years, exposing people to increased risk of cancer from their drinking water. A highly publicized example of well contamination occurred in Woburn, Massachusetts, where town wells were discovered to contain trichloroethylene, a suspected carcinogen. Many feared this contamination resulted in a cluster of childhood leukemias in the vicinity of the accused industries.

The potential for water pollution varies in type and scope throughout the U.S. and the world. A country's level of industry, agriculture, population, socioeconomic conditions, water delivery systems, and regulation all impact the degree of water contamination. Historically, water pollution in this country did not garner significant national attention until the 1960s and 1970s. Early legislation in the 1940s, 1950s, and 1960s allocated and expanded federal funding for municipal wastewater treatment plants. But it was not until the passage of the Federal Water Pollution Control Act in 1972 that the U.S. began a serious effort to mend its damaged and abused waterways. The Federal Water Pollution Control Act, renamed the Clean Water Act in 1977, established national standards for the nation's waterways and set limitations on allowed pollutant discharges. The Act's intent was to clean up the nation's waterways, including a return of navigable waters to a "fishable and swimmable" condition by July 1983 and a halt of pollutant discharges into waterways by 1985.[44] Subsequent amendments in 1977 and 1987 strengthened this Act, tightening regulation on pollutants. Despite this well-intentioned legislation, many rivers and lakes remain dangerously polluted.[45] The EPA estimates that public and private costs for water pollution treatment is $64 million a year.[46]

Water Quality

Water quality encompasses various characteristics of water, from taste and color to temperature and purity. Different levels of water quality are found regionally and throughout the world, based on available technology and economic resources. Water quality can vary depending on its intended use. High quality is needed for drinking water; lower quality is sometimes acceptable for irrigation purposes, as in wastewater reuse. Water use may also affect water quality, degrading its purity through pollution or a pH change. With pollution threatening the U.S. water supply, drinking water must be evaluated against certain standards to determine its safety and acceptability. Several different types of pollutants may contaminate a water supply. These can be categorized as physical, chemical, biological, and radioactive contaminants. Each category poses different types and levels of hazards to the consumer when they are present in the water supply. Consequently, in the U.S., public water supplies undergo testing to ensure its purity and safety to the consumer.

Routine biological testing involves the culturing of samples of 100 milliliters (ml) of tap water for the presence of indicator bacteria known as coliforms. The coliforms inhabit the intestines of warm-blooded animals including wildlife, birds, and humans. Their presence in the water supply indicates the possibility of fecal contamination and the potential for the presence of pathogens. The coliform test is used because the pathogens are difficult to grow. The absence of the indicator organisms does not assure the freedom of water from pathogens. Chlorination may kill the coliforms while permitting the resistant cysts of parasites such as *Cryptosporidium* or *G. lamblia* to survive. However, the coliform test has value as an indicator of potential contamination. The standard for water supplies is less than one coliform per 100 ml of water. Supplies exceeding this level require additional testing. If the supply continues to exceed the standard, then additional chlorine is usually added. In the event that the coliform remains high even after increasing the chlorine level, the town may issue a boil order requiring consumers to boil their water before consumption.

Tests for physical contaminants include odor, taste, color, turbidity, and total dissolved solids, referring to bicarbonates, sulfates, calcium, and other minerals and matter. These aesthetic water quality criteria impact the consumer's acceptance of the water. Foamability and excessive heat are also considered pollutants. Turbidity occurs when the water contains suspended clay or silt and may indicate other pollutants and stimulate bacterial growth. Color in the water results from organic matter, algae, metals, or colored industrial waste. Taste and odor signify organic material in the water and may also indicate the presence of pathogens or other pollutants. Foamability exists when detergents are present in the water supply. While not a health hazard, foam may indicate the presence of contamination.

Chemical contamination, both organic or inorganic, is tested using a spectrum of methods. Presently, more than 83 chemicals serve as the basis for testing in municipal water supplies — which are defined under the Safe Drinking Water Act (SDWA) as having 15 service connections used by year-round residents or regularly serving 25 or more people. The EPA has established standard or maximum contaminant levels (MCLs) for these 83 chemicals. The SDWA was reauthorized in 1986, requiring filtration for most public supplies having surface-water sources in order to remove pathogens such as *Cryptosporidium* and *Giardia lamblia*. Most of the public water supplies serving more than 10,000 people may require filtration under these requirements, and most of these systems have insufficient funding to carry out these federal mandates. Underfunding may also be the reason why a majority of small water supplies have not been adequately checked and may be operated by persons lacking training or qualifications, resulting in a large share of waterborne diseases in these communities. More than 60% of the nation's 60,000 community water supplies that serve less than 500 people are at increased risk to chemical and microbial contamination.[47] The contaminants of concern in drinking water include microbes, volatile organic chemicals, synthetic organic chemicals (including pesticides), inorganic chemicals (lead, mercury, and arsenic), and radionuclides.[22,52]

Right to Know

Consumers concerned about their tap water now have easier access to water quality information, thanks to the finalization by the EPA of a new law. This law requires public water systems to inform their consumers about the quality of their drinking water, any contaminants, and related health hazards. Since 200 million Americans receive their water from water utilities affected by this law, its impact will be significant. By October 1999, utilities had to send out written reports detailing the quality of the water. As these utilities already report this information to state Departments of

Environmental Protection, no additional water testing is necessary. This legislation empowers consumers through access to vital public health information.[48]

Types of Pollution

Inorganic Compounds

Inorganic compounds of many types may be present in the water systems. Of particular concern are lead, cadmium, mercury, arsenic, and copper. These substances can cause serious acute and chronic health problems. Lead, in particular, has been linked with brain damage and nervous system disorders. Mercury has also been implicated in central nervous system problems.[49] Also, Cadmium has been known to cause kidney damage.

Synthetic Organic Chemicals

Organic compounds in the water systems can present a serious health threat. The word "organic" refers to a substance containing the element carbon. Carbon can bond with four other atoms, which allows it tremendous variety in the types of compounds it forms. The types of synthetic organic chemicals in existence today number in the tens of thousands. Organic compounds can occur naturally in water, in the form of decaying vegetation and microorganisms. These natural organics may cause taste and odor problems in water. Of greater concern are the tremendous number of synthetic organic compounds that may pollute waterways. Sources of synthetic organic compounds include industry, agriculture, and even households. Each time a toilet is flushed or sewage sludge and manure is applied to the fields, large quantities of drugs like antibiotics, hormones, and pain killers, etc., are entering the environment and seeping into our waterways.[50] Disinfection with chlorine may also create organics in a water supply.

Synthetic organic compounds can be classified as volatile organic compounds (VOCs) or synthetic organic chemicals (SOCs). VOCs, as their name implies, are volatile and vaporize easily. VOCs migrate quickly in the ground and travel far from their site of introduction. VOCs present a threat to groundwater, where they are less able to vaporize and can accumulate. A dangerous VOC, carbon tetrachloride, has been found in well water. SOCs threaten surface waters through both accidental and purposeful discharges into waterways. Atrazine, chlordane, lindane, and toluene are several dangerous synthetic organic chemicals in use today.[51] Polychlorinated biphenyls or PCBs are other hazardous organic chemicals that have been found in some U.S. water supplies, even though their U.S. production ceased in 1977.

Radioactive Material

Radioactive materials from natural and man-made sources can migrate into water systems. The radiation emitted from decaying radionuclides can have dangerous, even fatal health consequences. The most common radioactive substances in water are radium, uranium, radon, and certain man-made radionuclides. While naturally occurring radionuclides appear mainly in groundwater, surface waters are more likely to contain artificial radionuclides from atmospheric fallout.[52]

Sources of Pollution

Overview

Pollution sources are labeled "point" sources or "nonpoint" sources. A point source of pollution stems from a single discrete location, such as an outflow pipe. This type of source can be monitored, measured, and controlled. Industrial discharges are generally considered point sources, as they enter waterways at a single location. The U.S. Environmental Protection Agency regulates point-source discharges through a permit system. A nonpoint source of pollution has a wide range of entries into a waterway and is difficult to measure and manage. Agricultural and livestock farming activities are considered nonpoint sources of pollution.[53] Nonpoint sources of pollution have now surpassed point sources as the main culprits of water pollution in the U.S.[44]

Underground Injection Wells

Underground injection wells (UIW) may pollute water. An underground injection well is any hole in the ground that is deeper than it is wide and into which liquids are injected. Five types of UIW's exist — Class I to Class V — and each is used for a different purpose. The most troubling are the Class V wells, which are shallow injection wells. These simply constructed wells may contain various pollutants, including industrial waste, agricultural drainage waste oil, and wastewater. All of these may threaten sources of drinking water. The likelihood of contamination depends on the proximity of the well to an aquifer, the quality of the well construction, and the type and amounts of fluid injected.[37]

Industrial Discharges

Unfortunately, part of creating a product includes creating waste. In the past, the liquid portion of this waste was dumped directly into our waterways or into poorly constructed landfills that leaked dangerous substances into the water. As knowledge about the degree and the risks of water pollution increased, public concern over the health effects of chemicals in our water systems also increased. Many people worried that chemical pollutants in our water system would cause cancer and other diseases. Today, restrictions are placed on what and how much can be dumped into waterways. The government has enacted laws, such as the Safe Drinking Water Act and the Clean Water Act, that require monitoring for specific chemical contaminants in water. The National Pollutant Discharge Elimination System (NPDES) regulates the discharge of wastewater into U.S. waterways. An NPDES permit stipulates the allowable levels of contaminants that may be discharged in an effluent. This level is called an effluent limitation. The EPA has set standard effluent limitations for specific industries, such as paper mills and chemical manufacturers. The permit also specifies the required tracking of discharged chemicals.[54] Despite the permit system, chemical contamination is still a growing problem. The vast number and volume of chemicals used by industry make it impossible to monitor water systems for each specific chemical. Even with the limited number of chemicals monitored, each year thousands of public water systems violate one or more of the thresholds due to the high cost of monitoring. Adding new chemicals to the hit list every few years increases the financial strain.[55]

Agriculture

Agriculture demands an average of 70% of the world's fresh surface waters, making it the largest consumer of fresh water. Not surprisingly, much of the pollution entering our waterways comes from agricultural sources. During rainfall, fertilizers and pesticides applied to crops can flow as runoff into surface waters or percolate through the ground to groundwater. In September 1999, North Carolina was flooded as a result of Hurricane Floyd. The floodwaters managed to shut down one third of the state, an area of 18,000 square miles affecting 2.1 million people. The floodwaters also succeded in causing huge amounts of hog feces and urine to flow into rivers and enter the groundwater. Drinking water was tainted by such overflow of man-made lagoons, causing an increase of fecal coliform bacteria levels to three times the normal average.[56] Increasing amounts of fertilizer and pesticides used for agricultural purposes also increases the opportunity for water pollution. Information on agricultural pollution in developing countries is limited. However, we do know that pesticide use in developing countries continues to increase, suggesting that incidents of water contamination will occur more frequently.[58] Another area of our country that is affected by agricultural runoff is an area in the Gulf of Mexico known to farmers and fishermen as the "dead zone." The source of the dead zone lies inland, stemming from the agricultural runoff of farms, which is then carried down the Mississippi River and out into the Gulf of Mexico. The principle cause of this "dead zone" is a result of nitrogen fertilizer and hog farm surface runoff. When the nitrogen reaches the Gulf, it stimulates growth just like it would on land, causing major algae growth. As the algae die and decay, they remove oxygen from the water, which results in the death of plant and animal species emigration.[59]

Pesticides

Although pesticides have improved the world's food supply, they do pose potential health risks if they enter the water supply. Some pesticides have been linked to an increased risk of cancer. The U.S. uses 1.1 billion pounds of pesticides annually. Pesticides have been found in stormwater runoff from both agricultural areas and urban areas.[55] Pesticide contamination of surface waters occurs when rainfall carries the chemicals in runoff into the waterway. Pesticides may also enter surface waters directly from the atmosphere. Pesticides have been found in U.S. surface waters since the commencement of water tests in the 1950s. Surface-water monitoring studies conducted throughout the U.S. have detected over 100 pesticides or related compounds. The types of pesticides found in surface waters change from region to region, seasonally, and over time. Organochlorines, most of which the U.S. banned in the 1970s, have been seen in decreasing concentrations in the water supply. Herbicide use has increased, as has the concentration of herbicides in the water supply, which regularly surpass the maximum contaminant level (MCL) standard.[59]

Fertilizer

Worldwide, fertilizers are increasingly found in water sources, both surface and groundwaters. Nitrate, a major component in fertilizer, has been detected in wells in many rural areas of the U.S. where fertilizer use is greatest. In fact, 9% of wells in agricultural regions of the U.S. exceed the 10 mg/l nitrate limit. Nitrates pose a health risk to pregnant women, infants, and young children. Babies are especially at risk from a condition called methemoglobinemia or "blue baby" syndrome, in which nitrates act to reduce the red blood cells' ability to transport oxygen. This condition can be fatal.[61,62] Fertilizers in water can also cause eutrophication, which is nutrient-overload of a water source. Eutrophication promotes algae growth, disrupts the ecosystem, deoxygenates water, and may result in fish kills. Nitrate contamination of groundwater poses a significant problem as it is difficult to clean up.[54]

Stormwater

Defined as "stormwater runoff, snow melt runoff, and surface runoff and drainage," stormwater poses a threat to drinking water sources. It may contain pathogens, nutrients, such as nitrates, synthetic organic chemicals, heavy metals, and salts. Groundwater contamination can occur when stormwater transports pollutants through the soil into the groundwater. Impervious surfaces, such as roads and parking lots, increase both the amount of stormwater and the level of contamination. Currently, stormwater discharges from industrial, municipal, and other small facilities are regulated under the Water Quality Act of 1987. Permits are required for industrial and municipal stormwater discharges to lessen the possibility of water pollution.[63]

Acid Mine Drainage

Water sources may be contaminated by drainage from mines. Acid mine drainage generally comes from active or abandoned underground mines. This acidic drainage enters streams and rivers, acidifying the water and harming the aquatic environment. Heavy metals may also be leached into the water during drainage formation. Sealing off abandoned mines can alleviate this pollution threat. Reducing water flow from the mines also can limit the amount of acid mine drainage. Treatment of drainage is another option for reducing its impact on waterways.[64]

Waterborne Disease

Historically, waterborne disease has been a significant threat to human health. About 3.3 billion people in the 127 countries of the developing world suffer water-related diseases. There are almost 6 million deaths from water-related diseases each year, and 50% of people on Earth lack adequate sanitation.[10] In the developing world, waterborne illness causes 900 million cases of diarrhea and claims 12 million children's lives annually.[65] Before methods of water purification were established in this country, diseases like cholera and polio were common health problems. With the development of chlorine treatments and other purification methods in the early part of the 20th century, the incidence of waterborne disease in this country decreased significantly. Despite extensive water treatment in the U.S., water may still carry waterborne disease. Over the past 10 years, several outbreaks of waterborne disease have occurred. In fact, in the U.S., waterborne illness causes over 900 deaths annually and more than a million cases of gastrointestinal illness. Two of the reasons for the increase in waterborne disease are inadequate purification and water contamination.[66]

Microbial contamination of water occurs when a pure source of water is polluted by raw sewage or animal feces. Since this water is thought to be pure, it is used for drinking water and subsequently makes people ill. This problem occurs when an undetected contamination source enters the water delivery system of a city or town, polluting the water. The usual culprits are wild animal feces or sometimes runoff from farms. Contaminated water could cause a host of gastrointestinal diseases, and dehydration from severe vomiting or diarrhea could be fatal in children, the elderly, and people with weakened immune systems.[67] The most common microbial contaminants are *E. coli*, *Giardia lamblia*, or *Cryptosporidium parvum*.

Over the past 15 years, the increased incidence of waterborne disease has forced us to examine our water purification methods. Even when water is chlorinated, it might still harbor disease-causing microorganisms. Certain microorganisms, in particular, *Giardia lamblia* and *Cryptosporidium parvum*, are very hardy and can tolerate the concentrations of chlorine used to treat drinking water. Unfortunately, when ingested, these organisms can cause disease and even death. A major U.S. outbreak of Cryptosporidiosis occurred in Milwaukee, Wisconsin, in 1993, when over 400,000 developed gastrointestinal illness from drinking water contaminated with *Cryptosporidium*. One hundred people died from the illness during this outbreak. As *Cryptosporidium* lives in the water supplies of approximately 45 million Americans, it represents a significant public health threat. Removal of this dangerous parasite from drinking water requires expensive filtration systems.[65] Other occurrences of waterborne illness in the U.S. demonstrate that our water is not free from biological contamination.

WATER TREATMENT

Municipal Water Treatment

In the U.S., most of the drinking water comes from 200,000 community water supply systems. These systems, approximately 140,000 small-scale suppliers and 60,000 municipal supply systems, supply water to 241 million Americans.[66] Most of this water requires treatment to remove contaminants before it can safely be used for drinking water. As mentioned previously, surface water is exposed to many sources of pollution and requires extensive treatment. Due to its protected nature, groundwater may require less treatment, though increasing reports of groundwater contamination may suggest additional purification.

Water treatment proceeds as a series of steps, each intended to remove additional unwanted material from the water and improve the water quality. The main steps of treatment are sedimentation, coagulation–flocculation, filtration, and disinfection. Other treatments include softening, fluoridation, taste and odor removal, algae control, and aeration. After this treatment, the water should be free of almost all contaminants. Sedimentation involves the settling out of large or weighty suspended objects from the water. A settling tank holds the water for 24 hours while the objects settle out. The next step in the process, sedimentation–flocculation, involves the addition of chemicals such as hydrated aluminum sulfate to facilitate large particle formation. These particles, or floc, settle out of the water when the water is allowed to sit. Filtration also removes suspended material but involves passage of the water through beds of sand or other porous material. While filtration is generally used in large water systems, it can also be used for individual water supplies. Filtration units may be attached directly to the faucet. Disinfection, discussed below, is the next crucial step. Other types of water treatment include aeration and water softening. Aeration is a technique used to remove volatile organic compounds (VOCs) from a water supply. An upward stream of air intersects with a downward flow of water. As the air rises through the water, it pulls the VOCs into the airstream and out of the water. Water softening involves removing the minerals calcium and magnesium from a water supply to prevent the build-up of mineral deposits on cooking equipment, washing machines, and water tanks.[52]

Disinfection

The most critical step in water treatment, disinfection, should destroy all organisms in the water supply. To ensure proper disinfection, organic matter and other material must be removed prior to disinfection. An important characteristic of an effective disinfectant is that it leave a residual to prevent regrowth of microorganisms. This residual maintains the purity of the water as it travels to the consumer. Chlorine is the major disinfectant used in the U.S.'s water systems today. It provides an inexpensive, relatively effective method of killing most waterborne microorganisms and protecting water as it travels through a delivery system. However, chlorine does have some drawbacks. When added to water containing certain organic matter, chlorine can combine with these pollutants to form dangerous disinfectant by-products called trihalomethanes or THM. These compounds have been linked with an incidence of bladder and rectal cancer. Despite the increased risk of cancer from trihalomethanes, chlorinated water is safer than drinking unchlorinated water due to the threat of waterborne disease. In the many countries throughout the world that lack purified chlorinated water, waterborne disease is a tremendous public health problem. The World Health Organization estimates that 9.1 million people die annually worldwide from disease caused by waterborne pathogens. Not chlorinating water to avoid the risks from trihalomethanes would increase the incidence of waterborne disease.[69,70]

Although chlorine is the most widely used disinfectant, other disinfection methods exist including ozonation, ultraviolet light exposure, and chloramination. Ozone can disinfect water 3100 times faster than chlorine and leaves no residual byproducts. However, because ozone does not provide a long-lasting residual, ozonated water is treated with a small dose of chlorine. Chloramination, which involves mixing chlorine and ammonia in the water to create chloramines, also disinfects the water and does not form trihalomethanes. But because chloramination lacks the disinfectant strength of chlorine, the water usually requires a second dose of chemical.[52]

Home Water Treatment

In some areas of the U.S., the available water might not meet purity or safety standards. In such situations, people have the option of buying home water treatment systems for their individual water supply. The water treatment market contains many options for the water-wary consumer, from the simplest of systems to the complex (and expensive). A relatively inexpensive option, activated carbon filters, can remove mercury, volatile organic compounds, trihalomethanes, and unpleasant tastes and odors. However, if these filters are improperly maintained, they may act as breeding grounds for bacteria. More expensive home treatment units, such as "reverse osmosis" or "distillation" units, remove most heavy metals and bacteria.[66]

REGULATIONS

Safe Drinking Water Act

In the U.S., drinking water received from public drinking water systems is regulated by the Safe Drinking Water Act (SDWA). The SDWA allows the EPA to set Maximum Contaminant Levels (MCLs) for water pollutants to protect the public health. The MCL is determined by multiplying the greatest dose of a chemical that generates no negative health effects in animal studies. Chemicals with an MCL also have a Maximum Contaminant Level Goal, which is the target level of a chemical in the water. While MCLs are required, MCLGs are not. MCLs have not been established for all the potential chemical contaminants in treated water. Enforcement of the SDWA is left to the individual states, with oversight provided by the Office of Groundwater and Drinking Water, a division of the USEPA. State standards for public water systems must be as strict as EPA standards. All public water systems in the U.S. must comply with the National Primary Drinking Water Regulations, which contain the MCLs. These water supply systems must test the water for specific contamination, including various inorganic and organic compounds, trihalomethanes (a specific organic), turbidity, coliform bacteria, and natural and synthetic radionuclides.[71]

The SDWA has evolved considerably since its inception in 1974. Presently, every five years the EPA administrator selects at least five chemicals from a list of contaminants not currently monitored, but which are known to be in the water supply. The administrator must decide whether to regulate these chemicals. A substance may also be added to the list if it is a CERCLA hazardous substance or a FIFRA pesticide. While new requirements provide improved water quality, these requirements place a financial strain on a community's water budget. To properly implement the current goals of the SDWA will require tens of millions of dollars. The EPA estimates that $138.4 billion will be required over the next 20 years to overhaul the nation's 55,000 aging community water systems.[72]

WASTEWATER DISPOSAL AND TREATMENT

Sewage

In many developing countries around the world, human waste pollutes the land and the water. People lacking any sanitation facilities simply defecate on the ground, leaving fecal matter to contaminate food and water sources. The developed world generally has ready access to sanitation in the form of septic systems or municipal sewers. In the U.S., wastewater from household water use is generally routed away from the house through pipes to treatment facilities or septic systems. This wastewater contains human wastes, food, detergents, and household chemicals. Sewage or wastewater is 99.9% water. Organic matter and impurities comprise the remaining 0.1%. Upon entering the ground or a waterway, this 0.1% of matter can drastically impact the quality of the groundwater or surface water. This organic material can serve as food for organisms living in the water.

Biological Oxygen Demand

As microorganisms decompose organic material in surface water, they use oxygen dissolved in the water. When a waterway has plentiful dissolved oxygen (DO) to meet the biological oxygen demand (BOD) of the organic matter, then the water can still support aquatic life. However, if a waterway is overloaded with biodegradable organic pollutants, this decomposition process can deplete the supply of dissolved oxygen. This deoxygenation diminishes the water's capacity to support aquatic life and can result in fish kills and algal blooms. The BOD of waste can be determined through a simple BOD test. Water reaerates through a natural process over time and distance from the waste input. The deoxygenation and reaeration of water can be presented graphically as a Sag Curve. A Sag Curve demonstrates the level of dissolved oxygen over time, showing the critical level where aquatic life dies (Figure 9.8).[4]

Types of Disposal

Pit Privies

In many areas of the world, human excreta and wastewater are dumped directly into waterways, due to lack of proper sanitation facilities. Such practices contaminate the water supply and increase the spread of waterborne disease. Rudimentary facilities

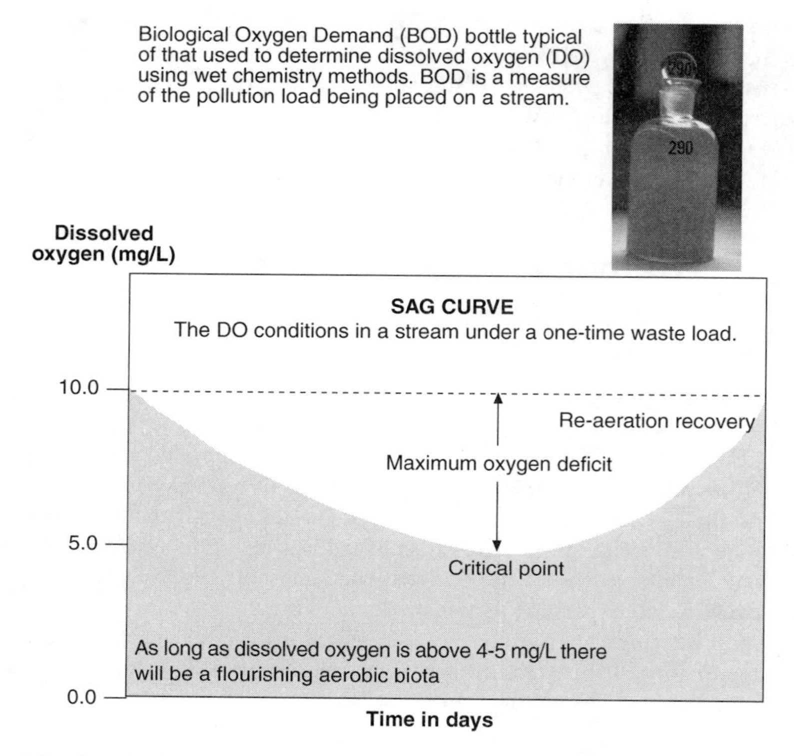
Biological Oxygen Demand (BOD) bottle typical of that used to determine dissolved oxygen (DO) using wet chemistry methods. BOD is a measure of the pollution load being placed on a stream.
290
Dissolved oxygen (mg/L)
SAG CURVE
The DO conditions in a stream under a one-time waste load.
10.0
5.0
0.0
Re-aeration recovery
Maximum oxygen deficit
Critical point
As long as dissolved oxygen is above 4-5 mg/L there will be a flourishing aerobic biota
Time in days

Figure 9.8 A typical sag curve resulting from the deposit of high BOD materials into a receiving body of water.

such as pit privies can protect water supplies from fecal contamination. A pit privy is simply a manually dug hole, with some type of seat placed over it and surrounding walls for privacy (Figure 9.9). Used mainly in developing countries, the pit privy isolates human waste from the water supply. However, if not properly maintained, the privy itself can become a source of contamination.[73]

Septic Systems

Septic systems provide a method of wastewater disposal to many people in the U.S., especially in rural areas or where no access to municipal sewage treatment exists. A septic system is a two-stage system, made up of the septic tank, usually a precast or reinforced concrete tank, and an absorption field, where the effluent flows into the soil. Sewage flows from the house into the septic tank, where it is held for 24 to 72 hours (Figure 9.10). During this time, the heavier materials settle to the bottom, where they form a layer of sludge. Lighter particles migrate to the top of the wastewater, forming a scum layer. The middle layer, consisting of wastewater, flows out of the tank into a series of perforated pipes that lie in the drainage field. The effluent flows out of the pipes and percolates through the soil. The soil in which the effluent flows must allow for adequate percolation. This capacity is crucial for

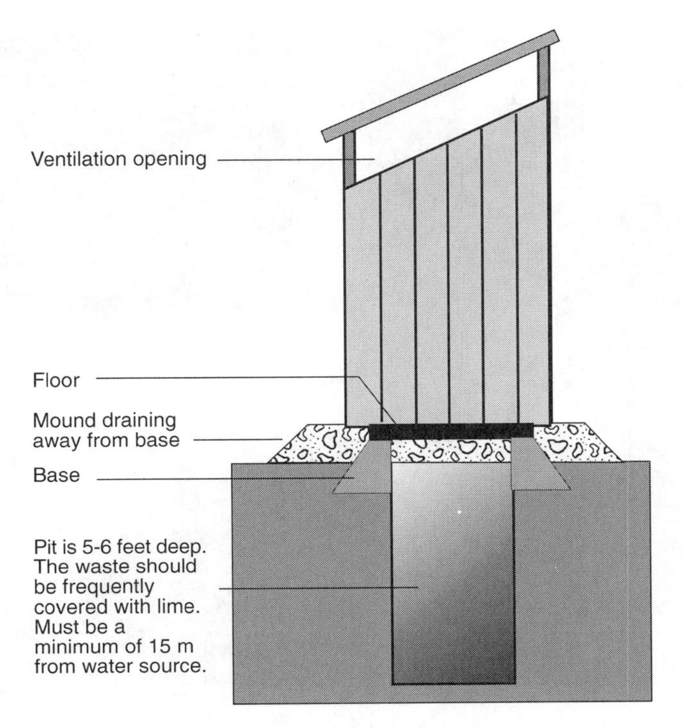
Ventilation opening
Floor
Mound draining away from base
Base
Pit is 5-6 feet deep. The waste should be frequently covered with lime. Must be a minimum of 15 m from water source.

Figure 9.9 A typical pit privy construction. (Adapted from Chanlett, E.T., *Environmental Protection,* 2, McGraw-Hill Book Company, New York 1979, chap. 3.)

an effective septic system. Soil permeability is determined through a perc test. Filtration through the soil further purifies the effluent as soil organisms break down the organic material into simpler inorganic forms. The minimum septic tank capacity is 750 gallons, extending up to 2000 gallons for large households. Tanks require periodic cleaning to remove the scum and sludge layer (Figure 9.11). A view of a typical installation of a septic system including the septic tank and leach filed is shown in Figure 9.12.[52]

Municipal Sewage Treatment

Sewage treatment provides a method to speed up and control water's natural process of purification through biooxidation, filtration, and settling. The end product of sewage treatment, called an effluent, should have significantly reduced BOD and pose little or no harm to the waterway or human health. Achieving these goals requires several stages of treatment, including primary, secondary, and tertiary treatments and sludge disposal (Figure 9.13).

Primary Treatment

Primary treatment is largely a mechanical process, concerned with the removal of solids. First, screens are used to catch large materials as the sewage flows through

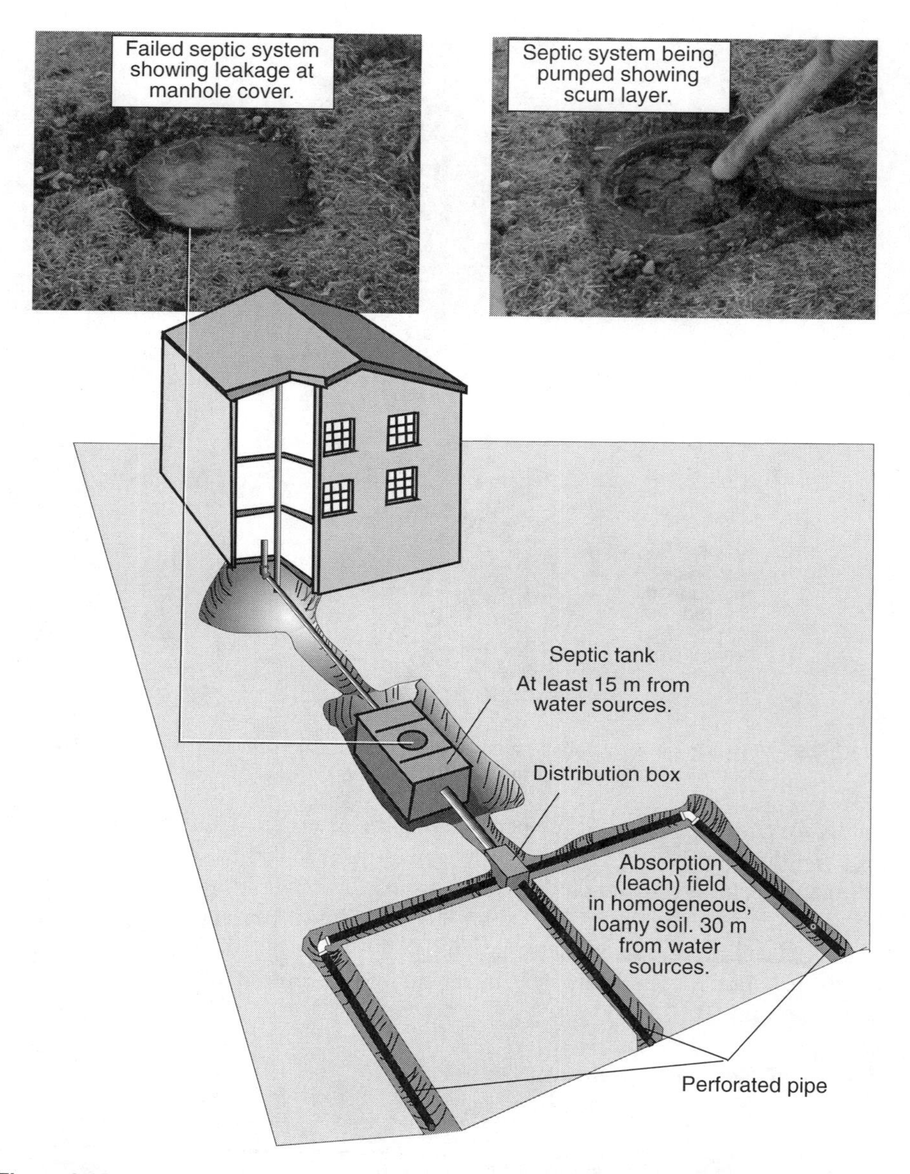

Figure 9.10 A typical septic system construction. (Adapted from Chanlett, E.T., *Environmental Protection,* 2, McGraw-Hill Book Company, New York 1979, chap. 3.)

them. Next, the comminuter, a grinder or a shredder, reduces the size of the solids in the water and removes any coarse objects (Figure 9.14). The sewage flows into a grit chamber, where heavy matter such as stones, sand, and glass particles can settle out. This prevents such materials from entering the pumps and destroying the mechanisms by abrasion. From the grit chamber, the sewage flows to the primary settling tank, called the primary clarifier.[71] Here the flow is slowed, the settleable

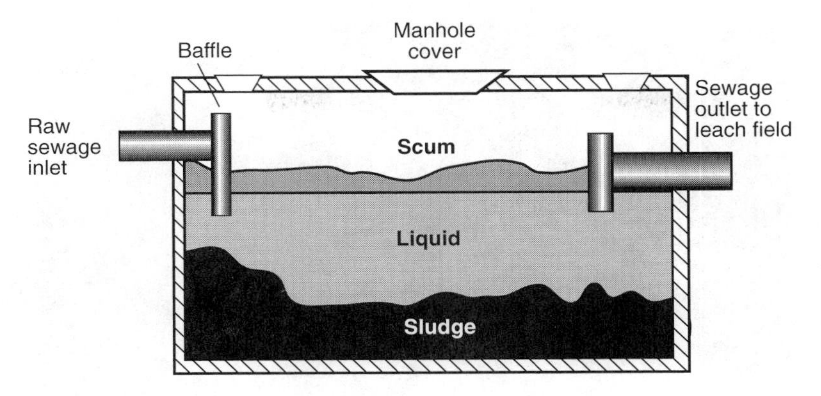

Figure 9.11 A cross-section schematic of a septic tank.

solids descend to the bottom where they form sludge that is pumped out for processing and drying. A skimmer removes floating debris, and the liquid portion of the sewage overflows into a receiving channel around the clarifier (Figure 9.14). At this stage, the effluent still contains substantial organic matter, causing a high BOD and numerous microbes. Many municipal systems throughout the U.S. still stop treatment at this primary stage. Chlorine is usually added at this point and the effluent is pumped to a receiving body of water, where the pollution load from this effluent will be high.

Secondary Treatment

Secondary treatment employs biological oxidation to break down pollutants into more stable inorganic or oxidized organic forms that resist further metabolism and do not contribute to the biological oxygen demand when released into a body of water. Trickling filters and activated sludge treatment employ bacteria to break down and digest organic material in the sewage. Trickling filters consist of a bed of coarse stones or plastic media, covered with biomass consisting of bacteria, protozoa, and fungi. When wastewater is applied to the stones, dissolved organic matter is adsorbed by the slime and digested by the microorganisms. Activated sludge treatment occurs in an aeration tank, where sewage is mixed with "active" or bacteria-rich sludge to create a "mixed liquor." This mixture is aerated for several hours while the sludge bacteria digest the organic material in the sewage (Figure 9.16). This mixed liquor is transferred to a settling tank, where the activated sludge settles out, most of which is routed back to the aeration tank. The effluent flows to a sedimentation tank, known as a secondary clarifier. An advantage of activated sludge treatment is that it can remove nitrates and phosphorus from sewage, whereas trickling filters do not.[74,75]

Tertiary Treatment

In some cases, additional treatment of the effluent is necessary to further reduce the BOD and suspended solids. A number of tertiary treatments or advanced wastewater treatment methods accomplish this goal, including air stripping by ammonia

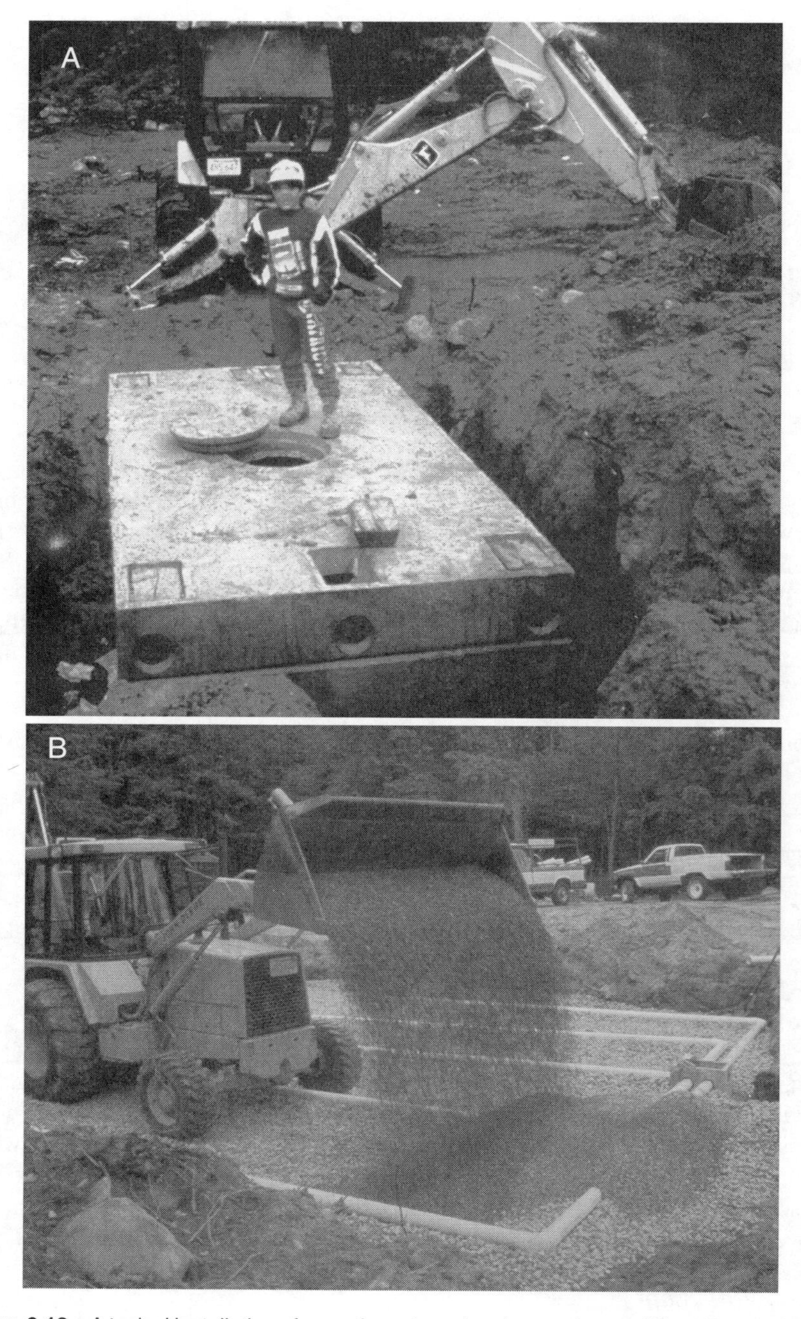

Figure 9.12 A typical isntallation of a septic system showing septic tank (A) and leach field (B).

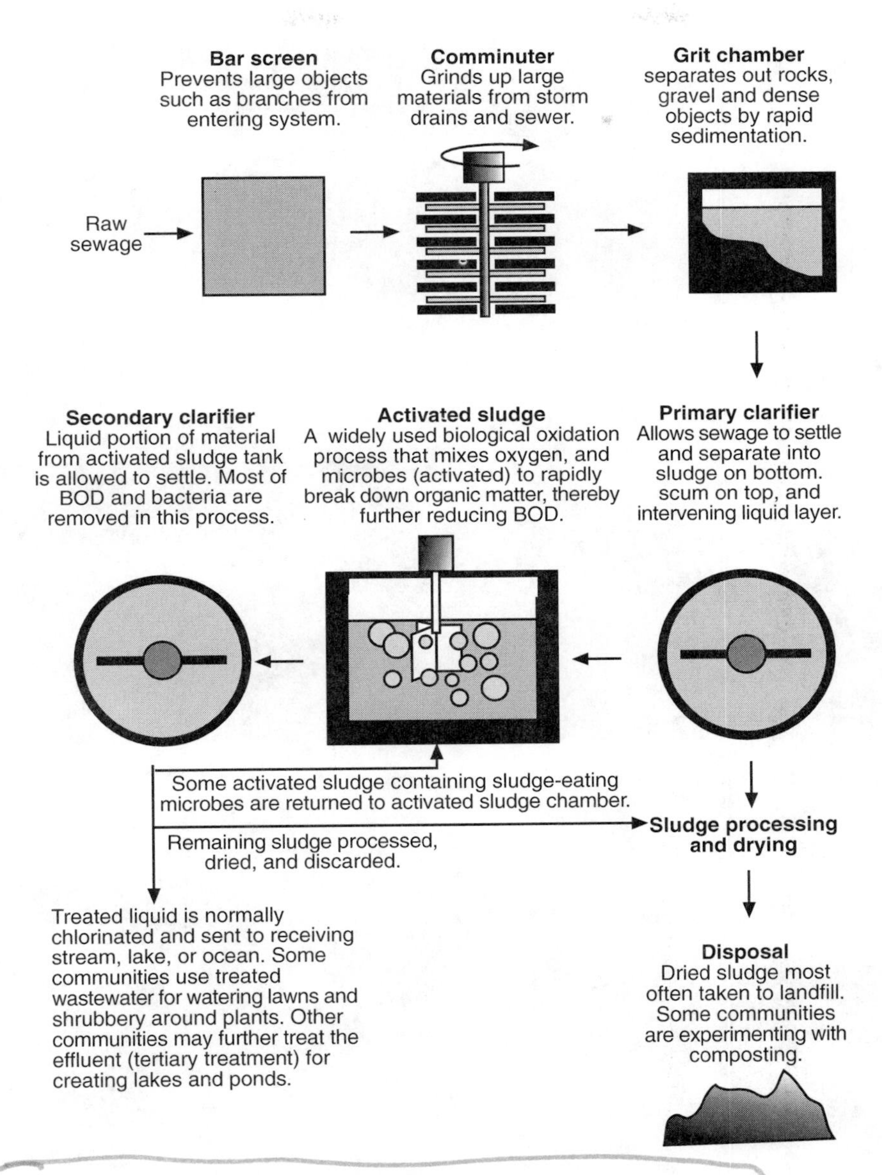
Bar screen
Prevents large objects such as branches from entering system.
Comminuter
Grinds up large materials from storm drains and sewer.
Grit chamber
separates out rocks, gravel and dense objects by rapid sedimentation.
Raw sewage
Secondary clarifier
Liquid portion of material from activated sludge tank is allowed to settle. Most of BOD and bacteria are removed in this process.
Activated sludge
A widely used biological oxidation process that mixes oxygen, and microbes (activated) to rapidly break down organic matter, thereby further reducing BOD.
Primary clarifier
Allows sewage to settle and separate into sludge on bottom. scum on top, and intervening liquid layer.
Some activated sludge containing sludge-eating microbes are returned to activated sludge chamber.
Remaining sludge processed, dried, and discarded.
Sludge processing and drying
Treated liquid is normally chlorinated and sent to receiving stream, lake, or ocean. Some communities use treated wastewater for watering lawns and shrubbery around plants. Other communities may further treat the effluent (tertiary treatment) for creating lakes and ponds.
Disposal
Dried sludge most often taken to landfill. Some communities are experimenting with composting.

Figure 9.13 A schematic of a typical municipal wastewater treatment plant.

and rapid granular filtration.[74] At this stage the effluent may or may not be disinfected before it is released into receiving waters. Controversy exists about disinfecting wastewater with chlorine. While the chlorine kills disease-causing organisms, it also may kill aquatic life when treated effluent is discharged into a waterway. Disinfection practice varies from state to state.[3]

Figure 9.14 Bar screen (left) and comminuter (right) at a municipal wastewater treatment plant.

Figure 9.15 A primary clarifier at a municipal treatment plant.

Figure 9.16 An activated sludge unit seen full and empty.

Sludge Treatment and Disposal

Sludge refers to the solids and liquids separated out of wastewater during sewage treatment. Wastewater treatment in the U.S. generates approximately 8 million dry tons of sludge annually. Sludge is generally composed of organic matter, pathogens, metals, and chemicals removed from wastewater. Treatment of sludge entails the

Studio dried
on huge
rotating barrel
that suctions
out moisture,
creating blanket
of dried sludge
Sludge used in composting
effort at the University of
Massachusetts

Figure 9.17 A sludge-drying apparatus and a sludge composting effort.

removal of liquids and digestion of the organic matter. The treatment process includes thickening to increase solid density, dewatering to remove liquid, stabilization to digest or neutralize the sludge, conditioning to promote larger particle formation, and disinfection. Sludge disinfection is a crucial step as it destroys pathogens to prevent the spread of disease. Digested sludge may be air-dried (Figure 9.17).[72,73]

In the past, sludge was either dumped in the ocean or buried in a landfill. Today, sludge disposal poses an increasing problem, since ocean disposal is no longer legal and landfill space is increasingly scarce. One solution to sludge disposal is to use it as fertilizer (Figure 9.17). However, some sludge may contain heavy metals. These should not be used for agricultural purposes. Such sludge may pose a risk to groundwater sources if improperly disposed in a landfill, as the metals or toxics may leach into the soil and water. Incineration presents another disposal option, although it generates air pollution, requires significant energy input, and creates ash waste.[74,76]

WATER POLLUTION AND HEALTH

The reality of polluted water can be both disruptive and devastating. In Westport, Massachusetts, wells contaminated with the cleaning solvents trichloroethylene and chlorobenzene forced homeowners to limit their exposure to their well water through drinking and cooking with bottled water, limiting shower time, and wearing gloves to wash dishes.[68] Increased risks of cancer and uncertainty about the future face consumers of polluted water. Unfortunately, only time can reveal the true dangers of contaminants in our waters, as little is known about the long-term effects of exposure to various water pollutants such as synthetic organic chemicals. Consumers armed with information about their water sources can protect themselves from exposure to polluted water. Worldwide, water pollution continues to plague many developing countries, inflicting immediate and long-term damage on the public's health.

FUTURE OUTLOOK

To adequately meet water demand in the future, we must recognize the value of this resource today. Increasing population, increased pollution, and competition for water challenge the goal of satisfying everyone's basic needs for water. Conflicting user groups, such as industry and agriculture, will vie for water rights in the face of scarcity. Effective water management, including conservation, accurate pricing, and education can aid efforts to use water efficiently. Throughout the world, the problems of pollution and scarcity will increase over time. The need to protect water from pollution will increase as fresh water supplies dwindle, both in the U.S. and worldwide. The notion of sustainable water use will continue to gain momentum in the face of a water crisis, encouraging conservation and preservation of our water resources. On a planet covered with water, yet plagued by waterborne disease, drought, and water mismanagement, we must promote conservation, efficiency, and frugality. Every drop counts.

REFERENCES

1. Turk, J. and Turk A., *Environmental Science,* 4th ed., Saunders College Publishing, Philadelphia, PA, 1988, chap. 17.
2. Grigg, N.S., *Water Resources Management,* McGraw-Hill, New York, 1996, chap. 1.
3. Nadakavukaren, A., *Our Global Environment,* Waveland Press, Prospect Heights, IL, 1995, chap. 14.
4. Chanlett, E.T., *Environmental Protection,* 2, McGraw-Hill Book Company, New York 1979, chap. 3.
5. Biswas, A., *Water Development and Environment Water Resources,* McGraw-Hill, New York 1997, chap. 1.
6. Johnson, B. K and Donahue, J. M., *Water Culture and Power,* Island Press, Washington, D.C., 1998, chap. 1.
7. Stille, A., The Ganges' next life, *The New Yorker,* Jan. 19, 58, 1998
8. Yates, L.D., A water supply development project in Quiescapa, Bolivia, *J. Environ. Health,* 3, 20, 1997.
9. World Health Organization, Implementation of the Global Strategy for Health for All by the Year 2000, 8th Report on World Health, 1993, chap. 6.
10. Gumpte, P., Water: Not a drop to drink, *Newsweek,* Mar. 29, 1999.
11. Anonymous, Thirsty planet, *Environ. Health Perspect.,* 104(l), 12, Jan. 1996.
12. Hutchings, C., Down to the last drop, *Geographical Mag.,* 71, 24, Aug. 1997.
13. Pinkayan, S., *Asian International Waters,* 4, Biswas, A.K. and Hashimoto, T., Ed., Oxford University Press, 1996, p. 1.
14. Verghese, B.G., Towards an Eastern Himalayan Rivers, in *Asian International Waters,* 4, Biswas, A.K. and Hashimoto, T., Eds., Oxford University Press, 1996, p. 29.

15. Nishat, A., Impact of Ganges Water Dispute on Bangladesh, *Asian International Waters,* 4, Biswas, A.K. and Hashimoto, T., Oxford University Press, 1996, p. 60.
16. Hinrichson, D., The world's water woes, *Intl. Wildl.,* 22, Jul/Aug, 1996.
17. Revkln, A.C., Chasing a deal on water with a few pitchers of beer, *New York Times,* Nov. 5, 41, 1995.
18. Grigg, N.S., *Water Resources Management,* McGraw-Hill, New York, 1996, p. 493.
19. Biswas, A. K., *Water for Sustainable Development of South and Southeast Asia in the Twenty-first Century,* 4, Biswas, A.K. and Hashimoto, T., Oxford University Press, 1996, chap. 2.
20. Hoffbuhr, J.W., Is there enough? *J. Am. Water Works Assoc.,* 6, Aug. 1999.
21. World Health Organization, Water Consumption and Sustainable Water Resource Management, Organization for Economic Co-operation and Development, 1998, p. 23.
22. Von Huben, H., *Water Sources,* 2nd ed., American Water Works Association, Denver, CO, 1995, chap. 5.
23. Winpenny, J., *Reforming Water Resources Policy,* Food and Agricultural Organization of the United Nations, Rome, 1995.
24. Lyman, P.L., Drought in Texas and Oklahoma stunting crops and economies, *The New York Times,* Aug. 12, 1998.
25. Von Huben, H., *Water Sources,* 2nd ed., American Water Works Association, Denver, CO, 1995, p. 98.
26. Easter, K.W., Becker, N., and Tsur, Y., *Economic Mechanisms for Managing Water Resources: Pricing, Permits, and Markets, Water Resources,* Biswas, A. K., McGraw-Hill, New York, 1997, p. 589.
27. Lindeburg, M.R., *Civil Engineering Reference Manual,* 6, Professional Publications, Inc., Belmont CA, 1997, p. 7.
28. Squandering the planet's lifeblood, *Intl. Wildl.,* 24, Sep/Oct 1999
29, Von Huben, H., *Water Sources,* 2nd ed., American Water Works Association, Denver, CO, 1995, chaps. 3 and 7.
30. World Health Organization, *Guidelines for Drinking Water Quality,* 2nd ed., World Health Organization, 1997, p. 93.
31. Koren, H. and Bisesi, M., *Handbook of Environmental Health and Safety,* 3rd ed., Lewis Publishers, Boca Raton, 1996, p. 273.
32. Koren, H. and Bisesi, M., *Handbook of Environmental Health and Safety,* 3rd ed., Lewis Publishers, Boca Raton, 1996, p. 257.
33. Whelan, E.M., *Toxic Terror: The Truth Behind the Cancer Scares,* Prometheus Books, Buffalo, NY, 1993, chap. 14.
34. USEPA, The Quality of our Nation's Water, Office of Water, www.epa.gov/enviro.
35. USEPA, *Manual of Individual Water Supply Systems,* U.S. Government Printing Office, Washington, D.C., 1973, p. 34.
36. Kahn, J., *Economic Approach to Environmental and Natural Resources,* Harcourt and Brace Co., New York, 1995, chap. 14.
37. USEPA, *Is Someone Contaminating Your Drinking Water?* Groundwater Protection Council, Oklahoma City, OK, 1997.
38. Uehling, M.D., How safe is your drinking water?, *Pop. Sci.,* 10, 63, 1996.
39. White, F Gilbert, The Environmental Affects of the High Aswan Dam, 1988.
40. Grigg, N. S., *Water Resources Management,* McGraw-Hill, New York, 1996, chap. 13.
41. Geyer-Allely, E., Water Consumption and Sustainable Water Resources, Organization for Economic Co-operation and Development, 1998, 28.
42. Pinholster, G., Drinking recycled wastewater, *Environ. Sci. Technol.,* 29, 174, 1995.

43. Asano, T., Wastewater Reuse, *Water Resources,* Biswas, A.X., Ed., McGraw-Hill, 1997, p. 383.
44. Frederick, K.D., Changing water resources institutions, *Sustaining our Water Resources,* National Academy Press, Washington, D.C., 1993, p.73.
45. Grigg, N. S., *Water Resources Management,* McGraw-Hill, New York 1996, chap. 14.
46. Palmer, T., *Lifelines: The Case for River Conservation,* Island Press, Washington, D.C., 1994, 117.
47. Nadakavukaren, A., *Our Global Environment,* Waveland Press, Prospect Heights, IL, 1995, chap. 15.
48. Pfarrer, S., Water quality mandate praised, *Daily Hampshire Gazette,* Northampton, MA, Aug. 13, 1, 1998.
49. Von Huben, H., *Water Quality,* American Water Works Association, Denver, CO, 1995, chap. 6.
50. Montague, Peter. Drugs in Water, Sep. 3, 1998.
51. Von Huben, H., *Water Quality,* American Water Works Association, Denver, CO, 1995, chap. 7.
52. Koren, H. and Bisesi, M., *Handbook of Environmental Health and Safety,* 3rd ed., Lewis Publishers, Boca Raton, 1996, chap. 3.
53. Ongley, E. D., *Control of Water Pollution from Agriculture, 55,* Food and Agriculture Organization of the United Nations, Rome, 1996, chap. 1.
54. Gallagher, L.M., *Clean Water Act, Environmental Law Handbook,* 14, Sullivan, T.F.P., Ed., Government Institutes, Inc., Rockville, MD, 1997, p. 114.
55. Tibbets, J., Water pressure, rewriting the Safe Drinking Water Act, *Environ. Health Perspect.,* 104, 930, Sep. 1996.
56. Kilborn, Peter T., Storm highlights flaws in farm law in North Carolina; a threat to water supply, *The New York Times,* Sunday Oct. 17, 1999.
57. Ongley, E. D., *Control of Water Pollution from Agriculture,* 55, Food and Agriculture Organization of the United Nations, Rome, 1996, chap. 3.
58. Krafter, C. R., Pesticides in Storm Runoff From Agricultural and Urban Areas in the Tuolumne River Basin in the Vicinity of Modesto, California, U.S. Geological Survey Water Resources Investigation Report 98-4017, 1998.
59. Annin, P., Down in the dead zone, *Newsweek,* Oct. 18, 60, 1999.
60. Larson, S.J., Capel, P.D., and Majewski, M.S., *Pesticides in Surface Waters,* 3rd ed., Ann Arbor Press, Inc., Chelsea, MI, 97, chaps. 1, 3, 4, 7.
61. Mitchell T.J. and Harding, A.K., Who is drinking nitrate in their well water?, *Environ. Health,* 11, 14, 1996.
62. Holton, C., Nitrate elimination, *Environ. Health Perspect.,* 104, 3 6, 1996.
63. Pitt, R., Clark, S., Parmer, Y., and Field, *R., Groundwater Contamination From Stormwater Infiltration,* Ann Arbor Press, Inc., Chelsea, MI, 1996, chaps. 1 and 2.
64. Kelly, M., *Mining and the Freshwater Environment,* Elsevier Science Publishers Ltd., Essex, England, 1988, chaps. 1 and 4.
65. Riley, M.B., Murphy, A.D., Rosado, M.A.M., A reversal of tides, *Water Culture, and Power,* Donahue, J.M. and Johnston, BR., Eds., Island Press, Washington, D.C., 1998, p. 237.
66. Wasik, J., Special Report: How safe is your water, *Consumers Dig.,* 35, 63, May 1996.
67. Thompson, E., Rotting carcasses sewage create mounting health in N.C., *The Associated Press,* Sep. 20, 1999.
68. Joyce, S., Capturing *Cryptosporidium, Environ. Heatlh Perspect.,* 104, 834, Aug. 1996.
69. Whelan, E.M., *Toxic Terror,* Prometheus Books, Buffalo, NY, 1993, chap. 11.

70. Pilkington, N.H., Disinfection: An overview, *Modern Techniques in Wastewater Treatment,* Kolarik, L.O. and Priestly, A.J., Eds., CSIRO Publishing, East Melbourne, Australia, 1995, p. 75.
71. Von Huben, H., *Water Quality,* 2, American Water Works Association, Denver, CO, 1995, chap. 1.
72. Williams, S.E., *Safe Drinking Water Act, Environmental Law Handbook,* 14, Sullivan, T.F.P., Ed., Government Institutes, Inc., Rockville, MD, 1997, chap. 6.
73. McGarry, Michael C., *Waste Collection in Hot Climates: A Technical and Economic Appraisal in Water Wastes and Health in Hot Climates,* John Wiley and Sons, New York, 1977, p. 243.
74. Barnes, D., Aerobic wastewater treatment systems, *Modern Techniques in Water and Wastewater Treatment,* Kolarik, L.O. and Priestley, A. J., Eds., CSIRO Publishing, East Melbourne, Australia, 1995, p. 92
75. Casey, T. J., *Unit Treatment Processes in Water and Wastewater Engineering,* John Wiley & Sons, New York, 1997, chap. 16.
76. Cheremisinoff, P., *Sludge Management and Disposal,* Prentice Hall, Englewood Cliffs, NJ, 1994, chap. 1.
77. Humphrey, P., Water fears contaminate Westport, *The Boston Globe,* Feb. 26, 41, 1995.

CHAPTER 10

Air, Noise, and Radiation

INTRODUCTION

The problems associated with air pollution are not recent. The obnoxious stench of burning coal was so annoying to Edward I and Edward II of Great Britain that those who caught burning coal were at risk of losing their lives. In fact, a man was hanged during the reign of Edward II for burning coal and fouling the air.[1] Such rigorous penalties have not been witnessed since then. However, air pollution has continued to be a problem that worsened with the combined effects of an increasing population and the industrial revolution. Generally, air pollution was considered to be an annoyance, resulting in respiratory irritation, deposits of dust and filth on surfaces, and reduced visibility. However, the potentially severe consequences of air

OBJECTIVES FOR THIS CHAPTER

A student reading this chapter will be able to:

1. **List and explain the reasons why air pollution is considered a national and global threat**
2. **Discuss and describe the chemical and physical components of the atmosphere, and explain the mechanisms of dispersion**
3. **Describe the regulatory efforts in the U.S. with emphasis on titles of the 1990 CAAA**
4. **Discuss the issues behind stratospheric ozone depletion and global warming**
5. **List and discuss the nature, sources, and health and welfare effects of the criteria pollutants**
6. **List, discuss, and describe the major sources of indoor air pollution, including health effects and methods of control**
7. **Define noise pollution and radiation. List the major sources and known health effects of noise and radiation**

pollution were thrust into the vision of the public eye in documented episodes that have become "classics" in the air pollution literature. These episodes began in the first week of December, 1930, in the Meuse River Valley, Belgium. This deep river valley was highly industrialized with manufacturing plants in steel, glass, lime, fertilizers, and sulfuric acid. That week in December the air became stilled and the continuously emitted pollutants in the valley were trapped. The residents complained of irritated throats, coughing, and tightness in the chest. More than 60 deaths were attributed to the episode, primarily among the elderly and those with preexisting cardiorespiratory disorders. Nearly 18 years later in Donora, Pennsylvania, similar conditions prevailed with steel manufacturing concentrated in a river valley. A stable, slow-moving air mass resulted in the formation of a dense, cool layer of air near the Earth with warmer air above in a phenomenon known as an inversion. The noxious pollutants of industry were trapped below the interface of the warm air above and the cool air below and began accumulating on Wednesday, October 28. The first death from this episode occurred on that Saturday, and nearly 20 people died by Sunday evening. Six thousand people, or 40% of the population, developed symptoms associated with the episode including nausea, vomiting, irritation to the eyes, nose and throat, headaches, and muscular aches and pains. The severest symptoms and the most deaths were among the elderly and those with preexisting respiratory and cardiovascular conditions. The lungs of those who died showed evidence of capillary dilation, hemorrhaging, purulent bronchitis, and edema.[2,3]

Excess deaths were also demonstrated in London, England, during several air-pollution episodes associated with stagnant air. These episodes occurred in 1952, 1956, 1957, and 1962. London's worst episode was in 1952, during which a thick yellow smog settled over the city for 5 days, resulting in the deaths of 4000 more people than the number expected in the absence of air pollution. The deaths were mostly among the elderly and those with preexisting respiratory or cardiac problems. The excess deaths spiked with the air pollution episode and subsided when the stagnant air was eventually swept away by more turbulent conditions.

Episodes of air pollution resulting in measurable levels of excess deaths have also occurred in Los Angeles and New York, primarily in the 1950s and 1960s. The air pollution disasters focused public attention on the severity of air pollution. It became clear that exposures to concentrated levels of ambient pollutants could cause severe symptoms and even death. Such awareness prompted greater attention to studying the factors associated with air pollution and then establishing regulations to control and limit the pollutants. Such efforts have made it unlikely that similar air pollution episodes will occur in industrialized countries. However, countries such as China, Mexico, and Eastern Europe are facing substantial air pollution problems as they strive for industrial growth in the absence of vigorous air pollution control policies.

Although progress in controlling air pollution in the developed nations has been made, problems of global concern remain. Government leaders and scientists are building toward a consensus in the midst of controversy that industrial emissions such as carbon dioxide and other greenhouse gases are contributing to global warming with potentially severe consequences. Emissions of nitrogen and sulfur oxides are thought to be major contributors to acid deposition — which is associated with the disappearance of fish and other biota in the tens of thousands of lakes and streams worldwide — and the decimation of high-altitude forested areas. Chlorofluorocarbons, once widely used in air conditioners, electrical solvents, and the production of foam cups have been linked to to the catalytic destruction of stratospheric ozone in a process known as ozone depletion. Such depletion is thought to result in higher levels of damaging ultraviolet radiation reaching the Earth. These atmospheric effects of air pollution have mobilized governments internationally to meet and attempt to resolve these issues. Actions that may be taken are likely to have far-reaching effects on economies, standards of living, and the livelihood of hundreds of millions of people. The quality of our lives, the stability of our environment, and the survivability of the planet have been brought forward as arguments in these controversial issues. Is there a global warming problem? Is stratospheric ozone being depleted? Is acid deposition killing forests internationally and destroying aquatic life? These issues are presented in this chapter in a scientific and objective framework for the reader to draw an independent conclusion.

Air pollution not only represents a perceived threat to global ecological damage but also exerts both acute and chronic threats to human health at exposures within the normal range of pollutants in our cities and towns.[4] These threats include respiratory disorders such as asthma, bronchitis, and emphysema. There are also increased risks of cardiovascular disease, cancer, and respiratory infections. Air pollutants have been implicated in direct injury to living plants; to materials including the corrosion of metal; soiling of buildings; and the degradation of paints, leather, paper, textiles, and dyes. These effects result in significant economic losses approaching several billion dollars annually.[5] Based on the overview given above, the arguments for controlling air pollution are substantial. The student of air pollution requires an understanding of its mechanisms, which must begin with a discussion of the atmosphere and methods of pollution dispersion.

THE ATMOSPHERE AND METHODS OF DISPERSION

Chemical Characteristics

The Earth's atmosphere consists of a mixture of gases, water vapor, and both solid and liquid particles. It is a dynamic system with most components leaving and entering on a continuing basis and with a wide variety of chemicals created in the atmosphere in the presence of sunlight, catalysts, and/or moisture. In spite of this continuing exchange of atmospheric ingredients, some of the gases remain relatively constant in their proportions. Nitrogen (N_2) represents a constant 78% of the 500 billion tons of air surrounding the planet, while oxygen (O_2) remains steady at 21% and argon (Ar) at 0.9%. These three gases collectively represent 99.9% of the atmosphere. The remainder of gases in constant proportion consist of traces of neon (Ne), helium (He), hydrogen (H_2), xenon (Xe), and krypton (Kr) (Figure 10.1). The atmosphere also contains substances that tend to be variable in their concentration and include water vapor, carbon dioxide, carbon monoxide, methane, ozone, ammonia, and nitrogen oxides. Many of these latter compounds enter the atmosphere from both anthropogenic and natural sources and include substances such as carbon monoxide, ammonia, and nitrogen oxides, which are regulated as pollutants. Additionally many other substances enter the atmosphere from human activities. Those that have the capacity to cause human disease and welfare effects resulting in disturbances to plant life and physical structures are collectively known as air pollutants.

Although the main components of the atmosphere, nitrogen and oxygen, have remained relatively constant for millions of years, current scientific thought suggests this was not always the case. The primitive atmosphere is thought to have been rich in carbon dioxide with little free oxygen. Nearly 2 billion years ago, green plants evolved that used carbon dioxide to grow by photosynthesis, producing energy-rich sugars and a waste by-product now known as oxygen. Organisms adapted to the presence of this waste oxygen and developed into the aerobic species of today,

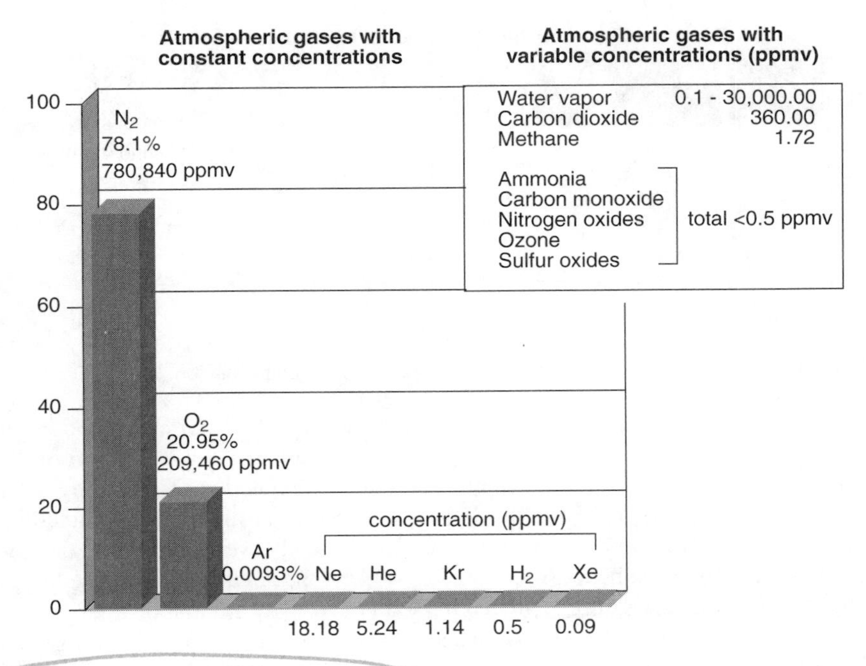

Figure 10.1 Atmospheric components.

including humans.[6] The Earth's atmosphere and climate have achieved a dynamic balance through the removal and recycling of carbon, nitrogen, oxygen, and other substances that have sustained plant and animal life for millions of years. Human technology and explosive populations could potentially alter this balance, causing changes in the Earth–atmosphere system that jeopardizes our sustainability.

Physical Characteristics

Solar Radiation

The life on Earth requires a continuing source of energy. That energy is supplied by the sun, which radiates enormous electromagnetic energy at effective temperature of 6000°F and travels in less than 10 minutes to the Earth at 186,000 miles per second.[7] More than 99% of the energy from the sun is within the spectral range of 150 to 4000 nanometers (0.15 to 4.0 μm). Fifty percent of the energy striking the Earth is in the infrared spectrum exceeding 0.7 μm (700 nanometers), while slightly more than 40% is in the visible spectrum of 400 to 700 nanometers (0.4 to 0.7 μm), and about 9% is in the energetic and shorter wavelengths of 150 to 400 nanometers (0.15 to 0.4 μm) (Figure 10.2).[7,8] The solar radiation reaching the Earth's atmosphere is absorbed, scattered, and reflected so that an average of less than 50% reaches the Earth's surface. The upper and lower atmosphere absorbs and scatters 15% to 20% of the total incident solar radiation. When skies are clear, carbon dioxide and water

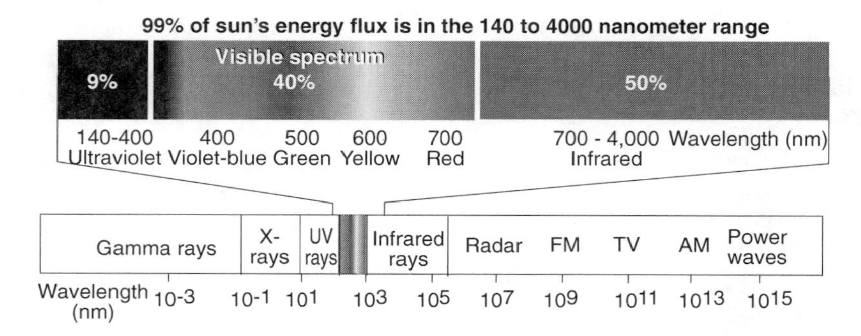

Figure 10.2 Electromagnetic spectrum and components of solar radiation.

vapor strongly absorb infrared radiation, while clouds and dust particles in the lower atmosphere scatter and diffuse visible light waves. Clouds cover a major part of the Earth at most times, causing significant scattering and reflection of incoming solar radiation. Solar radiation is also reflected back to space by ground surfaces including oceans and lands. The combined reflective ability of cloud cover and ground surfaces is known as the Earth's albedo. About 30% to 40% of the solar radiation is reflected by this albedo, with most of it attributed to cloud cover. The amount of solar energy absorbed must ultimately be radiated back to space in order to maintain a global radiation balance, otherwise the Earth would soon be overheated. This is not normally a uniform process, but on average the annual global temperature varies less than one degree centigrade. Solar energy that is absorbed by ground surfaces is radiated back as heat in longer, lower-energy, infrared wavelengths. You can observe this effect if you look down a long stretch of blacktop roadway on a hot summer day and see the waves of infrared energy rising from the pavement and distorting the scenery in the distance. The blacktop absorbed the higher energy (wavelengths of the visible range) and radiated back longer wavelengths (infrared radiation in the form of heat). If this reflected heat energy (infrared) is absorbed by infrared-absorbing gases such as carbon dioxide, chlorofluorocarbons (CFCs), methane, and water vapor, then it traps the warmth and reflects it back to the Earth's atmosphere in a process known as the greenhouse effect (Figure 10.3). If the greenhouse gases increase in concentration, it is logical that more heat energy will be absorbed and the average annual global temperature may rise, causing a global warming trend. This is the focus of world attention and a matter of discussion later in this chapter.

Vertical Temperature Differences and Atmospheric Regions

Imagine yourself in the basket of a hot air balloon lifting off the ground with a thermometer in your hand. You would normally experience a declining temperature as you gained altitude at a rate of about –6.5°C/km, or a loss of about 65°C over the zero to 10 km altitude range of this region known as the troposphere (Figure 10.4). Of course, when you enter levels of 6 km or higher, the air becomes so thin that you've entered the "killing zone," where supplemental oxygen is

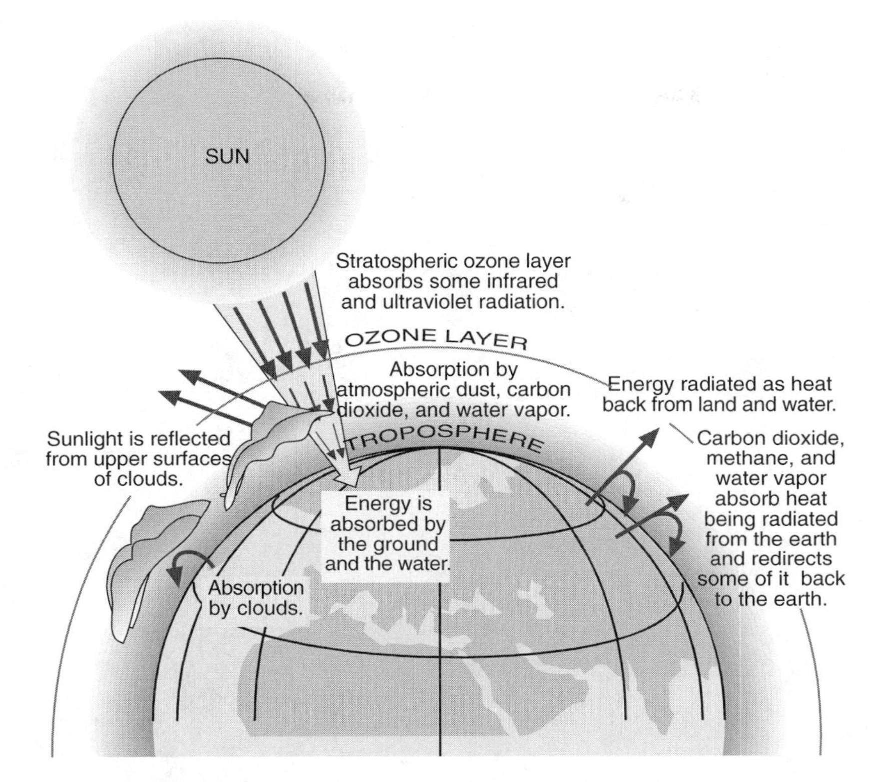

Figure 10.3 Mechanisms of reflection and absorption of solar radiation.

required. The thickness of the troposphere is greater over the equator (12 km), and lesser over the poles (8 km). The troposphere is where life activities occur and where the changes and manifestations in weather and climate are driven by variations in thermal energy, friction, and global rotation (Coriolis force). The end of the troposphere is marked by stationary or isothermal temperature as you increase in altitude from 10 km to 20 km in a region known as the tropopause. The temperature begins to increase to 0°C beyond this region to an altitude of about 50 km. This area is known as the stratosphere, and contains concentrations of ozone approaching 200 parts per billion (ppb) in the densest region at 24 km to 40 km (15 to 24 miles). Ozone (O_3) is an isotope of oxygen that absorbs infrared radiation, causing the elevated temperature of the stratosphere. Ozone also strongly absorbs ultraviolet radiation in the UB-B region of 230 to 320 nanometers, thus restricting damaging levels of UB-B from reaching the Earth and causing increased skin cancer in humans. The stratospheric ozone concentrations vary geographically and seasonally. Peak ozone levels are found between autumn and spring and are normally most dense over the poles. Levels above the stratosphere include the stratopause, mesosphere, mesopause, and thermosphere. These regions are in the outermost atmosphere where light molecules are lost to space, and the relatively few molecules of N_2 and O_2 are

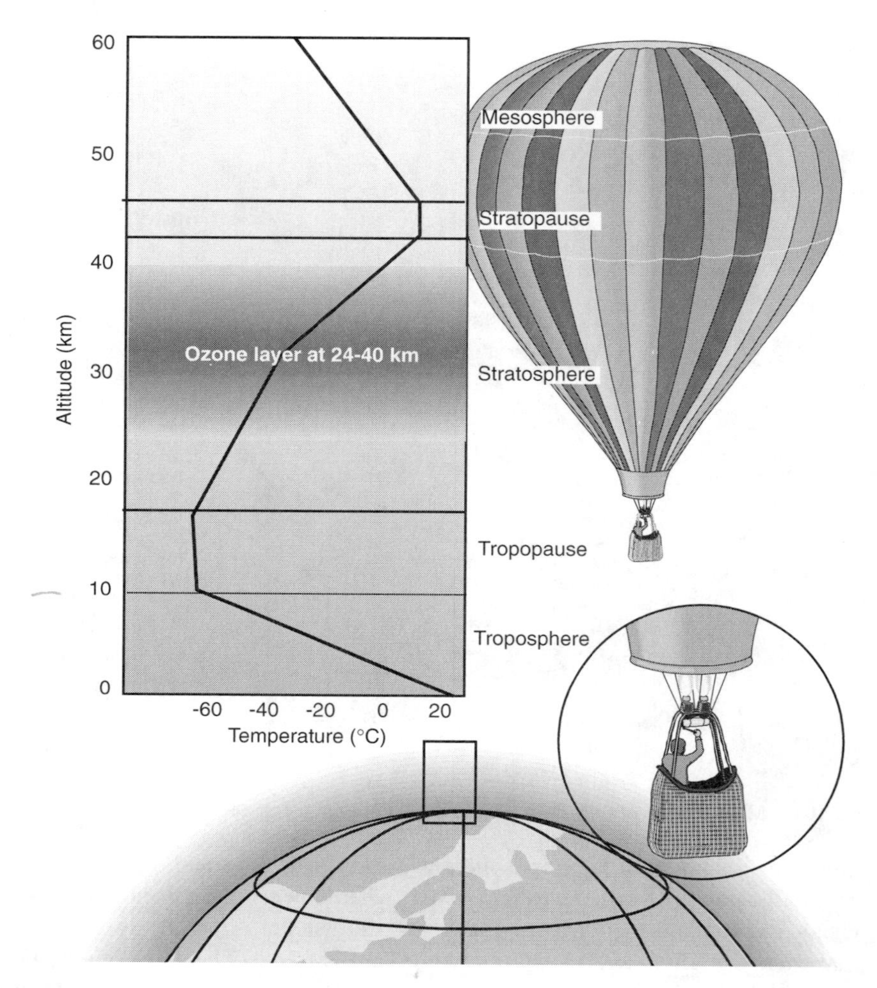

Figure 10.4 Atmospheric regions and temperature profile. (Adapted from Chamber, L., *Classification and Extent of Air Pollution,* Stem, A.C., Ed., 3rd ed., Academic Press, New York, 1976, chap. 1.

widely separated and may become photoionized. The region at 90 km to 400 km is also known as the ionosphere, where the Earth's magnetic field captures protons and electrons emitted by solar storms resulting in the aurora borealis (northern lights) and auroroa australis (southern lights).[7]

Atmospheric Pressure and Density

Gravity pulls the atmosphere to Earth, with the densest part of the atmosphere closest to the surface. About 99% of the atmospheric mass is below 30 km (18 miles); 90% is below 12 km; and 75% of the atmosphere is below 10 km (Figure 10.5). A column of air exerts a pressure on an object at sea level of 14.7 lbs/in2 (PSI). This

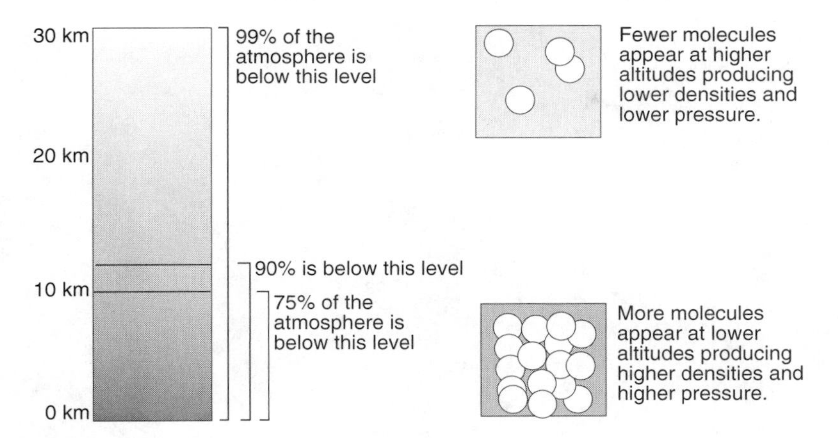

Figure 10.5 Atmospheric density as a function of altitude.

pressure can be demonstrated by creating a vacuum in a plastic bottle. The external atmospheric pressure will cause the bottle to collapse. Pressure is defined as the force per unit area and is related to the density of the air such that there is both low pressure and low density at higher altitudes. This can be demonstrated by boiling water first at sea level and then at high altitude. The temperature remains steady at 100°C as gases are released from the boiling water. Boiling the water atop a high mountain peak will occur at a lower temperature of perhaps 90°C, since the pressure is lower and gaseous vapors can escape more readily under lower pressure (Figure 10.6). Atmospheric pressure is measured by a barometer and normally expressed in millimeters or inches of mercury. The pressure at sea level is equal to 760 mm mercury (Hg), 29.92 inches of mercury, or 1010 millibars. Pressure and atmospheric density in the troposphere are not constant and vary in response to changes in thermal radiation and air movement. Air that is warmed rises in contrast to the cooler air around it. A hot-air balloon uses a propane burner to heat the air inside a fabric envelope, causing the balloon and its passengers to rise. The energy driving air movement in the atmosphere is the sun, which strikes the Earth at varying angles, warming the air unevenly. Warmer air becomes less dense, expands, and rises, thereby creating a column of air that is lower in pressure. Cooler air is more dense, settles, and is characterized by higher pressure. Differences in pressure cause air to move in both horizontal and vertical patterns. Air tends to move from regions of high pressure to low pressure. Therefore, as the sun heats a body of air and it rises, a region of low pressure is created into which flows cooler air. The greater the temperature differences, the larger the pressure differences will be, and the more rapid or violent the flow of air will be. If the air flows horizontally, this is known as wind. The direction of the wind is influenced by the rotation of the Earth (Coriolis force, Chapter 1), friction, and the differential warming provided by the sun. Such forces cause air to flow into regions of low pressure in a cyclonic motion and then to rise. This cyclonic motion is counterclockwise in the Northern Hemisphere and clockwise in the Southern Hemisphere (Figure 10.7). When cool air descends in the

90°C
Lower pressures at higher altitudes permit vapors from boiling water to escape more easily, thereby keeping the boiling temperature lower.
100°C
Higher pressures at lower altitudes restrict the escape of vapors from boiling water thereby keeping the boiling temperature at 100°C at sea level.

Figure 10.6 The demonstration of atmospheric pressure at low and high altitudes.

Northern Hemisphere, it is radiated outward in a clockwise motion known as an anticyclone (Figure 10.7). The direction is reversed to a counterclockwise motion in the Southern Hemisphere. These represent very large-scale air movements and play important roles in the dispersion of air pollutants. Low-pressure migrating cyclones usually are associated with inclement weather including precipitation, clouds, and windy conditions. High-pressure systems normally signal cooler, dry air with sunny conditions.

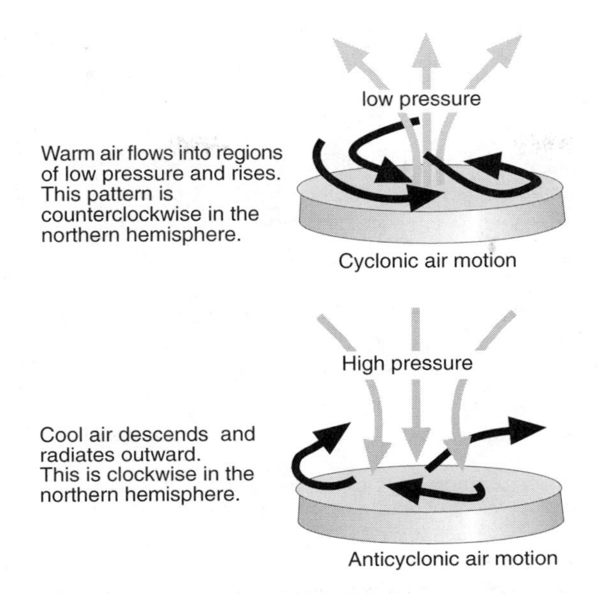

Figure 10.7 Cyclonic air motion in a low pressure system and anticyclonic air motion in a high-pressure system in the Northern Hemisphere.

Atmospheric Inversions

There is an ample supply of naturally occurring and anthropogenic pollutants in the atmosphere. They are normally removed by chemical reactions in the atmosphere, dry or wet deposition, and microbial conversions known collectively as sinks. Additionally, such pollutants are normally dispersed by wind and turbulence to very low concentrations that are unlikely to produce adverse health effects. However, sometimes atmospheric conditions become very stable and slow-moving, causing the pollutants to build in concentration. The concentration of pollutants is markedly influenced by meteorological conditions. The stability of the air directly impacts the mixing and transport of pollutants along with chemical reactions in the atmosphere that form new categories of pollutants. A stable air mass is slow-moving with little turbulence and permits the accumulation of pollutants. Certain meteorological conditions promote stability. Under normal conditions in the troposphere, air that is warmed rises as long as the air above and around it is cooler. The warmer the air mass in relation to its surroundings, the more rapidly it will rise. As the air rises, it will lose heat from such mechanisms as radiative cooling at an average of –6.5°C/km, which is termed the normal lapse rate. Under theoretical conditions a parcel of warm, dry air would cool at the rate of –10°C/km, and this is referred to as the dry adiabatic lapse rate. The actual change of temperature with altitude is referred to as the environmental lapse rate, and it is a direct measure of atmospheric stability. This stability can be measured by placing a thermometer at ground level and another several meters in the air to determine the environmental lapse rate (Figure 10.8). If the air cools with altitude at a rate faster than the dry adiabatic lapse rate of

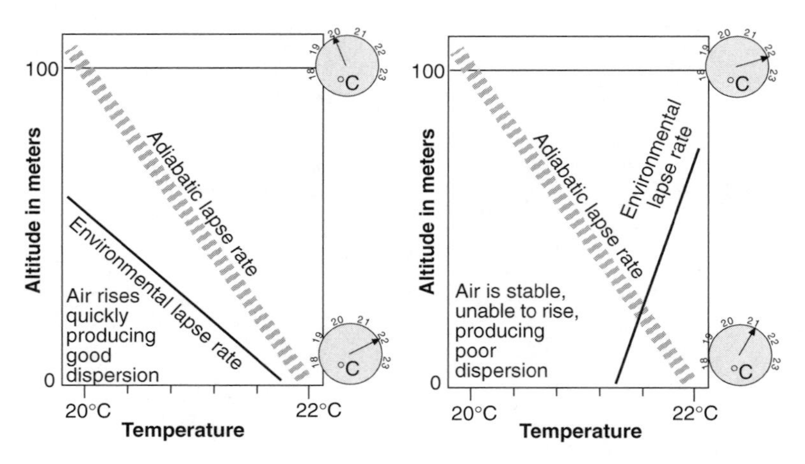

Figure 10.8 Dispersion characteristics of air having environmental lapse rates that are cooler (good dispersion) or warmer (poor dispersion) than the adiabatic lapse rate.

–10°C/km, then the vertical motion of the air will be quite rapid, permitting cooler air to sweep in and replace it, thereby creating turbulent conditions with good dispersion of pollutants. Atmospheric turbulence promotes pollution dispersion and results from mechanical obstruction and atmospheric heating. The higher a column of air can rise, the greater the maximum mixing depth (MMD) will be. The MMD varies with the time of day, the season, cloud cover, geography, and the presence of anticyclones. Mixing depths tend to be greater during the daytime, in summer, and in geographical areas where there is little ground cover or water to absorb the heat, such as deserts or the Great Plains. However, if the air temperature does not cool sufficiently with increasing altitude, or even gets warmer at higher elevations, the upward movement of air is prevented, stable conditions predominate, and dispersion is poor (Figure 10.8).[3,9] The warming of air with increasing elevation is a reversal of the normal tropospheric cooling process and is known as atmospheric inversion. It is the most common atmospheric condition associated with stability. The cooler, denser air near the ground is unable to rise through the warm layers above it and is therefore trapped until turbulent conditions can be reestablished. There are two primary types of inversions.

Radiation inversion normally occurs at night when heat absorbed by the Earth during the day radiates out quickly, leaving the air nearest the ground cooler than the radiated warmer air above. These inversions are normally not persistent and are easily broken up when the sun rises in the morning to cause convective currents. Such short-lived inversions seldom present a health hazard since there is little time to accumulate pollutants.

Subsidence inversion refers to the descent of air masses. Cool air masses (anticyclones) associated with Hadley cells descend, causing air molecules to compress in a layer above the ground. This layer becomes warmed by compression while the air at ground level remains unchanged and often cooler than the air above, producing an inversion layer at distances of 500 to 1000 meters above the ground (Figure 10.9).

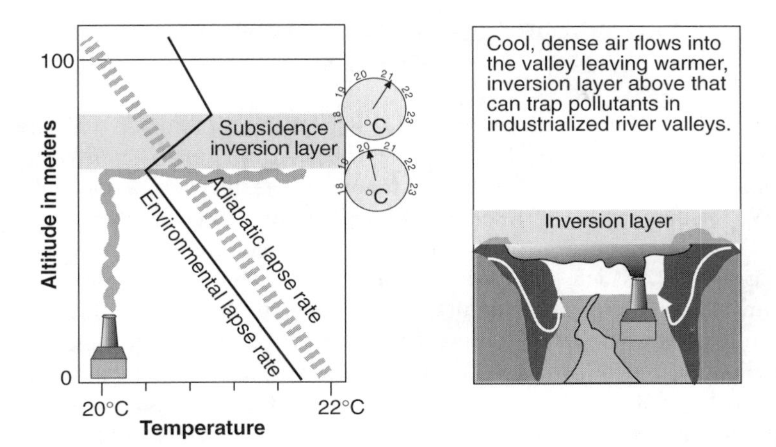

Figure 10.9 Environmental lapse rate conditions associated with subsidence inversion, and nocturnal inversion occurring in a river valley.

Such inversions may occur frequently along coastal areas such as California, which has documented inversions more than 300 days of the year. Such subsidence inversions occur most often during the fall and winter months and are particularly troublesome since they may persist for days and permit the build-up of potentially dangerous levels of pollution.

Nocturnal inversions also occur when the cool, dense air along mountain tops descends into the river valleys during the evening hours and accumulates under a layer of warm air above. The shaded valleys are often covered with mists or clouds, resisting penetration by sunlight until mid-day when the sun is directly overhead, where it can cause convection currents to break up the inversion layer. When such conditions combine with migratory anticyclones, the inversions become very persistent and hazardous. The stagnation of air produced by such migrating anticyclones was responsible for the classic air pollution disasters of Meuse Valley, Belgium, in 1930, Donora, Pennsylvania, in 1948, and others occurring in England, New York, and Los Angeles.[3,9,10]

Inversions are not alone in meteorological events that influence air pollution. Sunlight and humidity are important in forming chemical reactions in the atmosphere, producing secondary pollutants such as ozone created from nitrogen oxides in the presence of sunlight. Nitrogen oxides are considered primary pollutants since they are derived from identifiable sources such as smokestacks or automobile exhausts. Rain, sleet and snow serve as sinks to remove pollutants from air through absorption or adsorption, carrying pollutants to the ground or bodies of water where they may cause harmful effects. Pollutants emitted into an air mass may not widely disperse, and a migrating air mass may accumulate pollutants as it moves great distances in a process known as long-range transport. Much of the acidity from atmospheric deposition falling on the northeastern U.S. has been attributed to industries and power plants in the Midwest, burning high sulfur coal.

THE HISTORY OF AIR POLLUTION CONTROL IN THE U.S.

The control of air pollution was seen as a local control issue up until the 1950s, when a federal role began to slowly evolve. It was in 1955 that Congress authorized the Public Health Service in the Department of Health, Education and Welfare (DHEW) for the purpose of researching air pollution and conducting training programs. However, air quality continued to grow worse and Congress passed the Clean Air Act of 1963, which gave federal authority to develop air quality criteria for protecting the public health; conduct research on sulfur dioxide pollution; provide grants to establish state air pollution control agencies; and to provide more research, training, and technical assistance with regard to air pollution. Amendments to this Act in 1965 established the National Air Pollution Control Agency (NAPCA) and set emission standards for 1968 model light-auto motor vehicles. In spite of these regulatory initiatives, air pollution continued to worsen; and the 1963 Clear Air Act proved inadequate to deal with the problem. The response to Congress was to pass the Comprehensive Air Quality Act of 1967, which represented the first attempt to develop a regional approach for the control of air pollution through the designation of Air Quality Control Regions (AQCRs). However, states still kept the main responsibility for enforcement and were expected to develop air quality standards and plans for implementing them. Progress proved to be slow; and as Americans celebrated "Earth Day" for the first time on April 22, 1970, public attention to environmental concerns reached a high level, stimulating Congress to strengthen air pollution control initiatives. This time Congress passed the 1970 Clean Air Act Amendments (CAAA) which dissolved NAPCA; transferred air pollution control activities to the Environmental Protection Agency (USEPA) which the President created by executive order; increased federal enforcement authority; established uniform National Ambient Air Quality Standards (NAAQS) for six criteria pollutants; required states to develop plans for implementing programs to achieve NAAQS in the states and submit the State Implementation Plans (SIPs) to the USEPA; and gave American citizens the right to sue private and government entities to enforce air pollution requirements. Additional amendments were made in 1977 to the 1970 CAAA that postponed and extended many compliance deadlines for air-quality standards and auto-emission standards.[11]

However, the 1977 CAAA did authorize the USEPA to regulate chemicals suspected of destroying stratospheric ozone and also provided for the protection of clean-air areas by preventing further air pollution through the concept of PSD or Prevention of Significant Deterioration.[11] Despite these efforts and some success under pollution control, three major threats to the nation's environment and to the health of millions of Americans defied these regulatory initiatives. These threats included acid rain, toxic air emissions, and urban air pollution. Additionally, the threats to the global environment persisted in the form of global warming by greenhouse gases and the continued destruction of stratospheric ozone by ozone-depleting chemicals. President George H.W. Bush proposed major revisions to the Clean Air Act in June of 1989, which were passed by Congress and signed into law on November 15, 1990. These revisions are known as the Clean Air Act Amendments of 1990. Many of the provisions of this Act are still coming on line as there are many progressive and creative new directions included in the Amendments.[12] There are several titles to the Act that are intended to produce a healthy, productive environment combined with a sustainable energy policy and economic growth.

Titles of the Clean Air Act

Title I: Provisions for Attainment and Maintenance of the NAAQS

Building upon the 1970 and 1977 CAAA, the 1990 CAAA attempted to strengthen the provisions protecting the public against seven of the most widespread and common pollutants designated as criteria pollutants (Table 10.1). The criteria

Table 10.1 National Ambient Air Quality Standards

POLLUTANT	AVERAGING TIME	PRIMARY STANDARD	MAIN SOURCES
Carbon monoxide	8 hours	9.0 ppm	Transportation
Hydrocarbons (corrected for methane)	3 hours (6–9 am)	160 μg/m3	Transportation, Industrial processes
Nitrogen dioxide	Annual average	0.05 ppm	Stationary source fossil fuel combustion, Transportation
Sulfur dioxide	Annual average	0.03 ppm	Stationary source fossil fuel combustion
Particulates			Multiple sources including stationary source fuel combustion, industrial processes, and transportation; fine particles are associated with photochemicals products of fossil fuel combustion
PM_{10}	Annual arithmetic mean	50 μg/m3	
$PM_{2.5}$*	Annual arithmetic mean	15μg/m3	
	24 hr. average	65μg/m3	
Ozone	1 hour 8 hour*	0.12 ppm 0.08 ppm*	Secondary pollutant formed in presence of sunlight, NOx, and hydrocarbons; fossil fuel combustion is a major contributor
Lead	3 months	1.5μg/m3	Food, dust, older houses with lead paint

•July 1997 amendments to the 1990 CAAA.

pollutants are those for which standards have been set establishing maximum concentrations allowed in the ambient or outdoor air environment known as NAAQS. The USEPA promulgated standards for six classes of pollutants in 1971, and an air-quality standard for lead was promulgated in 1978. These standards are based upon Air Quality Criteria which document the relationship of these pollutants to effects on health and welfare, therefore providing a scientific basis for the standards.[11]

Over 100 million Americans live in cities that fail to meet the NAAQS for ozone, and many people continue to be exposed to unacceptable levels of carbon monoxide and particulates. The 1990 CAAA established these as nonattainment areas, which were first described in the 1970 CAAA and refer to those AQCRs as designated by the USEPA that exceed the levels of one or more of the criteria pollutants during the year. The USEPA ranked these areas according to the severity of an area's air pollution problem. A nonattainment status carries serious implications because new industries or sources that might contribute to the pollution would be prohibited and therefore limit economic development. This was economically very confining. The USEPA, therefore, developed an offset policy that allowed an industry to build a new facility if it achieved agreements with nearby sources to reduce their emissions so that no net increase in pollution levels occurred. Alternatively, new sources could be developed in a nonattainment region if the pollution emission rates could be reduced maximally by the best available control technology (BACT) as designated by the USEPA. Under the 1990 CAAA, the USEPA classified nonattainment areas for ozone, carbon monoxide, and particulates based on the extent to which the NAAQS is exceeded. The 1990 CAAA further establishes specific pollution controls and attainment dates for each control. Carbon monoxide and PM_{10} are classified as either moderate or serious nonattainment areas, while ozone has five nonattainment classifications ranging from marginal (Albany, NY) to extreme (Los Angeles, CA). The more severe the nonattainment problems, the more controls the area needs to apply and the more time it is allotted to do so.

Title II: Provisions Relating to Mobile Sources

Most automobiles emit up to 80% less pollutants than in 1960, but they still account for the greatest combined amount of criteria pollutants including carbon monoxide, hydrocarbons, and nitrogen-oxides. The main reason for this is the proliferation of vehicles on the nation's highways. More than one in every five households in the U.S. has three or more cars, and the miles traveled are projected to grow by 2% each year through 2010 as Americans no longer just travel from home to work but link their trips to shopping, after-school activities, day care, and many other activities. Efforts by government agencies to discourage the use of single-occupant vehicles (SOVs) has been generally discouraging since Americans value their travel independence and continue to enjoy the luxury of inexpensive fuel, affordable cars, and free highways.[13] To combat this problem, the 1990 CAAA reduced the allowable emissions for carbon monoxide, hydrocarbons, and nitrogen-oxides from vehicle tailpipes through tighter inspection and maintenance (I/M) programs. The amount of evaporation of gasoline while refueling at the pumps has also been reduced through the use of specially designed recapture nozzles. Programs requiring cleaner reformulated gasoline and/or oxygenated fuels are also being phased in throughout areas of the country in nonattainment areas for ozone and carbon monoxide, respectively.

Methyl-t-butyl ether (MTBE) is the biggest volume oxygenate used in gasoline and contains at least the 2.7% O_2 by weight required to reduce CO emissions. The additives must be used during the winter months of November through December when levels of CO peak. The amount of MTBE has increased from the 1.8 billion gallons used in 1992 since the introduction of a program to use reformulated gasoline to combat ozone depletion.[14] Additionally, 26 of the country's most polluted areas are regulating companies and agencies with centrally fueled fleets of 10 or more vehicles such that vehicle emissions are further reduced. This process began in 1998.

Title III: Air Toxics

Congress recognized that very hazardous chemicals other than the criteria pollutants were being emitted into the nation's air and directed the USEPA under the 1970 CAAA to establish a list of such chemicals along with standards and control techniques. Unfortunately, the USEPA acted very slowly over 20 years, listing only eight hazardous air pollutants (HAPs) including arsenic, asbestos, benzene, beryllium, mercury, radionuclides, vinyl chloride, and emissions from coke ovens. Congress was aware that more than 2.7 billion pounds of toxic air pollutants were being emitted annually in the U.S. and that exposure to such substances may be causing up to 3000 cancer deaths annually.[12] They obtained the information that industry was required to supply under the Superfund "Right to Know" rule (SARA, Section 313). This rule was passed in 1986 in response to the deaths of nearly 2000 people in 1984 at Bhopal, India, because of the accidental release of toxic chemicals from a chemical manufacturing facility. This heightened concern prompted Congress to list 189 toxic air pollutants for which emissions must be reduced. The USEPA was required to publish a list of source categories that omit certain levels of these pollutants and to issue standards based on the best demonstrated control technology within the regulated industry, known as the Maximum Achievable Control Technology (MACT). The source categories were to be controlled by the year 2000, although companies that voluntarily reduced emissions could obtain a 6-year extension.

Title IV: Acid Deposition Control

Emissions of nitrogen and sulfur oxides are partially converted in the atmosphere to nitric and sulfuric acids which return to the Earth in rain, snow, fog, and on dry particles to cause acidic deposition. Such acidic deposition harms non-buffered lakes and streams by killing fish and other aquatic organisms, causing forest destruction, corroding steel, damaging concrete and marble structures, reducing visibility, and even contributing to adverse health effects.[12] Despite progress in reducing emissions under the 1970 CAAA, nearly 20 million tons of O_2 were emitted annually in 1990, mostly from combustion of fossil fuels by electrical generating facilities. The goal of the 1990 CAAA was to reduce the total SO_2 emissions by half to 10,000 tons per year through a phased program requiring power plants to reduce emissions. The

1990 CAAA is the first air regulation to encourage the use of market-based principles, such as emission banking and trading through the use of allowances. It works as follows: coal-burning electric utilities and other industries will be allowed to emit nearly 9 million tons of SO_2, while close to 250,000 theoretical tons were set aside for auction. If a utility expected it could not reduce its SO_2 emissions to meet the new federal standard, it could purchase an "allowance" from other utilities whose emissions were lower than that required under federal regulations. An allowance is defined under 1990 CAAA as the right to emit one ton of sulfur dioxide. This is a market incentive approach where allowance, or the right to pollute, could be traded or banked on the market. This concept was originally proposed by members of the Environmental Defense Fund (EDF) as part of the effort to reduce SO_2 emissions 50% by the year 2010. According to EDF representatives, the program has been very successful with emissions, leading to reductions of 3.4 million tons in one year alone at unexpectedly low program costs.[15] However, the market for the sale of allowances has significantly diminished, and auction prices for a one-ton allowance have dropped from the 1990 expected price of $1000 to $250 in 1992, $140 in 1995, and $68 in 1996. Such inexpensive prices worry some people that utilities will find it less expensive to buy allowances and so pollute considerably more.[15]

Title V: Permits

The 1990 CAAA also introduced an operating permits program similar to that used on other federal legislation designed to protect water such as the Federal National Pollution Elimination Discharge System (NPDES). This updated the Clean Air Act to make it more consistent with other federal statutes. Under the permit program, air pollution sources regulated by the Act must obtain an operating permit through their state environmental program. State programs must be approved by the USEPA if they are to administer the permit system. The permit requires that all rules and regulations regarding air pollution control be listed on the one permit, making enforcement a much easier process. Additionally, the permit will list the permissible levels of pollutants and the required control measures. The source is required to file periodic reports showing its level of compliance with the permit obligations. They must also pay a fee for the permit that is intended to cover costs by the State in administering the program.

Title VI: Stratospheric Ozone and Global Climate Protection

Stratospheric ozone is most concentrated in the region of 24 km to 40 km, although it is among the least concentrated of naturally occurring atmospheric gases. Ozone is formed when ultraviolet rays split oxygen molecules to release an oxygen atom that combines with an oxygen molecule to form ozone (O_3). Although created by UV light, ozone also absorbs UV energy in the range of 200 to 300 nm, protecting plant and animal life from the harmful effects of intense ultraviolet radiation. In the mid-1970s, concern about the integrity of the ozone layer grew when scientists Rowland and Molina, from the University of California, reported that chlorofluorocarbons were attacking and destroying the ozone layer. The CFCs were widely used

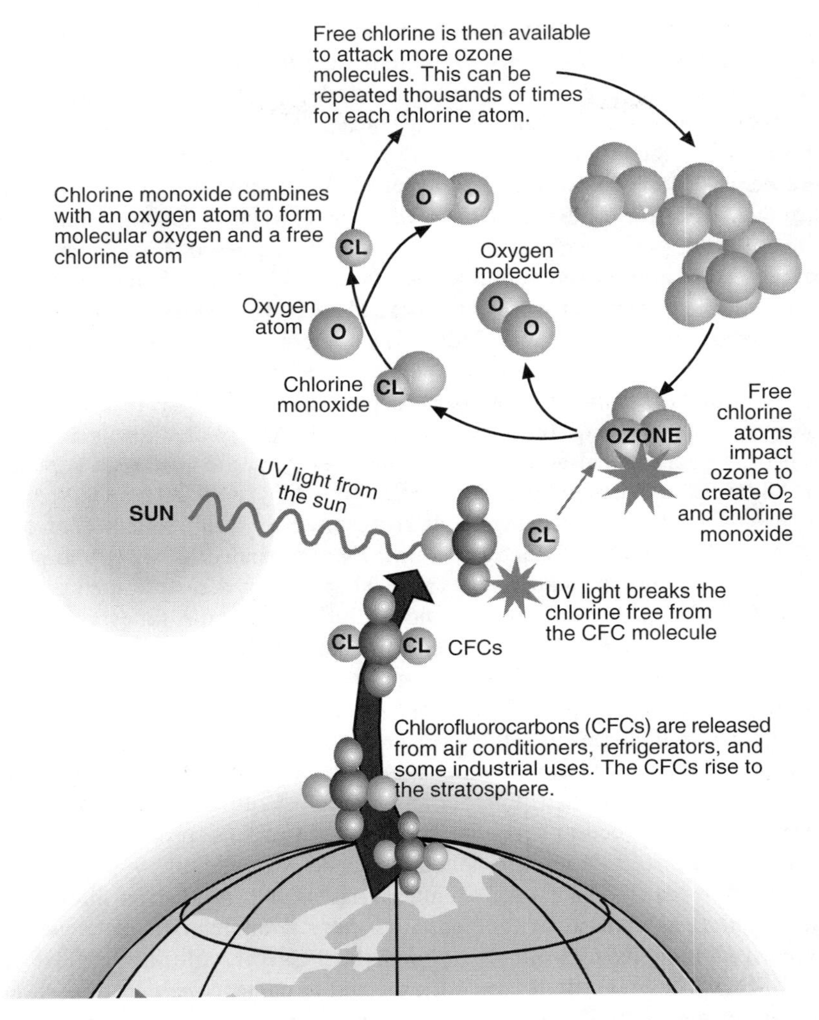
Free chlorine is then available to attack more ozone molecules. This can be repeated thousands of times for each chlorine atom.
Chlorine monoxide combines with an oxygen atom to form molecular oxygen and a free chlorine atom
O O
CL
Oxygen molecule
O
O
Oxygen atom
O
Chlorine monoxide
CL
OZONE
Free chlorine atoms impact ozone to create O2 and chlorine monoxide
UV light from the sun
SUN
CL
UV light breaks the chlorine free from the CFC molecule
CL
CL
CFCs
Chlorofluorocarbons (CFCs) are released from air conditioners, refrigerators, and some industrial uses. The CFCs rise to the stratosphere.

Figure 10.10 The catalytic destruction of ozone in the atmosphere by CFCs according to the Rowland-Molina hypothesis.

as refrigerants, aerosol propellants, and foam-blowing agents in styrofoam containers. The scientists began their work in the 1970s and showed that CFCs could be photolytically destroyed in the atmosphere to release chlorine atoms that could then catalytically destroy O_3 molecules in the stratosphere (Figure 10.10). Mario Molina, Sherwood Rowland, and Max Planck received the Nobel Prize for chemistry in 1995 for their work in establishing that CFCs were destroying the ozone layer.[16] The work of Rowland and Molina erupted from the narrow view of a few esoteric scientists to international attention in 1985 when British scientists discovered a large hole in the ozone over the Antarctic during September to mid-October. This was subsequently confirmed by the National Aeronautics and Space Administration (NASA).[17]

The Antarctic hole has continued to grow since then, while also lasting longer. In the early 1990s, NASA scientists reported that very low concentrations of stratospheric ozone were appearing over the middle latitudes of 20 to 60 degrees. While some ozone loss was partially due to volcanic dust and gasses hurled into the atmosphere by the eruption of Mount Pinatubo in the Philippines, losses in ozone were being observed over the U.S., Japan, Europe, Russia, and China before and since then. The World Meteorological Organization reports that the ozone hole over Antarctica is approximately as large as the combined area of Canada and the U.S., peaking at 7.7 million square miles and lasting for 50 days. Scientists predict the ozone will likely grow larger over the next 15 to 20 years. Despite global efforts to halt the production of ozone-depleting CFCs, there remain substantial amounts of CFCs to be released, and CFCs may persist in the stratosphere for many years.[18]

The efforts to control and halt ozone-depleting chemicals has been controversial. However, representatives from 29 nations met in Montreal, Canada in 1987 and reached an international agreement (Montreal Protocol) to freeze CFC production at 1988 levels, and then cut production in half by the year 2000. Mounting evidence that the ozone layer was being depleted resulted in a subsequent conference in London in 1990, and the Montreal Protocol was amended to hasten the phase-out of CFC production. The amendments also called for a phase-out of ozone-depleting halons (used in fire extinguishers) and CCL4 in the same time period. Methyl chloroform is to be phased out completely by the year 2005. More than 80 countries signed the 1990 agreement in London. The 1990 CAAA required that Class I ozone-depleting chemicals be phased out on a schedule specified in the 1987 Montreal Protocol. Class II chemicals (HCFCs) were to be phased out by 2030. However, as stratospheric ozone depletion became progressively worse and became apparent over North America and Northern Europe, representatives of more than 100 nations met in Denmark and agreed to speed up the phase-out of CFC production. Production of CFCs for use in the U.S. was halted by January 1, 1996.[8,11]

The efforts to find substitute chemicals has usually resulted in inferior performance and the requirement to modify existing equipment to make use of the alternatives such as hydrofluorocarbons (HCFCs) or chlorine-free HFC-134a (CF3CFH2). These substitutes are not expected to damage the ozone layer, but are still infrared absorbers and contribute to the greenhouse effect. Therefore, the amendments to the Montreal Protocol target many CFC substitutes for phase-out by 2040 through voluntary efforts.[11] However, there was concern that many LDCs would be unable to afford the expense of modifying equipment to accept the substitute refrigerants. The Montreal Protocol allowed for continued production of CFCs in developed nations for use in developing countries in order to prevent them from building their own CFC-producing factories. American factories may produce 53,500 tons each year for export until 2005. Unfortunately, there has been evidence that CFCs have been returning to this country illegally, where a 330-pound cylinder that costs $70 in Europe is sold for $242 in the U.S. This is still less costly than the CFCs recycled in the U.S. Officials believe as much as 10,000 tons entered the U.S. in 1995 from Russia and other countries in the former Soviet Union.[19,20] Despite these troublesome and controversial events, overall production of CFCs is down

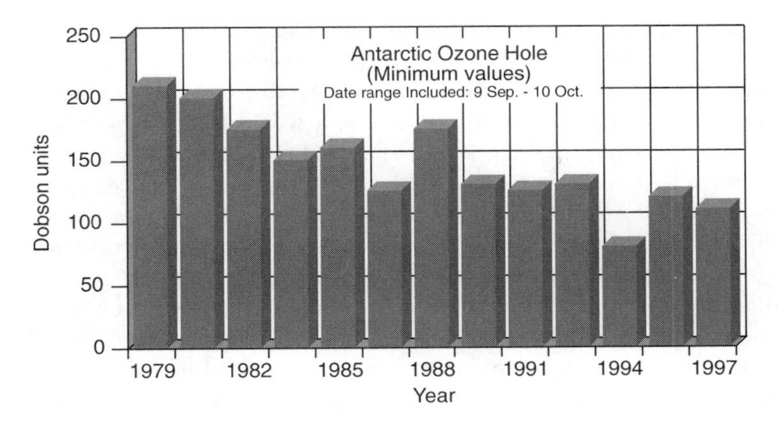

Figure 10.11 Antarctic ozone hole minimum vaues, 1979–1997. (Adapted from USEPA, The Size and Depth of the Ozone Hole, written by USEPA's Stratospheric Protection Division, June 10, 1998, hhp://www.usepa.gov/ozone/science/hole/size.html.)

markedly; and the National Oceanic and Atmospheric Administration (NOAA) has reported a reduction in the rate at which CFCs are accumulating in the atmosphere.[21] A graph is shown featuring the lowest recorded ozone during September/October of the years 1979 to 1997 as measured by satellite using total ozone mapping spectrometers (TOMS) (Figure 10.11). The minimum value of 88 Dobson units was recorded in 1994.[22] Although there appears to be a slight increase in stratospheric ozone compared to 1994, the recovery of stratospheric ozone is expected to take decades or up to a 100 years.[21]

Title VII: Provisions Relating to Enforcement

Since the start of the clean act movements in the middle of the 20th century, the laws and regulations have changed with relationship to ongoing research and public policy. In view of the fact that the laws have changed, it is also important to keep the enforcement ability up to date. Included in the 1990 Clean Air Act are a broad array of authorities to make the law more readily enforceable, thus bringing it up to date with the other major environmental statutes (EPA 1990 Clean Air Act Amendments from air quality handout).

Through the Clean Air Act of 1990, the EPA has gained authority to issue administrative penalty orders up to $20,000 and field citations up to $5000 for lesser infractions. Contained in the Clean Air Act of 1990 is a clause that actually enhanced civil judicial penalties. Criminal penalties for known violations are upgraded from misdemeanors to felonies, and new criminal authorities for known and negligent endangerment will be established.

In addition, polluting sources must remain within compliance and annually certify their compliance with the EPA. The EPA has also gained the ability to administer subpoenas for compliance data. EPA will also be authorized to issue compliance orders with compliance schedules of up to one year.

Health and Welfare Impacts of Ozone Depletion

If the stratospheric ozone that protects the Earth from UV-B radiation (280 to 320 mm) were brought to sea level pressures, it would be about ⅛ inch thick. The destruction of this very delicate layer could lead to increased UV-B radiation that is thought likely to result in increases in basal and squamous cell skin cancer (such cancers now affect 500,000 people annually along with 25,000 new cases each year of malignant melanoma); suppress or weaken the human immune response system; damage the cornea and conjunctiva of eye; reduce plant leaf size, total dry weight, and stunting of plant growth; and decrease amounts of phytoplankton and zooplankton. Studies showed that phytoplankton production was reduced by 6% to 12% under the ozone hole. This lack of productivity can have several negative impacts. Since phytoplankton photosynthesis is reduced, less carbon dioxide will be removed from the atmosphere, further enhancing the greenhouse effect. Also, since phytoplankton are at the base of the food chain, a reduction could threaten the food source of the world's oceans.[23,24] It needs to be mentioned that pilot studies in Southern Chile, where people have been periodically exposed to high levels of UV-B, found no convincing evidence to support the reported anecdotal acute adverse health effects due to ozone thinning.[24] The authors conclude that exposure to UV-B at current levels in Punto Arenas, South America, were not likely to produce adverse health effects, but cautioned that increased levels may be of concern if they should occur.

The reduction of CFCs is important in the protection of stratospheric ozone. CFCs also contribute to global warming. The greenhouse gases contribute the following amounts to global warming: CO_2 (55%), CFCs (7% to 24%), methane (15%), and nitrous oxide (6%).[8] The CFCs are strong infrared absorbers and are up to 20,000 times as efficient as carbon dioxide in producing warming effects. Efforts at reducing CFCs are doubly important to protect the stratospheric ozone layer and to reduce the impact on global warming.

Other Titles to the 1990 CAAA

The Clean Air Act Amendment of 1990 contains many new provisions with regard to tougher enforcement, new research, and development in air monitoring. There are also unemployment benefits through the Job Training Partnership Act for those workers who may have lost their jobs because industries had to comply with the new provisions of the 1990 CAAA. Efforts under the Act also attempt to reduce the problem of pollution-created haze areas at national parks and other parts of the country.

Revised Ozone and Particulate Standards

On July 17, 1997, the USEPA announced new NAAQS for ground-level ozone and particulates. Ground-level ozone is a major component of smog that is photochemically produced as a secondary pollutant of the stratosphere from the interaction of sunlight, nitrogen oxides, and hydrocarbons. Ground-level ozone will be discussed in more detail later in the chapter. The previous ozone standard was last revised in 1979 and set at 0.12 ppm for one hour. However, based on more than 3000 new studies on the ecological and health effects of ozone published since 1980, it has become clear that the standard did not adequately protect the public health; and the USEPA replaced the previous standard with an 8-hour standard set at 0.08 ppm.[25]

Similarly, the USEPA announced a new standard for particulate matter (PM) under the NAAQS, based on the review of hundreds of scientific studies. USEPA has added a new annual PM 2.5 standard set at 15 micrograms per cubic meter (ug/m^3) and a new 24-hour PM 2.5 standard set at 65 ug/m^3. The USEPA is retaining the current annual PM10 standard set at 50 µg/m.[3,26] The health effects of particulates are discussed in detail in a following section in this chapter.

These changes to the standard were met with significant opposition from the auto industry, oil corporations, and from some state governors. Areas of the country producing more smog and soot than considered healthy under the proposed standards would face economically strangling restrictions including controls on truck operations, mandates to reduce pollution-causing auto emissions to zero, requirements that vehicles be run on fuels other than gasoline, strict controls on power plants and factories, and increased emission checks on autos. The new standards are considerably tougher than those developed under the 1970 CAAA and 1990 CAAA, which are still unmet by many states across the country.[27]

The Issue of Global Warming

The Hot Air Treaty, Kyoto, Japan

The global warming treaty completed in December 1997 (Kyoto, Japan) asked Western nations to reduce greenhouse gases to pre-1990 levels by 2010. The U.S., if it were to comply, would need to reduce greenhouse emissions to 7% below the level of the year 1990 or by about 33%, one third of its otherwise projected use in the year 2010.[28] Western European nations and Japan are assigned similar goals. According to the agreement, reductions would not start taking effect until 2008. Even if the nations complied, total world production of CO_2 would be reduced by only 1.1 billion tons by the year 2010. None of the developing nations have agreed to the Kyoto Treaty, even though they are soon expected to produce more greenhouse gases than the West. The LDCs have the urgent problem of providing for exponentially increasing populations, and further reducing use of fossil fuels is unlikely. Inexpensive labor and unregulated industry could attract business from the West, thereby shifting jobs, machinery, and CO_2 production to the LDCs with no net overall decrease in greenhouse gases, while greatly disturbing Western economies. The U.S. Congress has already voted down any treaty that does not include the LDCs, and if the accord is not ratified, global emissions of greenhouse gases could become double their preindustrial level by the year 2050.[28] As an example, China, with its heavy reliance on coal, will be the largest, single producer of climate-warming carbon dioxide by 2010 to 2030. Despite rising health problems among its citizens because of increased air pollution, there appears to be no slowing in China's coal-burning practices as it races to feed its billions of people and enter the world's industrial marketplace.[29]

Global Warming: The Controversy

The question of global warming has been raised by climatologists since the early 1970s. It has been argued that human activities have upset the balance of atmospheric carbon dioxide through the combustion of fossil fuels that releases carbon oxides; the clearing of forests that produces CO_2 and removes a vital consumer of CO_2; and the destruction of phytoplankton by pollution of the oceans. Since the preindustrial age a steady in incresase in CO_2 in the atmosphere from 280 ppm to 370 ppm has alarmed specialists about what to do with excess CO_2.

Greenhouse gases such as carbon dioxide, methane, ozone, CFCs, and nitrous oxide are strong absorbers of infrared radiation. When the sun's energy reaches the Earth, much of it is absorbed by ground cover and water and reflected back as lower energy infrared radiation. An increasing blanket of carbon dioxide around the planet absorbs some of this infrared energy, trapping it and causing the Earth to warm in a process known as the greenhouse effect (Figure 10.12). This concept of global warming caused by greenhouse gases is undergoing tremendous scientific debate, and came to public attention in 1995 because icebergs the size of small states broke off the Antarctic ice shelf; the annual average global temperature has risen by about 0.5C (1°F) since the 19th century;[30] U.S. and French scientists report that satellite measurements show the global sea has risen faster (3 minutes/year) in recent years than in previous decades;[31] climate scientists reconstructed average annual temperatures back to the year 1400 and found that 1990, 1995, and 1997 were the warmest

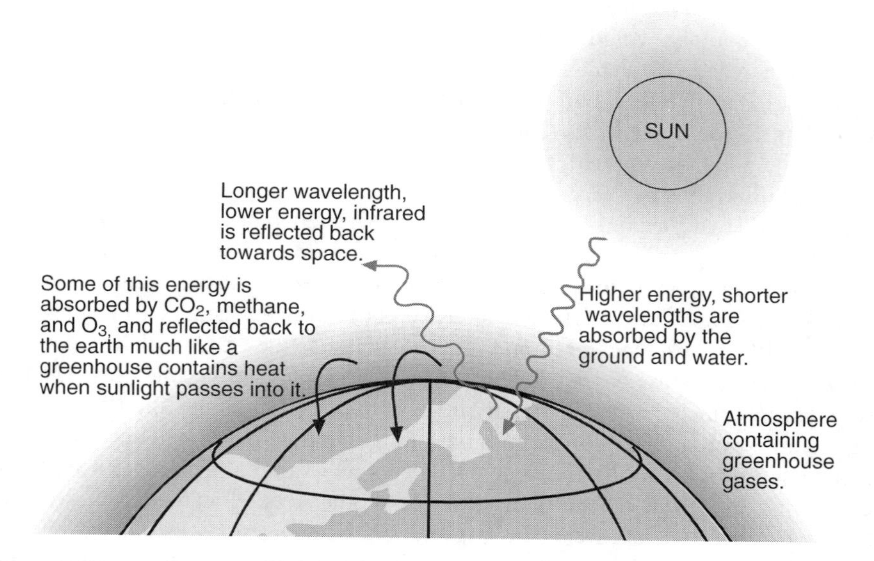

Figure 10.12 The absorption of reflected infrared energy to produce the greenhouse effect.

Figure 10.13 Glaciers at Gross Glockner in Austria have receded rapidly in the last few years. This is a view of the visitors' center overlooking the receding glacier.

in 600 years;[32] and mountain glaciers around the world are rapidly retreating such as the one at Austria's highest mountain peak, Gross Glockner (Figure 10.13), while those at Glacier National Park in Montana have shrunk 50% since 1850.

There have been various measures proposed for combating greenhouse gas increases. Alternative methods such as other sources of energy including wind, solar and nuclear power; increased numbers of trees and woody plants in deforested areas to natural CO_2 sinks; and the newest method of carbon sequestration takes CO_2 from stationary sources, such as a chemical factory or power plant, and stores it for thousands of years either underground or in deep oceans. The ocean contains at least 50 times more carbon than the atmosphere does, making it a more efficient reservoir for carbon dioxide, yet having less effect in altering the chemistry of the water. It will take hundreds of years for the CO_2 to reach the surface of the ocean, rendering it ineffective as a greenhouse gas.[33]

Researchers at the Max Planck Institute for Meteorology in Hamburg, Germany, say that it is unlikely that natural forces (less than a 5% chance) caused the temperature rise.[30] Other reputable scientists are not ready to point the finger to anthropogenic sources such as greenhouse gases as the source of global climate change. In fact, while international leaders battle over treaties to resolve a perceived global threat from greenhouse gases, scientists still line up along a mark in the sand as to who supports or rejects the role of human-induced global warming. Recent developments favor the theory that greenhouse gases play a role in increasing global temperatures, but uncertainties remain in the studies aimed at showing this, and a definitive answer may not be possible.[30] The scientific community will likely never depend on a single study or observation to accept the concept of greenhouse warming, but will gradually accept the concept when there is a consistent accumulation of evidence.[30] The combined search for such proof encouraged paleoclimatologists at the University of Massachusetts to reconstruct temperatures over the past 600 years using a method of indirect indicators (proxy climate indicators) including ice core samples, coral records, tree rings, and historical temperature measurements. They compared estimated temperatures to factors that affect climate including the atmosphere's carbon dioxide supply and discovered CO_2 to be the dominant factor over the past several decades.[32] Many factors influence climate as mentioned in Chapter 1, and global climate is likely influenced by many complex factors much greater in their effect than anthropogenic greenhouse gases. Climate is affected by increases in atmospheric gases that absorb energy, changes in the Earth's orbital geometry, changes in ocean temperatures, volcanic activity with increased atmospheric and reduced sunlight penetration, and variations in solar radiation.

While the meaning of the rise in global temperature and effects such as increasing ocean levels continue to be disputed, the spread of certain diseases has been clearly documented. Those diseases have been found at latitudes where they have not been previously identified. The increased incidence of hanta virus, malaria, cholera, and toxic algal blooms suggest a change in ecological balance and predator–prey relationships that can influence the risk of disease in humans. Regardless of whether there is a proven relationship between global warming, climate changes, and the emergence of new patterns of infectious disease, the public health community is obligated to investigate these new trends and determine what additional analysis and intervention are necessary.[34]

Factors Affecting Global Climate Change

Orbital Geometry as a Factor Affecting Climate

You may recall the frigid weather of 1994 that plummeted temperatures to all-time lows of –20°F to –30°F in cities around the U.S. This caused people to question whether the greenhouse effect was real, and some considered that glaciers may be returning. In the 1840s, the idea that glaciers covered much of the Northern Hemisphere was very controversial, but scientists have shown abundant evidence that thick ice sheets covered much of the North American continent, gouging out lakes, taming mountain ranges, and recreating the landscape. However, recent discoveries are even more surprising. Using core samples from sea-floor sediments and ice core samples from Greenland and Antarctica, scientists have shown that carbon dioxide rises in positive, almost perfect, correlation with rising global temperatures; and climate changes are normally abrupt and often extreme over the history of the planet.[35] The records show that mean global temperatures fluctuated widely with transitions from warm to cold, often measured in decades (Figure 10.14). The interglacial period we are now in has lasted 10,000 years, and this stable period of warmth is very unlike the unstable climate history of the planet. It is probably no coincidence

WEB PICK **Go to http://www.unix.oit.umass.edu/~envhl565/index.html** **Click on "Chapter Web Links" to find Air Pollution — USEPA Office of Air and Radiation on the World Wide Web under Chapter 10**	
Chapter 9	Air Quality and Radiation
Whose Site?	U.S. Environmental Protection Agency, USEPA's Office of Air and Radiation (OAR)
URL	http://www.epa.gov/oar/
What's Here?	The USEPA's Office of Air and Radiation (OAR) provides information on air pollution, clean air, and air quality. OAR develops national programs, technical policies, and regulations for controlling air pollution and radiation exposure. OAR is concerned with pollution prevention, indoor and outdoor air quality, industrial air pollution, pollution from vehicles and engines, radon, acid rain, stratospheric ozone depletion, and radiation protection. There are specific links and information on acid rain, ozone depletion, global warming, air quality where you live, transportation and fuels, toxic air pollutants, indoor air quality, visibility, and off-road equipment. There is a hot-topics section called AIRlinks that includes information on regional ozone transport, the new ozone and particulate standards, clean-burning gasoline, MTBE, mercury, and energy. Check out the "maps" link under Tools and Technical Info. This **AIRGraphics** website (*AGWeb*) gives you access to maps and charts of air pollution information. You specify criteria, such as pollutant name and amount; AIRGraphics produces a map and displays it in your browser. AIRGraphics displays two types of data: air pollutant emissions estimates and air quality measurements. You should also try linking to the New England Region 1 EPA site on ozone at http://www.epa.gov/region01/eco/ozone/. Click on the "ozone mapping system" and look at near-real-time information on ozone concentrations in the New England area.

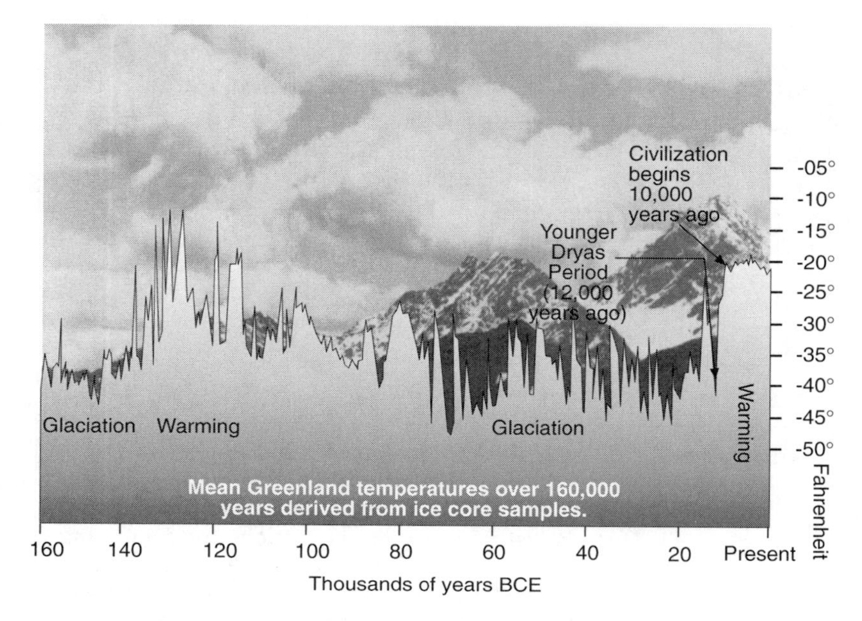

Figure 10.14 Record of average annual temperatures in Greenland derived from ice core samples. (Adapted from Lemonick, M.D., The ice age cometh? *Time,* Jan. 31, 79, 1994.)

that human population marked its beginning just about 10,000 years ago with the climate stability. Humans were certainly not sufficient in number or technology to affect global climate over the previous 160,000 years, so what events caused such drastic changes? Among the more widely accepted theories is that of Milankovitch, a Yugoslavian astronomer who theorized such climate shifts were the result of subtle changes in the Earth's orbit around the sun (eccentricity) and slight variations in the Earth's tilt (obliquity), which produce climactic cycles (ice ages) of Earth at 23,000, 41,200 and 100,000 years. The Milankovitch theories gained support from examination of oxygen isotopes in ocean floor sediments revealing climate conditions over eons of time. Heavy oxygen isotopes indicate warmer weather, while lighter ones predominate in cooler weather.[17,35] This Milankovitch theory does not explain all ice ages, nor the rapid shifts in global temperature evidenced by the Greenland ice core samples.

Changes In Ocean Temperature

Wallace Brodkerad and George Denton from the University of Maine have advanced one of the more interesting theories about climate changes involving ocean currents. They propose that a very salty, dense stream of water flows northward under the Gulf Stream carrying heat from the tropics. It rises to the surface in the far north as water above is swept away by harsh weather and deposits its heat to warm the arctic, cools, sinks, and returns to the tropic. This continuing process keeps the arctic free of glaciation. Disturbances to this process could lead to a new period

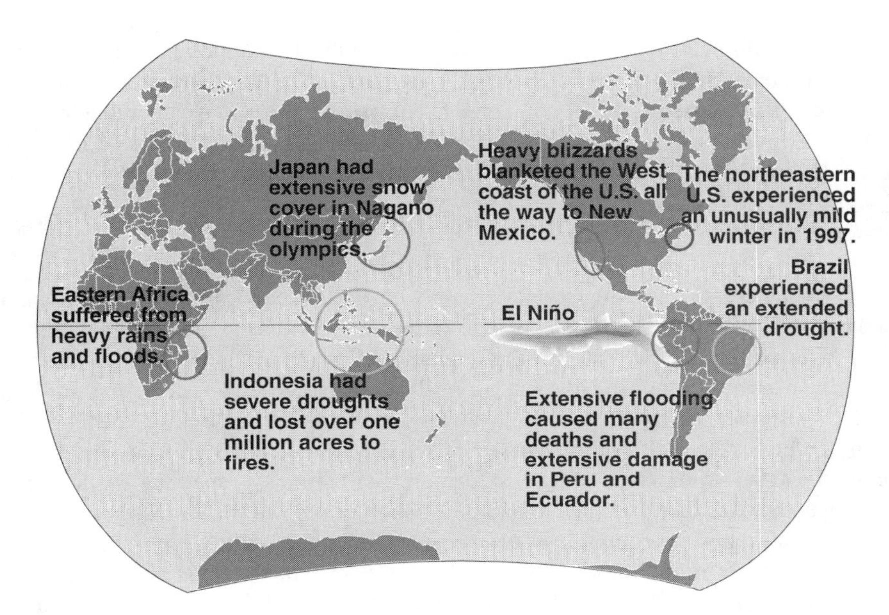

Figure 10.15 Global effects of El Niño. (Adapted from Noshi, J.M., The fury of El Niño, *Time,* Feb. 16, 67, 1998.)

of glacier formation. This was thought to occur during the period of glaciation known as the Younger Dryas period, about 12,000 years ago. During this period, huge quantities of fresh water from the North American continent emptied into the North Atlantic for over 100 years.[35]

Some of greatest extremes in weather have been attributed to ocean currents known as El Niño (the baby). El Niño is a change or shift in ocean temperatures along with atmospheric conditions in the tropical Pacific that changes weather patterns all around the world (Figure 10.15). Scientists believe that high atmospheric pressure in the Eastern Pacific sends tradewinds westward, pushing Pacific water ahead of them. Consequently, the water on the Asian and Australian coasts is about one to five feet higher than in the Americas. When the pressure drops, the water moves east again forcing downward a cool layer of water underneath called a thermocline, while permitting the warmer, unmixed water on top to carry its heat and moisture toward the Americas and away from Indonesia.[36] Because of this process, Indonesia faced its worst drought in 50 years with fires that consumed more than one million acres of forest and threatened hunger for many, while the western coasts of the Americas suffered heavy precipitation, powerful storms, and incredible winter blizzards. Torrential rain fell in East Africa, powerful hurricanes swirled in the Pacific, flash floods killed hundreds in Peru and Ecuador wiping out entire villages, and typhoons smashed into Japan. The ocean currents associated with El Niño produced remarkable global effects, and scientists are concerned that El Niño is changing dramatically from its 2-million-year-old cycle, although somewhat chaotic, to recur far more frequently and persist much longer. Some climatologists argue that the increase in severity and frequency of El Niño is a result of global warming.[37] When the waves associated with El Niño bounce off the Americas and return, then bounce back off of Asia, they lift up the thermocline, causing a cooling effect that ends the El Niño effect. This causes shifts in weather that include a cold and wet north, and dry, hot south that is opposite of the effects of El Niño. This shift in Pacific Ocean temperature is known as La Niña.

Volcanic Activity

Volcanic eruptions in the modern era may have extreme localized effects on land and may cause short-term global changes in weather patterns as sunlight is inhibited by a layer of particles thrust into the atmosphere. However, ancient volcanoes are thought to have erupted in Siberia 250 million years ago and lasted for a million years. This coincided with the Great Dying, occurring at the end of the Permian period, which killed 95% of life in the oceans and more than 70% of the vertebrates on land. Ocean levels dropped by 300 feet, and acid rain enveloped the Earth. Some scientists theorize that this prolonged volcanism blocked out the sun, causing extensive glaciation and poisoning the atmosphere with sulfur oxides.

Scientists claim that other factors may have contributed to this mass extinction which developed the events leading to the extinction of the dinosaurs some 65 million years ago. But accumulating evidence points toward extended volcanic activity as having dramatic effects on the Earth's climate and its inhabitants.[38]

Solar Radiation

Scientists have recently discovered that the sun is not constant in its solar output but varies with changes in the sun's luminosity, which is in turn associated with sunspot activity. Reduced sunspot activity appears to result in a cooling effect, whereas increased activity increases warming of the Earth. The sunspots show cycles of 11 and 22 years that correlate with nearly half of the global warming evidenced over the last 100 years.[17]

THE CRITERIA POLLUTANTS

Introduction

The USEPA has reported substantial progress in improving the air quality in many states including the heavily polluted areas of Southern California, Hawaii, and regions in Arizona and Nevada. Despite major increases in population and acute travel over the past decade, air pollution has decreased by about one third overall in these regions. The greatest reductions have been recorded for lead (93%), followed by carbon monoxide (35%) and particulate matter (26%). While the majority of pollutants are being steadily reduced, nitrogen oxides, ground-level ozone, and fine particulates remain a problem in many areas.[39–41]

The total emissions of criteria pollutants fell 32% since 1970 in the face of a 29% increase in the U.S. population, a 121% escalation in vehicle miles traveled,

Table 10.2 Percent Changes in Ambient Concentrations of Air Pollutants from 1970 to 1996 (Adapted from USEPA, National Air Quality and Emission Trends Report, 1996, USEPA Office of Air Quality Planning & Standards, USEPA document 454/R-97-013 1/13/98, WYSIWYG://http:flwww.usepa.gov/oar/aqtmd96/trendsfs.hmfl.)

CRITERIA POLLUTANTS	PERCENT CHANGE IN AMBIENT CONCENTRATIONS FROM 1970-1996
Carbon monoxide	-31%
Nitrogen dioxide	+8%
Sulfur dioxide	-39%
Particulates (PM_{10})	-73%
Ozone	-38%
Lead	-98%

and a gross domestic product increase of 104%.[39] Still, nearly 46 million people live in counties that fail to meet the air quality standards for one or more of the criteria pollutants (Table 10.2).[39] Fine particulates from power plants, motor vehicles, and photochemical reactions in the atmosphere kill as many as 64,000 Americans each year from lung and heart disease.[42] Similarly, the American Lung Association released a survey of 13 U.S. cities linking exposure to ground-level ozone (smog) to as many 50,000 emergency room visits and 15,000 hospital admissions.[43] The health and welfare problems associated with criteria pollutants have not been completely resolved and many issues remain. A summary of criteria pollutants sources, health and welfare effects is presented in Table 10.3.

Particulate Matter (PM)

Particulate pollutants include airborne particles in liquid or solid form that range in size from visible fly ash greater than 100 μm to particles 0.005 μm in size (Table 10.4). Particles may be produced naturally, such as pollen or sea spray, or by human activities such as industrial processes, agricultural activities, fossil fuel combustion, and traffic. Particulates include dust, smoke, soot, (carbon), sulfates, nitrates, trace metals, and condensed organic compounds. Particulates produce a number of effects adverse to human interests including respiratory and cardiac health hazards to humans; the deposit of grime and soot on buildings; the reduction in sunlight, thereby causing a regional or global cooling effect; and reduced visibility in areas of extensive smoke or particulate pollution.

The characteristics for coarse particles (2.5 to 10 μm) are very different from fine particles (<2.5 μm) and relate to the adverse effects they produce. Coarse particles originate from wind-blown dust coming from deserts, unpaved roads over which vehicles travel, and agricultural fields, pollen, mold spores, and plant debris. These particles tend to be mostly minerals consisting of aluminum, silicon, potassium, and calcium, which are chemically basic. Fine particles (<2.5 μm) are mostly

Table 10.3 Sources and Health and Welfare Effects of Criteria Pollutants

CRITERIA POLLUTANTS AND SOURCES	HEALTH EFFECTS	WELFARE EFFECTS
Carbon monoxide Incomplete combustion of fossil fuels as in vehicles, kerosene heaters, boilers, and furnaces; cigarette smoking, forest fires, and biological decomposition	Interferes with oxygen transport in blood by binding with hemoglobin; causes headaches, fatigue, cardiovascular disease, and central nervous system disorders	Effects on plants or materials are not evident
Nitrogen dioxide Emitted from the combustion of fossil fuels in vehicles, industrial boilers, and electric generating utilities	Causes increased risk of respiratory infections and aggravates symptoms in persons with asthma and chronic bronchitis	Produces a reddish-brown haze over cities which reduces horizon visibility, causes leaves to yellow, and is a precursor to acid deposition and tropospheric ozone
Sulfur dioxide Fossil fuel combustion especially in coal-burning electric power utilities, metal smelters, oil refineries, and industrial boilers	Causes irritation of the throat and lungs and aggravates symptoms in persons with asthma and chronic bronchitis	Causes corrosion and deterioration of metals, brittleness of paper, paint discoloration, damages textiles and leaves of plants, and is a precursor to acid deposition
Particulates PM_{10} and $PM_{2.5}$ Fossil fuel combustion emissions, industrial processes, photochemical reactions in atmosphere, mechanical abrasion	Aggravates asthma, heart disease, and chronic lung disease; alters lung's natural cleansing mechanisms; smaller particles associated with the most severe symptoms	Causes soiling of materials, grime deposits and reduced visibility; particulates from volcanic eruptions can reduce solar energy and produce temporary cooling effect
Ozone A product of NOx emissions from motor vehicles, power utilities, and industries burning fossil fuels, combined with hydrocarbons and sunlight in the atmosphere	Causes breathing difficulty, irritation to mucous membranes, and increases risk to respiratory infections; acute exposures causes respiratory pain, bronchoconstriction, lung edema, and abnormal lung development	Corrodes rubber, paint; weakens fabrics, rubber, and produces leaf damage and retardation of plant growth
Lead Historically emitted from vehicles burning leaded gasoline; emissions have been reduced by 98% since 1974; most lead exposures in U.S. today are not airborne	Damage to nervous system, blood forming tissues, kidneys; evidence of neurobehavioral disorders inlcuding learning disabilities and antisocial behavior	No known effect on vegetation or materials

emitted from fossil fuel combustion in industry, residences, and motor vehicles. They are also created in the atmosphere from gases of nitrogen and sulfur oxides and volatile organic compounds forming sulfates, nitrates, and aliphatic and aromatic hydrocarbons

Table 10.4 Sources and Characteristics of Coarse and Fine Particles

PARTICULATE SIZE	PARTICULATE SOURCES	CHEMISTRY
Coarse Particles These are coarse particles from 1-100 μm including the course fraction PM_{10} (2.5-10 μm)	Industrial and mechanical processes such as fragmentation of matter and atomization of liquids, agricultural and forestry activities, and dust from unpaved roadways, mold spores, and wood ash	Silicon, aluminum, iron, potassum, and calcium are common components; the coarse particle samples tend to be alkaline
Fine Particles These are fine particles less than 1 μm, but including much of the mass of the $PM_{2.5}$ fraction; the $PM_{2.5}$ fraction consists of fine particles less than 1.0 μm and some coarse particles in the 1-2.5 μm range	Industrial and residential combustion of fossil fuels; secondary particles produced by direct, catalytic, and photochemical oxidation of nitrogen and sulfur compounds, and volatile hydrocarbons to produce sulfates, nitrates, and oxyhydrocarbons	Elemental and organic carbon such as from fuel combustion (soot), sulfates, nitrates, condensed organic compounds, oxyhydrocarbons, and trace metals; the fine particle samples tend to be acidic

including aldehydes, ketones, phenols, esters, terpenes, and phenylacetic acids.[26,44] Coarse particles can build up in the respiratory system causing respiratory disorders such as asthma in women. Smaller particles are more apt to be associated with increasing lung function in persons who have preexisting conditions such as asthma; increasing premature death among the elderly and those with cardiopulmonary disease; deterioration of respiratory defense mechanisms and adverse changes in lung tissue and structure; and increased respiratory disease and symptoms.[26,40,45] The recent community studies provided support for issuing new standards for particulates in order to protect the public health and welfare. The new standard of $PM_{2.5}$ is set at 15 μg/m^3 annual arithmetic mean and 6.5 μg/m^3 for a 24-hour average since these fine particles are more closely linked to mortality and morbidity effects than the previous PM_{10} standard.

Ozone and the Photochemical Oxidants

There are two types of ozone known as *good* and *bad*. Good ozone is a layer in the stratosphere previously discussed which absorbs UV-B radiation and protects the Earth from excess amounts of these damaging rays. The bad ozone is formed on the troposphere (nose-level) by a complex series of reactions including sunlight, nitrogen dioxide, and volatile hydrocarbons. The reaction follows what is outlined in Figure 10.16.

The photochemical oxidants resulting from such reactions include primarily ozone, nitrogen oxides, and alkyl peroxy radicals (RO_2). The alkyl peroxy radicals

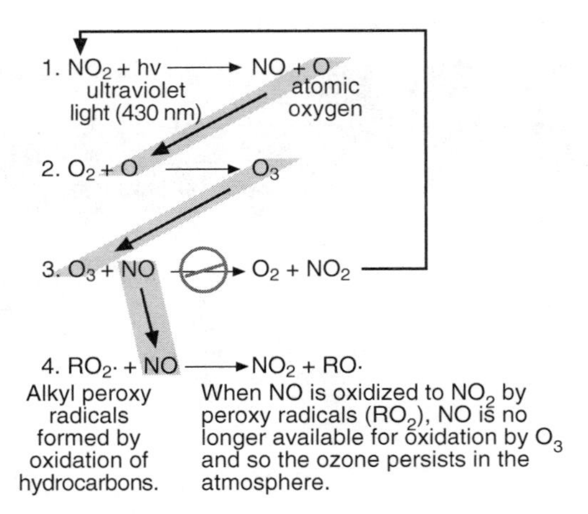

Figure 10.16 Proposed mechanism for the production and maintenance of tropospheric ozone. (Adapted from Godish, T., *Air Quality,* 3rd ed., CRC/Lewis Publishers, Boca Raton, FL, 1997, chap. 2.)

are produced by oxidation of hydrocarbons. An oxidant is a substance that readily gives up an oxygen atom or removes hydrogen from a compound. Photochemical refers to the initiation of these reactions by sunlight. This mixture of photochemical oxidants is often referred to as photochemical smog and is most concentrated in areas with heavy traffic, intense sunlight, and stable air conditions. Such conditions are especially prominent in Southern California and, to a lesser degree, along the northeast coast, especially from May through September. Ozone is the primary indicator of smog and in concentrated form is a dense (1.6 times as heavy as air), violet blue gas with a sharp metallic odor. It is a powerful oxidant capable of breaking molecular bonds and rapidly degrading many structured materials and plant and human tissues.

The ozone standard was revised in 1979 to be 0.12 ppm for one hour. However, more than 3000 studies since the 1980s show evidence that adverse health effects occur at lower exposures than the previous standard and that these symptoms are aggravated by longer exposures. Exposures to ozone are associated with increased hospital admissions from asthma and accounted for up to 20% of summertime admissions of respiratory distress. Respiratory infections and inflammation are aggravated by ozone as well. Children and adults who play and work outside during summer months are at elevated risk to ozone exposure and may experience coughing, chest pain, and a reduction in lung function. Exposure to ozone over long periods has been shown to cause inflammation of the lungs, impairment of lung defense mechanisms, premature aging of lung tissue through irreversible damage, chronic bronchitis, and emphysema. Consequently, the USEPA has set a new 8-hour standard at 0.08 ppm based on the decision that this level will protect the public health.[46]

Carbon Monoxide

Carbon monoxide is a colorless, odorless, tasteless gas produced from the incomplete combustion of fossil fuels with the greatest contribution by motor vehicles. In 1993, motor vehicles accounted for nearly 75% of anthropogenic carbon monoxide emissions.[44] Between 1986 and 1996, carbon monoxide concentrations in ambient air decreased 37% while overall emissions decreased by nearly 20%. These improvements were made while vehicle miles traveled increased 28% during the same period.[47] Carbon monoxide enters the bloodstream through the lungs and combines with hemoglobin of red blood cells to form carboxyhemoglobin. Carbon monoxide has a stronger affinity for hemoglobin than oxygen. Therefore, as levels of carboxyhemoglobin rise, the adverse effects associated with oxygen deficiency increase. The health threat from chronic low-level exposure is most serious for people who have preexisting cardiovascular disease.[48] Inhalation of slightly elevated levels of carbon monoxide can produce symptoms ranging from dizziness and headaches to visual impairment, reduced work capacity, poor learning ability, lowered manual dexterity, and difficulty in performing complex tasks. Higher levels of CO inhalation can lead to paralysis of motor function and death.[47]

Lead

Lead is a systemic heavy metal poison that enters the human body through inhalation of contaminated air and ingestion in food, water, soil, or dust. Lead accumulates in the bones, blood, and soft tissues with target organs being the central nervous system, the blood-forming tissues, and the gastrointestinal tract. Lead exposures have been associated with neurological impairments, mental retardation, and behavioral disorders. Even at very low doses, children and fetuses suffer from central nervous system damage. The association of lead with behavioral problems and reduced intellectual ability caused lead to be placed on the list of criteria pollutants in 1977 when the Clean Air Act was reauthorized. Since automobiles burning leaded gasoline were the largest contributor to lead exposure, the phase-out of leaded gasoline has been the predominant control strategy. Lead emissions from highways have decreased 99% since 1987, while overall ambient lead concentrations decreased 95% since 1977, and 75% since 1987.[39]

Sulfur or Oxides

Health and Welfare Effects

Sulfur, phosphorous, carbon, oxygen, hydrogen, and nitrogen are the main components of most living things and consequently are found in fossilized forms including the fossil fuels such as coal and oil. The combustion of fossil fuels results in the oxidation of sulfur to produce sulfur dioxides. They react with moisture, oxidizing agents and reactive hydrocarbons and other substances in the atmosphere to produce sulfates, sulfites, and sulfuric acid. The primary source of emissions is electric utilities. These compounds may produce a variety of adverse health effects. Sulfur oxides are also involved in corroding metals; disintegrating marble, limestone, and dolomite; attacking fabrics; reducing visibility; and impairing the growth of plants and forests. The most prominent health concerns associated with SO_2 include respiratory illness, effects on breathing, a reduction in lung defenses, and aggravation of existing cardiovascular disease. Persons with asthma, cardiovascular disease, and chronic lung disease are most sensitive to the effects of sulfur dioxide. Children and the elderly are also at increased health risks from the inhalation of sulfur oxides.[47] Sulfur oxides are among the main precursors to acid deposition, with nitrogen-oxides being the second greatest contribution.

Acid Deposition

Acid deposition is a term that represents a more inclusive understanding of the acidification process than the term acid rain, since acidity may be found in rain, sleet, snow, fog, clouds, and adsorbed to particles. Consequently, the term acid rain is being replaced by the term acid deposition although the two are presently used synonymously. The problems associated with acid deposition were first recognized about 30 years ago in Sweden when changes in the plant and animal life of freshwater lakes was associated with increased acidity from acid deposition.[17] Initially, the revelation of acidic deposition was not widely accepted by the world scientific community; but as measurements of rainfall pH demonstrated greatly expanding areas of increasing acidity, the problem became the subject of domestic and international debates over acid rain controls. The U.S. passed the 1990 CAAA with major provisions for sulfur and nitrogen-oxide controls, although atmospheric acidity still presents a major environmental problem.

Rain, snow, clouds, and sleet are normally somewhat acidic because of the presence of CO_2 in the atmosphere, producing carbonic acid with pH levels of nearly 5.0 in the absence of pollutants. The pH scale is logarithmic and is based on the negative log of the hydrogen ion concentration. Consequently, a pH of 3.0, as has been found in precipitation in the Ohio Valley, is 100 times the acidity of rain with a pH of 5.0. About 70% of sulfur-oxide emissions originate from large electric utilities burning high sulfur coal, while other major contributors to NOx and SOx include industrial boilers, metal smelters, and automobiles.[17] Sulfur acids have contributed about 65% to the acid deposition problem while nitrogen oxides contributed 35%. However, as regulatory controls became more effective, the levels of sulfur oxides in the atmosphere have decreased by 39% since 1970 while ambient concentrations of NO_2 increased 8% over the same period.[47] Consequently, the emissions of NOx from automobiles and electric power generators are taking on greater importance.

Effects of Acid Deposition on Ecology

Acid deposition is a product of fossil fuels with most SOx occurring from electric utilities burning coal or oil. Many of these plants employ the strategy of building tall stacks so the discharged gases will be carried out of the vicinity. Although locally effective, the gas streams are carried by prevailing winds over great distances to be deposited far from the origin in a process known as long-distance transport. Heavily industrialized areas of the Ohio River Valley, upper Midwest, and areas of Illinois, Pennsylvania, West Virginia, and Indiana are often considered to be the source of significant acid plumes that travel eastward to deposit all along the northeast coast off the U.S. and parts of Canada. The heavily industrialized areas of Great Britain, Central Europe, and parts of China create similar conditions for polluting distant locations. Sweden and Norway point to central Europe as a major cause of the acidity falling on them. Although long-distance transport contributes significantly to acidic deposition, local sources are also important contributors. Just how much each contributes to the problem has been strongly contested by those allegedly producing the pollutants and those who receive them.

The New England areas, including upstate New York, the mountainous areas of West Virginia, Pennsylvania, Virginia, Kentucky, Wisconsin, Minnesota, parts of Canada, and western mountainous areas of North America are sensitive to acidity because these areas are deficient in magnesium and calcium carbonates. Therefore, the soils and water have low acid neutralizing capacity and are referred to as acid-sensitive ecosystems (Figure 10.17).[17]

Figure 10.17 Acid sensitive regions of North America and Canada. (Adapted from Godish, T., *Air Quality,* CRC/Lewis Publishers, Boca Raton, FL, 1997, chap. 4.

Aquatic Ecosystems

When lakes and streams receive acid deposition in acid-sensitive areas, they are unable to neutralize or buffer the acidity. Consequently aquatic systems in the Adirondacks or New York, New England, and other sensitive areas become chronically and progressively acidified or receive sudden massive doses of acidity in spring thaws. This results in a process known as shock loading. As the acidity of a body of water increases, there is a reduction in the diversity of species and a shift in species composition. The smallest of the organisms, unicellular phytoplankton, disappear, followed by the benthic invertebrates. These organisms serve as important food sources and nutrient recyclers (decomposers) so that large aquatic creatures are adversely affected by their disappearance. Acidification can lead to mobilization of aluminum and possibly other toxic metals, which, in combination with the increased acidity, leads to reductions in fish populations. At pH levels below 5.5, fish populations decline with the most sensitive, juvenile fish first disappearing. The eggs, larvae, and juvenile fish fail to thrive in a process called recruitment failure, often leaving acidified streams and lakes with a diminishing population of older, larger fish. Eventually, entire lakes and streams become devoid of fish, such as in the Adirondacks, mid-Atlantic mountains, New England, Ontario, Quebec, Sweden, and Norway.

Effects on Forests and Plants

The causal role of acid deposition in the decline of forested areas in North America, Germany, and central Europe has been a matter of considerable investigation and debate. Acidic deposition at levels of pH 4.0 to 5.0, which are most common, do not appear to cause widespread adverse effects on forest ecosystems.[17] However, conifer forests such as Red Spruce on mountain tops in New Hampshire, Vermont, and the Appalachians have been more than 80% decimated at the cloud line. Severe damage is also evidenced in central Europe in such places as the Czech Republic and Poland, where 60% to 70% of forests show evidence of damage associated with sulfur and nitrogen disposition. The mechanisms to such destruction are not immediately obvious but may be attributed to combinations of ozone and acid clouds (as low as pH 2.2). This results in damaged leaves; mobilized toxic metals in soil such as aluminum, which adversely effect roots; leaching nutrients from soil; and overstimulation of plants from excess nitrates, which aggravates deficiencies of other nutrients. These factors may combine to increase forest susceptibility to insect and fungal pathogens.[49]

Current Directions in SOx Control

Ambient concentration of SO_2 decreased 37% since 1987 while emissions decreased by more than 14%. In October 1999, the EPA conducted a study to determine whether emission regulations impacted the recovery of lakes and streams in North America and Northern Europe, which had suffered years of acid deposition. The study concluded significant declines of 1% to 6% sulfate per year in many lakes and streams. However, reduced emissions do not automatically translate to immediate improvement in streams, lakes and rivers. For example, some waters in Maine continue to acidify even though sulfuric acid concentrations in rain have decreased. The scientists concluded that several factors contributed to continued lake acidity — variations in climate, increasing levels of nitrogen compounds in precipitation such as nitric acid, declines in the ability of a watershed to neutralize an acid, and the short length of time that data has been collected.[50]

Reductions in SO_2 emissions are due mostly to controls implemented under USEPA's Acid Rain Program.[45] This program features a number of options including switching to low sulfur fuel; using scrubbers that remove SO_2 from the stack emissions; washing coal, which removes up to 50% of the sulfur content; and using advanced combustion technologies such as fluidized-bed combustion to remove sulfur and nitrogen oxides.

Nitrogen Oxides

Nitrogen and oxygen are combined in the high-temperature confines of industrial boilers and automotive engines to produce nitric oxide. This is oxidized in the atmosphere to produce nitrogen dioxide, an orange-yellow to reddish-brown gas that can often be seen enveloping cities from a distance. Nitrogen oxides are an important precursor for ozone and acidic deposition (nitric acid). Nitrates arising from NOx emissions fall from the sky to stimulate algae growth (i.e., red tide) causing toxic conditions for aquatic life in bays and estuaries. Nitrogenous compounds also contribute significantly to the formation of fine particulates. Nitrogen dioxide irritates lung tissues and reduces resistance to respiratory infections, while possibly increasing the incidence of acute respiratory disease in children.[47]

HEALTH IMPLICATIONS OF AIR POLLUTANTS

The criteria air pollutants have been implicated in a number of adverse health problems including reduced lung functions, exacerbating problems such as asthma, increasing risk of chronic bronchitis and emphysema, and aggravating acute respiratory infections. Carbon monoxide, sulfur dioxide and particulates have also been shown to contribute to cardiovascular disease.[40,42,45,51] It is calculated that 125 million Americans breathe unhealthy air resulting in approximately 70,000 premature deaths in the U.S. alone.[52] Fine particulates from motor vehicles and power plants are reported to kill some 64,000 Americans a year and may be a major contributor to the epidemic of childhood asthma sweeping the country and developing nations worldwide.[42]

Asthma occurs in persons with hyperactive airways when the bronchial tubes respond to allergens, pollution, irritants, cold air, and even exercise by small-muscle contraction, dilation of blood vessels, and mucous secretion. The result is suffocating paroxysms where the asthma sufferer struggles for each breath through closed airways clogged with mucous (Figure 10.18).[53] Asthma has increased by 42% since the mid-1950s. In 1998 alone, 17.3 million people suffered from severe asthma,

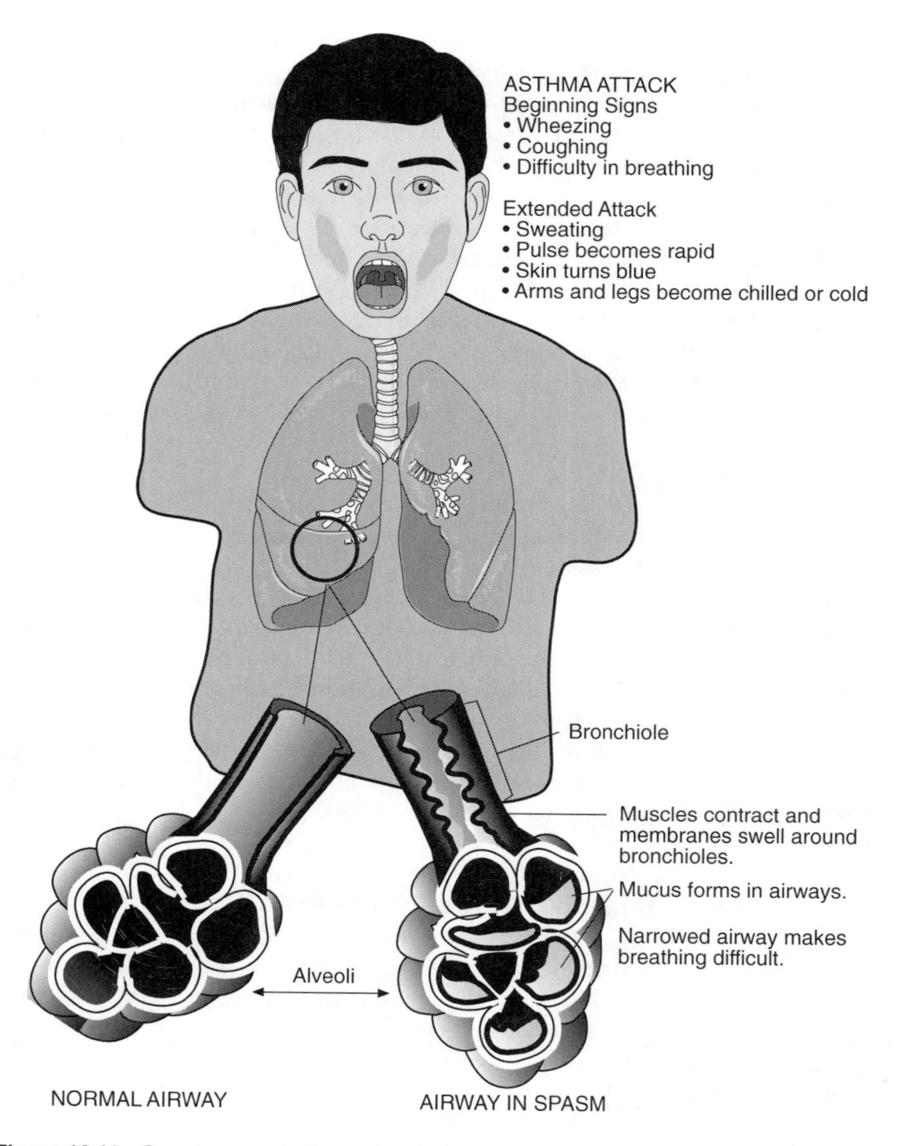

Figure 10.18 Symptoms and effects of asthma.

4.8 million of which were children.[54] According to the Centers for Disease Control (CDC), asthma rates among children below 17 years of age increased 72% between 1982 and 1994.[55] In reality, there has been a two-fold increase in the number of children who have had asthma since 1980. This statistic alone has spurred concern among public health researchers.[56] It has become more severe, and research shows that children, minorities, and urban poor are at greatest risk. Asthma now tends to be concentrated in urban pockets where children live under poor conditions and are

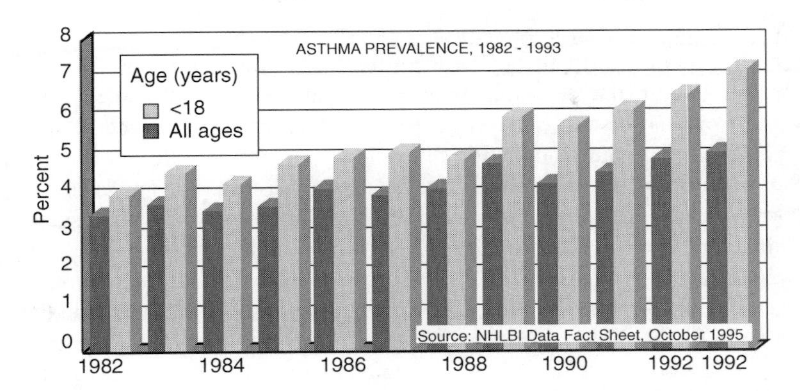

Figure 10.19 Trends in the prevalence of asthma from 1982 to 1996. (Adapted from Friebek, E., The attack of asthma, *Environ. Health Perspect.,* 104:1, 22, 1996.)

frequently exposed to allergens such as mites and cockroaches[57] and to episodes of air pollution involving fine particulates and ozone (Figure 10.19).[58] Air pollutants, particularly ozone and fine particulates, were found to be consistently associated with asthma exacerbation in a prospective cohort study of children with asthma.[59] However, this link to specific air pollutants remains controversial. The prevalence of asthma is rising in both industrialized and developing countries, and the two countries with the greatest increase in asthma prevalence are Australia and New Zealand, neither of which has extensive air pollution.[60] Serious questions remain to be answered on this controversy. Have all the responsible gases been identified? Is there an interaction of gases and particulates? What are the mechanisms by which pollutants exacerbate asthma?

Due to the rapid increases in severe asthma cases, alternative plans and programs have been started to aid asthma research. For example, President Clinton included $50 million in grants for his 2000 budget. These grants will be used for states to test methods for improving the health of children with asthma who are enrolled in the Medicaid program. In addition to the asthma grants set aside by Clinton, the National Institute of Health (NIH) plans to spend $118 million on asthma research. Furthermore, the National Institutes of Occupational Safety and Health (NIOSH) has begun a nationwide study to look into the occurrences of asthma in children who attend public schools.[54]

Main Mechanisms of Air Pollutant Effects on Respiratory System

Air is generally laden with dust, microbes, and gaseous emissions that may create adverse effects to human health. However, the human respiratory system has a number of defense mechanisms to ward off such attacks. The first line of defense includes nasal hairs that trap larger particles. A combination of mucous and cilia line the bronchi and bronchioles that trap and move smaller particles out of the respiratory tree. Irritating dusts and gases may cause the smooth muscles of the bronchioles to constrict, restricting heavier exposure in the short term. However, prolonged or exaggerated constriction is often combined with mucous secretions leading to breathing difficulties. Particles that reach the deeper portions of lung, such as the alveoli, may be confronted by the alveolar macrophages, which can ingest, destroy, and discard many microbes. Pollutants may produce their adverse effects by inhibiting and inactivating mucociliary streaming; killing or neutralizing alveolar macrophages; constricting airways; causing vasodilation and excess mucous secretion; or causing changes in alveolar cell-wall structure through abscesses and thickening which causes scar formation. The formation of scar tissue is characteristic of emphysema, in which the walls of alveoli disintegrate and the compartments become unnaturally enlarged. Elasticity of the living tissue diminishes and the lungs are unable to recoil and expel air, reducing the efficiency of the lungs in enhancing oxygen and carbon dioxide. Victims ultimately may become wheelchair-bound, gasping for each breath, and unable even to expel sufficient air to blow out a candle. There is no cure for emphysema.

INDOOR AIR POLLUTION

Imagine suffering through episode after episode of pneumonia within a few short months, or suffering an attack of breath-restricting asthma every time you enter the basement of your home, or developing nausea and violent headaches upon entering your place of employment each day. These are examples of afflictions associated with indoor pollution in private homes and public places.[61] At one time, most people were aware of the potential threats of outdoor air pollution, but today many experts are cautioning against the effects of indoor pollutants. There has been a growing body of scientific literature in the last several years that points to indoor air as being more polluted and potentially more hazardous than outdoor or ambient air.[62] Additionally, people spend an average of 90% of their time indoors while some at-risk subgroups such as the elderly, very young, and chronically ill may spend nearly all their time indoors.

Sources of Indoor Air Pollution

Sources of indoor pollution come from several major categories that include combustion sources such as acid, coal, gas, oil and kerosene; household cleaning products; furnishings such as cabinetry or furniture made of pressed wood; newly installed carpeting that emits volatile organic carbons (VOC from adhesives); building materials and paints that off-gas volatile compounds; products for personal care and hobbies; cooling and humidification systems; external sources such as radon and tobacco smoke; and moisture that supports the growth of molds and other biological contaminants (Figure 10.20). Sources such as air fresheners, paneling, cabinets, and rugs may emit pollutants at a fairly continuous rate for some time, while other pollutant sources or activities release pollutants intermittently such as cigarette smoking, use of household cleaners, or pesticide sprays. How long pollutants remain in

1. Moisture
2. Pressed wood furniture
3. Contaminated humidifier
4. Moth repellants
5. Personal care products
6. Room air fresheners
7. Chemical cleaners and disinfectants
8. Pressed wood cabinets
9. Unvented gas stove
10. Household chemicals
11. Tobacco smoke
12. Wood stove
13. Wood paneling
14. Asbestos pipe insulation
15. Unvented dryer
16. Radon infiltration
17. Pesticides
18. Auto exhaust
19. Auto cleaners and additives
20. Paints, thinners, and stains

Figure 10.20 Sources of potential indoor air pollutants. (Adapted from Anonymous, Carbon monoxide-heart failure link, *Environ. Health Perspect.*, Vol. 104, 21, 138, 1996.)

the air and their relative concentration depends on several factors. Elevated heat and humidity increases the release of some pollutants into the air such as a formaldehyde. Concentrations are greatly influenced by the amount of ventilation as well.

Air enters a house three ways:

1. *Natural Ventilation*: Air enters through open windows and doors, ridge vents, the roof, ventilation pipes, or other portals designed for air access.
2. *Infiltration*: Air is drawn into the house through cracks around doors and windows, spaces in construction joints, or spaces or cracks in the foundation.
3. *Mechanical Ventilation*: Air may be drawn out of the home through externally vented fans or into the structure by whole-house fans in air handling systems.

The rate at which air is replaced in the structure by external air is termed the air exchange rate. This average rate for an American home is 0.7 to 1.0 air changes per hour. Houses that are tightly sealed without provisions for an exchange may experience as little as 0.2 air changes per hour, while drafty houses may approach up to 10 times that amount.[48] More recently constructed office buildings and nonresidential structures may have sealed windows that restrict air infiltration. While ventilation and cooling systems are intended to maintain comfort, air that is recirculated in a building without sufficient outside makeup air may exhibit increased levels of CO_2 and pollutants. Prior to 1973, at least 15 cubic feet per minute (CFM) per occupant was required for make-up air. However, the standard was reduced after 1970 to 5.0 CFM; but this level may be insufficient in providing a safe and healthier environment.

Signs of Indoor Pollution

Signs of indoor air pollution may include physical or health symptoms or both. Physical symptoms may include heating or cooling equipment that is dirty and/or moldy; moisture condensation on walls and windows; air that has a stuffy or unpleasant odor; and signs of water leakage anywhere in the building with the growth of molds. Health indicators of indoor air pollution may include immediate or acute effects such as eye irritation, dry throat, headaches, fatigue, sinus congestion, sun irritation, shortness of breath, cough, dizziness, nausea, sneezing, and nose irritation. These symptoms are vague and may originate from many sources. Some people recognize that the symptoms occur when they enter a structure and diminish when they leave. In many cases, however, there is not a readily identifiable pattern or illness. Additionally, chronic illnesses such as emphysema, cancer, and heart disease may not appear for years; and any association with exposure to substances in a particular building is usually not recognized. When a number of occupants of a building display acute symptoms without a particular pattern and the varied symptoms cannot be associated with a particular source, the phenomenon is often referred as sick building syndrome (SBS). Further, maladies may not be the cause of indoor air pollution but may be related to psychological stress, inadequate lighting, excessive heat or cold, or even noise.[63]

When well-defined illnesses occur in a building and they can be traced to specific building problems, the illnesses are referred to as building-related illnesses (BRI).

The well-defined illnesses normally include asthma, hypersensitivity pneumonitis, humidifier fever, or Legionnaire's disease. The problems associated with SBS or BRI have apparently increased as concerns about energy conversation have escalated, resulting in the construction of airtight buildings with thicker insulation, magnetically sealed doors, sealed windows (office buildings) or triple-glazed windows that serve to limit the exchange of air with the outside. Office buildings are at risk of indoor air pollution from poorly designed or improperly used or maintained ventilation systems; the presence of pollution sources such as harsh chemicals, improperly vented photocopiers, or biological contaminants from dirty ventilation systems; and uses of the building for which the building was not designed, such as a paint shop or automotive repair service in the basement of an office building.

Common Sources of Indoor Air Pollution

The most common sources of indoor pollution include environmental tobacco smoke, radon, biologicals, nitrogen dioxide, carbon monoxide, organic gases, formaldehyde, respirable particles, and pesticides.

Environmental Tobacco Smoke (ETS) and Other Combusted Materials

Although smoking has been under vigorous attack on the U.S. and the opportunities for smoking in public buildings has greatly diminished, smoking still contributes to nearly 500,000 deaths each year in the U.S.[64] The smoker is exposed to nearly 4700 compounds in mainstream smoke through direct inhalation. The sidestream smoke that originates from the burning end of the cigarette and exhaled smoke make up environmental tobacco smoke (ETS) to which non-smokers may be exposed in a form referred to as "passive" or "second-hand" smoke. Such involuntary exposures have prompted numerous legal actions to further restrict the public places where smoking is allowed. However, people continue to be involuntarily exposed to ETS in the home and many commercial establishments, increasing the risk of respiratory disorders. Second-hand smoke has been linked to the rise of lung cancer risk by an average of 30% in non-smoking spouses.[61] Other sources of combustion products include gas and kerosene space heaters that are unvented, and improperly vented wood stoves, fireplaces, and gas stoves. Pollutants from these sources include carbon monoxide, nitrogen-oxides, particulates, and a variety of aromatic hydrocarbons and acid aerosols. Unintended deaths occur every year from the buildup of CO in the home because of improper ventilation of home space heaters or furnaces. Lesser concentrations may lead to headaches, dizziness, and aggravate preexisting heart disorders. Nitrogen-oxide exposure can increase the risk of respiratory infection, emphysema, and may aggravate respiratory disorders such as asthma.

Radon

Radon is a colorless, odorless gas that occurs naturally by the decay of radium-226. It is found most everywhere but is highest in concentration in uranium and

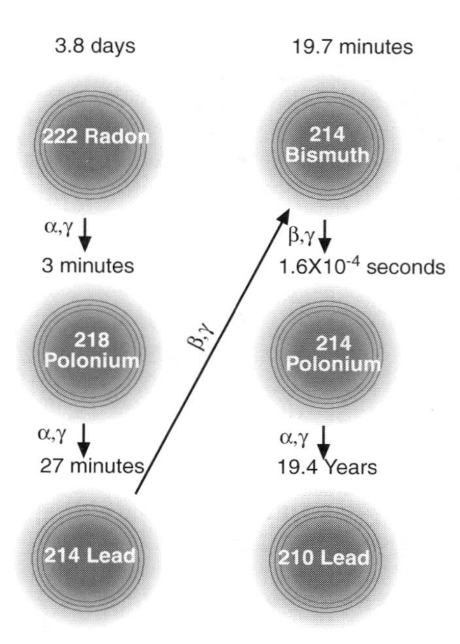

Figure 10.21 Radioactive decay of products of radon.

phosphate ores. Areas with lesser concentrations of radon are found in limestone, shale, and granite. Radium-226 is ubiquitous in low concentrations and does not present a health issue until it is trapped and concentrated in areas such as the basement of a building or house built in soils where uranium is in the soil or rock.

As the uranium naturally radioactively decays, it releases radon gas that further decays into short-lived radon daughters and gamma rays (Figure 10.21). The radon daughters posing the greatest risk are the alpha-emitters polonium-218 and 214. The radon progeny quickly attach to particles in the air, which may then be inhaled to lodge in sensitive areas of the upper respiratory tree. Once lodged in human tissue, the radioactive materials increase the risk of lung cancer, causing from 5000 to 20,000 excess cancer deaths a year in the U.S.[62]

Recent evidence released by the National Research Council points to radon as a significant public health problem causing as many as 21,800 deaths per year, with the greatest risk to people in homes where cigarettes are smoked. Radon is considered to be the second-leading cause of lung cancer exceeded only by cigarette smoking.[65] Radon in soil normally enters structures under slight negative pressure through cracks in concrete walls and floors, sumps, floor drains, dirt floors, joints, and crawl spaces. Radon may also be present in well water or even foundation materials. It is estimated by the USEPA that as many as 8 million homes in the U.S. have elevated radon levels. Although some regions are at greater risk than others, buildings and households should be checked since even adjacent houses may show remarkably different levels of radon. Most hardware stores have radon detection kits for sale. The kits are either alpha track detectors or chemical canisters. The kit instructions usually suggest hanging the detector in a closed cellar at 3 to 4 feet above the ground for several days to a month. The kits are returned by mail to the manufacturer for analysis, and a report is usually returned within 2 to 3 weeks to the homeowner or building manager. The federal action guidelines recommend that corrective actions be taken at 4pCi/L or above. Radon concentrations are reported as picocuries per liter (pci/L). A picocurie is the equivalent of one trillionth of a curie representing slightly more than 2.0 radioactive disintegrations per minute. The average concentration of radon in ambient air products is about 0.25 pci/L.

Some of the highest levels of radon have been measured in a region called the Reading Prong, located in an area covering Southeastern Pennsylvania and parts of Connecticut and New Jersey.[66] The highest level ever recorded in a household of 2600 pci/L was identified in Southeastern Pennsylvania. The Northeast has higher mean levels than most other sections of the country, but the presence of radon in a home is unpredictable. Households must be tested to determine if they exceed the recommended action level of 4 pci/L. Surveys have shown that one in every 15 homes in the U.S. exceeds the action guideline. At this level, the estimated number of lung cancer deaths to radon exposure would be 13 to 50 per 1000 for an average lifetime exposure. This risk is equivalent to smoking five cigarettes a day.[62,67]

The mitigation of radon may involve a number of actions such as installing sub-slab or basement ventilation; covering sump holes; sealing cracks and other openings in floors, joints, and walls; ventilating crawl spaces; installing air-to-air heat exchanges; or installing a drain tile suction system around the outer foundation walls. In some cases, more than one method may be used.

Biological Contaminants

The most common biological contaminants include molds, mildew, viruses, bacteria, dust mites, cockroaches, pollen, animal dander, and cat saliva. The sources for these agents are numerous, including people-transmitted viruses and bacteria, household pets, contaminated air-handling or ventilation systems, moist surfaces on which organisms can grow, dried fecal material from cockroaches and other insects, and soil and plant debris. The major threat to the biological contaminant of the home is moisture. Water may enter through a porous basement floor or walls, a leaky roof, or remain in unventilated bathrooms or as a reservoir in dirty humidifiers. In these situations, water stimulates the growth of mold, dust mites, and other infectious or allergenic organisms.[68]

Surveys have shown that 30% to 50% of homes surveyed in the Northern U.S. and in Canada have problems with moisture, while structures in the South may be at higher risk. Wet surfaces, moist carpeting (often associated with concrete floors), water-damaged materials, and standing water can all support the growth of molds, mildew, bacteria, and insects, such as house and dust mites. Dust mites are microscopic insects that feed on shed human skin, thrive in moist environments, and are powerful allergens. Molds are also potent allergens, and they can provide nutrients to bacteria, spread viruses, emit chemicals such as cancer-causing phenols and trichloroethylene, increase susceptibility to a host of diseases by weakening the immune system, and initiating aggravating asthma.

Possible symptoms of illness caused by biological contaminants include running nose, colds, flu-like symptoms, headaches, unexplained fatigue, and digestive problems. Allergic rhinitis, hypersensitivity pneumonitis, asthma, and other serious allergies are common. Some biological contaminants transmit infectious illnesses such as influenza, measles, and chicken pox while molds and mildews release disease-producing toxins.[62,68] Homes that have water stains or other signs of moisture, leaking, plumbing, mold, mildew, or dust remaining after cleaning are signs of a possible water problem.

Reducing Exposure

Since moisture always migrates to colder surfaces, taking ample steps such as using fans, warming basements, ventilating attics and bathrooms, and insulating exterior walls can check the unwanted movement of water. Gutters may be used to drain water away from buildings. Other methods of controlling biological vectors include washing bedding frequently in hot water, maintaining humidity between 30% to 50%, using a high-efficiency particulate filtering vacuum cleaner, enclosing pet litter boxes, discarding wet carpeting and other water-damaged materials, and practicing good housekeeping procedures.

Carbon Monoxide, Nitrogen-Oxides and Respirable Particles

These substances are manually emitted from unvented kerosene and gas heaters, wood stoves, fireplaces, environmental tobacco smoke, automobile exhausts in attached garages, and other combustion sources. These have been discussed previously under combustion sources.

Organic Gases and Pesticides

Most households and many businesses have a wide variety of products that can emit potentially hazardous materials including paints, strippers, disinfectants, cleaners, repellants, automotive products, hobby supplies, volatile office supplies, and pesticides. Such substances often exceed levels found outside the home by a factor of up to five times, while some activities such as paint stripping may produce concentrations of 1000 times outdoor levels.[62,69] The potential health consequences are numerous based on the extraordinary variety of chemicals available in the home and business. These symptoms range from initiation to mucous membranes, multiple organ damage, headaches, nausea, cancer, and many other maladies. Exposures to such substances can be limited by properly disposing of unneeded products; following instructions on a label and using the chemical in a well-ventilated area if so advised; and using non-toxic alternatives where possible.

Formaldehyde (HCHO)

Formaldehyde is found in pressed wood products such as cabinets and furniture made from plywood, particleboard, wall paneling, and fiberboard. It is also found in urea formaldehyde foam insulation (UFFI), some textiles, and environmental tobacco smoke. However, formaldehyde-omitting products are now being manufactured with much less HCHO emission potential, and emissions from pressed wood products are presently 10% of those used 20 years ago.[66] Since older products and structures have off-gased most of the volatile HCHO, and new products have very little HCHO, the significance of formaldehyde as a public health issue has been markedly reduced. Exposures still can occur, which is aggravated in the presence of a confined or poorly ventilated space, combined with high heat and humidity. Adverse health effects include irritation to the mucous membranes, severe allergic reactions, fatigue, wheezing, and coughing.

NOISE

Introduction

People spend an average of 90% of their lives indoors and are exposed not only to chemicals and biologics but also to sounds that enter their homes in the form of heavy traffic, descending or disembarking airliners, trains, and loud radios. Outside they are confronted with the additional sounds of snowmobiles, jet skis, and motorcycles in areas previously thought to be quiet reserves. Sound itself is not a pollutant; but when it interferes with tasks, when it distracts, annoys or disturbs, or when it causes losses in hearing or alters physiology in a negative way, then it becomes unwanted sound, or noise. Noise may be considered a pollutant whose primary medium of transmission is air, so it is considered in this chapter as another form of air pollutant. Organizations such as Noise Network have been created for people who cannot endure sources of noise disturbance any longer. The organizations help clients relocate from an area of high noise to one of a lower frequency of occurrence. Tragedies have been the result of normally noisy areas. For example, a 78-year-old farmer committed an unthinkable act after months of enduring noisy neighbors. He reached his breaking point, shooting two of the neighbors, killing one, and wounding the other. He was later convicted of manslaughter and is presently serving two consecutive life sentences. This is an uncommon effect of noise pollution; however, it is true and could be a cause for alarm down the noisy road.[70]

The Physics of Sound

Sound is a form of energy that is produced by the vibration of objects which compress and expand air, water, or solids to produce waves. When the vibration occurs in a car, variations are produced in the normal atmospheric pressure which are carried to the eardrums and cause the membranes to vibrate. These vibrations are passed to the inner ear and ultimately to the brain, which interprets these vibrations of acoustical energy as sound. Because sound depends on the movement of molecules causing positive and negative pressure waves, it cannot travel in a vacuum. As an object vibrates, waves radiate outward in an expanding sphere much like the ripples produced when a stone is dropped in water. Waves are produced

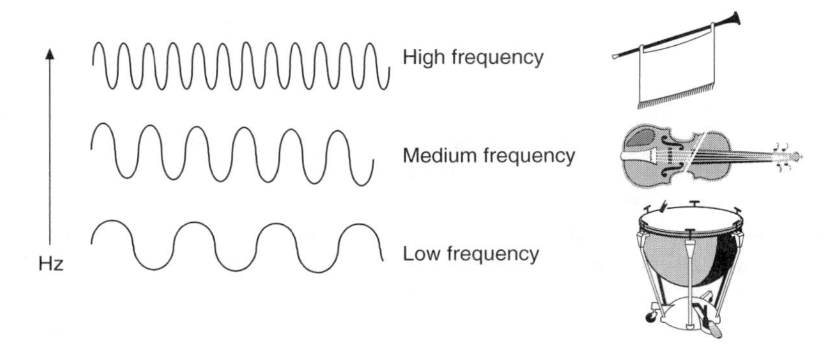

Figure 10.22 Frequency differences of sound. Higher frequencies are associated with more cycles per second (hertz), and are perceived as higher pitch.

which consist of alternating peaks and troughs that may vary in their length, peak, and amplitude. These wave characteristics produce changes in frequency (or pitch) and amplitude, which translates to loudness. Frequency is directly related to the speed of vibration. The faster the vibration, the more frequent the waves are per unit time and the higher the pitch (Figure 10.22). Frequency is expressed as cycles per second, and the standard unit of expression is hertz (Hz), which is equivalent to one wave per second passing over a particular point. The normal range of human hearing is in the frequencies of 50 to 20,000 Hz. Sound waves also vary in the height or amplitude and reflect the amount of energy formed in the sound wave (Figure 10.23). Soft sounds have a low amplitude while loudness is characterized by large amplitudes. This amplitude intensity or loudness is measured in decibels (dB). The scale of dB ranges from zero (0) dB, which is the threshold for hearing among the normal population, to 194, which is the threshold for pure tones.[71,72]

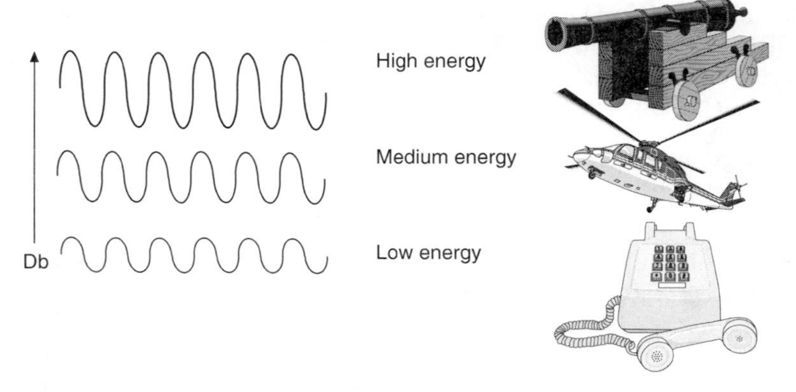

Figure 10.23 Amplitude differences of sound. Higher amplitudes or wave heights are associated with greater energy and are perceived as loudness.

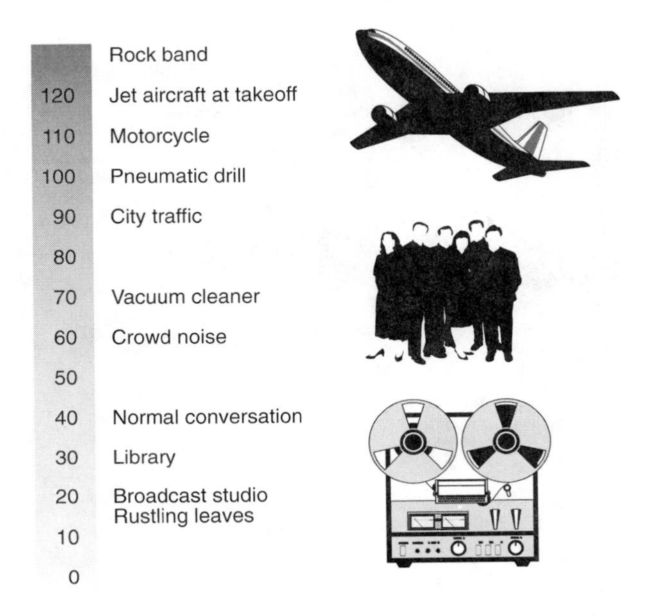

Figure 10.24 Sound levels of various activities.

The perception of sound (hearing) is not uniform across frequencies (Hz) and amplitudes (dB). Human hearing is most acute on the range of human speech, which is about 400 to 4000 Hz. At frequencies on either side of the range, hearing is less sensitive and greater amplitudes are required to hear these frequencies. At amplitudes below zero dB, humans cannot detect sound pressures although they exist, and as sound increases above 70 dB it becomes increasingly difficult to participate in conversation. Sounds above 90 dB may lead to hearing loss. The pain threshold for noise begins at 120 to 130 dB for most people. Since the perception of loudness doubles with each 10 dB increase, this is not a linear relationship, and small differences in dB readings are heard as very large changes.[72,73] The relationship of decibels to common sounds is shown in Figure 10.24.

Physiology of Sound and Health Effects

Sound compression waves are directed by the external auditory canal (meatus) where they strike the tympanic membrane, causing it to vibrate in concert with the waves. The tympanic membrane is connected to a series of three very small bones in the middle ear known as the malleus, incus, and stapes, which transmit the vibration to the oval window of a small-shaped structure called the cochlea (Figure 10.25). The transmitted vibration causes corresponding waves in the liquid (perilymph), which under appropriate conditions causes the basilar membrane to

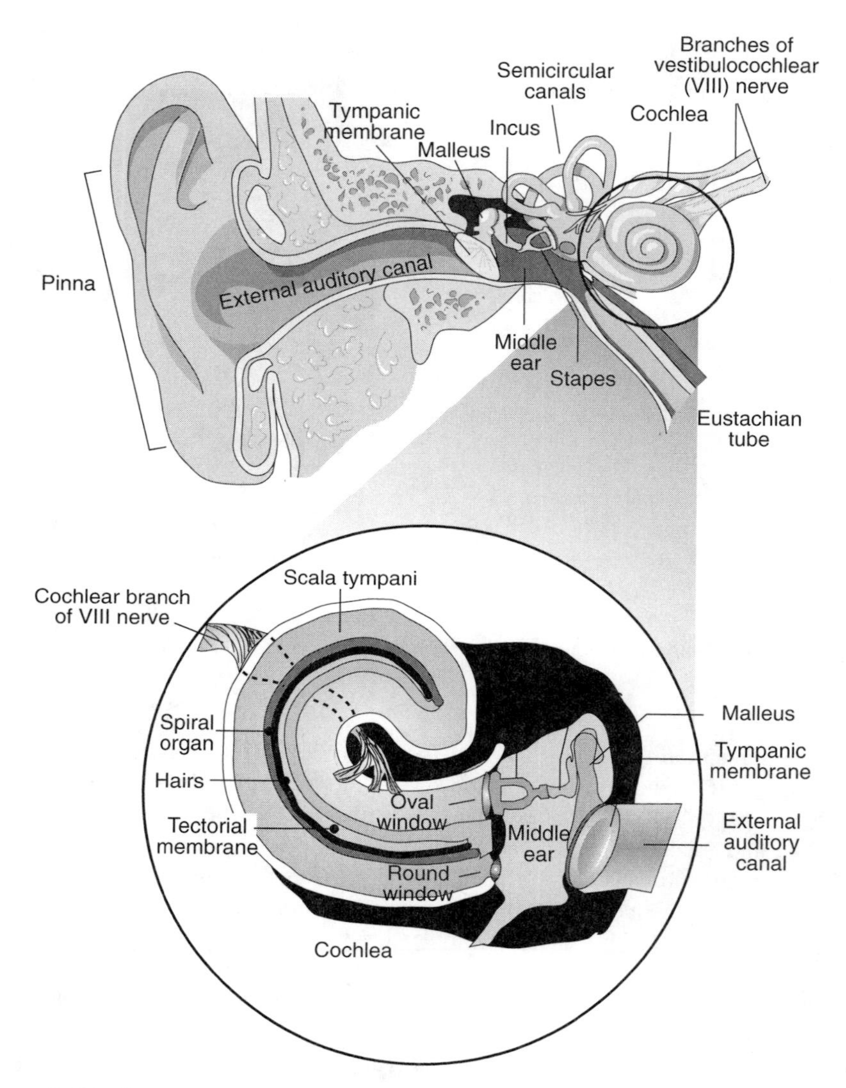

Figure 10.25 Structure of the ear and hearing apparatus.

vibrate. This in turn causes hair cells of the spiral organ (organ of Corti) to move against the tectorial membrane. The hair cells convert a mechanical force (pressure) into a nerve impulse (electrical signal) which is ultimately transmitted to the auditory area of the temporal lobe of the cerebral cortex.[73] Excessive sound pressure (loud noises) can destroy the delicate hairs in the spiral organ. This process usually begins with the hairs at the base of the cochlea most sensitive to high-pitched sounds and with continued exposure to high dB sound. The damage extends to other hairs and [illegible] 2 Study Guide

[illegible] the book [illegible] What does [illegible]?

[illegible] things [illegible] familiar with [illegible] for Quiz 2:

- What is [illegible]? Who are the [illegible] and the history [illegible] relevant [illegible] post-911 [illegible] What [illegible] emergency response [illegible]

- [illegible] alpha particles [illegible] gamma [illegible] ionizing and non-ionizing radiation [illegible] types of protection [illegible] safety [illegible] for radiation [illegible]

- [illegible] does the [illegible] What [illegible] permanent damage [illegible] protect [illegible]

Q1 92 .15

Q2 95 .15

Q3 94.3 .15

FWU 98 .2

Research .15

Final .2

EH 2000: Quiz 3 Study Guide

Quiz 3 will draw from both the book and my PPT slides. 2 hours max.

These are some specific things to be familiar with and know for Quiz 3:

- What is the ICS? What are the elements of it, and the hierarchy? Why does it exist? How has it become so much more relevant to EH in the post-911 world? What are some functions of the EH professional in emergency response efforts?
- Describe: radioactivity, isotope, alpha particle, beta, x-ray, and gamma radiation; distinguish between ionizing and non-ionizing radiation; what type of radiation is UV radiation (e.g., from the sun)?; what are the major safety controls used for radiation?
- How does the human ear work? What are the parts and when permanent damage results, what specific part is irreparably destroyed? How do we protect workers from hearing damage?

Who are the most significant utilizers of water in the world? What percentage of the world lacks decent water? Who dies from water pollution globally, and in what numbers? Who or what causes the most pollution, both globally and/or in the USA alone? What is the major source of material put into domestic landfills?

- Describe some of the broad environmental regulations for waste or hazardous materials control and use, including: MSW, TSCA, EPCRA, CWA, RCRA, CERCLA, MSDS, CFCs, O3 (ozone), NPDES.

- What is a Criteria Air Pollutant? Be familiar with the major categories of air pollutants, their sources, and toxic effects (i.e., why do we consider them hazardous air pollutants?). Understand an atmospheric inversion, and be familiar with significant inversions or other air pollution events from the past, both in the USA and especially in Europe.

- What is a hazardous waste? Who defines hazardous wastes? Who regulates it? Are radioactive wastes hazardous wastes, in the legal sense? What is pH and how is it related to waste categories? What is a landfill, an incinerator, a deep injection well, etc.?

corresponding lower pitches. Such continued exposure can lead to increased damage of hair cells until 30% to 50% of the hair cells are destroyed, at which time measurable hearing loss becomes evident. Such damage resulting in loss of hair cells produces permanent hearing loss and is known as permanent threshold shift (PTS). Permanent hearing loss can result from prolonged exposures to sound levels of 70 dB to 85 dB, which are at or above the level of vacuum cleaners or city traffic. Higher sound intensities require much less time to cause damage, and being on a jet plane landing field without ear protection could cause measurable hearing losses in less than an hour. Noise can also result in a temporary threshold shift (TTS), which is a reversible hearing loss normally lasting from moments to hours, although longer periods of temporary hearing loss are possible. Very loud explosions or gunfire can cause acoustical trauma, which can rupture or damage eardrums, middle ear bones, and even the organ of Corti.

The measurement of hearing loss is usually performed on an instrument called an audiometer, whereby a person is placed in a soundproof room and listens on earphones to sounds produced at different pitches and decibels by a trained audiologist. Persons with hearing problems will be unable to detect sounds at normal threshold levels, and the intensity will have to be increased. If normal values are 0 to 25 dB, then hearing the same sounds at 26 to 40 dB would indicate a mild hearing loss. A severe hearing loss would be 71 to 90 dB.

Noise has also been associated with adverse health effects beyond hearing loss. Noise may be an annoyance. The annoyance level appears to escalate as a function of intensity (loudness) and pitch. Therefore, loud, high-pitch sounds (above 5000 Hz) tend to be the most annoying.[73] Related to this annoyance is the creation of stress and associated physical effects such as increased blood pressure, elevated heart rate, headaches, colitis, and ulcers. Such responses may seem unreasonable, but in the presumably quieter realm of historic man, sudden loud sounds prompt immediate physiological responses to the presence of danger as a mechanism of survival. Unfortunately, prolonged exposure resulting in continued, exaggerated physiological response leads to damaging health consequences.

Noise can lead to interference with speech communication. Most of us have been in the presence of a loud band or noisy crowd where we had difficulty hearing a person yelling at us within a few inches from our ears. Noise ratings have been developed that measure its ability to interfere with speech communication or a scale known as the preferred octave speech-interference level (PSIL). The PSIL is useful in determining what changes in PSIL are necessary to making communication easier.[74] Noise levels have also been implicated in decreased learning ability, work performance, increased safety hazards, and sleep disruption.[69,71]

Regulation of Noise

The regulation of sound requires that it be measured according to a standard. The measuring device most frequently used is a sound pressure level meter that converts sound captured by a microphone to electrical signals which can deflect a needle on an indicating meter (Figure 10.25). Most instruments contain A, B, and C scales. The A-weighing network is biased to high-frequency noise and best correlates with annoyance effects and hearing damage in humans. Thus, the sound reading on this scale would be expressed in dBA units. This is the scale most frequently used in assessing noise exposure. Exposures are regulated for the occupational environment by the OSHA (Occupational Safety and Health Administration) Department of Labor with a permissible exposure limit of 90 dBA for an 8-hour day, 40-hour work week. Permissible exposure times decrease half for each five dB increase.

Federal agencies that regulate noise include the Federal Highway Administration (FHA), Housing and Urban Development (HUD), OSHA, Federal Aviation Agency (FAA), and the U.S. Environmental Protection Agency (USEPA). Each of these agencies is responsible for regulating and controlling noise to specific operations or segments of society under their direction. OSHA regulations are designed to protect workers from occupational noise-induced hearing losses. The USEPA is mandated under the Federal Noise Control Act of 1972 and Quiet Communities Act of 1978 to abate noise in the ambient environment and in communities through identifying levels of environmental noise considered protective of human health; identifying major noise sources requiring regulation; and providing financial assistance to state and local governments to investigate noise problems and develop control measures in the communities. Many programs were initiated in the 1970s, but the elimination of federal funding and associated programs in the 1980s all but did away with state and local noise control programs throughout the country. Many federal regulations under the NSAD 1972 noise abatement were never promulgated, and it is unlikely that significant regulatory attention will be paid to issues of noise until there is public outcry focusing on the increasing noise pollution in our environment.[72]

RADIATION

Introduction

Marie and Pierre Curie termed the mysterious phenomenon "radioactivity" and isolated two radioactive substances, polonium and radium, from uranium ore samples. These discoveries, along with the discovery of x-rays, quickly led to the birth of a whole new field of scientific inquiry with far-reaching implications in physics, biology, medicine, and, unfortunately, warfare.

Ionizing Radiation

Early investigations of radioactivity revealed that the observed emissions were of several different kinds, consisting of subatomic particles — protons, neutrons, or electrons, which are released when atoms decay, later becoming known as particulate radiation. The particulate radiation includes alpha and beta particles. Atoms are the basic units of elements and consist of a small, dense center, called a nucleus,

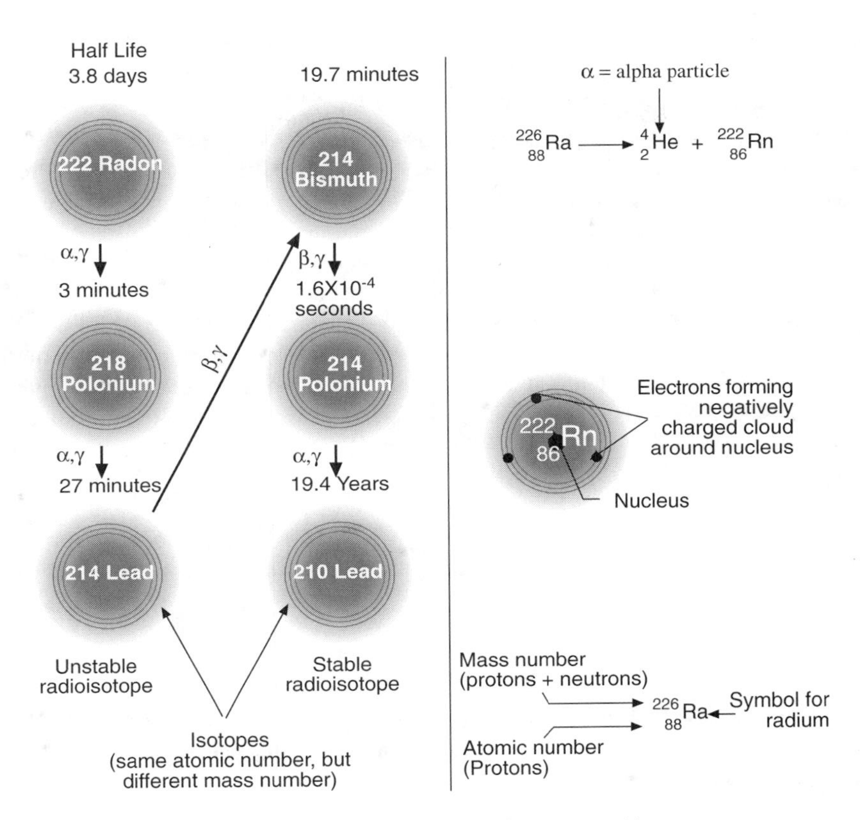

Figure 10.26 Radioactive decay of products and nuclear composition.

surrounded by a cloud of negatively charged electrons (Figure 10.26). The positive charge in the atom is attributed to the protons or atomic number. The protons account for approximately half of the mass of the atom, with the other half composed of neutrons that carry no net electrical charge and are therefore neutral. The total number of protons and neutrons is equal to the mass number of the atom. A particular element may have more than one mass number. Lead may have mass numbers of 214 or 210, although they have the same atomic number. Such atoms are called isotopes, and the difference in mass number results from different numbers of neutrons. Many elements have stable and unstable isotopes. Unstable isotopes or radioisotopes (Figure 10.26) spontaneously emit energy in the form of radiation that was originally detected by their ability to expose photographic film in the dark. When a radioisotope decomposes, it releases energy in the form of electromagnetic radiation (g- or x-rays), and energy of motion from particles (a or b). Naturally occurring gamma rays (g) consist of highly energetic short wavelengths of electromagnetic radiation, a form of energy which also includes ultraviolet light, visible light, infrared waves, and microwaves.

Radioisotopes

Ionizing radiation includes alpha (a) and beta (b) particles, x-rays and gamma rays because the particles or rays involved are sufficiently energetic to dislodge electrons from the atoms or molecules they encounter. Ionizing radiation's ability to destroy chemical bonds gives it special significance. An atom that is missing one or more electrons is referred to as an ion, and energetic radiation capable of doing this is called ionizing radiation. X-rays and gamma rays are included in this category and are capable of breaking chemical bonds and ionizing molecules. When critical molecules are disrupted, the cell will undergo degenerative changes or even die. A critical molecule in human and animal tissue is DNA. Injury results from the ionization of water molecules in tissue that form ion pairs (free radicals), which instantly react with the cell components to produce cross-linking of macromolecules causing changes in the tertiary, or 3-dimensional, structure of the molecule. These changes may disrupt the function of proteins, deoxyribonucleic acid (DNA) and other molecules, causing mild reactions to cellular death. The disruption of the DNA molecule may cause genetic errors to occur that will be evidenced in abnormal growth and maintenance of the affected cells, and death of the cells is often the ultimate outcome.

Radiation Exposure

Every individual comes into contact with ionizing radiation from three general sources: naturally occurring; naturally occurring but enhanced by human actions; and human generated.

Natural Sources

The Earth, air, water, and food all contain traces of radioactive materials that constitute a source of "background" radiation. The following are sources of background radiation.

Cosmic Radiation

These are high-energy particles composed primarily of protons and electrons that stream toward the Earth from outer space and the sun following episodes of solar flares. Exposure to cosmic radiation is considerably less at sea levels than at high altitudes. For this reason, residents of Denver receive about twice as much cosmic radiation than the residents of LA or Miami.

Radioactive Minerals on the Earth's Crust

Uranium, thorium, potassium, and radium are all found in soils and rocks in many parts of the world. Inhabitants of areas such as the Rocky Mountains are exposed to higher levels than those on the east coast, where radioactive minerals are much less abundant.

Radionuclides in the Body

A number of radioactive substances enter the body by ingestion of food, milk, or water, or by inhalation, which are incorporated into body tissues where their concentration may be sustained at a steady rate or may increase with increasing age.

Enhanced Natural Sources

The impact of human activity on levels of exposure would be much lower under other conditions. Examples of enhanced, natural sources include uranium mill tailings, phosphate mining, and jet airline travel (due to the increased exposure to cosmic radiation at high altitudes).

Human-Generated Sources

These include medical applications, nuclear weapons fallout, nuclear power plant emissions, and consumer products. Medical applications include medical x-rays and radiopharmaceuticals for both diagnostic and therapeutic purposes. Nuclear weapons fallout from the first atomic bomb to the end of atmospheric testing of nuclear weapons by several major powers of the world, primarily the U.S. and the former Soviet Union, resulted in significant amounts of radioactive fallout worldwide. Very small amounts of exposure are received from home smoke detectors, static eliminators, airport security checks, TV receivers, and tobacco products.

Health Impacts on Ionizing Radiation

During the early years of work with radioactivity, the medical community itself, particularly radiologists, saw a much-increased incidence of leukemia. This was due to their enthusiastic use of new machines that exposed them and their patients to large amounts of radiation on a daily basis.

Dosage

Since ionizing radiation can neither be seen nor felt, human exposure is measured in terms of the amount of tissue damage it causes. Under the International System of Units, the term gray (GY) is the unit of absorbed dose, used to quantify the amount of energy from ionizing radiation absorbed per unit mass of material. There is little evidence that present average levels of exposure to ionizing radiation are having a serious adverse effect on the health of the general public. However, higher doses can be injurious to living tissues. The International Congress of Radiation Research has been involved in a long-running argument over what is considered a safe dose of radiation. U.S. radiation experts from the University of California (UC) at Davis have stated that there are safe values of exposure to radiation (i.e., threshold) that the body can tolerate without the development of cancer. Conversely the International Commission on Radiological Protection stands behind the linear, no-threshold theory, stating that the indicated safe dose proposed by the UC Davis researchers does not exist. These arguments about the validity of the linear, no-threshold theories are important because regulators use the theory when setting radiation exposure standards.[75]

Dose Rate

The time span during which a given amount of radiation is delivered is thought to be very significant in determining the extent of tissue damage incurred, perhaps even more important than the total dose.

Radiation-Induced Mutations

Exposure to low levels of ionizing radiation can result in gene mutations. There is a linear relationship between increased radiation dose and increased numbers of mutations. Nuclear power plant workers and medical personnel who receive small doses of radiation on a daily basis have been shown to exhibit a marked increase in the frequency of mutations in their chromosomes.

Radiation and Birth Defects

A number of congenital abnormalities, as distinguished from defects caused by mutant parental genes, are known to result from x-ray exposure *in utero*. Sensitivity to radiation is greater during fetal development.

Radiation-Induced Cancer

The potential for health damage as a result of radiation exposure was first recognized in connection with the development of skin cancers among early radiologists, of whom nearly 100 fell victim to this disease within 15 years following the discovery of x-rays. Leukemia is the classic example.

Radiation and Nuclear Power Generation

Most attention is focused on nuclear power plants due to their perceived threat. Nuclear power production involves a number of steps, referred to as the nuclear fuel cycle. These steps include mining the uranium, crushing and processing or milling the uranium, converting yellowcake to gaseous uranium hexafluoride, enriching uranium to increase the concentration of the isotope from its original, fuel fabrication, power production, reprocessing, and waste disposal.

Although highly controversial, nuclear power is the cleanest fuel known to man. The problems with nuclear power controversiality do not deal with the toxic gas or water production as other fuels sources do, but to the production of the waste. The waste for a nuclear reactor, depending on the isotope used for production, has varying half-lives. These half-lives correspond to the time it takes for the isotope to decay to one half of its original limit. Nuclear power is used throughout the U.S. for the production of public electricity. Another major source of nuclear power is through the naval defense departments of the world. International navies harness the nuclear energy for use in submarines and other seagoing vessels. The use of nuclear power allows submarines to stay submerged for longer periods of time, allowing a more efficient use of time and federal money.

In 1982 Congress enacted the Nuclear Waste Policy Act, which delegated responsibility for high-level radioactive waste management to the federal government and designated the U.S. Department of Energy as the lead agency to coordinate the effort to site, construct, and operate the nation's first permanent repository for such wastes.

ULTRAVIOLET RADIATION

Wavelengths of the electromagnetic spectrum ranging between 40 and 400 nanometers in length are categorized as ultraviolet (UV) light. Certain portions of this range are strongly absorbed by living tissues, particularly by DNA, which constitutes the major target of UV damage. Injury to the hereditary material of cells is the reason for the lethal or mutational effects that excess UV exposure can provoke in living organisms. Light in the ultraviolet range does not penetrate beyond the upper epidermis, but the lack of melanin in people without color places them at increased risk to skin cancers.[71]

Three major types of skin cancer account for over 700,000 new cases of the disease diagnosed in the U.S. each year. Basal cell carcinoma and squamous cell carcinoma represent the vast majority of skin cancers, accounting for nearly one-third of all cancers occurring the U.S. today. Chronic exposure to sunlight (especially the UV-B component of sunlight) is recognized as the cause of over 90% of these two cancers. Malignant melanoma is much less common but far more deadly. Americans born in the 1990s are 12 times the risk of developing melanoma as those born 50 years ago and twice as likely as those born just 10 years ago. Unlike the nonmelanoma skin cancers, malignant melanoma appears to develop as a result of occasional severe sunburn rather than prolonged low-dose exposure to sunlight.

The beneficial effects of ultraviolet light include the production of vitamin D that prevents the formation of bone-deforming rickets, and the germicidal properties of UV that are found to reduce bacterial and viral infections. This has been demonstrated in operating rooms and in the treatment of bacterial skin diseases including acne and boils.[71]

REFERENCES

1. Chamber, L., *Classification and Extent of Air Pollution,* Stem, A.C., Ed., 3rd ed., Academic Press, New York, 1976, chap. 1.
2. Morgan, M.T., Gordon, L., Walker, B. et al., *Environmental Health,* 2nd ed., Morton Publishing Co., Englewood, CO, 1997.
3. Whelan, E.M., *Toxic Terror,* Prometheus Books, Buffalo, New York, 1993, chap. 10.
4. Godish, T., *Air Quality,* 3rd ed., CRC/Lewis Publishers, Boca Raton, FL, 1997, chap. 5.

5. Liu, B. and Xu, E., Physical and Economic Damage Functions for Air Pollutants by Receptors, USEPA/600/6-76-011, 1976.
6. Gribbon, J., Carbon dioxide, ammonia, and life, *New Scientists,* 94:1305, 413, 1982.
7. Godish, T., *Air Quality,* CRC/Lewis Publishers, Boca Raton, Florida, 1997, chap. 1.
8. Nadakavukaren, A., *Our Global Environment: A Health Perspective,* Waveland Press, Inc., Prospect Heights, IL, 1995, chap. 11.
9. Godish, T., *Air Quality,* CRC/Lewis Publishers, Boca Raton, FL, 1997, chap. 3.
10. Nadakavukaren, A., *Our Global Environment: A Health Perspective,* Waveland Press, Inc., Prospect Heights, IL, 1995, chap. 12.
11. Godish, T., *Air Quality,* CRC/Lewis Publishers, Boca Raton, FL, 1997, chap. 8.
12. USEPA, The Clean Air Act Amendments of 1990: Summary Materials, USEPA Office of Air and Radiation, Reproduced by Library of Congress, Congressional Research Service, Nov. 15, 1990.
13. Kessler, J. and Schroder, W., Meeting Mobility and Air Quality Goals: Strategies That Work, USEPA, Office of Policy Analyses, Oct. 13, 1993.
14. Anderson, E.V., Health Studies Indicate MTBE is safe gasoline additive, *C&EN,* Sep. 20, 9, 1993.
15. Bradley, J., Buying high, selling low, *E/The Environmental Magazine,* Jul/Aug 1996, hhp://www.emagazine.com/2curr3.
16. Chandler, D.L., MIT scientist shares Nobel for identifying ozone damage, *The Boston Globe,* Oct. 12, 101V, 1995.
17. Godish, T., *Air Quality,* CRC/Lewis Publishers, Boca Raton, FL, 1997, chap. 4.
18. Associated Press (AP), Ozone hole grows to record Size, *CNN Interactive,* Nov. 2, 1996.
19. Wald, M.L., Smuggling of polluting chemicals is resolved, *NY Times National,* Sep. 17, 1, 1995.
20. Begley, S., Holes in the ozone treaty, *Newsweek,* Sep. 25, 70, 1995.
21. Stevens, W.K., Scientists report an easing in ozone-killing chemicals, *New York Times,* Aug. 26, 1993.
22. USEPA, The Size and Depth of the Ozone Hole, written by USEPA's Stratospheric Protection Division, June 10, 1998, hhp://www.usepa.gov/ozone/science/hole/size.html.
23. Suitil, K., Holey war, *Discover,* Jan., 75, 1993.
24. Schein, O., Vincencio, C., Munoz, B., et al., Ocular and dermatologic health effects of ultraviolet radiation exposure from the ozone hole in Southern Chile. *Amer. J. Public Health,* 85:4, Apr., 546, 1995.
25. USEPA, USEPA's Revised Ozone Standard, Fact Sheet, USEPA, Office of Air and Radiation, Office of Air Quality Planning and Standards, Jul. 17, 1997, http//ttnwww.rtpnc.usepa.gov.
26. USEPA, USEPA's Revised Particulate Standard, FACT sheet, USEPA Office of Air & Radiation, Office of Air Quality Planning & Standards, Jul. 17, 1997, http://ttnwww.rtpnc.usepa.gov.
27. Gerstenzank, J., Clean-air plans fuel backstage U.S. fight, *Los Angeles Times: Science,* Sun., Apr. 26, 1997, http://www.latimes.com/home/news/science/environ/x000035587.html.
28. Easterbrook, G., Hot air treaty, *U.S. News & World Report,* Dec. 22, 46, 1997.
29. Tyler, P., China's inevitable dilemma: coal equals growth, *New York Times,* Nov. 29, A I, 1995.
30. Monastersky, R., Dusting the climate for fingerprints: Has greenhouse warming arrived? Will we ever know? *Sci. News,* Vol. 147, June 10, 362, 1995.
31. Monastersky, R., Satellite detects a global sea rise, *Sci. News,* Vol. 146, Dec., 388, 1994.

32. Mann, M.E., Bradley, R.S., and Hughes, M.C., Global scale temperature patterns and climate forcing over the past six centuries, *Nature,* Vol. 392, Apr. 23, 779, 1995.
33. Schneider, D., Burying the problem, could pumping carbon dioxide into the ground forestall global warming? *Sci. Amer.,* Jan. 1998, p. 21.
34. Wiant, C., Global warming: public health and the debate about science and policy, *J. Environ. Health,* Mar. 1998 Vol. 60, n7, p. 31(2).
35. Lemonick, M.D., The ice age cometh? *Time,* Jan. 31, 79, 1994.
36. Noshi, J.M., The fury of El Niño, *Time,* Feb. 16, 67, 1998.
37. Pfarrer, S., Expert fears El Niño global warming link, *Daily Hampshire Gazette,* Fri., Apr. 17, 9, 1998.
38. Nash, M., When life nearly died, *Time,* Sep. 18, 95, 1995.
39. USEPA, National Air Quality and Emission Trends Report, 1996, USEPA Office of Air Quality Planning & Standards, USEPA document 454/R-97-013 1/13/98, WYSIWYG://http:flwww.usepa.gov/oar/aqtmd96/trendsfs.hmfl.
40. Suplee, C., Dirty air can shorten your life, study says. *Washington Post,* Mar. 6, Al, 1995.
41. News and Analyses, Air pollution in the U.S., *Sci. Amer.,* Apr., 27, 1997.
42. Allen, S., More dying of dirty air than in cars, study finds, *Boston Globe,* May 9, 14, 1996.
43. Washington (CNN), America's dirty little secret: smog still a health problem, *CNN Interactive,* June 21, 1996.
44. Godish, T., *Air Quality,* 3rd ed., CRC/Lewis Publishers, Boca Raton, FL, 1997, chap. 2.
45. Raloff, J., Hearty risks from breathing fine dust, *Sci. News,* Vol. 148 Jul. 1, 5, 1995.
46. USEPA, USEPA's Revised Ozone Standard, FACT Sheet, USEPA Office of Air Quality Planning and Standards, Jul. 17, 1997, http://ttnwww.rtpnc.usepa.gov.
47. USEPA, National Air Quality and Emission Trends Report, FACT Sheet, USEPA, Document Number 454/e - 97-013, USEPA Office of Air Quality Planning and Standards, 1, 23, 1998, wysiwyg://l lhttp://www.usepa.gov/out/aqtmd96/trendsfs.htrffl.
48. Anonymous, Carbon monoxide-heart failure link, *Environ. Health Perspect.,* Vol. 104, 21, 138, 1996.
49. Flynn, J., The following forest, *Amicus J.,* 15:4, 1, 1994.
50. Shaw, R., Environmental News Network, Recovery Begins From Acid Rain Damage, Mon. Oct. 18, 1999.
51. News & Analyses, Air pollution in the U.S., *Sci. Amer.,* Apr. 7, 27, 1997.
52. Reese, A., Blue skies, *E/The Environmental Magazine,* Nov/Dec 1999.
53. American Lung Association, Life and Breath: Answer and Questions, item, #1245, 15, Jan. 1990.
54. Stolberg, S.G., Poor are fighting baffling surge in asthma, *The New York Times,* Oct. 18, 1999.
55. Cole, C.L., High rate of asthma inspires new study, *Boston Sunday Globe,* Nov. 24, 1, 1999.
56. Shell, E.R., Does civilization cause asthma? *Atlantic Mon.,* 285, 90, 2000.
57. Rosenstreich, D.L., Eggleston, P., Kaltan, M., et al., The role of cockroach allergy and exposure to cockroach allergen in causing morbidity among inter-city children with asthma. *N. Engl. J. Med.,* 333, 1356, 1997.
58. Friebek, E., The attack of asthma, *Environ. Health Perspect.,* 104:1, 22, 1996.
59. Thurston, G.G., Lippman, M., Scott, M.B. and Fine, J.M., Summertime haze air pollution and children with asthma, *Amer. J. Respir. Crit. Care Med.,* 155, 154, 1997.
60. Barner, R. I., Air pollution and asthma, *Postgraduate Med. J.,* 70, 319, 1994.
61. Lyons, B., Something in the air, *Ladies Home J.,* Feb., 102, 1995.

62. USEPA, The Inside Story: A Guide to Indoor Air Quality, USEPA Office of Air and Radiation, and the Consumer Product Safety Commission, U.S. Government Printing Office, LM/4uc/1-88/004, 1988, 32 pp.
63. Menzies, D., Tumbly, R.M., Nunes, R, et al., Exposure to varying levels of contaminants and symptoms among workers in the office buildings, *Amer. J. Public Health,* 86: 1629, 1996.
64. Godish, T., *Air Quality,* 3rd ed., CRC/Lewis Publishers, Boca Raton, FL, 1997, chap. 6.
65. Leary, W., Research ties radon to as many as 21,800 deaths each year, *The New York Times National,* Feb. 20, A13, 1998.
66. Godish, T., *Air Quality,* 3rd ed., CRC/Lewis Publishers, Boca Raton, FL, 1997, chap. 11.
67. USEPA, A Citizens Guide to Radon, U.S. Government Printing Office, USEPA-86-CC4, 1986.
68. Hager, M.G., Mites and molds, *Self,* May, 167, 1997.
69. Holusha, J.A., Study calls household materials especially toxic, *The New York Times,* Mar. 15, 1995.
70. Geary, James, Mad about the noise, *Time,* Jul. 27, 1998.
71. Nadakavukaren, A., *Our Global Environment: A Health Perspective,* Waveland Press, Inc., Prospect Heights, IL, 1995, chap. 13.
72. Godish, T., *Air Quality,* 3rd ed., CRC/Lewis Publishers, Boca Raton, FL, 1997, chap. 12.
73. Toroza, G.J. and Anagnostakos, N.R., *Principles of Anatomy and Physiology,* 6th ed., Harper and Row Publishers, New York, 1990, chap. 17.
74. National Bureau of Standards (NBS), Fundamentals of Noise: Measurement, Rating Schemes and Standards, NTIA Publication No. 300-15, 1971.
75. Birchard, K., Experts still arguing over radiation doses, *Lancet,* 354, 400, 1999.

CHAPTER 11

Solid and Hazardous Waste

INTRODUCTION

It was just another day on the Yangtze River of China when Zhoa Bangying, 21 years old and a uniformed attendant of a Russian-built hydrofoil boat, threw a large bag of trash out of the window and into the river. China is entering an era of economic development complete with fast food, single-use disposable items, and a throwaway culture. Throwing trash out of a window into the river or onto the streets is so commonplace that no one is surprised. Garbage and rubbish float down China's rivers, choke its canals, fly out of passenger train windows, and create mountains of trash.[1] China is not alone in this disregard for proper disposal. The constant influx of rural

OBJECTIVES FOR THIS CHAPTER

A student reading this chapter will be able to:

1. **Discuss and explain the consequences of improper solid waste disposal**
2. **List and characterize the typical municipal waste stream**
3. **Describe and discuss the methods of reducing the solid waste stream through reuse and recycling efforts**
4. **Describe the methods of collection and disposal of municipal solid wastes including the benefits and problems associated with landfills and incinerators**
5. **Differentiate the types of hazardous waste and discuss reasons for proper disposal, giving some case examples**
6. **List and describe the various methods of hazardous waste control emphasizing waste reduction, volume or hazard reduction, and long-term storage and disposal options**
7. **Discuss the positive and negative aspects of clean-up efforts under "Superfund" and some of the major concerns regarding its future operation**

people to the city of Nairobi has doubled its population in the last 10 years and overwhelmed its ability to provide basic services such as rubbish collection. Littering the streets with solid waste is commonplace, and heaps of uncollected, stinking, and fly- and rodent-infested garbage appear everywhere.[2] Once renowned as "The Green City in the Sun," Nairobi is now riddled with heaps of uncollected garbage and is called by many "The Stinking City in the Sun."[3] The failure to properly manage the solid wastes in Nairobi are thought to be contributing to the increased incidence of shigellosis, paratyphoid, and cholera.[4] Such diseases are easily spread as accumulating garbage has clogged drainage systems, flooding the cities, polluting water sources, and converting the Nairobi River into an open sewer devoid of higher life forms.[5]

There are a number of reasons given for the indiscriminate dumping, including poor management of finances and resources, a lax attitude of city employees, and little concern among residents for the cleanliness of their city.[2] The value systems of people moving to Nigeria, or Wuhan, China, or many other areas of the world, change in the face of economic development and industrialization. China was once a culture of thrift, where rags and cloth diapers were reused and where national cleanliness was monitored and enforced. Now it is giving way to a hurried, money-driven, throwaway society.[1] This is not unusual for cities. Cities have historically been centers for filth and disease. Cities in England and Western Europe were disease-ridden centers of filth up until the 19th century. It was the norm to cast garbage, rubbish, and body wastes out of windows onto the streets below, or left on the floor about the living space. The consequence of increasing refuse, filth, insects, and rodents resulted in disease outbreaks of enormous consequences such as the bubonic plague or "black death" that ravaged Europe in the 13th and 14th centuries. The relationship between disease outbreaks and accumulated filth gradually clarified in the eyes of municipal leaders so that by the early 1900s efforts were undertaken in Europe and North America to improve sanitation. One of the major efforts was directed at promptly collecting urban wastes and dispersing them at some distance from the population to reduce the probability of disease. This normally meant dumping the collected refuse on the outskirts of town. Unfortunately, much of the world continues to follow this pattern of refuse disposal, which invites the proliferation of rodents and insects, becomes a source of contamination to groundwater, pollutes ambient air when combusted, facilitates the spread of debris around the dumping site, lowers property values about the site, and encourages the spread of disease from microorganisms and toxic chemicals. It has only been within recent years that the U.S. has adopted waste management policies that recognize the need for safer and more effective methods of collecting, storing, transporting, and disposing of the unuseable or unwanted materials generated by society. This author recalls living as a child in a rural area of Massachusetts and helping to load the domestic trash from our household onto a neighbor's truck. The trash was then discarded at the town dump located in a local wetlands area reached by an unpaved road. There was no soil cover and no barrier to the soil beneath, and so it was called an open dump. Rats, flies, and mosquitoes were rampant. Periodically, the dump was set on fire to reduce the volume and temporarily discourage the proliferation of pests.

A large percentage of municipal solid waste (MSW) was thrown into open dumps up through the early 1970s, with very little recycling occurring. Larger communities with diminishing open land for dumping opted to construct incinerators, and there were more than 300 municipal incinerators operating in the U.S. during the 1960s. The introduction of strict air quality regulations in the 1970s caused many of the incinerators to close as they represented major sources of air pollution. These same regulations imposed restrictions on the burning of open dumps. During the mid-1970s, incineration capacity declined, open dumps disappeared, and the generation of MSW increased dramatically. During the same period, the recovery of materials, such as from recycling, grew very slowly.[6] The federal government passed regulations in 1976 that forbade open dumping while introducing the concept and use of the sanitary landfill. This law is known as the Resource Conversation and Recovery Act (RCRA; pronounced "rickra") of 1976 (P. L. 94-580). There are three distinct programs under RCRA. Subtitle D encouraged states to develop comprehensive plans for the management of solid wastes with emphasis on those of a nonhazardous nature, such as household wastes. Subtitle C was designed to control the improper disposal of hazardous waste through a manifest process known as the "cradle-to-grave" approach. Subtitle I was designed to minimize the contamination of groundwater from underground storage tanks through leak detection requirements, mitigation and prevention of leaks, and new performance standards for underground storage tanks. Subtitle D was soon found to be inadequate despite regulations that dictated the conditions for the proper siting, construction, and management of sanitary landfills. Nearly 94% of 17,000 land disposal sites surveyed in the mid-1970s failed to meet minimum requirements.[7] The consequences included substantial odors and debris, contamination of groundwater, proliferation of pests, and other environmental pollution problems.

Seeing the need for additional control over the disposal of solid wastes, Congress passed the Hazardous and Solid Waste Amendments Act of 1984 (P.L. 98-616). This provided strict requirements for the proper siting of landfills to minimize surface and groundwater pollution. However, in an effort to achieve an improved and sustainable waste management program, the USEPA promulgated new RCRA Subtitle D landfill requirements that went into effect in October of 1993. The landfills were to be constructed with double liners, systems for collecting the liquids percolating through the wastes (leachate), groundwater monitoring wells, and methane venting and detection systems. The costs for siting and operating landfills escalated dramatically as the number of landfills diminished in the face of rapidly growing amounts of MSW. Not only has the population of the U.S. increased dramatically since the 1960s, but a person in the U.S. today generates more than 4.3 pounds of refuse per day compared to 2.7 pounds in 1960. This exceeds 1500 pounds per person per year, more than any other country.[8] The problem of diminishing refuse disposal capacity has been further exaggerated by the community opposition to landfill siting which has become known as the "not in my back yard" (NIMBY) syndrome. The cost for a truck to unload its waste at a landfill (tipping fee) has approached $65 to $100 per ton in the northeast U.S., making other disposal alternatives more attractive. Although sanitary landfilling continues to be among the most widely used method of disposal, the USEPA's Agenda for Action endorsed the concept of integrated waste management by which MSW is reduced or managed through several different practices that include source reduction (including reuse of products and backyard composting of yard trimmings); recycling of materials (including composting); and waste combustion (preferably with energy recovery) and landfilling.[6]

The concepts of MSW and MSW management are discussed in the following sections.

DEFINITION AND CHARACTERIZATION OF MUNICIPAL SOLID WASTE

Definition of MSW

The USEPA's Office of Solid Waste uses a materials flows methodology to estimate the amount of municipal solid waste (MSW) generated. This methodology

Table 11.1 Sources of MSW and Example Products from Each Source

SOURCES	EXAMPLES OF PRODUCTS
Residential, including single and multifamily houses	Nondurable paper items such as magazines, newspapers, and advertising flyers, plastic and glass bottles, aluminum and steel cans, packaging, food wastes, yard wastes
Institutional, including schools, hospitals, prisons, and nursing homes	Food wastes, papers from classrooms and offices, disposable tableware and napkins, paper towels from restrooms, and yard trimmings
Commerical, including retaurants, office buildings, and stores (wholesale and retail)	Food wastes, paper products from offices, restrooms, and serving tables, disposable tableware, corrugated and paperboard products, and yard wastes
Industrial packaging and administrative wastes	Wooden pallets, office paper, corrugated and paperboard, plastic film, and food wastes (from cafeterias)

Adapted from USEPA, Characterization of Municipal Solid Waste in the United States: 1997 Update, Prepared for USEPA by Franklin Assoc., Ltd., Prairie Village, KS, Report No. EPA 530-R-98-Oct.-May, 1998, chap. 1.

is based on production data (by weight) for both materials and products in the waste stream and is adjusted for imports and exports. Materials in MSW include paper and paperboard, yard trimmings, food wastes, plastics, glass, metal, and wood wastes.

Municipal solid wastes may also be categorized by the products generated, which include durable goods (e.g., appliances), nondurable goods (e.g., newspapers), containers and packaging, food wastes and yard trimmings, and miscellaneous organic wastes. Such wastes generally come from residential, commercial, institutional, and industrial sources. Examples of the types of MSW in each of these categories are listed in Table 11.1.

Characterization of MSW

Contrary to popular belief, MSW does not include everything that is landfilled in RCRA Subtitle D landfills (RCRA Subtitle D), but excludes municipal sludge, industrial nonhazardous waste, construction and demolition waste, agricultural waste, oil and gas waste, and mining wastes. (Figure 11.1). The term "generation" refers to the weight of materials and products as they enter the waste management system from residential, commercial, institutional, and industrial sources and before materials are recovered or combusted. The total amount of MSW materials generated (thousands of tons) from 1960 to 1996 are shown in Figure 11.2. Total MSW generation peaked in 1994, declined slightly in 1995 to 211,460 tons, and again in 1996 to 209.7 million tons.[9] Paper and paperboard materials represented the largest component at 13%, followed by food wastes (10.4%) and plastics (9.4%) (Figure 11.3). The amount of MSW generated in 1996 and categorized as products

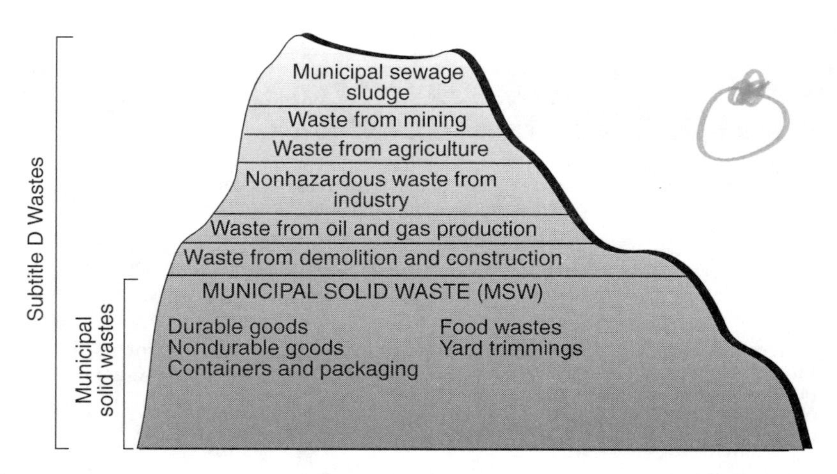

Figure 11.1 Municipal solid wastes as a component of RCRA Subtitle D wastes. (Adapted from USEPA, Characterization of Municipal Solid Waste in the United States: 1997 Update, Prepared for USEPA by Franklin Assoc., Ltd., Prairie Village, KS, Report No. EPA 530-R-98-Oct.-May, 1998, chap. 1.)

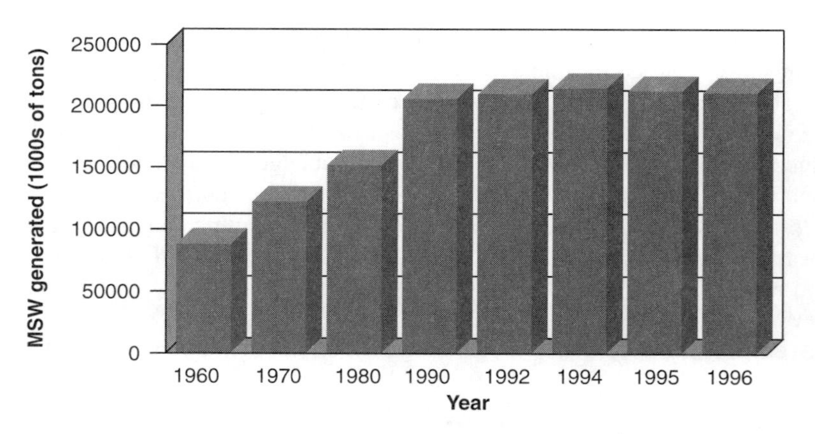

Figure 11.2 The total amount of MSW generated (thousands of tons) from 1960 to 1966. (Adapted from USEPA, Characterization of Municipal Solid Waste in the United States: 1997 Update, Prepared for USEPA by Franklin Assoc., Ltd., Prairie Village, KS, Report No. EPA 530-R-98-Oct.-May, 1998, chap. 1.)

is shown as Figure 11.4. The major proportion of generated waste is categorized as containers and packaging, representing 33% or 69.2 million tons. Nondurable goods represented the second largest amount (26.5%) followed by durable goods as 15.1%.[9]

The composition and quantities of MSW in the U.S. have changed since 1960. Total materials as products were reduced by 200,000 tons for 1995 to 1996 and over 1.5 million tons from 1994 to 1996. Paper and paperboard products in MSW exhibited the largest decline, while plastic products had the largest increase, growing by nearly one million tons in a single year.

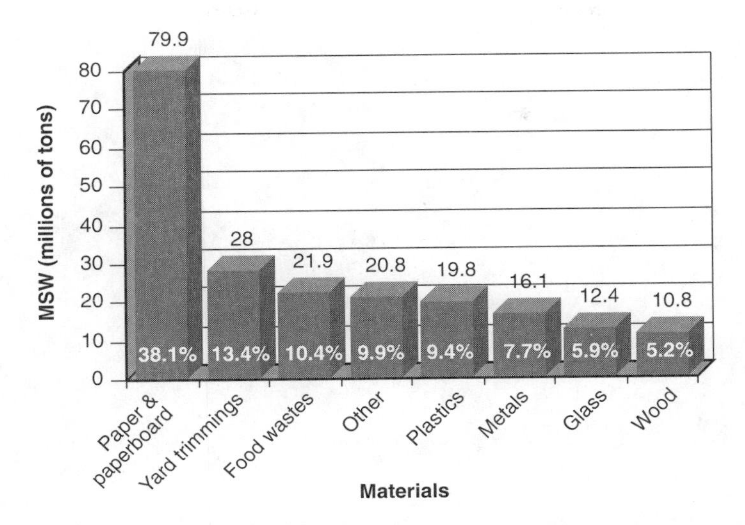

Figure 11.3 Major material components of MSW by percent and weight (millions of tons): 1996. (Adapted from USEPA, Characterization of Municipal Solid Waste in the United States: 1997 Update, Prepared for USEPA by Franklin Assoc., Ltd., Prairie Village, KS, Report No. EPA 530-R-98-Oct.-May, 1998, chap. 1.)

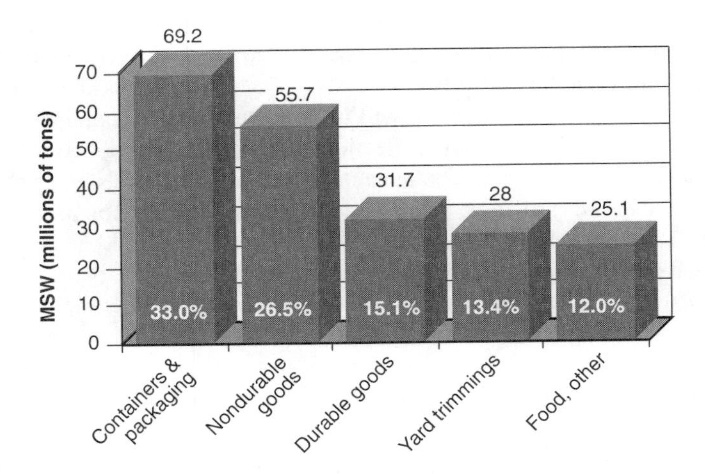

Figure 11.4 Products generated in MSW by weight and percent, 1996. (Adapted from USEPA, Characterization of Municipal Solid Waste in the United States: 1997 Update, Prepared for USEPA by Franklin Assoc., Ltd., Prairie Village, KS, Report No. EPA 530-R-98-Oct.-May, 1998, chap. 1.)

COLLECTION AND DISPOSAL OF SOLID WASTE

Collection of MSW

Whether collected by private companies under contract or by municipal employees, the collection and disposal of solid waste is the responsibility of government,

Figure 11.5 Waste-compacting collection vehicle.

which must develop and enforce regulations that protect the public health. The collection of MSW consumes $7.5 billion of the $10 billion annual cost for the collection and disposal of solid waste. This effort requires nearly 170,000 employees distributed among public and private sectors.[10] Factors that must be considered in collection include whether the waste will be collected by the municipality or under private contract; the frequency of collection; what materials will be salvaged; the types and location of collection containers or bags; what materials will be forbidden from routine disposal (i.e., tires, automobile batteries, mercury-containing batteries, fluorescent lights, durable white goods); whether wastes should be separated prior to collection as required by certain recycling efforts; and what type of equipment may be required. Collection vehicles often include manually loaded, compacting bodies, which increase the size of the load that can be carried and facilitate emptying at the disposal site (Figure 11.5). Communities that offer recycling programs and curbside collection will usually employ specialized trucks that have several compartments or sections necessary for the segregation of collected recyclables (Figure 11.6).

Many rural communities do not offer a collection service but require the residents to bring their refuse and/or recyclables to a transfer station. A transfer station is a site where solid waste is concentrated before taken to a processing facility or a sanitary landfill (Figure 11.7). The concentration most often involves compaction by placing the waste into a metal channel where a ram compresses the waste into a roll-off collection container (Figure 11.8). The container may then be hauled onto a tractor bed and transferred for processing or landfilling. Some transfer stations also feature recycling bins where separated items such as metal cans, plastic milk bottles (HDPE), paper and cardboard, and glass containers may be collected and periodically removed for further processing and resale (Figure 11.9).

Figure 11.6 Compartmentalized truck assigned for pickup of recycled materials.

Figure 11.7 Transfer station where waste compaction and separation take place.

Management of MSW

Once waste is collected, it must be processed in some way for final disposal. Presently, most states use some combination of waste management that involves one or more components of landfilling, combustion, or recycling. Nationally, more than

Figure 11.8 Close-up view of compaction ram at a transfer station.

Figure 11.9 Recycling bins at a transfer station.

half (55%) of the total MSW collected is landfilled. About 27.3% is recovered for recycling (or composted), and 17.2% is combusted with most combustion systems employing energy recovery (Figure 11.10).

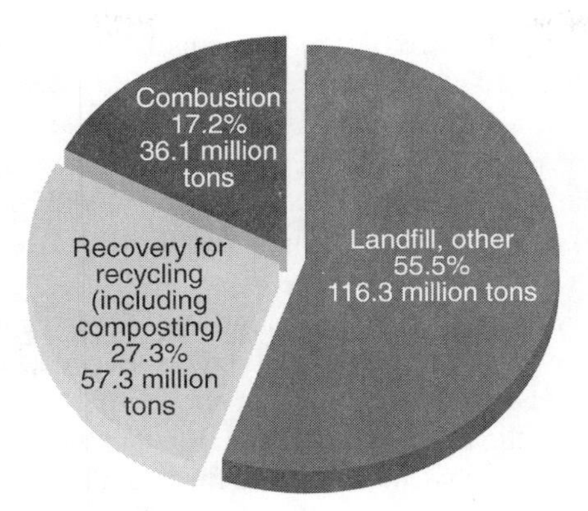

Figure 11.10 How MSW is managed in the U.S. (Adapted from USEPA, Characterization of Municipal Solid Waste in the United States: 1997 Update, Prepared for USEPA by Franklin Assoc., Ltd., Prairie Village, KS, Report No. EPA 530-R-98-Oct.-May, 1998, chap. 1.)

Landfills

Design

Landfills have historically been sources of pollution to streams and groundwater; methane gas, which can leak into basements in potentially explosive amounts; odors; and breeding grounds for insects, rodents, and gulls. Stricter requirements for the siting and construction of landfills under RCRA Subtitle D mandate that all new landfills, or ones that are to be expanded, must be constructed with double liners consisting of impervious layers, leachate collection systems, monitoring wells to detect potential migration of pollutants from the landfill site, and methane detection systems also designed to detect lateral migration of methane. (Figure 11.11).

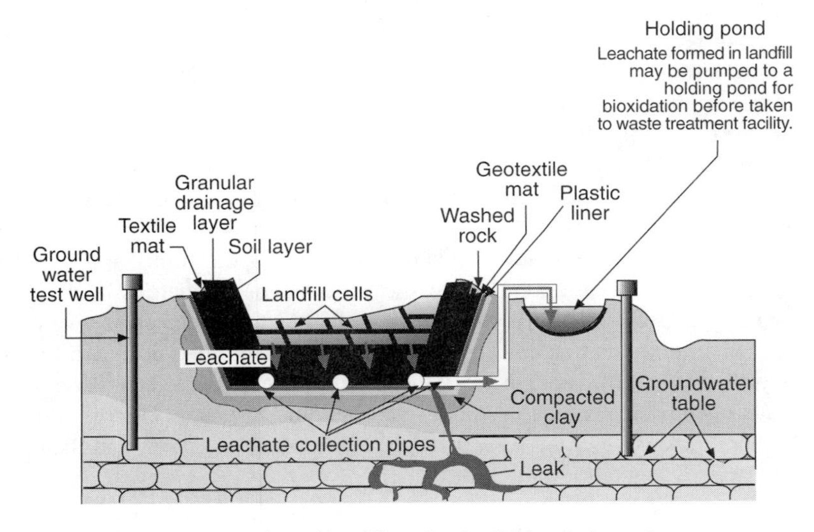

Figure 11.11 Typical construction of landfill and potential leachate routes.

The reason for such extensive design concerns is that a landfill is a depression in the ground (or constructed on the surface to resemble a stadium) which is designed to hold accumulated wastes. Its purpose is to contain wastes and prevent pollutants from reaching surface or groundwater. Landfills have been described as bathtubs that can leak from the bottom or spill over the top.[11] Landfills with two liners are like a bathtub within a bathtub, designed to further restrict the migration of pollutants.

Solid wastes that are deposited in the landfill tend to mix and degrade chemically and biologically to produce a variety of gaseous (methane), liquid (leachate), and solid (metal oxide) products. The amount of moisture, oxygen, temperature, and contents of the fill affect the level of decomposition and the products produced. The major products of concern include leachate and methane. Leachate is usually a dense, dark liquid produced when surface water and groundwater infiltrate through the solid waste and pick up solid matter, chemicals, and waste products of microbial decomposition that serve as a potential serious source of contamination to surface and groundwater. Methane gas is produced when organisms anaerobically decompose organic waste such as a food waste and sewage sludge disposed in the landfill. Methane gas can migrate laterally to enter the cellars of nearby structures and produce potentially explosive concentrations.[10] Although biodegradation does occur in a limited amount in landfills, the marketing of biodegradable products to the public may be overstated. Archeologist William Rathje has been excavating and researching landfills for years. His discovery of 25-year-old hotdogs, guacamole, steaks, and even older, in-tact newspapers has Rathje theorizing that very little biodegradation takes place in a landfill. He concludes that most garbage is "mummified" and takes up space indefinitely since microbes have insufficient moisture and air to decompose the bulk of landfill ingredients.[12]

The critical elements of a landfill include a bottom liner or double bottom liner; a leachate collection system; a cover; the natural hydrogeologic setting; and monitoring systems for detecting methane and leachate plumes. (Figure 11.11). The national hydrogeologic setting should be away from surface water and aquifer with impervious material below (fractured bedrock should be avoided), while also having a geology simple enough to permit the installation of wells to intercept escaping pollutants. The bottom liners may be composed of one or more layers of clay or a synthetic flexible membrane (or some combination of both) (Figure 11.12). The synthetic liners are often constructed of high-density polyethylene (HDPE). However, other materials may be used for plastic liners (referred to as Flexible Membrane Liners or FMLs). Leachate seeps to the bottom of the sloped landfill and is collected by a system of pipes that are laid along the bottom. The leachate is intended to be periodically pumped and transported to a wastewater treatment plant where solids are removed and liquids treated. The cover cap of the landfill is designed to restrict the entry of water from precipitation and usually is constructed of several layers sloped away from the center to promote the shedding of water. The first layers are usually membrane covered by a layer of permeable sand or gravel, and then a layer of topsoil where vegetation may root.

Figure 11.12 Installation of a bottom plastic liner in a landfill.

Landfills Are Not Secure

Landfills are vulnerable to leakage from several different sources. Landfill liners, even constructed of HDPE, can be expected to leak more than 20 gallons per acre per day through tiny holes created during manufacture and those created when the seams are welded together during installation.[13] Furthermore, such liners are also permeable to a variety of dilute solutions including xylenes, toluene, trichloroethylene (TCE), and methylene chloride, which pass through the solid matrix of 100 ml-thick HDPE liners in less than 2 weeks. Such liners also become brittle and will fracture or crack from stress and chemical assault in under 2 years.[14] Even benign chemicals such as margarine, shoe polish, vinegar, and peppermint oil may attack plastic liners.[11]

Leachate collection systems can fail as the polyvinyl chloride pipes fail from chemical assault;[14] are crushed by the weight of the landfill; or become clogged from silt, mud, microbial growth, or precipitation of minerals. Once clogged, leachate accumulates, causing increased pressures that lead to leaks or overflows.[11]

Landfill caps or covers are meant to be impermeable but are disturbed by burrowing and soil-dwelling mammals, insects, worms, and reptiles; erosion by precipitation, freeze–thaw cycles, and wind; roots of vegetation that may penetrate the cover; uneven settling that causes rips and tears on the membrane or cracks in the clay surface; upward migration of chemicals or piercing objects; and sunlight on exposed surfaces. Caps that are not maintained permit precipitation to accumulate, causing leachate overflow or penetration through the bottom liner.

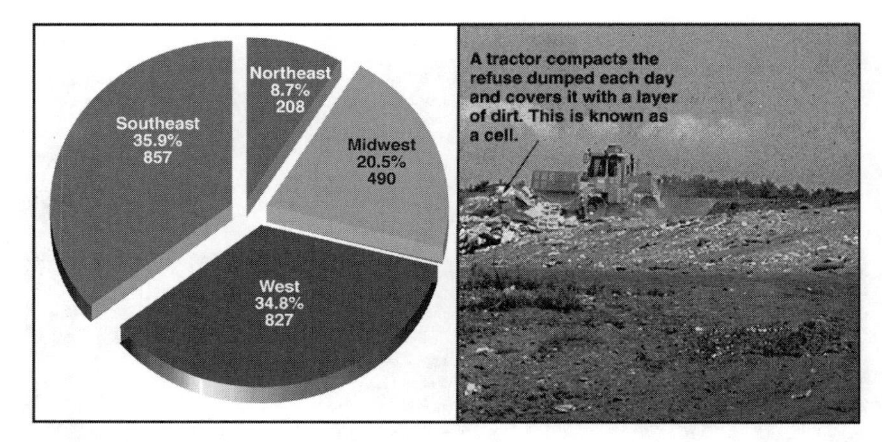

Figure 11.13 Landfill capacity by region in the U.S. Also shown is the formation of a cell in a typical landfill operation.

Leachate collection systems, bottom liners, and caps are ultimately expected to fail and cause the contamination of groundwater, which is likely to remain contaminated for thousands of years.[14] Since groundwater is a diminishing resource, alternative methods of waste management must be employed. The combination of increased costs for landfill construction, the difficulty in siting landfills, and the moratorium on landfill development and expansion by some states has resulted in a significant decline in the number of landfills since a peak of 8000 community landfills in 1988.[17] There are less than 2400 municipal solid waste landfills today, although the total landfill capacity has not declined considerably. Newly constructed landfills are much larger. The northeast U.S. has the fewest landfills (208) and the fewest remaining years of landfill capacity (Figure 11.13).[6]

One of the northeastern states, Massachusetts, has adopted a Solid Waste Master Plan designed to reduce alliance on municipal landfills and to enter the next century with a waste reduction and recycling-based integrated solid waste management system in place.[15] This plan calls for a moratorium on new landfills and the closure of existing unlined ones, with the anticipated result being significant reduction in leachate (Figure 11.14).[15] As landfills are closed, Massachusetts seeks to increase its efforts in recycling and source reduction, while nonrecovered waste is diverted to energy facilities and composting facilities.

Source Reduction

Activities that reduce the amount of waste toxicity prior to entering the waste stream are referred to as source reduction. Source reduction activities may include:

1. Products package reuse — the use of glass bottles has declined in favor of plastic containers, but glass beer or soft drink bottles represent an excellent example of containers that can be reused for their original purpose. Some chain grocery stores offer reusable cloth shopping bags. Clothing, furniture, and appliances are often

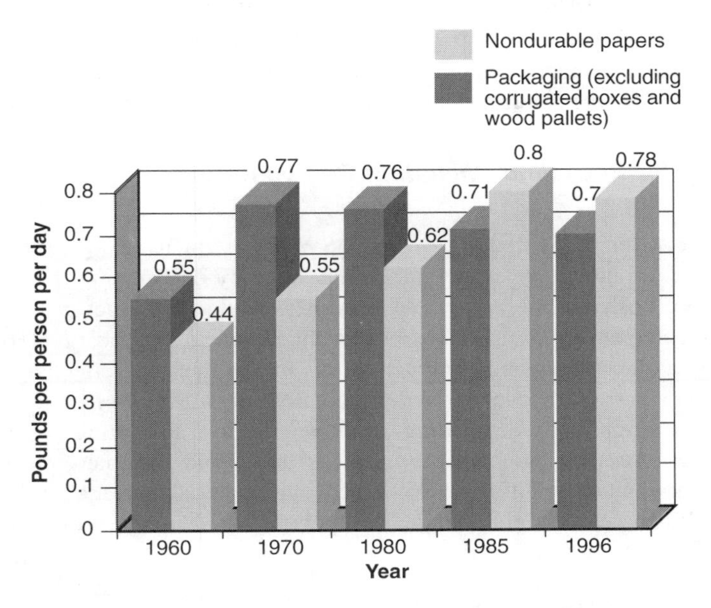

Figure 11.14 Generation of packaging, and non-durable papers from 1960 through 1986 (pounds per person per day). (Adapted from USEPA, Characterization of Municipal Solid Waste in the United States: 1997 Update, Prepared for by Franklin Assoc., Ltd., Prairie Village, KS, Report No. EPA 530-R-98-Oct, May, 1998, chap. 3.)

informally reused among family and friends, sold at yard sales, or donated to agencies that will redistribute them.

2. Package or product redesign that reduces material or toxicity — containers may be redesigned with fewer materials such as aluminum beverage cans, which are nearly 20% lighter than they were in 1976, or reducing the thickness of plastic liners used in cereal boxes.[7] The consumer may choose to purchase less toxic materials such as citrate-based cleaners or nontoxic pesticides. Manufacturers may use less toxic packaging by avoiding the use of highly bleached white corrugated boxes in favor of brown corrugated boxes printed with nontoxic inks.
3. Reducing use by modifying practices — consumers have a wide latitude to reduce material use. Many stores now offer the opportunity to purchase foods or other materials in bulk, thereby minimizing packaging. Paper can be used for copying on both sides, while electronic mail (email) reduces the need for hardcopy mail. Many products are now offered in concentrated form, including juices, detergents, and fabric softeners, that can be reconstituted in a larger and reusable container.

Trends in source reduction are difficult to estimate since the sharing or passing of goods among family and friends cannot be quantified. However, trends in the use of nondurable papers (newspapers, office papers, etc.) and packaging containers (excluding corrugated boxes) have been used as indicators. The generation of packaging on a per-person basis from 1960 through 1996 shows an increase in 1970 followed by slight decreases in the early 1980s and a relatively flat response thereafter (Figure 11.14). The generation of nondurable paper wastes increased through the late 1980s and early 1990s, and has decreased only slightly in recent years

(Figure 11.14). The advent of computers has not diminished the per capita generation of newspapers, magazines and other nondurable papers as expected. The generation of nondurable papers contiinues to be a matter of concern in recycling efforts.

Recovery for Recycling (Including Composting)

The act of removing materials from MSW for productive use is referred to as recycling or resource recovery. It can be measured by determining the amounts of purchases of postconsumer recovered materials including the net exports of such materials. The recovery of materials for recycling is one of the most preferred waste management alternatives and continues to be among the most effective as well.[7,9] Most citizens perceive recycling as desirable and are willing to participate in programs that are minimally demanding on time and effort. Recycling is perceived to conserve resources by reducing the need for virgin and nonrenewable materials; reduce the amount of pollution by using secondary materials that require less energy to process (using recycling aluminum cuts pollution and energy uses by more than 90% compared to the use of virgin materials); and save energy by using recycled materials since less is required for processing. The USEPA considers the recycling of materials in MSW among the more important waste management strategies, and many states have adopted recycling into their integrated waste management plans. These states have set recycling goals, including the percent of MSW to be recycled and the dates by which this will be accomplished. The state of Massachusetts now recycles 33% of the MSW, but has not met the ambitious goal of 46% recycling.[15]

Recyclable materials must be collected before they can be processed. Collection normally involves curbside collection, drop-off programs, container deposit programs, and buy-back operations. Curbside collection requires residents to separate their refuse into recyclables and discards and to place the separated recyclables in collection containers at the curb. Some states have adopted unit-based pricing systems, also known as "pay-as-you-throw" programs. A typical program would require the resident to purchase plastic trash bags at selected community vendors, identified with the name of the town on it. The cost may be $.50 to $1.00 each. Discards would go into the bags and the resident would have a financial incentive to recycle and avoid filling the bags quickly. Such programs have been shown to be the greatest single stimulus to increasing residential recycling.[15] Nationally, slightly more than half of the U.S. population (about 135 million people) have access to one of the 9000 curbside collection programs, although most are not based on "pay-as-you-throw" programs. The northeast serves more than 43 million people or over 80% of its population with a curbside recyclables program (Figure 11.15).[6] There are in excess of 10,000 locations where citizens may drop off materials for recycling. There are currently nine states that have container deposit systems in place. These are sometimes referred to as "Bottle Bill" states, which require consumers to pay a refundable deposit on soft drink and beer containers. Such programs have documented a marked increase in recycling of these containers, with over 35% of all recycled beverage containers occurring in these nine states.

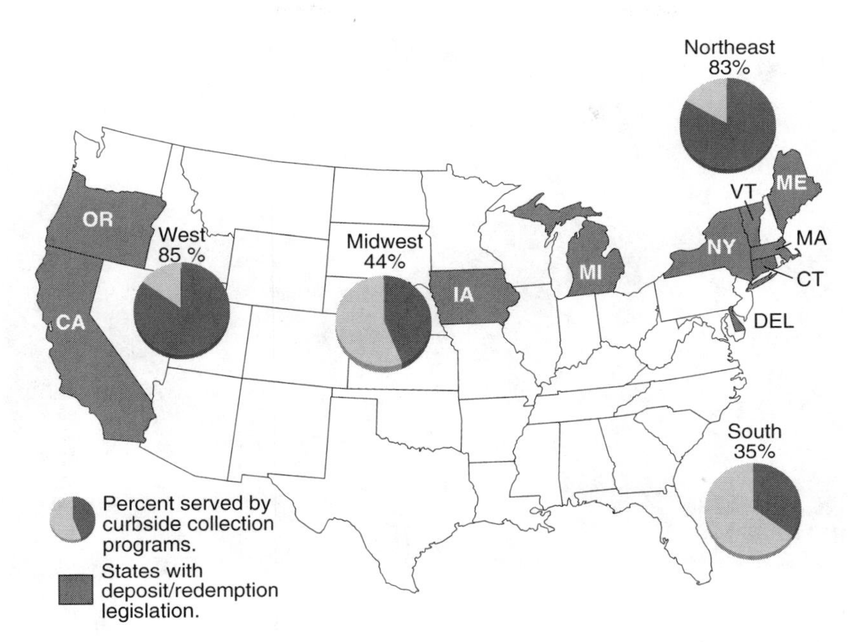

Figure 11.15 States with deposit/redemption legislation and regions served by curbside collection programs. (Adapted from USEPA, Characterization of Municipal Solid Waste in the United States: 1997 Update, Prepared for by Franklin Assoc., Ltd., Prairie Village, KS, Report No. EPA 530-R-98-Oct, May, 1998, chap. 3.)

Once the recyclable materials are collected, they must be further processed and separated to create large amounts of homogeneous secondary materials that are attractive to manufacturers as alternatives to virgin materials. The facilities that perform the function of preparing recyclables for marketing are referred to as materials recycling facilities (MRFs) (pronounced "murfs"). In a typical MRF, recyclables are dumped to a cement floor (Figure 11.16), picked up by a front-end bucket loader, and placed on conveyor belts that carry the material past magnetic separators to remove steel cans and steel objects, and then past a line of people who manually remove recyclable materials from the stream (Figure 11.17). The materials often require further processing. Glass bottles are separated into clear, brown, and green glass, and ground into small smooth pellets called cullet. An example of green glass cullet is shown in Figure 11.18. There are more than 380 MRFs operating in the U.S. processing nearly 30,000 tons of material per day. Some of these MRFs employ more sophisticated technologies to sort recyclables, including magnetic pulleys, optical sensors, eddy currents, and air classifiers.

Some MRFs process mixed waste that would normally go to a landfill complete with food scraps, dirty diapers, and mustard jars. There are only 58 such facilities in the country, most of them in the West. These "dirty MRFs" tend to be popular in areas of the country where administrators have opted not to involve citizens in recycling efforts. Theoretically, the percent of material recycled would be very high because all municipal refuse would be delivered to such facilities.

Figure 11.16 Recyclables awaiting processing at a typical MRF.

Figure 11.17 Manual separation of recyclables at a typical MRF.

Trends In Resource Recovery

Materials recovered from MSW have been measured by the purchases of post-consumer recovered materials plus net exports of the materials. Since 1960 there has been a consistent increase in the weight of MSW recovered and in the percentage of total MSW recovered (Figure 11.19).

Figure 11.18 Green glass cullet as a recyclable product after processing.

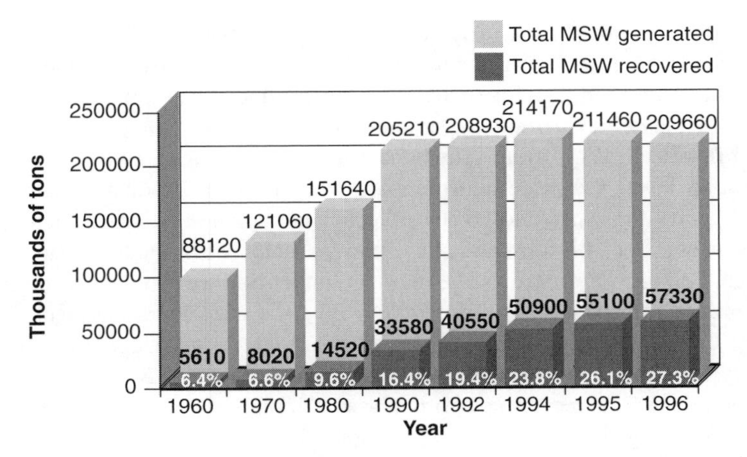

Figure 11.19 Recovery of MSW: 1960–1996 (in thousands of tons and percent of generation). (Adapted from USEPA, Characterization of Municipal Solid Waste in the United States: 1997 Update, Prepared for USEPA by Franklin Assoc., Ltd., Prairie Village, KS, Report No. EPA 530-R-98-Oct.-May, 1998, chap. 1.)

Recovery of materials from MSW remained relatively flat from the 1960s through the mid-1980s. However, escalating costs of landfilling, concerns about environmental pollution, and legislative initiatives motivated a renewed interest in recycling, which was once strong during the World War II era. Recovery rates increased significantly from the mid-1980s to the present national level of over 27% (Figure 11.19). Most of the recovery appears to be in paper and paperboard, accounting for nearly 57% of the material recovered. This is followed by yard trimmings and food wastes that are composted (19.7%), and metals at 11.1% (Figure 11.20).

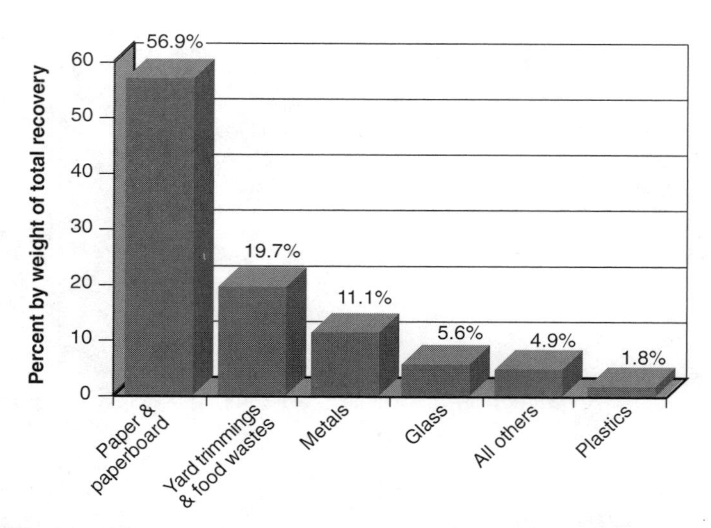

Figure 11.20 Materials recovery in 1996 (by weight of total recovery). (Adapted from USEPA, Characterization of Municipal Solid Waste in the United States: 1997 Update, Prepared for USEPA by Franklin Assoc., Ltd., Prairie Village, KS, Report No. EPA 530-R-98-Oct.-May, 1998, chap. 1.)

Both prices and markets have varied considerably over the past several years, making further overall gains in recycling more problematic. Downturns in global economies have reduced the demand for many recycled products. The amount of recovered materials, especially paper and paperboard, has increased dramatically, contributing to a fall in prices. Very often, virgin material capacity such as virgin pulp will be sufficient to lower the price and compete with recycled products. Manufacturers usually specify the types and levels of acceptable contamination in such materials as glass, paper, steel, plastics, and aluminum. In many cases, the levels of acceptable contamination are very narrow. The market competition for products also influences recyclability. For instance, the production of glass containers has declined since 1993 by 1500 tons per year, causing recycling efforts for glass containers to remain relatively level. Long transport distances for recycled materials also increase costs. This, combined with recycling efforts abroad, diminishes export markets. Another factor influencing the recycling effort is the changing capacity to use recovered materials. There are numerous plastic resin types, but most recycling of plastics is directed at unpigmented HDPE (high-density polyethylene) bottles for milk, water, and juice, or PET (polyethylene terephthalate) used in soft drink bottles. The recycled PET is used in the manufacture of fibers for carpets and fiberfill for garments, while recovered HDPE is used to manufacture new bottles, drainage pipe, trash bags, and plastic lumber. Most other plastic wastes have a limited market because the technology to recover and reuse them has not been adequately developed. Research is continuing to find new methods for the processing and reuse of mixed plastic waste and selected individual resins. Another factor influencing the recycling market is legislation that fosters recycling. Federal agencies and many state governments mandate a minimum recycled content for targeted products such as newspapers, or for products they purchase such as office paper. The state of Massachusetts has increased its purchase of recycled products from \$8.2 million in fiscal year 1994 to nearly \$34 million in fiscal year 1997.[15] President Clinton signed an executive order in 1993 requiring all federal agencies to purchase paper containing a minimum of 25% recycled paper as of the year 2000.

Composting

There are approximately 2200 municipal composting programs nationally directed at yard trimmings. Most of these programs are located in the northeast and midwest and handle more than 25,000 tons per day.[9] There are 14 mixed waste composting facilities (eight in the midwest) that remove large items and metal from the mixed waste and compost the combined paper, food waste, yard trimmings, and wood. Composting is a controlled process of degrading organic matter by microorganisms into a humus-like material. The composted material is often low in plant nutrients but is useful for conditioning soil by improving soil porosity and aeration and increasing water retention. Some communities are experimenting with adding domestic sludge to the mix and co-composting the sludge and biodegradable refuse together. Unfortunately, public acceptance of this co-composted material is not universal because of the real or perceived dangers of pathogens and/or toxic agents surviving in compost. The composting process is normally performed by removing inorganic materials such as glass, metal, and tires, and shredding the organic material. In some sophisticated operations, the entire process from shredding to aeration, curing, and finishing are performed in an in-vessel composting system (Figure 11.21). Most composting use a windrow process in which rows of material are placed next to each other outdoors and periodically mixed to incorporate air into the mix (Figure 11.22). This permits the organisms to aerobically digest the materials,

Figure 11.21 In-vessel composting system. This system is used to grind mixed food and plant wastes, aerobically digest them, and provide a finished compost product.

Figure 11.22 Compost piles arranged in rows at different stages of digestion and curing.

increasing the heat of the compost pile to temperatures sufficient to kill most pathogens and insect eggs. When the materials have been left long enough to cure so that cellulose and lignin are broken down, the material is normally screened to size and either packaged or used in bulk for soil enhancement.

Compost that is consistently high in quality and blended to meet the needs of end users is successfully marketed. Compost is used for wetlands mitigation, land reclamation, storm filtrates, soil amendments, mulches, and low-grade fertilizers.[16] The majority of composted material is used in landscaping, agriculture, and for landfill cover. Many of the markets have seen little penetration. However, as the quality and consistency of compost improve, there is likely to be higher consumer acceptance when combined with education on the benefits of compost use.

Combustion

The treatment of MSW by combustion includes the combustion of mixed MSW, the preparation of refuse-derived fuel (RDF), or a separated component of MSW (such as rubber tires) with or without the recovery of energy.[9] Generally, combustion with the production of energy is called waste-to-energy (WTE), while combustion of MSW without energy recovery is called incineration. The burning of unsegregated wastes (mass burn) in incinerators was very popular prior to the 1970s when air quality regulations made the operational costs prohibitive. However, improved methods in pollution control and monitoring, combined with escalating costs of alternative refuse disposal, have encouraged the growth of the refuse combustion industry. It is estimated that 17% or 35 million tons per year of MSW are combusted. Most

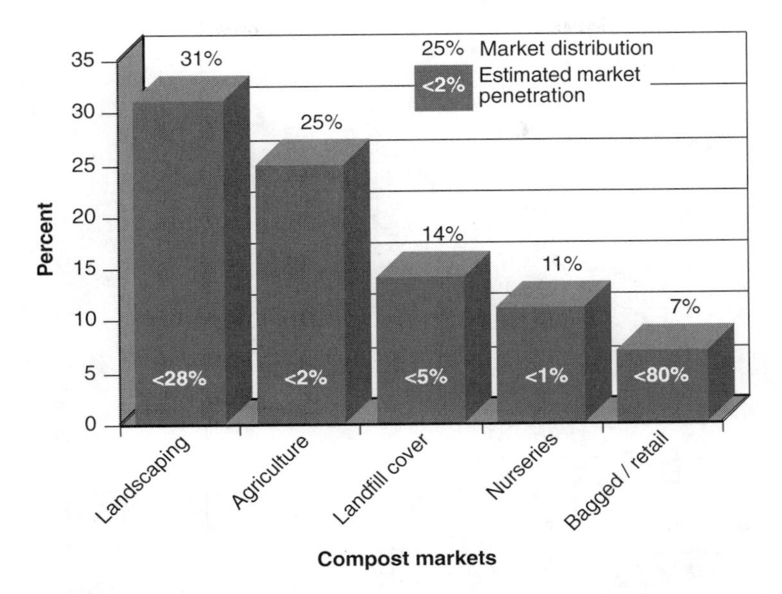

Figure 11.23 The market distribution and estimated market penetration for compost. (Adapted from USEPA), Characterization of Municipal Solid Waste in the United States: 1997 Update, Prepared for by Franklin Assoc., Ltd. Prairie Village, KS., Report No. EPA 530-R-98-Oct.-May, 1998, chap. 4.)

facilities combusting MSW also incorporate recovery of energy as steam or electricity or both. The sale of energy helps to defray operational costs. These WTE plants generally mass burn, although many incorporate recycling activities that remove non-combustible items such as metals and glass prior to burning. A state-of-the-art WTE facility integrates with a statewide master plan to combust only waste that cannot be recycled or composted. An example of such a facility is operated by Springfield Resource Recovery, Inc. (SRRI) in Springfield, Massachusetts. The facility is designed to minimize emissions in stages. The first stage incinerates MSW by controlled combustion that minimizes combustion pollution. The second stage recovers energy and converts it to useable steam and electricity, and the third stage minimizes or neutralizes pollutants coming from the waste (Figure 11.24). Independent testing firms use EPA-approved testing methods to evaluate the levels of emissions, including dioxins, furans, lead, mercury, beryllium, trace metals (cadmium, chromium, copper, manganese, nickel, tin, zinc), inhalable particles (PM_{10}) and total suspended particulate matter (TSP). This is necessary because of increasing concerns about possible toxic air emissions from these facilities. The USEPA approved tighter emissions controls in 1991 for combustion facilities in accordance with the Clean Air Act Amendments. Systems now employ specialized reactors, bag houses, computerized combustion control systems, gas scrubbers, and continuous monitoring systems to control and evaluate emissions. The fly ash captured from the bag house may be mixed with the bottom ash to reduce the probability of it being designated as hazardous waste.

There were 110 WTE facilities in 1996 with a total design capacity of 100,000 tons per day. Most of the facilities are located in the northeast and southern U.S. Another

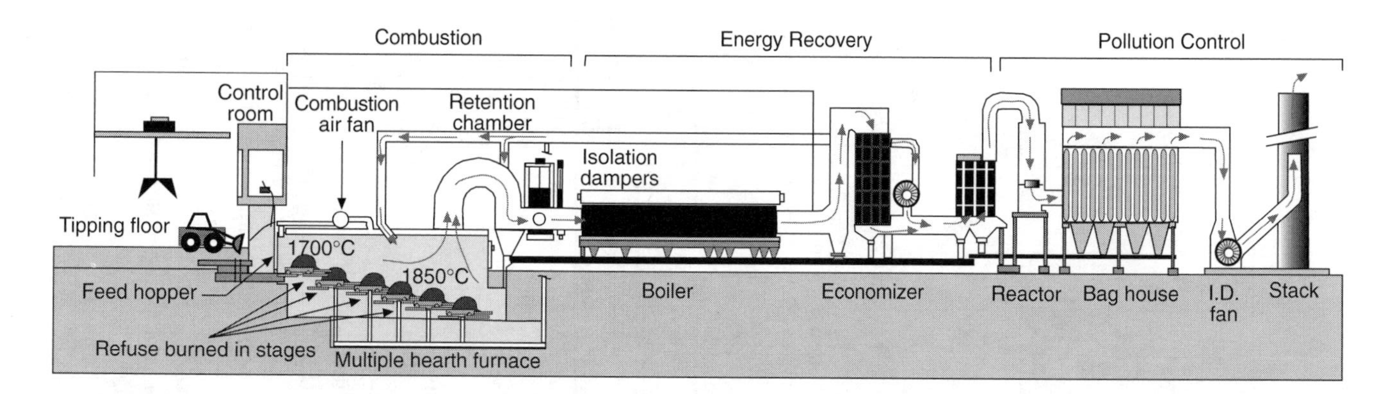

Preseparated MSW is deposited to the tipping floor. MSW is swept to the incinerator feed hopper and burned in a multiple hearth furnace in stages for 3 hours until fully combusted. A lean atmosphere and controlled temperatures reduce the formation of CO, NOx, and dioxin emissions.

Hot gases from the incinerator pass through water tubes in the boiler which heat water to create steam. The steam expands through turbines generating electricity.

Economizers extract heat and recycle it for plant uses. Acid gases are neutralized in a dry absorption reactor. Particulates are removed by filters in the bag house. Fly ash and bottom ash are collected and deposited in a monofill. All systems are continuously monitored for CO, CO2, NOx, SO2, and opacity. (Opacity is an indirect measurement of smoke indicated by the degree of obstruction of a beam of light.)

Tipping floor

Control room

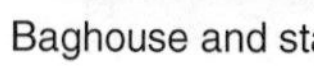

Baghouse and stack

Figure 11.24 Stages and operations of a modern WTE facility. This diagram is based on the design of the WTE facility in Springfield, MA.

5400 tons a day are prepared as refuse derived fuel (RDF), which is a pelletized fuel derived from MSW that can be mixed with coal and burned in a standard boiler. The future of constructing additional WTE facilities is in some doubt as operational costs rise and combustible products are diverted to increasing recycling efforts. Courts have overturned flow-control ordinances passed by communities to guarantee minimum tonnage of combustibles to WTE facilities and incinerators. Therefore, the overall operating capacity for WTE facilities is less than 85% of design capacity.[6]

HAZARDOUS WASTES

Background

During the spring of 1997, heavy rains raised the groundwater level and turned the Love Canal region in Niagra Falls, New York, into a muddy swamp mixed with some of the 19,000 tons of toxic wastes buried in the abandoned canal by Hooker Chemical Company. The chemically-laced ooze seeped into basements, onto playgrounds, and over lawns, creating a storm of protests, accusations of severe health effects, costs of over $37 million to relocate families from the area, and another $30 million to aid families. More than $150 million and 12 years have been spent on clean-up efforts. Love Canal became a major turning point in directing public attention to other chemical wastelands scattered across the country in abandoned warehouses, manufacturing facilities, landfills, and processing plants.[17] In 1982, scientists recommended a government buyout of Times Beach, Missouri, after waste oil dealer Russell Bliss sprayed the dirt roads in 1971 with dioxin-laced oil to keep the dust down to a minimum.[18] The events at Love Canal and Times Beach are not unique. The Escambia treating company near a Pensacola, Florida, neighborhood closed its doors years ago; but now a toxic mess piled in a 60-foot mound by the USEPA has sat in the neighborhood since 1991, covered only with a polyethylene bag. This is a black, working-class neighborhood, and the citizens felt that their back yards were being contaminated by dioxin and chemical preservatives such as creosote and pentachlorophenol. The USEPA announced in 1996 it was planning to spend $18 million to relocate people from 158 houses and 200 apartments, making it the largest move since Love Canal and Times Beach, which took place more than a decade earlier. Agency officials acknowledged that they had to consider environmental justice in the decision. The idea of environmental justice developed from studies indicating minorities were being subjected to environmental pollutants disproportionately. This prompted President Clinton to issue an executive order in 1994 that decreed agencies musts address the notion of environmental justice when considering the disposal of hazardous wastes. Additionally, the order required minorities be given a voice in environmental regulation and that the clean-up of the worst toxic waste sites in minority areas be given priority.[19]

In the early and mid-1970s, cases of acute lymphocytic leukemia began to surface in the town of Woburn, Massachusetts. Although no conclusive link was established between discarded organochlorine chemicals appearing in the drinking water and elevated leukemia levels, the involved companies agreed to pay nearly $70 million in clean-up costs. The story, as written by Jonathan Harr, became a best seller titled *A Civil Action*, and was made into a film starring John Travolta. These events point to increased public awareness and concern about hazardous wastes as one of the major environmental challenges. The USEPA estimates that in excess of 270 million tons of hazardous waste are generated every year in the U.S., with most of it produced by the petrochemical industry.[20] Discarded or released hazardous substances have contaminated surface and groundwater, soils, and air. Drinking water has been contaminated, food sources polluted, and wildlife habitats destroyed or compromised.

What is a Hazardous Waste?

The proliferation of hazardous wastes, industrial sources, and community outcry prompted Congress to pass the Resource Conservation and Recovery Act of 1976. This was the first federal regulatory program targeted at controlling present and future hazardous waste. This Act was intended to prevent the improper disposal of hazardous waste using a manifest tracking system that attempts to follow the waste from its point of generation to its final treatment and disposal. The process is called the "cradle-to-grave" system and is meant to protect public health by defining what wastes are hazardous; tracking wastes to the point of disposal; assuring that treatment, storage, and disposal (TSD) facilities meet minimum national standards; and making certain that TSDs are properly maintained after closure and that facility owner/operators are financially responsible for hazardous waste releases that may occur at their facility.

Everyone who handles the waste, including the generator, the shipper, and the manager of the TSD facility, must sign a multipart manifest with copies going to the state and federal regulatory agencies and back to the generator after various signatures are applied. The final disposal of wastes has become expensive, and persons dump the waste along roadways during rainstorms, down storm sewers, or on privately owned land may realize significant profits by failing to report these activities at the outset. However, the penalties are severe and include criminal prosecution and hefty fines. The definition of hazardous waste in RCRA includes any discarded material that may pose a substantial threat or potential danger to human health or the environment when improperly handled. The USEPA has established that a waste be considered hazardous if it meets either of following qualifications. First, the substance should be listed as one of the 400 wastes or waste streams under Title 40, Parts 261, 31-33 of the Code of Federal Regulations (CFR). Such substances include wastes generated by chemical manufacturers, vehicle maintenance shops, paper industries, metal manufacturing, cleaning agents and cosmetics manufacturers, the construction industry, the printing industry, and many others. These are called listed wastes.

Second, wastes are considered hazardous if they possess certain properties or characteristics, such as if they are toxic, ignitable, corrosive, or reactive according to the specifications outlined in RCRA. A substance is considered toxic if it is potentially poisonous to humans and can cause acute or chronic health problems.

Wastes are considered toxic if they produce sufficient amounts under the Toxic Characteristic Leachate Procedure (TCLP) described by the USEPA. This process requires acidified liquid to be percolated through waste in a laboratory-created landfill condition for 24 hours. Samples of the leachate are tested for the amounts of the target substance released. Substances are ignitable if they have a flash point below 60°C or can cause fires spontaneously by friction, moisture, or chemical change. Certain plastics, oils, solvents, and paints fit this category. Wastes are corrosive if they have a pH below 2.0 or above 12.5, and include battery acid, drain cleaners, and other compounds that will dissolve living tissue, metals, and other materials. Reactive wastes may include discarded metallic sodium or wastes from the manufacture of certain explosives that can react vigorously with air or water to generate toxic gases or even explode.

Hazardous Waste Regulations

The method of dealing with hazardous wastes in the late 1970s and early 1980s was based on the premise that such wastes are inevitable and acceptable methods of treatment and disposal were the answer to protecting public health. However, the rapidly increasing generation of hazardous wastes, the discovery of tens of thousands of hazardous waste sites, and growing public concern demanded an alternative to this "end of pipe" approach. Consequently, amendments were made to RCRA in 1984 that significantly expanded its regulatory powers. The amendments are known as the Hazardous and Solid Waste Amendments (HSWA), and they are focused on protecting groundwater by restricting the treatment, storage, and disposal of hazardous wastes in land management facilities; mandating stricter requirements for landfills accepting hazardous waste; requiring a schedule for determining if the landfilling of untreated hazardous waste should be phased out; creating a new program for the detection, control, and management of hazardous liquids (primarily petroleum) in underground storage tanks; and increasing the numbers of people who fall under RCRA regulations by including small-quantity generators. Small-quantity generators would include those that produce 100 kg (220 lb) per month of hazardous waste, or 1kg (2.2 lb) per month of extremely hazardous wastes. Previously, only large-quantity generators that produced more than 1000 kg (2220 lb) per month of hazardous waste were subject to RCRA provisions. Congress was concerned about the reliance on land disposal of hazardous wastes and wanted such disposal to be minimized or eliminated. Land disposal, particularly landfills and surface impoundments, should be the least favored method of handling hazardous wastes. There was concern that all land disposal facilities, including those that are designed to be secure, will ultimately leak.[21] The same HSWA amendment called for the reduction of hazardous wastes generated at the source, and such activities should have priority. This effort was further strengthened by the passage of the Pollution Prevention Act of 1990 that emphasized a federal policy of preventing the formation of hazardous wastes; followed in priority by the recycling or reuse of hazardous materials; and the last activity in the hierarchy is treatment and disposal in an environmentally safe way. The USEPA strongly committed to this policy by making waste minimization

WEB PICK Go to http://www.unix.oit.umass.edu/~envhl565/index.html Click on "Chapter Web Links" to find Scorecard Home on the World Wide Web under Chapter 11	
Chapter 9	Solid and Hazardous Waste
Whose Site?	Environmental Defense is a national nonprofit organization headquartered in New York City, representing more than 300,000 members. Since 1967, it has linked science, economics, and law to create innovative, equitable, and cost-effective solutions to the most urgent environmental problems.
URL	http://www.scorecard.org/
What's Here?	Environmental Defense sponsors several publications and websites including Scorecard, a comprehensive online tool for monitoring and taking action on chemical releases nationwide; and ForMyWorld, a noncommercial web site that delivers detailed, zip code-specific environmental information for every community in America. Type in your zip code under Scorecard and retrieve information specific to your selected area on water quality, air quality, potential sources of land contamination, lead hazards, toxic chemical releases from manufacturing facilities, environmental justice, and agricultural pollution.

its highest priority and requiring hazardous waste generators to certify their efforts at waste minimization on the hazardous waste shipping manifests.[22] Similar requirements exist for owners and operators of TSD facilities. The USEPA has created The Waste Minimization National Plan, which is a long-term national effort to reduce the quantity and toxicity of hazardous wastes. The goals of the plan are to achieve a 50% reduction in the most persistent, bioaccumulative, and toxic (PBT) chemicals in the nation's hazardous waste by the year 2005 as compared to the baseline year of 1991. Bioaccumulative (B) refers to chemicals that tend to concentrate in animal and plant tissues. Many chlorinated organics and heavy metals tend to fit in this category; prevent the transfer of chemical releases from one medium (air, water, or land) to another; and give priority to source reduction (reducing waste prior to its being generated) and environmentally sound recycling over waste treatment and disposal.[23]

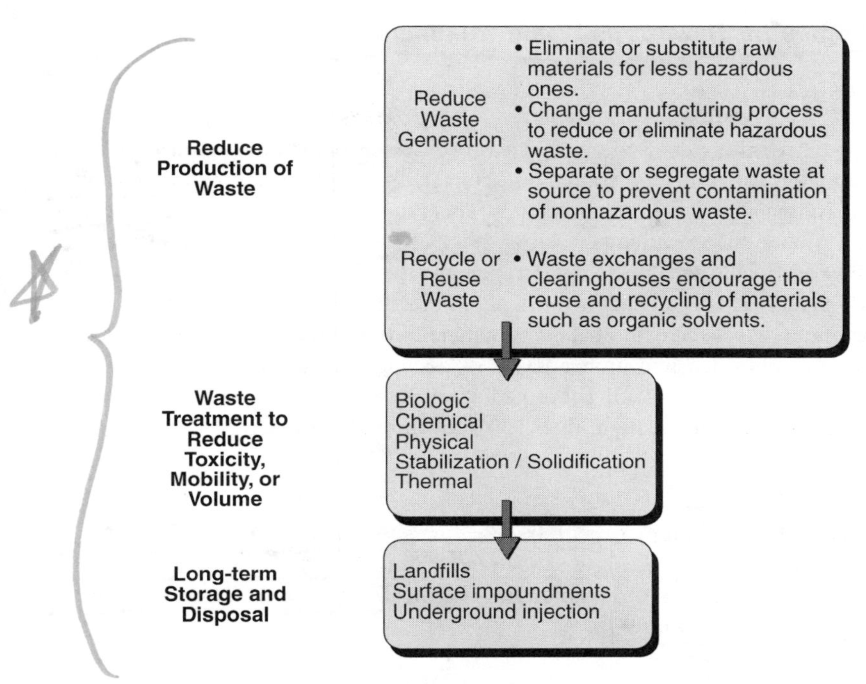
Reduce Production of Waste
Reduce Waste Generation
• Eliminate or substitute raw materials for less hazardous ones.
• Change manufacturing process to reduce or eliminate hazardous waste.
• Separate or segregate waste at source to prevent contamination of nonhazardous waste.
Recycle or Reuse Waste
• Waste exchanges and clearinghouses encourage the reuse and recycling of materials such as organic solvents.
Waste Treatment to Reduce Toxicity, Mobility, or Volume
Biologic
Chemical
Physical
Stabilization / Solidification
Thermal
Long-term Storage and Disposal
Landfills
Surface impoundments
Underground injection

Figure 11.25 A desirable hierarchy for the management of hazardous wastes. (Adapted from Stephens, R.D., Hazardous Wastes, in: *Principles and Practice of Environmental Medicine,* Tarcher, A B., Ed., Plenum Medical Book Co., New York, 1991.)

The Management of Hazardous Wastes

The three major options for managing hazardous waste include: (1) reducing the production of waste by reducing the amount generated or recycling/reusing the hazardous material after its generation; (2) reducing the volume and/or hazard of the waste; and (3) long-term storage or disposal (Figure 11.25).[20]

Generation Reduction of Hazardous Waste

The highest priority item in the management of hazardous waste is to reduce the production of waste at the source. The generation of hazardous waste can be minimized by eliminating or substituting raw materials for less hazardous ones; changing the manufacturing process to reduce or eliminate hazardous waste; and separating or segregating waste at the source to prevent the contamination of non-hazardous waste. Recycling and reusing hazardous materials before they enter the waste stream is also strongly supported by the USEPA and most state regulatory agencies. Waste clearinghouses and waste exchanges have been organized to facilitate the transfer of discard materials from one industry to another that may use it as raw material in their process. Organic solvents are among the most widely recycled and reused materials.

Technologies for Hazardous Waste Treatment

Biological

Microorganisms are used to break down or degrade organic compounds as they use waste materials for a food source and convert them into water, carbon dioxide, and simple inorganic and organic molecules. This process is known as bioremediation and has been used successfully to clean up oil spills and other organic wastes where soil permeability allows good oxygen penetration. Waste degradation may be enhanced by adding oxygen and/or nutrients to biostimulate the waste-eating organisms. Treatment may also occur in aboveground facilities or reactors where conditions may be better controlled and higher levels of decontamination can be achieved on wastes fed through the system. Bioremediation is being extensively tested under USEPA or state oversight at more than 150 sites throughout the U.S.

Chemical

Treatments are applied that neutralize acid or alkaline wastes, detoxify, dechlorinate, oxidize, coagulate, precipitate, or concentrate hazardous wastes. Chemical treatment is used most often to change the chemical structure of waste and render it nontoxic or less hazardous.[24]

Physical

Nonchemical methods that separate components of the waste stream are termed physical methods. Such methods include activated carbon to remove organic materials from liquid waste, distillation, filtration, coagulation, centrifugation, reverse osmosis, and ion exchange.[24] Once separated, the hazardous components may be reused or more efficiently destroyed or disposed.

Solidification/Stabilization

Both physical and chemical methods are used to form a solid, inert substance that limits the migration of hazardous materials from a landfill. Compounds amenable to fixation include organic salts, heavy metals such as lead, mercury, and zinc, and many inorganic compounds. Binding agents may include organic polymers, silica, or cement-based materials that are used to encapsulate materials in a water-resistant matrix, which can then be more safely disposed in a landfill. Stabilization can also be achieved by inserting electrodes into the contaminated soil and passing huge quantities of electricity through it, which melts and fuses the soil into a glass-like block. This process is called *in situ* vitrification (ISV), and such stabilized blocks are often left in place since the danger of pollutant migration is largely diminished.

Thermal Treatment

Incineration accounts for less than 2% of the hazardous wastes managed although it offers the advantages of completely destroying many forms of hazardous waste, reducing the volume of waste, and recapturing energy in the form of steam or electricity. Thermal treatment can occur via incineration or pyrolysis. Pyrolysis combusts materials in an oxygen-starved atmosphere and is not as widely used as incineration, which is conducted in an oxygen-rich environment. Incineration is normally conducted at temperatures of 425°C to 1650°C (800°F to 3000°F) in a turbulent atmosphere of sufficient duration. This ensures good mixing and effective destruction of the waste approaching levels of USEPA's current standard of 99.99%.[20,25] Incinerators have been used successfully to treat waste paints, plastics, mineral oils, pesticides, solvents, sludge, resins, greases, and waxes. The threat of hazardous emissions, including acids, dioxin, and heavy metals, has led to strict regulations on emissions from incinerator facilities.[26] The materials generated in an incinerator include carbon dioxide, water, sulfur, nitrogen-oxides, acid gases such as hydrogen chloride, and ash. Gas streams must be monitored, and hazardous materials must be removed from the emissions by various scrubbers, filters, and electrostatic precipitators. This has prompted very high construction and maintenance costs but has become a useful alternative as the options for land disposal continue to decrease.

Hazardous Waste Disposal

Even after extensive treatment, some amount of hazardous wastes will remain in the form of ash, chemically-treated residue, physically separated materials, or even in original waste form. Such waste may be rot, but it may be economically reusable or recyclable and is usually dispersed to landfills, surface impoundments, or underground injection wells. Nearly 80% of hazardous wastes are disposed to land with 96% of these wastes treated or disposed at the site where they are generated.[20,27] Historically, such wastes were deposited to unlined pits and lagoons (surface impoundments) that presented extensive environmental risk from overflows, leaching, and volatilization.

Landfills

A common method for hazardous waste disposal is the secure chemical landfill as designated in the 1984 HSWA of RCRA. Secure landfills require groundwater monitoring, double liners, a leachate collection system, and financial guarantees for postclosure activities (Figure 11.26). Just as with MSW landfills, current technologies almost guarantee the leakage of hazardous materials from commercial hazardous waste landfills.[25] Increased costs of construction, operation, and liability have curtailed the number of active commercial operations to under 20 landfills. These landfills are further restricted by legislation to accept mostly pretreated wastes that have been reduced in toxicity and/or treated to prevent leaching. The difficulty in siting for commercial hazardous waste landfills in New England and New Jersey, because of public resistance and other factors, has resulted in long-haul shipments at very high costs for businesses, schools, and commerce in these areas.

Deep-Well Injection

The process of deep-well injection involves pumping hazardous wastes through wells with double or triple casings into porous rock formations at depths of 1000 to

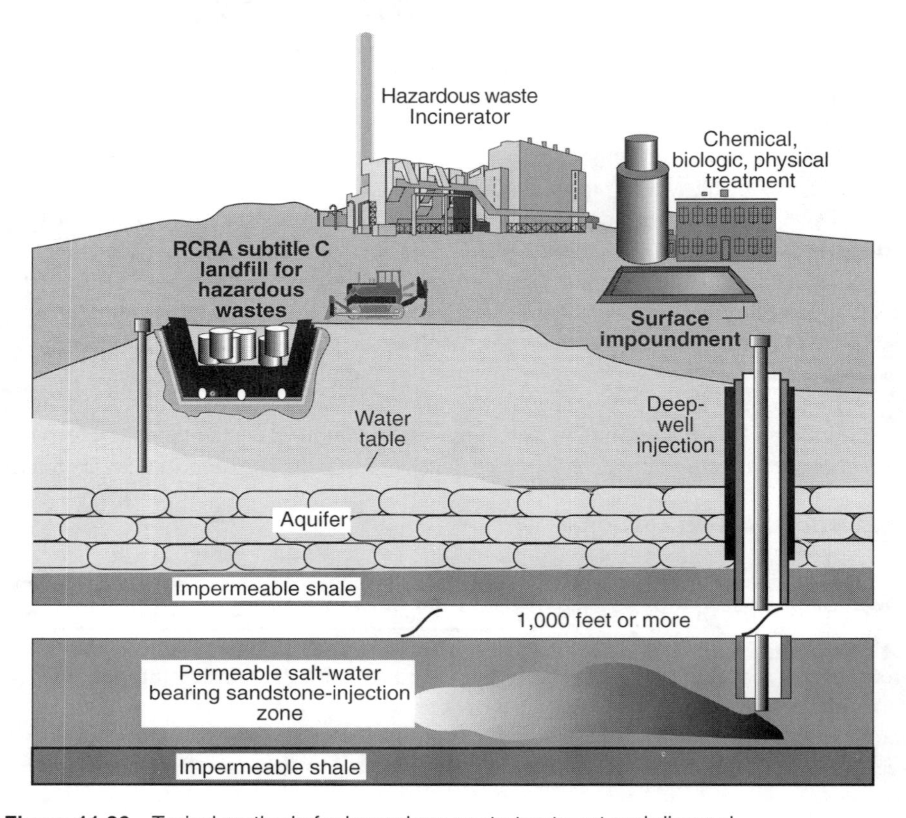

Figure 11.26 Typical methods for hazardous waste treatment and disposal.

10,000 feet below the surface, well below drinking water aquifers (Figure 11.26). Such wells are used by petroleum refineries and petrochemical plants located in the midwest and Texas, which pump nearly 30 million tons of hazardous waste into the ground each year. Despite legislation prohibiting the construction of injection wells within a quarter mile of a drinking water source, there remain concerns about such wells. Detection of leaks and methods of correcting them are severely restricted by the depth. Further, the Earth is a dynamic entity with continents drifting, tectonic plates shifting, and local geology under variations in pressure that combine to produce fractures and fissures that may release injected wastes to combine with groundwater at some future time.

Surface Impoundments

Surface impoundments are depressions in the ground or areas with dikes around them that contain liquid wastes. They vary in size and shape, ranging from several hundred square feet to hundreds of acres (Figure 11.26). Most of these pits, lagoons, or basins were unlined and located on site at manufacturing facilities. Legislators recognized the high risk of groundwater contamination; and regulations now require

Figure 11.27 Stored (and discarded) drums of hazardous waste.

such impounds to be equipped with double liners, leachate collection systems, and groundwater monitoring wells. However, just as with landfills, such impoundments will eventually leak and attention will focus on reducing hazardous waste through reduction, reuse, and recycling efforts.

Cleaning Up

Before the 1980s, it was common practice for industries and municipalities to haul the wastes to a depression in the ground, dump them, and cover them. In previous years, many local landfills were not even covered with soil. In many cases, drums of toxic wastes were simply stored in piles on site (Figure 11.27). Many of these sites were abandoned, forgotten, and even built upon as neighborhoods expanded with the pressures of a rapidly increasing population. Then the buried toxic sludge began to appear as rusted 55-gallon drums seeped their contents into the soil to contaminate groundwater or worked their way to the surface where the true drama of poisonous mud and rusted containers could be seen. Driven by the horrors of Love Canal, Chem-Dyne, Swartz's Creek, and dozens of now-historical hazardous waste sites, Congress acted in December of 1980 by passing the Comprehensive Environmental Response, Compensation, and Liability Act (CERCLA). This Act, also known as "Superfund," authorized the federal government to spend $1.6 billion over a 5-year period for emergency clean-up activities. The Superfund program also provided for response to the release of hazardous substances from leaking underground storage tanks, spills from transportation accidents, and pesticide-contaminated groundwater.[28] The Superfund program also provided for activities to oversee long-term containment of hazardous sites. Despite the term

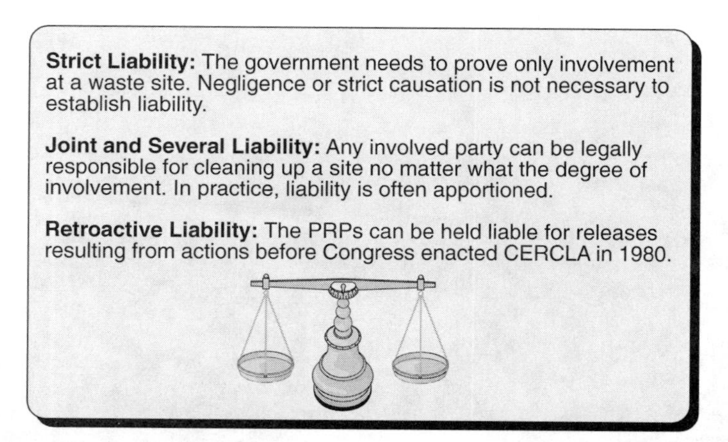

Strict Liability: The government needs to prove only involvement at a waste site. Negligence or strict causation is not necessary to establish liability.
Joint and Several Liability: Any involved party can be legally responsible for cleaning up a site no matter what the degree of involvement. In practice, liability is often apportioned.
Retroactive Liability: The PRPs can be held liable for releases resulting from actions before Congress enacted CERCLA in 1980.

Figure 11.28 Definitions of liability under Superfund. (Adapted from U.S. Congress, Congressional Research Service, Superfund Reform: Cleaning Up America's Toxic Waste Sites, *Congr. Dig.,* Mar., 68, 1998.

"compensation" in the law, no funds were provided for victims whose health may have suffered from exposure to toxic waste sites. The term "Superfund" is attributed to the fact that the bill created a trust fund financed primarily by excise taxes on chemicals and oil and an environment tax on corporations. The USEPA was charged with the responsibility of identifying those hazardous waste sites in most need of cleaning up. Superfund requires that for each site, the USEPA identifies Potentially Responsible Parties (PRPs) who are considered to be any entity, however large or small, that has anything to do with the site. The liability rules applied include retroactive liability, joint and several liability, and strict liability (Figure 11.28). This may force individuals, companies, and organizations (entities) to pay for clean-ups that have had little or even no part in causing the damage. In practice, state or federal agency personnel will attempt to identify entities with the largest pockets (most financial resources) as PRPs. Such rules have been held responsible for \$.36 to \$.60 of every dollar spent in Superfund going to legal and other transactional costs.[30]

There are presently nearly 33,000 hazardous waste sites that have come under Superfund authority. More than 1300 of these sites are on the USEPA's National Priority List (NPL). The NPL was established because sites vary in size, hazardous substance content, containment, and degree of threat to the surrounding populations and environment. The USEPA established a hazard ranking system (HRS) based on the estimated hazard potential of the hazardous waste site. The factors used to make this estimate include the waste characteristics (toxicity, quantity, solubility, reactivity); the distance to the local population; and the proximity to surface, groundwater, and drinking water supplies. Other factors include the level of containment such as the condition of the containers (i.e., where the drums are rusted), and whether there are leachate collection systems or liners. The NPL is updated annually.

Environmental groups, buttressed by a large contingent of concerned citizens, charged that RCRA was not being vigorously enforced. These actions caused Congress to respond by passing the Superfund Amendments and Reauthorization Act of

1986 (SARA) which increased the program's funding to $85 billion and provided new and stricter standards. A series of delays, lawsuits, and accusations followed, demanding that the Superfund program be repaired. Superfund had spent 13.8 billion of taxpayers' money and $12 billion of private sector money since 1980 to clean only 291 sites. This cost was expected to escalate to $42 billion in the next 70 years to clean up an estimated total of 2300 NPL sites.[30] Some $6 billion in the Hazardous Substance Superfund Trust Fund used for clean up is supported by taxes on domestically produced and imported oil; a tax on feedstock chemicals; a corporate environmental income tax; and general revenues, reimbursements, penalties, and interest on the trust fund.[17]

The reauthorizing of the Superfund law has been very contentious as evidenced by a flurry of reform bills introduced in both houses. The major arguments center around the issue of liability. Pending bills exempt small businesses from liability for actions that occurred prior to the Superfund law and provide liability relief for parties that contribute only small amounts of waste. The bills stress more cost-effective clean-up strategies by the USEPA and give greater control to states. Additionally, the amount of damages that can be recovered from harming natural resources is limited.[17] Proponents of the bills argue that liability issues discourage the development of Brownfields, which are abandoned and underused commercial and industrial properties. Opponents feel the bills will permit businesses to pollute with reduced liability and will shift economic liability to the taxpayers. Less stringent clean-up requirements would result in great exposures and risks to the public and permit natural resources to remain polluted. Regardless of the outcome of pending legislation, current efforts of Superfund are widely supported, and there has been significant progress in recent years cleaning up sites.

The USEPA listed 1405 sites which were distributed over six different stages involved in the clean-up process. Nearly 500 of the 1405 sites listed were completing construction, and nearly 55 sites had not yet started remedial assessment (Figure 11.29).[31]

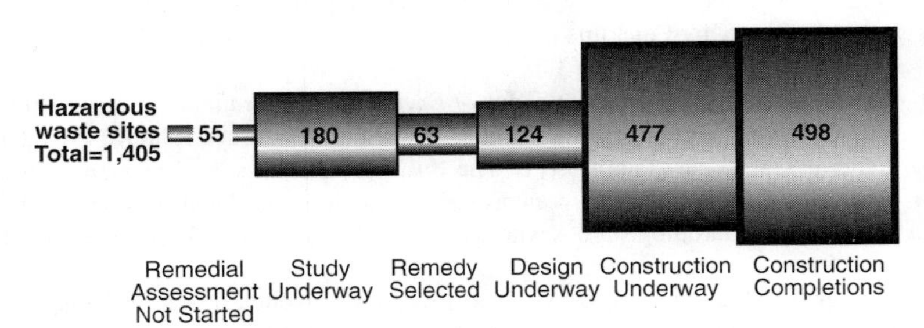

Figure 11.29 Site remediation progress at NPL sites. (Adapted from The Congressional Institute Inc., *Superfund* July 17, 1, 1996, http://www.conginst.org/cidoc-l.nsf/e74d.,167f730.d9718525636coo79ao93?Open Document.)

According to Carol Browner, former director of USEPA, as of September 1998, 1100 sites were undergoing or completed clean-up construction, with 400 of the worst sites completed. Two thirds of the completed cleanups were accomplished within the previous 4 years, and there is a commitment by the USEPA to clean up an additional 500 sites by the year 2000. The most recent listing (February 10, 1998) shows there are 1191 sites on the NPL, reflecting the removal of clean-up completions. If you would like to know of Superfund sites near you, you can search for local Superfund sites with the "Site information" page at www.epa.gov./superfund/sites/index.htm. To locate Superfund sites across the country, try www.epa.gov/superfund/sites/npl.htm.

REFERENCES

1. Tyler, P. E, When it comes to trash, Chinese just say throw, *The New York Times International,* Sun., Dec. 8, 5, 1996.
2. Mwanthi, L. 0., Nyabola, E., and Tenambergen E., Solid waste management in Nairobi City: Knowledge and attitudes, *Environ. Health,* 60 (5), 23, 1997.
3. Schernding, V.Y and Yack, D., Environmental health impacts associated with rapid urbanization in South Africa, *Public Health Rev.,* 9 (1-4), 295, 1992.
4. Waiyaki, P. G., Cholera: Its history in Africa with special reference to Kenya and other East African countries, *East Afr. Med. J.,* 73 (1), 40, 1996.
5. Otieno, F. A. 0., Quantity and Quality of Runoff in Nairobi: The Wasted Resources, *Proceedings of the Sixth International Conference on Rainwater Catchment Systems,* Nairobi, Kenya, 21, 1993.
6. United States Environmental Protection Agency (USEPA), Characterization of Municipal Solid Waste in the United States: 1997 Update, Prepared for U.S. Environmental Protection Agency, Municipal and Industrial Solid Waste Division, Office of Solid Waste, by Franklin Assoc., Ltd., Prairie Village, KS, Report No. EPA 530-R-98-Oct, May, 1998, chap. 3.
7. Nadakavukaren, A., *Our Global Environment,* 4th ed., Waveland Press, Inc., Prospect Heights, IL, 1995, chap. 16.
8. Grove, N., Recycling, *Nat. Geogr.,* 186: 1, Jul., 92, 1994.
9. United States Environmental Protection Agency (USEPA), Characterization of Municipal Solid Waste in the United States: 1997 Update, Prepared for U.S. Environmental Protection Agency, Municipal and Industrial Solid Waste Division, Office of Solid Waste, by Franklin Assoc., Ltd., Prairie Village, KS, Report No. EPA 530-R-98-Oct.-May, 1998, chap. 1.
10. Koren, H. and Bisesi, M., *Handbook of Environmental Health and Safety: Principles and Practices,* Vol. 11, 3rd ed., Lewis/CRC Publishers, Boca Raton, 1995, chap. 2.
11. Pellerano, M.G, and Parkhurst, B. The Basics of Landfills, Environmental Research Foundation, 1995, http://www.enviroweb.org/enviroissues/landfills/index.html.
12. Anonymous, Garbologist goes where few scientists dare, *Los Angeles Times,* Jul. 24, 1, 1993.
13. Bonoparte, R. and Gross, B. A., Field Behavior of Double-Liner Systems, in Waste Containment Systems: Construction, Regulation, and Performance, Bonaparte, R., Ed., Geotechnical Special Publication, No. 26, American Society of Civil Engineers, New York, 1990, chap. 2.

14. Lee, G. F. and Jones, A., Municipal Solid Waste Management in Lined, "Dry Tomb" Landfills: A Technologically Flawed Approach for Protection of Groundwater Quality, EL Macero, Calif., G. Fred Lee and Associates, March, 1992. Available from: G. Fred Lee and Associates, 27298 East El Macero, Drive, El Macero, CA, 67 pgs.
15. Commonwealth of Massachusetts, Department of Environmental Protection, Executive Summary: 1997 Solid Waste Master Plan Update, Executive Office of Environmental Affairs, Department of Environmental Protection, 1997. http://www.magnet.state.ma.us/dep/bwp/dswm/dswmpubs.htn#swmp.
16. United States Environmental Protection Agency (USEPA), Characterization of Municipal Solid Waste in the United States: 1997 Update, Prepared for U.S. Environmental Protection Agency, Municipal and Industrial Solid Waste Division, Office of Solid Waste, by Franklin Assoc., Ltd. Prairie Village, KS., Report No. EPA 530-R-98-Oct.-May, 1998, chap. 4.
17. Congressional Research Service, Superfund reform: Cleaning up America's toxic waste sites, *Congr. Dig.,* Mar., 68, 1998.
18. Howlett, D. and Tyson, R., Toxicity of Times Beach: EPA verifies Mo. town's worst fears, *USA Today,* Sep. 13, 10A, 1994.
19. Nassiter, A., Villain is dioxin: Relocation is response, *New York Times,* Oct. 20, 1996.
20. Stephens, R.D., Hazardous Wastes, in: *Principles and Practice of Environmental Medicine,* Tarcher, A B., Ed., Plenum Medical Book Co., New York, 1991.
21. Council on Environmental Quality (CEQ), Environmental Quality, 15th Annual Report, 1984. U.S. Government Printing Office, Washington, D.C., 1986.
22. United States Environmental Protection Agency, Safer Disposal for Solid Waste: The Federal Regulations for Landfills, Report No. EPA/530-SW-91-092, Washington, D.C., 1993.
23. United States Environmental Protection Agency, Waste Minimization National Plan: Reducing Toxics in Our Nation's Waste, EPA Solid Waste and Emergency Response, EPA Report No. 530-F-97-010, Washington, D.C., 1997.
24. Senkan, S.N. and Stauffer, N.W., What to do with hazardous waste, *Technol. Rev.* 84, 34, 1987.
25. Harris, R.H., English, C.W., and Highland, J.H., Hazardous waste disposal: Emerging technologies and public policies to reduce public health risks, *Am. Rev. Public Health,* 6, 269, 1985.
26. Rose, J. Incineration regulations set to tighten, *Environ. Sci. Technol.,* 28:12, 512A, 1994.
27. U.S. Congress, Office of Technology Assessment, Technologies and Management Strategies for Hazardous Waste Control, OTA-M- 197, U.S. Government Printing Office, Washington, D.C., 1983.
28. U.S. Congress, Office of Technology Assessment (OTA), *Assessment Superfund Strategy,* OTA-ITE-282, U.S. Government Printing Office, Washington, D.C., 1985.
29. U.S. Congress, Congressional Research Service, Superfund Reform: Cleaning Up America's Toxic Waste Sites, *Congr. Dig.,* Mar., 68, 1998.
30. The Congressional Institute Inc., *Superfund* July 17, 1, 1996, http://www.conginst.org/cidoc-l.nsf/e74d.,167f730.d9718525636coo79ao93?Open Document.
31. United States Environmental Protection Agency, Superfund Cleanup Figures, Apr. 1, 1998. http://www.epa.gov/superfund/whatissf/p/mgnirpt.htin.

CHAPTER 12

Assessing Human Risk

INTRODUCTION

A farmer stands in his field, staring in horrified awe at the stream of black waste flowing onto his land from an adjacent river. This scene occurred on April 29, 1998, when a waste reservoir in southern Spain burst, releasing toxic sludge into waterways and onto surrounding farmland. The reservoir served as a holding area for zinc, lead, iron, and cadmium wastes from a nearby Canadian-owned zinc mine. The dangerous waste posed both an immediate and long-term threat to the health and livelihoods of locals and to the environment. Crops were ruined and land was contaminated by

OBJECTIVES FOR THIS CHAPTER

A student reading this chapter will be able to:

1. **Define risk and discuss the uncertainties associated with environmental risk**
2. **Discuss the characteristics of risk**
3. **Define risk analysis and describe the tools used to perform risk anaylsis**
4. **Explain the concepts of dose, extrapolation, and accepable daily intakes (ADI)**
5. **List and discuss the process of risk analysis including hazard identification, dose–response assessment, exposure assessment, and risk characterization**
6. **Outline and discuss the major components of risk management and risk communication**

the toxic mess. Locals complained of burning eyes and throats. Ten tons of dead fish and shellfish were carted away from the acidic waters. With shock and dismay, people wondered at the odds of this unexpected event. A shift in the Earth apparently sent the reservoir wall tumbling, releasing millions of cubic feet of metal waste into the surrounding region. This event typifies the multitude of environmental risks facing the world today. These types of risks are uncertain and unexpected, and the consequences can be severe, even deadly.[1]

Risk is a integral part of life, permeating every aspect — from eating, breathing, sleeping, socializing, working and playing. Risk can be of a financial, personal, social, health, and environmental nature. Some industries, such as insurance, gambling, and finance, are based on risk. The word *risk* describes a range of activities, situations and concepts, from drinking a glass of red wine daily to skydiving to chemical exposure. Risk can imply chance consequence, danger, and opportunity.[2] In everyday language, risk is commonly used to describe types of people or situations. For example, the term *risky business* implies shady dealings, and a *risk-taker* suggests a brave, confident, perhaps foolhardy individual. *Risk-averse* describes someone who shies away from risk. *Risk-free* describes an activity that contains no uncertainty or negative consequences. A comparison of some of life's risks are shown in Figure 12.1.

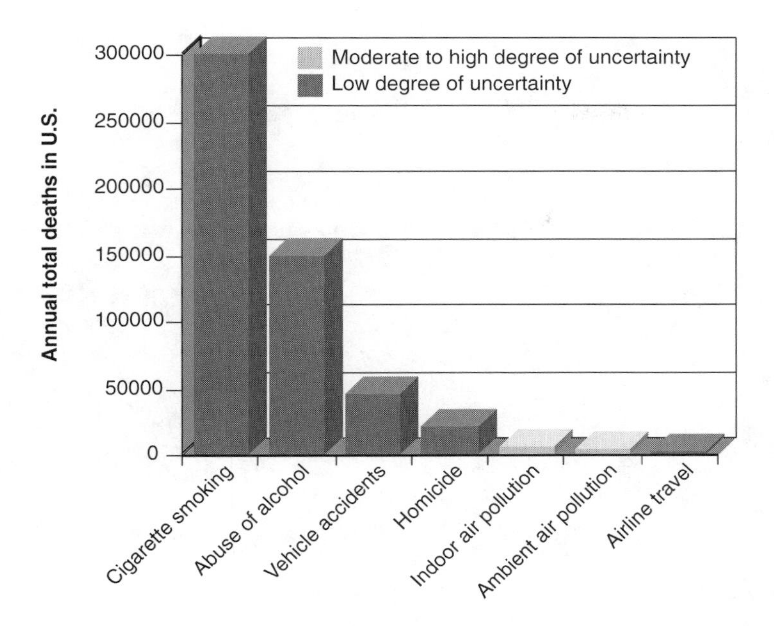

Figure 12.1 A comparison of some common risks. (Adapted from Moeller, D.W., *Environmental Health,* revised ed., Harvard University Press, Cambridge, MA, 1997, chap. 16.)

Many of the risks in our lives are voluntary, such as our diet, sunbathing, and smoking. People accept certain risks because they enjoy the benefit they receive from the behavior or activity. People risk money in the stock market in the hopes of making a profit; people live in earthquake- and hurricane-prone areas in exchange for the lovely climate often found in those regions. We accept many risks in our lives because they seem both mundane and remote. We cross streets, take pharmaceuticals, drive in cars, and fly in airplanes under the assumption that these activities are safe. However, even these everyday activities carry some level of risk. Because these activities are familiar and routine, they do not appear threatening. However, not everyone is content with life's daily dose of risk. Some people seek out extraordinarily high levels of risk, engaging in skydiving, bungee jumping, rock and ice climbing, and other extreme sports (Figure 12.2). People choose these activities because they enjoy the element of danger from the risk and the adrenaline rush. They have defined their own acceptable level of risk, which might be too high for most people who reach their risk threshold more easily.

ENVIRONMENTAL RISK

Some risks are well understood, such as driving and smoking. The vast quantity of information and statistics on automobile accidents allows us to draw conclusions about the risk of dying in a car crash. Similarly, voluminous data on smoking and

Figure 12.2 Extreme sports are becoming popular because of the excitement associated with high risk. (Source: Corel sampler, limited edition CD-ROM.)

disease demonstrates that smoking poses a health risk. However, some risks remain shrouded in uncertainty. Many environmental risks fall into this category. Environmental risk is a reality of today's world. The seemingly endless supply of synthetic chemicals, consumer goods, energy, and waste create new risks through chemical contamination, pollution, and environmental degradation. Environmental disasters such as chemical spills or explosions threaten millions of people living in the vicinity of manufacturing or storage facilities. The uncertain risks of global warming and ozone depletion loom ahead. A central factor of environmental risk is that it is usually involuntary. People do not choose to ingest chemical pollutants such as pesticides or industrial solvents in their food and water, to undergo workplace exposures to dangerous chemicals, to breathe polluted air, or to experience radiation exposure from nuclear fallout or faulty nuclear power plants. These environmental risks pose a unique problem to regulators charged with protecting the public's health. Limited information may be available on the health effects of these risks. Consequently, in an effort to protect the public's health, various government agencies study these potential hazards to determine the levels of risk they pose. This effort to understand these risks, and to quantify their impact on human health, is the field of risk analysis.

Risk Characteristics

Risk has several defining characteristics. The two characteristics that most embody risk are uncertainty and the unknown. Risk can be defined as the likelihood of an unwanted occurrence coupled with an element of uncertainty about when the risk might occur.[4] Many other definitions of risk exist, all varying slightly, but focused on the concepts of uncertainty and the unknown. Other risk characteristics may include dread, voluntary or involuntary, immediacy or latency, catastrophic potential, threat to future generations, and exposure to the unknown. Risk often encompasses the ideas of morbidity and mortality — that is, the possibility of sickness or death.[5] Certain risk characteristics, such as involuntary, unknown, and dread, generate greater fear in people than risks that are known or voluntary. As many environmental risks have the characteristics of unknown and dread, they cause people anxiety and concern about the consequences of exposure to the risk. Risk analysis is one method to address this fear and anxiety, through developing knowledge about a risk, understanding its potential health effects, and devising methods to limit and manage the risk.

Development of Risk Analysis

Risk analysis is the process of reviewing information about a hazard to characterize that hazard's impact on human health. This process involves a review of scientific studies, an understanding of the properties of a risk, an assessment of levels of human exposure and dose, and a conclusion about the likelihood, impact, and extent of a risk. By employing these methods, risk analysis allows researchers to develop conclusions on the severity and consequences of environmental risks. Risk analysis developed over the past 30 years out of a need to understand modern technologically-based risks. It began with an increased interest in the environment from the government, scientific, and public sectors. Rachel Carson's 1962 book *Silent Spring* spurred concern about environmental issues. Environmental crises such as Love Canal, Three Mile Island, and the chemical accidents in Seveso, Italy, and Bhopal, India, encouraged research into the effects of these environmental risks. The purpose was to assess the potential health impact of these hazards and to develop methods to cope with these risks.

The field of risk analysis developed as a response to several changes in the modern world. First, the number of risks society faces today has increased dramatically compared to times past. For example, over 70,000 synthetic chemicals are used today in industrial and manufacturing processes and in many consumer products and household items. Little detailed risk information is available for most of these substances.[6] Second, technology has improved our ability to measure risk. Through the continued development of risk analysis we have begun to examine the hazards of all consumer goods and even products that have been taken for granted as being good for our health. Recent studies have been conducted looking at the effects of cow's milk on the human body. The main focus of the study examines intestinal irritation, intestinal bleeding, anemia, and allergic reactions in infants and children, as well as infections of salmonella. In a paper written by Robert M. Kradigan, M.D., risk analysis of cow's milk examines the possibilities of viral infection and possible health hazards associated with milk.[7] Today, exposures are measured in the parts per million, billion, and even trillion, greatly increasing our capacity to accurately quantify exposure levels. Third, the number of government agencies directly or indirectly monitoring environmental risk have increased, such as the Food and Drug Administration, the Environmental Protection Agency, and state-run Departments of Environmental Protection. Even towns and cities have agencies regulating local risks to public health. Fourth, the number of laws and regulations governing the environment and protecting public health have increased dramatically over the past 30 years. Some of these laws drastically changed the management of risks and dissemination of information. The Superfund Amendments and Reauthorization Act of 1986 (SARA), also known as the Emergency Planning and Community Right-to-Know Act, ensured that communities were prepared for chemical emergencies and had access to information about chemicals at sites in the area. Other examples of regulations include the Food, Drug and Cosmetic Act, the Toxic Substances Control Act, and the Clean Water Act. Fifth, an increase in public interest in environmental risks has spurred research in quantifying risks. Special interest groups and grassroots movements have involved the public in environmental issues at both local and national levels.[8,9] The development of risk analysis has indeed led to the enlightenment of environmental hazards and risks associated with our lifestyles. Strides have been made in enlightening the masses on issues of risk analysis. Recent environmental risk analysis has led us to begin examining the effects of the population's negligence toward the environmental responsibility of this generation. A recent article written by Radim J. S(breve)ram for the Institute of Experimental Medicine in the Czech Republic discusses the effect that air pollution is having on the reproductive health of the world population. The risk analysis conducted concurs with other current research that pollutants in the air are a causative problem in reproductive health. For both men and women, polycylic aromatic hydrocarbons are the most harmful and common pollutants and are caused by both industrial and chemical processes polluting the air. A growing awareness has developed over environmental issues because of the development of risk analysis.[10]

A detailed qualitative and quantitative description of a risk can be created by gathering the relevant study information, considering the important factors, and developing a conclusion about the risk. Risk analysis has many applications across a variety of fields. This risk description can be used in decision making and the development of environmental and public health policy. For instance, risk analysis can be used to estimate the likelihood of developing cancer from exposure to radiation, to determine the side effects of a medicine awaiting FDA approval to ascertain whether a synthetic food product, such as Olestra or saccharin, is safe for human consumption, or if transgenically altered animals and plants pose a health risk. Risk analysis allows public groups to make informed decisions and weigh the risks and benefits in their community.

Tools of Risk Analysis

Risk analysis employs several scientific disciplines in its goal to characterize a risk (Figure 12.3). Significant amounts of information, both quantitative and qualitative, are needed to develop an adequate picture of a risk. Toxicology, epidemiology, clinical trials, and cellular studies provide some of the information for a risk analysis.

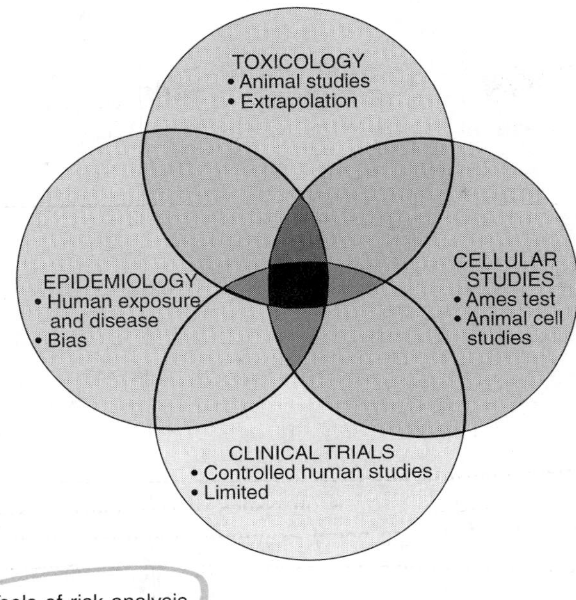

Figure 12.3 Tools of risk analysis.

The types of information reviewed include characteristics of the chemical, type and length of exposure, dose, animal response, and human response. These disciplines are discussed in greater detail below.

Toxicology

Toxicologists study chemicals to determine their physiological and health impacts on humans. Risk analysis relies heavily on toxicological studies and findings. A particular branch of toxicology — regulatory toxicology — aims at guarding the public from dangerous chemical exposures. For example, the health effects of nicotine, saccharin, and benzene have been explored and defined through toxicology studies. Toxicological studies usually involve controlled laboratory animal studies. The animals are evaluated for their responses to different doses of a substance. These controlled research conditions are crucial for procuring accurate information. Prior to a toxicology study, little may be known about a chemical, such as what amount causes minimal adverse health effects or death. In a study, scientists follow a series of steps to learn about a substance's activity in a living organism. A study can delineate both the lower and upper limits of a chemical's potency, that is, the amount that causes no effect in any animal and the amount at which all animals die. After these limits are established, scientists can work within this range to discover additional information about a chemical's health effects. As a chemical may have myriad health effects, the researchers will focus their interests on certain health responses. These specific physiological changes are called *endpoints*. Some common endpoints

WEB PICK **Go to http://www.unix.oit.umass.edu/~envhl565/index.html** **Click on "Chapter Web Links" to find Superfund-Risk Assessment Home Page on the World Wide Web under Chapter 12**	
Chapter 9	Risk Assessment
Whose Site?	U.S. EPA Office of Emergency and Remedial Response (OERR)
URL	http://www.epa.gov/superfund/programs/risk/index.htm
What's Here?	The U.S. EPA Office of Emergency and Remedial Response (OERR) developed this site to provide a consistent framework to evaluate and communicate the risks posed by hazardous waste sites. Risk assessors and risk assessment data play an important role in the characterization and clean-up of Superfund sites. The risk assessor ensures that the planning and work plan development tasks incorporate risk assessment data needs. Appropriate sampling will lead to comprehensive documentation of possible exposure pathways and help achieve standardization in risk assessment planning. There are a number of risk-assessment tools at this site including the Integrated Risk Information System (IRIS), Soil Screening Calculations, Exposure Factors Handbook, Johnson and Effinger Air Model, Guiding Principles for Monte Carlo Analysis, Proposed Guidelines for Carcinogenic risk Analysis, and much more.

frequently studied include the effective dose of a chemical, named the No Observable Effect Level (NOEL), the No Observable Adverse Effect Level (NOAEL), the Lowest Observable Adverse Effect Level (LOAEL), and death. All of these endpoints provide valuable information on the doses and effects of a chemical. This information can be incorporated into a risk assessment.[6,9]

Dose

When studying a chemical agent, toxicologists use different doses to elicit different animal responses. They ascertain the threshold dose, which is that dose at which no effects are seen in the study animals. As the administered dose increases above the threshold amount, animals will begin to show adverse effects and some animals will die (Figure 12.4). The dose at which 50% of all the test animals die is called the lethal dose or the LD50. Similarly, the ED50, or effective dose, is the

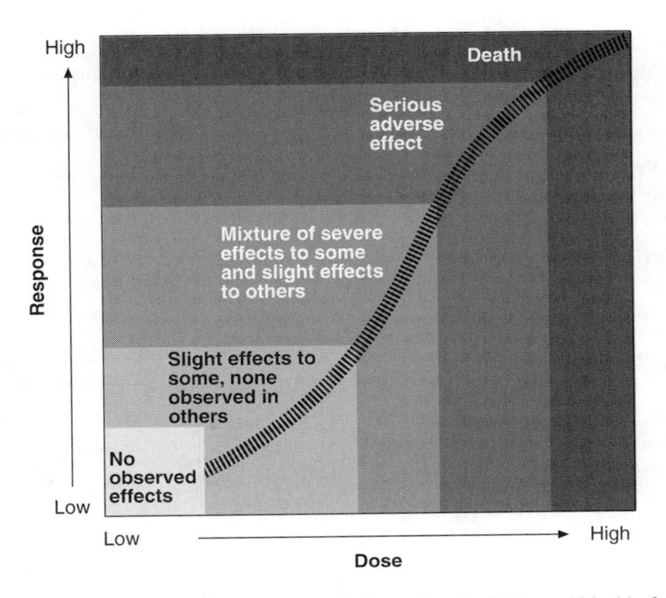

Figure 12.4 The dose response curve. (Adapted from Covello, V.T., and Merkhofer, M.W. *Risk Assessment Methods,* Plenum Press, New York, 1993, chap. 1.)

dose at which 50% of the animals demonstrate a response to the chemical. A very toxic chemical will have a low LD50, signifying a low dose sufficient to kill 50% of the test animals. A less toxic chemical will have a higher LD50. The Maximum Tolerated Dose (MTD) is a common level of chemical exposure in animal studies. At this dose, researchers expect to see only up to a 10% loss in weight and no death, clinical toxicity, or pathologic lesions in the animal population. Some critics of the MTD believe it is too high a dose. Chemicals that cause cancer at the MTD in test animals may not cause cancer at lower doses.[10]

Animal studies occur under controlled conditions. This control refers to exact doses, and specific lengths and times of exposure. By controlling the amount, the timing, and the duration of exposure, scientists can obtain increased quantity and quality of information on a chemical's health effects. For example, altering the amount, timing, or duration of a dose in a study can alter the chemical's opportunity to inflict damage. Large doses may cause one type of response; small doses, another. Timing of a dose may also be critical. For example, if exposure is timed to occur during a "window of vulnerability" in a fetus, minuscule amounts of an agent may cause severe damage. Outside this "window," this amount would be harmless to the fetus. Clearly, control of the dose provides tremendous influence over the types of health effects seen. In addition, this control allows for variability in study design. Toxicology studies can occur over short, medium, or long periods of time. Short-term, or acute exposure studies, occur over 2 weeks and generally involve high doses of the chemical substance. Medium-length or subchronic studies involve lower chemical doses over a longer time period, from 5 to 90 days. Long-term or chronic studies can last 2 years and use much smaller doses of chemicals. This variability improves understanding of a chemical's activity under a different conditions.[11] For comparison purposes, animal studies establish a control group not exposed to the agent under consideration. This group provides the researchers a level of certainty that the physiological responses seen after exposure are in fact due to exposure and not to some unrelated, uncontrolled variable.[12]

Extrapolation

Extrapolation can be defined as using known information to infer something about the unknown.[13] In risk analysis, the term extrapolation refers to the use of animal data to predict human response to chemical exposure. Extrapolation is widely used in toxicology to deduce the risks humans face The use of extrapolation is a vital tool of toxicology since human exposure studies are not widely accepted. Different types of animals can be used during laboratory studies, depending on the agent under consideration. While dogs and monkeys have been used for animal studies, the most common laboratory animal is the rat. During laboratory studies, animals generally receive high doses of the chemical of interest for a measured span of time, often their lifetimes. Laboratory animals have shorter life spans than humans, allowing studies to document a full lifetime of exposure, from infancy to maturity to death. The results from these high-dose, short-duration studies are used to extrapolate human response to the longer-term, lower-level exposures we generally receive. The difference in dose level and duration of exposure generates some doubt about the validity of extrapolation. Furthermore, biological and metabolic differences between the test animal and humans may undermine the validity of extrapolations. For instance, the test animal may have one method for detoxifying a substance, whereas a human may use a different metabolic pathway. Alternatively, a test animal may lack an adequate detoxification pathway, whereas humans may be very effective at detoxifying a substance. Because of these differences, a substance that causes cancer in one animal may not cause cancer in another animal, or vice versa. This fact limits the accuracy in extrapolating from animal data to human response.[11] Despite these drawbacks, extrapolation is a crucial method in studying potential human outcomes based on the results of animal studies.

Acceptable Daily Intakes

The purpose of toxicology studies is often to establish an acceptable level of exposure or dose of a substance that is considered safe. This level, which poses little risk, is termed the acceptable daily intake (ADI). In extrapolating from animals to human studies, scientists incorporate safety factors when determining acceptable exposure (dose) levels. The No Observable Effect Level (NOEL) established by the threshold dose in test animals is translated into the ADI or the reference dose for humans. These levels include safety factors from 10 to 10,000, depending on the protected group. The safety factor is added by dividing the threshold dose by some factor of 10, thereby reducing the acceptable level of exposure and risk from exposure and increasing safety. Again, which factor of 10 is used depends on the level of protection sought, the uncertainty inherent in the study data, and who is being protected. Lower safety factors are used when high-quality human data is available. Higher safety factors, up to 10,000, are used as the uncertainty level goes up and the quality of the data diminishes. Although the ADIs are based on scientific studies and include a margin of safety, these numbers are not foolproof. As more data becomes available about a substance, the ADIs may change, either lower or higher. For example, as we have gained knowledge about the dangers of exposure to lead, the minimum level of exposure considered harmful to human health has decreased, as adverse health effects are seen at extremely low concentrations of lead.[4] We also have had to incorporate these ADIs into our diets. The nature of the environment has changed so that we must examine dietary ADI to set safety standards for dosages of daily allowances. Recent studies have examined ADI soy-based powders and liquids available as alternatives to milk. Although soy-based formulas have been found to offer most of the nutrients of breast milk, recent studies are showing that infants on soy formula are two to three times more likely to develop thyroid disease. With studies of this sort, we are better able to access risk factors involved with exposure to substances that may pose safety hazards.[13]

Epidemiology

What Is It?

Epidemiology is "the study of the distribution and determinants of disease frequency in the human population."[14] In an attempt to determine causality between an exposure and a disease, epidemiologists study human response to various environmental agents, such as tobacco smoke, pharmaceuticals, exercise, and even hair dye. These are not controlled laboratory studies, but rather occur in the outside world, subject to fluctuating conditions. When an increase in the incidence of a disease is reported, epidemiologists try to account for many factors that can be related to its occurrence. This was true when high incidences of morbidity were discovered among inner-city children with asthma, and it was hypothesized that this could be related to the exposure to cockroach allergen. Children from eight major inner-city areas were studied. It was concluded that that this variable alone could not account for this situation. Many variables are introduced into these studies, such as personal choice, socioeconomic factors, age, demographics, and other factors. These factors, called confounders, can bias or skew the results of the study either toward more of an association than actually exists or toward less of an association than actually exists, depending on the confounding factor. Consequently, in designing a study, scientists will try to "control" these confounding factors.[15]

Study Types

Several different types of epidemiological studies exist, including cross-sectional, cohort, and case-control studies. These study types differ in design by the methods categorization of study subjects and the timing of information gathering. Each type of study has its own strengths and weaknesses in finding an association between an exposure and a disease. A cross-sectional study is a type of descriptive study design in which both a subject's exposure and disease are determined at the same time. This type of study provides general information on the disease, such as who and where. Cohort studies and case-control studies are types of analytic studies focused more on answering the question of why a particular disease develops. Cohort studies first group people by exposure and then follow them over time for disease development. This type of prospective study is useful for studying rare exposures. Case-control studies first group people by disease, then look retrospectively for a history of exposure. This type of study is useful for studying rare diseases. All of these study types provide valuable information on disease occurrence in human populations.[12,15]

Bias

Epidemiology involves the study of humans. Consequently, it avoids the extrapolation problem seen in animal studies. However, it is not without drawbacks. A significant problem in epidemiology is bias. Bias is a type of error introduced into epidemiological studies through differential treatment of the cases and controls in a study, either during the selection process or information gathering. Selection bias, information bias and nondifferential misclassification bias may lead to inaccurate study results. Researchers can address the problem of bias through careful study design. Despite this and other limitations, epidemiology remains an important tool for evaluating the link between exposure and disease.[12,15]

Clinical Trials

Clinical trials provide a direct means to evaluate human response to a substance under more controlled conditions than epidemiology studies. This type of risk assessment provides detailed information on dose response, toxicity, efficacy, and side effects. In clinical drug trials, individuals volunteer to be exposed to a substance or to ingest a drug and are assessed for their health response. The types of health responses measured are generally minor, short-term, and reversible, since testing human subjects involves ethical considerations and the trials are limited in duration. Volunteers must be informed of the risks, must give their consent, and may receive monetary compensation for their participation. Despite the volunteer status of subjects in human clinical trials, dissent exists about the ethics of human studies. Past studies on prisoners were halted due to ethical considerations. Recent controversy erupted when several pesticide manufacturers announced human testing of pesticides to determine safety. The companies resorted to human trials to demonstrate the safety of their products in the face of strict food quality legislation passed in 1996. Opponents felt the companies were taking advantage of volunteers, enticing them with significant monetary compensation, downplaying the inherent uncertainty of the exposure and misleading them by calling pesticides "drugs." Clinical trials are used widely by pharmaceutical companies prior to releasing drugs to the public. Although they are limited in scope, depth, and duration, clinical human trials yield crucial information on humans.[10,16] Clinical trials provide us with information on the safety and efficacy of drugs. Human trials can provide great risk to participants, but they also offer a great service to humanity. The promise of drugs that may eradicate the flu, the AIDS virus, or the common cold cause a stir at the mere thought of the possibilities and lives that may be saved. The National Cancer Institute investigated the drug tamoxifen in 1992 as a possible cure for previously treated patients and as a preventative agent for breast cancer. The drug was conducted as a placebo-controlled study on 13,388 healthy women. Half of the women were given tamoxifen and the other half was given the placebo, with varying results. The clinical trial found that there were very satisfactory results from the study with fewer cases of breast cancer than the placebo group of women. However, the women who took the drug had two to three times the incidence rate of uterine cancer. Only through clinical trials can we ascertain the health hazards associated with drugs that may be helpful and harmful. Their importance is crucial in the discovery of new drugs and their effect on the human body, both long and short term.[17]

Cellular Testing

Cellular studies are used to test chemicals for mutagenicity and carcinogenicity. These tests can be done on bacteria or animal cells. A widely known bacteria test is the Ames test, which utilizes Salmonella bacteria that lack the ability to produce histidine, a vital growth chemical. These bacteria are exposed to the chemical in question and assessed after 2 to 3 days for signs of growth. Indications of growth suggest a mutation has occurred in the bacteria's genetic information, conferring the ability to produce histidine. Although relatively inexpensive and quick, cellular studies are not always accurate in identifying carcinogens. This problem relates to the difficulty in using single-celled organisms to predict human response to a substance. Other cellular studies include animal cellular studies, which are used to identify potential carcinogens. In these tests, the chemical is applied to rodent or hamster cells. After a 1- to 3-week period, these cells are assessed for changes in growth. Unlike the Ames test, this type of study is fairly accurate in predicting carcinogenesis.[11]

THE PROCESS OF RISK ANALYSIS

With this arsenal of tools, the process of risk analysis can begin to identify and enumerate the risk of a particular hazard. This process has four steps: (1) hazard identification; (2) dose–response evaluation; (3) exposure assessment; and (4) risk characterization (Figure 12.5). These four steps describe the risk through a qualitative and quantitative evaluation.

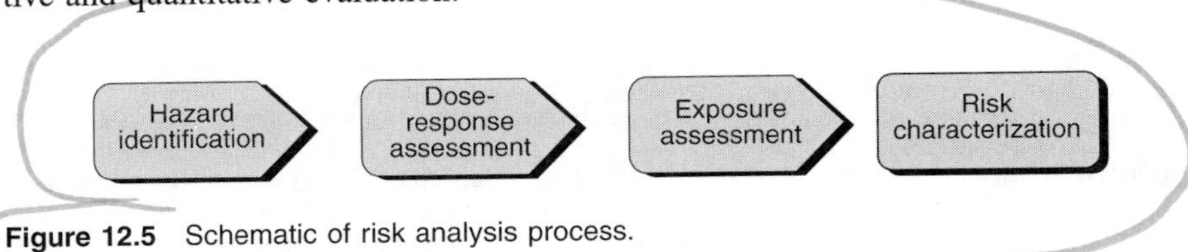

Figure 12.5 Schematic of risk analysis process.

Hazard Identification

The initial step in risk analysis — hazard identification — involves identifying chemicals that present a risk to human health. This step entails performing a qualitative assessment of a chemical's potential for negative health impacts on humans. A thorough evaluation of scientific research about the risk determines the likelihood of adverse effects on humans. A chemical's activity in living organisms is researched through a review of animal studies, cellular studies, epidemiological studies, and human trials. In addition, to ascertain the possibility of human exposure, a substance's boiling point, volatility, and solubility are considered. If a chemical has been identified as a potential threat to human health, the process of risk analysis continues.[3,18]

Dose–Response Assessment

The next step, dose–response evaluation, provides a quantitative view of the risk. This step also involves a review of scientific studies and data. In this case, the magnitude of response is correlated with the dose (Figure 12.6). The health response is reviewed at each dose to note the differences in response. This correlation provides valuable information about levels of risk at different levels of exposure. In addition, epidemiological studies are assessed to determine the agent's strength, potency, and capacity to produce negative health effects in humans. Although human studies are

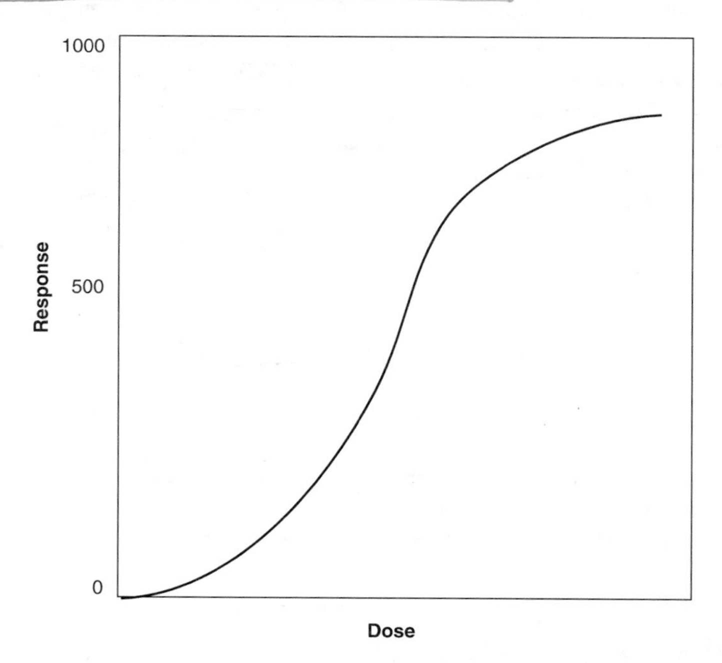

Figure 12.6 A typical simple dose–response curve. (Adapted from Cohrssen, J.J. and Covello, V.T., Risk Analysis: *A Guide to Principles and Methods for Analyzing Health and Environmental Risks,* The Council on Environmental Quality, 1989, chap. 3.)

preferable for this evaluation, such data is not always available. In such an instance, animal data is used.[3,18]

Exposure Assessment

After the dose–response assessment, an exposure assessment is performed. The purpose of the exposure assessment is to measure or estimate people's level of exposure. To accomplish this, the length of exposure, the amount of exposure, the route of exposure, the duration of exposure, and the number of people exposed are considered. Exposure is different from dose in that exposure refers to the amount of a substance in the environment, while dose is the level of a substance actually taken in by an organism. Dose can be influenced by many factors, such as how the substance enters the body, whether absorbed through the skin, ingested with food, or inhaled (Figure 12.7). This discrepancy between the amount of a substance measured and the actual dose is included in an exposure assessment. To perform an exposure assessment, specific information must be considered. The properties of the substance, such as its volatility, water-solubility, ability to become airborne, or migrate through soil will influence its relationship with the environment and affect its routes to human exposure. Besides considering the substance's properties, the assessment should specify the group exposed, with special attention given to those particularly vulnerable, such as pregnant women, children, the elderly, and the health-compromised. Finally, the type of exposure must be noted, whether continuous,

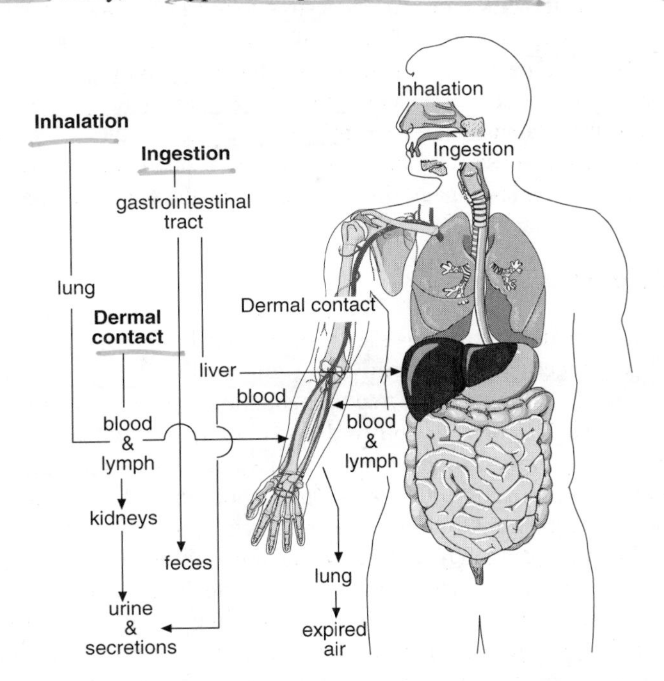

Figure 12.7 Routes of chemical exposure. (Adapted from Cohrssen, J.J. and Covello, V.T., Risk Analysis: *A Guide to Principles and Methods for Analyzing Health and Environmental Risks,* The Council on Environmental Quality, 1989, chap. 3.)

Deaths per million people in a selected population (nation).

Deaths per million people within a certain radius of the released pollutant.

Deaths per weight (tons or pounds) of the toxic substance(s) released.

Loss of life expectancy associated with exposure to the toxic material(s).

Figure 12.8 Types of risk estimates. (Adapted from Stern, P.C. and Fineberg, H.V., *Understanding Risk,* National Academy Press, Washington D.C., 1996, p. 50.)

intermittent, short-term, long-term, or chronic.[3,9] As is the case with studies looking at incinerators and the health risks associated with exposure to their emissions, cumulative effects should also be considered, taking into account the effect of all the sites in a given region. This is especially important with this particular hazard due to its ability to "travel long distances and persist in the environment."

Risk Characterization

The final step of risk assessment is risk characterization. Risk characterization provides a picture of the risk that addresses its severity, likelihood, and consequences. All of the previous information from the hazard identification, the dose–response evaluation, and the exposure assessment are incorporated into the risk characterization. The risk characterization includes an estimate of the negative effects to exposed individuals, such as the number of cases of cancer or deaths per 100,000 people (Figure 12.8). Other types of risk estimates include an individual lifetime risk, such as a one-in-a-million chance of cancer, a loss of life expectancy, as in days or years lost from exposure to a substance, and relative risk, that is, the risk of the exposed population vs. the unexposed. Probabilities of these effects are also provided. Any drawbacks, limitations, and uncertainties inherent in the estimate are incorporated into the risk characterization.[3,11]

Limitations of Risk Analysis

The conclusions drawn from risk analysis are not foolproof. Limitations of risk analysis include uncertainty, variability, and effect of multiple exposures. Uncertainty exists in all estimates and recommendations about environmental risks. Generally, "zero risk" is not a possibility, due to high degrees of uncertainty involved in risk assessment. Although people want to be assured about a safe level of exposure to a chemical, sometimes too little is known to about the substance to provide any real assurance. A frightening example is represented by pharmaceuticals pulled from the market. Although a drug may undergo years of laboratory testing and even human trials, sometimes rare side effects are not seen until the drug is approved and released to the public. This scenario occurred recently with the popular diet drug Fen-Phen and the fast-acting analgesic Duract. In the case of both drugs, dangerous and even deadly side effects appeared in human users shortly after the drugs were available by prescription. Both drugs were subsequently taken off the market.[19] Even with rigorous laboratory testing and careful risk analysis, a degree of uncertainty remains.

This uncertainty factor can erode the purpose of recommended doses, as people may doubt that any safe exposure level exists. In addition, interpersonal variability may strongly affect an individual's risks. Some people in a population are simply more sensitive to exposures than others and may be more likely to experience adverse health effects from a "safe" dose. Therefore, the built-in safety factors may not be enough to protect all individuals from harm. In addition, risk analysis rarely considers the fact that people are exposed to multiple risks at the same time. Rather, risk analysis focuses on exposure to a single agent. Any synergistic or additive effects are ignored, providing an inaccurate picture of the true risks of real-world exposures. Despite these limitations, risk assessment is still a valuable tool for exploring and understanding the risks of the modern world.[10]

RISK MANAGEMENT AND COMMUNICATION

Management

Risk assessment provides us with tremendous amounts of information about risks. Risk management involves merging the results of risk analysis with various social factors such as socioeconomic conditions, political pressures, and economic concerns (Figure 12.9). Three avenues of risk management are educational, economic, and regulatory. These are not mutually exclusive but can be used together to manage a risk. Educational risk management can include use of the media to inform the public about risk or the strategic placement of warnings to provide people with risk information, as on drugs and other chemical products. Economic risk management is accomplished through pollution taxes and permits that emphasize risk reduction through monetary incentives or disincentives. Regulatory risk management, also called command and control, is evident in the numerous laws governing potential environmental and health hazards in the U.S. The Clean Water

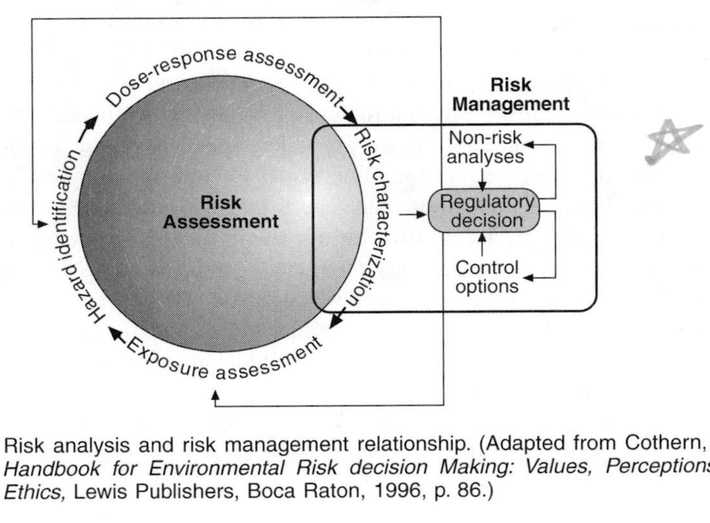

Figure 12.9 Risk analysis and risk management relationship. (Adapted from Cothern, C.R., *Handbook for Environmental Risk decision Making: Values, Perceptions and Ethics,* Lewis Publishers, Boca Raton, 1996, p. 86.)

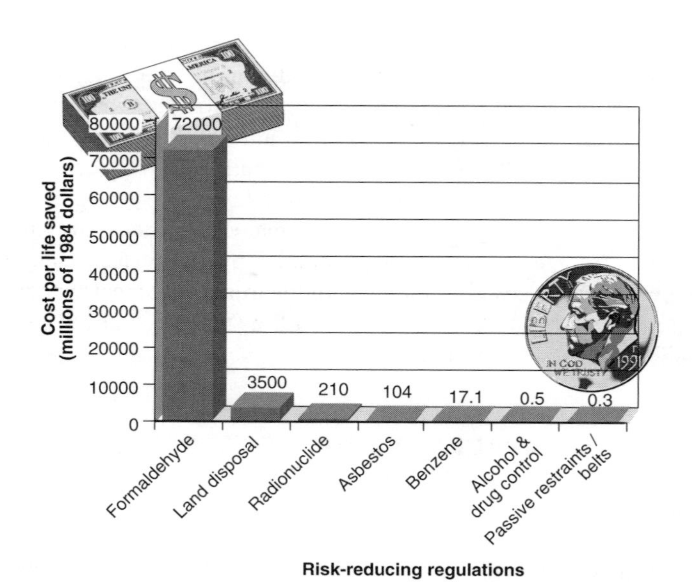

Figure 12.10 Comparative costs of risk management. (Adapted from Viscusi, W.K., *Rational Risk Policy,* Clarendon Press, Oxford, 1998, p. 98.)

Act, the Safe Drinking Water Act, and the Toxic Substances Control Act are examples of regulatory risk management, where laws restrict and limit pollution, chemical exposures, and releases.[21]

Risk management generally involves comparing the risk to some other factor such as cost, or reducing the risk or the benefit gained from the risk (Figure 12.10). Several analytical methods include cost–benefit analysis, risk–benefit analysis, and contingent analysis. These comparisons allow regulators and agencies to determine the best course of action for managing or controlling a risk. Economics are a critical factor in risk management. Consequently, the best course of action is not always the one that reduces the most risk, but rather is the most economically feasible option, reducing the greatest amount of risk per dollar spent. Risk management decisions do not solve all problems surrounding a risk or satisfy everyone's concerns. A decision that benefits one group may cause harm to another. An example is the process of selecting the location of an ash dump. A risk management decision is made to site the dump in a sparsely populated, rural area. This decision solves the problem of where to dispose of incinerated waste. However, the residents of the rural community receive little or no benefit from the dump. Instead, they face the risk of land and water pollution from the ash, a decrease in aesthetic value of their land from the existence of the dump, and the noise of waste trucks through a previously low-traffic area. Risk management also encompasses the idea of an acceptable level of risk and an acceptable number of cases. We cannot eliminate all risk, but we can hopefully reduce risk to some acceptable level. This idea of an acceptable number of cancer cases or an acceptable number of deaths is an uncomfortable but inherent aspect of risk management. Clearly, risk management is a complex undertaking that requires careful consideration of the risk characterization, political, economic, and social factors.[23,24] At times the multilevel process of risk management can result in indecision, delayed action, or in some incidences no action due to the absence of funding. An example of this occurred in 1998 when a "panel convened by the National Institute of Environmental Health Sciences decided there was enough evidence to consider the invisible waves called electromagnetic fields, like those generated by power lines and electric appliances, a 'possible human carcinogen.'" Because no consensus could be reached regarding issues like safety thresholds and funding was withdrawn, nothing has been done with this information.

Risk Communication

Risk communication occurs when a dialogue occurs between two or more parties involved in a risk. These parties may be government agencies, industry, and citizen groups. Methods of risk communication include public hearings, emergency hotlines, and information pamphlets. The goal of risk communication is to effectively relay risk information developed through risk analysis to various interested groups. The World Resource Report can be used as an example of risk communication. It is a publication that is released jointly by the World Resources Institute, United Nations Environment Programme, the United Nations Development Programme, and the World Bank. Their collaborative message places an emphasis on prevention and further support "coordination and communication among agencies that don't often interact" (1998, NEHA). Risk communication can be challenging, as it requires addressing people's different risk perceptions, biases, scientific knowledge, educational backgrounds, even race and gender. Many of these areas require significant sensitivity and tact. Not all risk communication is effective. Certain factors are required for effective risk communication. Most importantly, trust and credibility are needed. As trust and credibility are influenced by the related factors of openness and honesty, knowledge and expertise, and concern and care, these factors also influence the success of risk communication. The public often mistrusts industry, so industry must act to earn the public trust. On the other hand, industry often perceives the public as lacking scientific knowledge, so the public must strive for a clear understanding of scientific principles. In addition to requiring trust, risk communication must often address differences in perception and meanings. Successfully addressing these differences allows for open, effective communication among parties subject to the risk.[25]

Other obstacles that hamper effective risk communication are the scientific terms and expressions used to discuss and describe risks. The public's vs. the experts' understanding and interpretation of risk information often differ significantly. For example, the phrase "one in a million" can be interpreted in two ways: the expert states the statistic to address society's risk as a whole. However, the individual, interested in his own chance of developing cancer, applies the probability to himself and attempts to determine his chance of being that "one in a million." In addition,

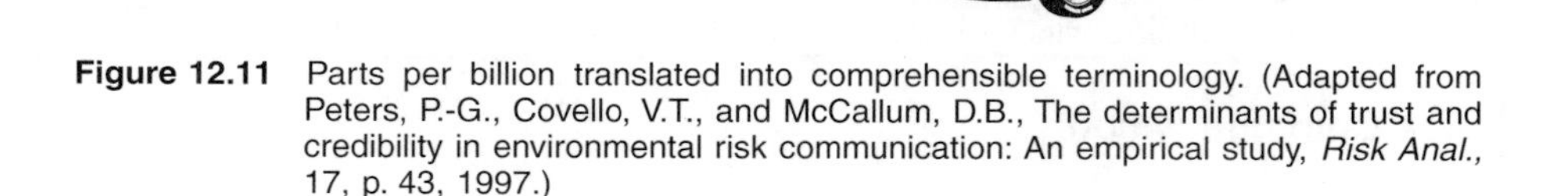

TIME: One second in 6.4 months.
LENGTH: 1.56 inches in 25,000 miles (the circumference of the earth at the equator).
AREA: One sq. yard in 324 square miles.
VOLUME: Approximately a half drop of water in a 17,000 gallon pool.
WEiGHT: A particle of dust (<0.01 grams) on a 2,000 pound automobile).

Figure 12.11 Parts per billion translated into comprehensible terminology. (Adapted from Peters, P.-G., Covello, V.T., and McCallum, D.B., The determinants of trust and credibility in environmental risk communication: An empirical study, *Risk Anal.,* 17, p. 43, 1997.)

scientific language with which the expert is comfortable may be foreign and meaningless to a lay person. A term such as "parts per billion" is difficult to interpret and may render any risk information useless to someone who cannot relate to the expression. Translating terms such as "parts per billion" into comprehensible and applicable measurements, such as one second in 6.4 months, or 1.56 inches in 25,000 miles, or one half drop of liquid in 17,000 gallons, can facilitate comprehension of scientific language and increase risk communication (Figure 12.11). Risk communication involves building bridges among opposing groups, so that all parties can understand the relevant concepts and issues, although they may not agree on them. Risk communication is an essential aspect to managing environmental risks today.

REFERENCES

1. Simons, M., Big sludge spill poisons land in southern Spain, *New York Times,* May 2, 1998, p. A3.
2. Jardine, C.G. and Hrudey, S.E., Mixed messages in risk communication, *Risk Anal.,* Vol. 17, 1977, chap. 1.
3. Moeller, D.W., *Environmental Health,* revised ed., Harvard University Press, Cambridge, MA, 1997, chap. 16.
4. Covello, V.T., and Merkhofer, M.W. *Risk Assessment Methods,* Plenum Press, New York, 1993, chap. 1.
5. Derby, S.L. and Keeney, R-L., Understanding "How safe is safe enough?," *Risk Anal.,* 1, 217, 1981.
6. Beck, BX, Rudel, R., and Calabrese, E., *The Use of Toxicology in the Regulatory Process, Principles and Methods of Toxicology,* Hayes, A.W., Raven Press, Sacramento, CA, 1994, chap. 1.
7. Kradjian, M.D, Robert M., The Milk Letter: A Message to Parents, Aug. 1998, http://www.afpafitness.com/milkdoc.htm.
8. U.S. Department of Health and Human Services, Task Force on Health Assessment, *Determining Risks to Health,* Auburn House Publishing, Co., Dover, MA, 1986, p. 3–4.

9. Halbeib, W.T., *Emergency Planning and Community Right-to-Know Act, Environmental Law Handbook,* 14th ed., Sullivan, TER, Government Institutes, Inc., Rockville, MD, 1997, pp. 481.
10. S(breve), Radim, Impact of Air Pollution on Reproductive Health, Environmental Health Practices, Nov. 1999, pp. 107–111, http://ehpnet1.niehs.nih.gov/docs/1999/107-11/editorial.html.
11. Malachowski, M.J., *Health Effects of Toxic Substances,* Government Institutes, Inc., Rockville, MD, 1995, chap. 4.
12. Beck, B.R., Rudel, R., and Calabrese, E., *The Use of Toxicology in the Regulatory Process, Principles and Methods of Toxicology,* Hayes, A.W., Raven Press, Sacramento, CA, 1994, p. 31.
13. Cohrssen, J.J. and Covello, V.T., *Risk Analysis: A Guide to Principles and Methods for Analyzing Health and Environmental Risks,* The Council on Environmental Quality, 1989, chap. 3.
14. Kerlin, Katherine, Soy Baby Blues; Can We trust Alternatives to Milk-Based Formulas, *E/Entertainment Magazine,* Nov./Dec. 1999, http://www.emagazine.com/november-december_1999/1199gl_health.html.
15. Schettler, T., Solomon, G., Burns, P., and Valenti, M., *Generations at Risk — How Environmental Toxins May Affect Reproductive Health in Massachusetts,* MASSPIRG, Martin VanderLoop Associates, 1996, chap. 3.
16. *The Merriam Webster Dictionary,* Merriam-Webster, Inc., Springfield, MA, 1995, p. 183.
17. Hennekens, C.H. and Buring, J.E., *Epidemiology in Medicine,* Little Brown and Company, Boston, 1987, chap. 1.
18. Rosenstreich, D.L., Eggleston, P., Kattan, M., Baker, D., Slavin, R.G., Gergen, P., Mitchell, H., Mortimer, K.M., Lynn, H., Ownby, D., and Malveaux, F., The role of cockroach allergy and exposure to cockroach allergen in casuing morbidity among inner-city children with asthma, *New Engl. J. Med.,* 1997; 336:1356–63.
19. Pappert, Ann, *TamoxiSpin* 39–42. Oct./Nov. 1999.
20. Hennekens, C.H. and Buring, J.E., *Epidemiology in Medicine,* Little Brown and Company, Boston, 1987, chap. 2.
21. Stecidow, S., New food quality act has pesticide makers doing human testing, *Wall Street Journal,* Sep. 28, 1998, p. Al.
22. U.S. Environmental Protection Agency, *Risk Assessment,* Government Printing Office, 1992, p. 3.
23. Kearney, W., O'Neill, M., Publication Announcement: New Waste Incinerators Safer, but Some Emissions and Health Concerns Need Further Study, National Academy of Sciences, Oct. 13, 1999, http:/www4.nationalacademics.org/…370D385256809-0070428E?OpenDocument.
24. Stern, P.C. and Fineberg, H.V., *Understanding Risk,* National Academy Press, Washington D.C., 1996, p. 50.
25. Sharpe, R., s20 Painkillers: How a drug approved by the FDA turned into a lethal failure, *Wall Street Journal,* Sept. 30, 1998, p. Al.
26. Cothern, C.R., *Handbook for Environmental Risk Decision Making: Values, Perceptions and Ethics,* Lewis Publishers, Boca Raton, 1996, p. 86.
27. Somers, E., Perspectives on risk management, *Risk Anal.,* 15, 677, 1995.
28. Viscusi, W.K., *Rational Risk Policy,* Clarendon Press, Oxford, 1998, p. 98.
29. Morgan, M.T., et al., *Environmental Health,* revised ed., Morton Publishing Co., Englewood, CO, 1997, p. 279.

30. Hallenbeck, W.H., *Quantitative Risk Assessment for Environmental and Occupational Health,* 2nd ed., Lewis Publishers, Boca Raton, 1993, p. 2.
31. Gross, L., Current Risks: Experts finally link electromagnetic fields and cancer, *Sierra,* May/Jun., 30, 1999.
32. New global health report warns about health risks of environmental degradation, *J. Environ. Health,* Sept. 1998, 61, 31–33, http://web7.infotrac.gale-group…959208w3/ 62!xrn_69_0_A21125877.
33. Peters, P.-G., Covello, V.T., and McCallum, D.B., The determinants of trust and credibility in environmental risk communication: an empirical study, *Risk Anal.,* 17, p. 43, 1997.
34. DiNardi, S.P., *Calculation Methods for Industrial Hygiene,* Van Nostrand Reinhold, New York, 1995, p. 76.

CHAPTER 13

Environmental Laws and Compliance

INTRODUCTION

Environmental laws are a maze of confusing and overlapping regulations that now cover several thousand pages of federal documents. It is not within the purpose or capability of this brief chapter to clarify or detail all of the environmental regulations, but to outline some of the major environmental legal initiatives and compliance issues. In pursuing this objective, the following areas will be covered:

OBJECTIVES FOR THIS CHAPTER

A student reading this chapter will be able to:

1. **Discuss how a law is made and describe the system of environmental laws**
2. **List and describe the major components of the major federal environmental laws including RCRA, CERCLA, EPCRA, SARA Title IIII, Pollution Prevention Act, CAA, CWA, SDWA, stormwater regulations, pesticide regulations, and underground storage regulations**
3. **Describe and discuss the major components of environmental compliance**

- Environmental laws — some fundamentals
- Resource Conservation and Recovery Act (RCRA)
- Comprehensive Environmental Responsibility, Compensation and Liability Act (CERCLA)
- Emergency Planning and Community Right-to-Know Act (SARA Title III), transportation of hazardous materials and wastes
- Pollution Prevention and Improved Waste Management Programs, regulation of underground storage tanks
- Pesticide regulation
- Air quality control
- Water Quality Control and Stormwater Management Safe Drinking Water Acts

ENVIRONMENTAL LAWS — SOME FUNDAMENTALS

The Making of a Law

The average time for an environmental law to be passed at the federal level may take several years. Every federal law starts as a bill and passes through a complex system of checks and balances in Congress, originally created by the framers of the constitution as a way of dispersing power and preventing tyrants from abusing that power. The bill is first introduced into the House. If it is passed, it goes to the Senate. These bills are then forwarded by the leadership to a standing committee (subcommittee) for review and support. Nearly 90% of bills never make it through this process. Recommended bills are brought forward through hearings for comments and opinions. The committee then meets to mark up (discuss and amend) the bill and to vote on it. If the subcommittee and the parent committee in the Senate vote favorably for the bill, it is sent (reported) to the full chamber for debate and a vote. Before a bill is sent to the House, it goes first to the Rules committee, where a time limit is set for debate; and it also indicates whether floor amendments are allowed. This is not so in the Senate, where riders are often attached to popular bills that may have no relationship to the original bill but often are passed along with it, forcing the president to accept it or veto the entire bill.[1]

If both the Senate and the House pass a bill with some differences (as is usually the case), then they go to a conference committee to resolve these differences. Should these differences be resolved, and the same version approved by both House and Senate, the bill moves forward to the President, who may sign it into law or veto it. Although Congress may override a veto by a two thirds majority vote in each chamber, it is very difficult to do so.

Environmental Laws Are Part of a System

Environmental law is more than a collection of unrelated regulations; it is a system of statutes, regulations, executive orders, factual conclusions, case-specific interpretations, and accepted environmental principles that have evolved into "Environmental Law."[2] Understanding the unifying principles of this law system is both a challenge and the key to understanding its complexity. Environmental law is a system that uses all of the laws in our legal system to minimize, prevent, punish, or remedy the consequences of actions that damage or threaten the environment and public health and safety. Environmental law encompasses all the environmental protections that originate from the U.S. Constitution and state constitutions; federal, state, and local statutes; regulations published by federal, state and local agencies; presidential executive orders; court decisions interpreting these laws and regulations; and the common law.[2]

Virtually all of the environmental laws require some form of compliance that falls within one or more of the following areas:[2] notification requirements; discharge or waste controls; process controls and pollution prevention; product controls; regulation of activities; safe transportation requirements; response and remediation laws; and compensation requirements.

Notification requirements are meant to deal with any intended or accidental releases of pollutants or hazardous wastes that need to be reported immediately to local and state and/or federal authorities. Failure to do so can significantly impact the environment, endanger public health, and cause significant liability to the persons associated with the release. Discharge or waste controls are regulations designed to prevent or minimize the discharge, release, or disposal of wastes and pollutants into the environment. *Process controls and pollution prevention* are regulations that promote minimizing waste generation, reducing the quantities produced and released, and preventing the release or discharge of pollutants into the environment. *Product controls* can reduce the generation and disposal of solid waste by altering the design or packaging of a product, along with the process by which it may be produced. Some regulations promote the use of less hazardous materials and reduced packaging to achieve these goals. The *regulation of activities* that threaten natural habitat, fragile ecosystems, and vital resources may be necessary to provide protection for resources, endangered species, or habitat. Some of those activities include mining, lumber harvesting, oil refining, and construction. *Safe transportation requirements* recognize that accidents and spills along our highways occur at an enormous rate, and laws have been established to reduce the risks associated with the transport of hazardous materials. These laws provide for special packaging, labeling, placarding, operator training, and methods of response. *Response and remediation laws* have been promulgated that regulate the clean-up of pollutants and hazardous wastes released to the environment. Laws such as CERCLA have specific provisions for the payment of clean-up costs. These same laws provide for *compensation requirements* requiring responsible parties to pay for clean-up costs for damages done to the health or environment of the private sector. Compensation for damages done to public assets may be recovered by representatives of the "public interest."[1,2]

The system of environmental law includes federal statutes, executive orders, state laws, tax laws, local and municipal laws, business and regulatory law, environmental law and judicial decisions, common law, and torts.

Federal statutes include regulations — such as the Clean Air Act or Clean Water Act — that establish federal and state regulatory programs allowing states to enact and enforce laws that meet federal minimum standards and achieve the regulatory objectives established by Congress. Most states have taken over regulatory programs as allowed under federal regulations and are often more stringent than the federal counterpart. These states are subject to federal intervention only when they fail to enforce the regulations. *Executive orders* are issued by the president and require federal facilities to provide leadership in protecting the environment and to comply with federal statutes and policies regarding the protection of the environment. They were first used in 1970 and include more than 18 executive orders, including such areas as federal compliance with right-to-know laws, pollution-prevention requirements, federal actions that address environmental justice in minority populations, and low-income populations.[3]

Tax laws are being used to create incentives for environmentally safe products and activities and disincentives against environmentally detrimental products and activities. Among the approaches that have been adopted or seriously considered are recycling tax credits, taxes on use of virgin materials, taxes on hazardous waste generation, and excise taxes on various products. Many state attorney generals use their business regulatory authorities to control environmental claims made for products. Full disclosure of environmental liabilities in statements and reports has been required by the Securities and Exchange Commission for some time. Local and municipal laws include zoning and noise control ordinances, nuisance laws, air emission requirements, landfill restrictions or closures, local emergency planning, and product recycling incentives. These are not trivial ordinances and regulations and must be incorporated into any compliance program. Such laws have been effectively used to control the operation and location of facilities.[2] Court interpretations of these laws and regulations are dynamic and changing; therefore, judicial decisions have important impacts on the manner in which environmental laws are applied. The environmental system of laws also incorporates *common law*, which is a body of rules and principles that pertain to the government and the security of persons and property. These basic rules were originally developed in England and were then brought to the American colonies. After the American Revolution, these basic rules were formally adopted and enforced by the states.[2] Civil suits, in which the plaintiff seeks to remedy a violation of his rights, are common law actions. Under common law, *tort* is a private wrong or wrongful act for which the injured party can bring forth a civil action. There is a general legal duty and responsibility to avoid causing harm to others by acts of omission or commission. When a person's rights have been violated by acts of carelessness, the injured party may seek compensation or restitution by means of a lawsuit. The 1990s have been called the era of "toxic torts" due to the filing of tens of thousands of lawsuits involving asbestos and toxic chemical litigation cases. The three most common types of torts encountered in environmental law include:

1. *Nuisance.* Generally, a person may use his land or personal property as he sees fit with the limitation that the owner use his property in a reasonable manner. Whenever a person uses his property to cause material injury or annoyance to his neighbor, his actions constitute a nuisance.
2. *Trespass.* An invasion of another's rights is the general definition of trespass. A more limiting definition of trespass is an injury to the person, property, or rights of another immediately resulting from an unlawful act. Unlawful intent is not necessary to constitute trespass.
3. *Negligence.* This is the portion of the tort law that deals with acts not intended to inflict injury. A case becomes one of criminal law if an intent to inflict injury exists. The degree of care and caution that would be taken or used by an ordinary person in similar circumstances is defined as the standard of care required by law in negligence cases. A defendant is liable if his actions cause injury, and without those actions the end result would not have happened. This involves a natural and continuous sequence of actions, unbroken by intervention. Persons harmed by the careless and improper handling or disposal of hazardous wastes can file suit to recover damages for their losses even if there has been full compliance with all government regulations and permit conditions.[2]

FEDERAL ENVIRONMENTAL LAWS

Managing Hazardous Waste

Hazardous waste management has become a crucial issue. Human health has been threatened and water supplies have been polluted by the improper handling, release, and disposal of hazardous waste. Most hazardous waste is banned from landfilling. Just a few commercial hazardous waste disposal facilities exist, and their total capacity is limited. Liability and costs of waste disposal are increasing. Federal regulations require a cradle-to-grave approach for managing hazardous waste. These regulations establish a system to determine which wastes are hazardous, track wastes from generation to disposal, operate disposal and temporary storage facilities, and set up state hazardous waste programs.[4] An environmentally sound and economic

WEB PICK **Go to http://www.unix.oit.umass.edu/~envhl565/index.html** **Click on "Chapter Web Links" to find Environmental Law Net on the World Wide Web under Chapter 12**	
Chapter 9	Environmental Regulation
Whose Site?	1997–2001 David S. Blackmar and © 2001 Greenberg Traurig, LLP. All Rights Reserved Worldwide. The content and arrangement of hyperlinks contained herein are the intellectual property of David S. Blackmar and Greenberg Traurig, LLP
URL	http://lawvianet.com/
What's Here?	Environmental Law Net is the creation of David S. Blackmar, who is an environmental lawyer with the international law firm of Greenberg Traurig, LLP. This site offers reliable primary legal resources that are indispensable to environmental lawyers and compliance managers. This site delivers those resources through a uniquely functional and efficient format that allows users to make informed decisions while saving time and money. The features of this site exist within two primary domains: (1) legal information libraries and (2) community resources. Some of the most useful resources provided on this site are typically not available in law firm libraries or through expensive proprietary networks. This site also delivers valuable original content and special features designed to make the Internet more informative, easy, and enjoyable to use. This site has sections on laws and regulations, court and agency decisions, agency documents and data bases, compliance, enforcement and litigation, and transactions. There are also useful sections of daily news articles relevant to the environmental field, an online forum, and a section on conferences and events.

approach to meeting legal requirements for hazardous waste management is reducing the amount of hazardous waste generated by the facility. Hazardous waste minimization is now the law. An integrated hazardous waste management program must be implemented with initiatives for reducing the use of hazardous waste in order to protect the facility and its employees.[5]

Resource Conservation and Recovery Act (RCRA)

The federal hazardous waste laws evolved after a series of amendments to the Solid Waste Disposal Act (SWDA) of 1965. The Resource Conservation and Recovery Act (RCRA) was passed as an amendment in 1976 to the SWDA. General guidance in meeting hazardous waste requirements was provided to the USEPA by RCRA. However, the USEPA did not issue key hazardous waste regulations until May 1980. The discovery of hazardous waste problems at Love Canal, New York, in 1979 to 1980, and at many other locations thereafter, increased pressures for faster regulatory action.[6] The foundation for RCRA waste management regulations can be divided into three distinct programs or subchapters.

1. Subchapter C establishes a system for controlling hazardous waste from the point of generation to the ultimate disposal.
2. Subchapter D establishes systems for controlling solid or nonhazardous waste, such as household waste.
3. Subchapter I regulates underground tank storage of toxic substances and petroleum products.

Subchapter C of the RCRA is concerned with the everyday environmental management of hazardous wastes and involves a "cradle-to-grave" system that tracks hazardous waste from its origin through its ultimate disposal (Figure 13.1). Facilities must comply with regulations if they generate, transport, store, treat, and dispose of hazardous waste in amounts specified under RCRA. Hazardous waste generators can be separated into three principal categories. State regulations may provide somewhat different designations and responsibilities to three groups:

1. Large-quantity generators (LQG) produce 1000 kilograms or more of hazardous waste (2200 pounds or more) or more than 1 kilogram (2.2 pounds) of acutely hazardous waste in a calendar month.
2. Small-quantity generators (SQG) produce 100 to 1000 kilograms of hazardous waste and less than 1 kilogram of acutely hazardous waste in a calendar month.
3. SQGs that are conditionally exempt include those that generate no more than 100 kilograms per calendar month of hazardous waste, or no more than 1 kilogram per month of acutely hazardous waste (AHW). These generators may be exempted from RCRA requirements providing their wastes are properly managed (e.g., sent to a solid waste disposal facility).

The administration of RCRA is done by the USEPA, and state programs must be consistent with federal RCRA regulations. Comprehensive cradle-to-grave regulation of hazardous wastes is provided for in most state laws. The lead agency in implementing RCRA is usually the state environmental regulatory agency. In general, most state hazardous waste regulations closely parallel federal provisions of RCRA.[5] Within some states, special problems are addressed by the addition of more stringent requirements. Hazardous waste management has been changed in many ways. There is a ban on land disposal of certain wastes, small-quantity generators

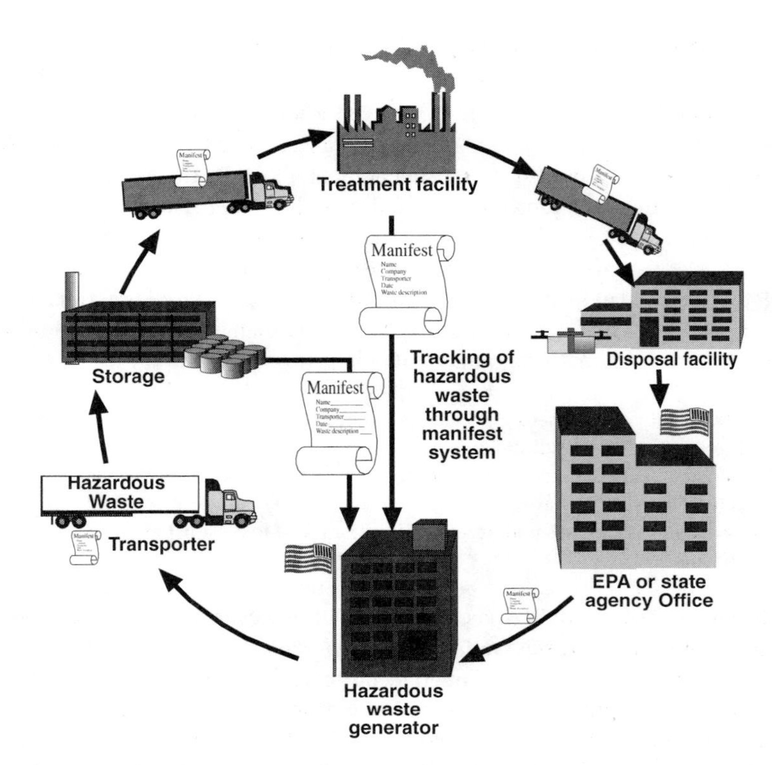

Figure 13.1 The RCRA "cradle-to-grave" system tracks hazardous waste from its origin through its ultimate disposal by a manifest system.

are under tighter controls, and regulation of used oil as a hazardous waste has gone into effect in some states.

Identifying a Hazardous Waste

There are two types of hazardous wastes: listed and characteristic. Listed wastes include hundreds of hazardous wastes including wastes from nonspecific sources (F Wastes), manufacturing processes (K Wastes), and discarded chemical products (P and U Wastes), which are listed in the RCRA regulations. The Code of Federal Regulations, Title 40, Part 261, Subpart D, contains these lists of hazardous wastes. Products not listed may be characteristically hazardous. Under the direction of RCRA, the USEPA developed generalized criteria for evaluating the potential hazard produced by a waste. The four criteria include:

1. *Ignitability* — easily catches fire, with a flash point of less than 140°F
2. *Corrosivity* — easily corrodes materials or human tissue, very acidic or alkaline (pH less than 2 or greater than 12.5)

3. *Reactivity* — explosive, reacts with water or acid, unstable
4. *Toxicity* — causes local or systemic damages that result in adverse health effects in an organism — that is, a poison (such as oil), an asphyxiant (such as carbon monoxide), a mutagen (a chemical that alters DNA), teratogen (one that causes birth defects), or a carcinogen (one that causes cancer); toxicity is determined by a test known as the Toxicity Characteristics Leachate Procedure (TCLP)

The USEPA developed lists of hundreds of hazardous wastes using these criteria. Generators of nonlisted waste must analyze the waste to determine if it exceeds the USEPA standards. If the waste exceeds any of the USEPA's four criteria, it is considered a "characteristic hazardous waste."

Tracking Hazardous Waste

All small- and large-quantity generators, transporters, and TSD (treatment, storage and disposal) facilities must obtain a USEPA identification number as required by most state environmental agencies and RCRA. These ID numbers are used by state agencies and the USEPA to track the flow or whereabouts of hazardous waste. This tracking is accomplished by using a paper trail created by shipping manifests. RCRA assures uniform record keeping for all shipments of hazardous wastes. A "Uniform Hazardous Waste Manifest" must be used for all shipments of RCRA-regulated hazardous wastes. State manifest forms, adapted from the USEPA form, supplement this documentation in a number of states. Each shipment of hazardous waste requires a USEPA manifest form.[6]

Other Requirements Under RCRA

Accurate record keeping is essential to a successful hazardous waste management program. Complete and up-to-date records are necessary to ensure compliance with environmental regulations. Shipments of hazardous waste need to be properly labeled, marked, and placarded according to the U.S. Department of Transportation (DOT) requirements. The facility managers remain responsible for the proper management of waste after it leaves the property, and thus great care must be taken in selecting a waste hauler and a treatment, storage, or disposal facility.[7]

Comprehensive Environmental Responsibility, Compensation and Liability Act (CERCLA)

The clean-up of inactive and abandoned hazardous waste sites is provided by the Comprehensive Environmental Response, Compensation, and Liability Act of 1980 (CERCLA or Superfund). CERCLA and RCRA have the same basic objectives. However, the purpose of RCRA is to prevent hazardous waste problems at active sites, whereas CERCLA is designed to provide clean-up of existing problems at inactive sites. CERCLA also provides a mechanism of establishing financial liability to any or all parties responsible for the release of hazardous substances. These potentially responsible parties (PRPs) are liable for the costs of cleaning up those substances. These strict joint and liability provisions of CERCLA can result in costly investigative and clean-up activities at waste disposal sites. According to the USEPA, the average cost per Superfund site is $35 million, and it is therefore important to manage wastes in a way that reduces the potential for exposure to these costs.[6]

The Superfund Amendments and Reauthorization Act (SARA) of 1986 amended and strengthened CERCLA, the basic Superfund law. The Regional USEPA is the state's lead agency for implementation of the Superfund. Although crude oil, petroleum products, and natural gas products are excluded from this list, oil spills are addressed by the National Oil and Hazardous Substances Pollution Contingency Plan.[6]

Steps in Superfund: Find, Prioritize, and Clean

The first step in Superfund was to find the hazardous waste sites. The identification of hazardous waste sites was and is accomplished in three major ways. First, the USEPA required all owners and operators (including former owners and operators) of hazardous substance TSD facilities to report the existence of these facilities by June 11, 1981. Second, the USEPA already regulated TSD facilities under RCRA, and these reported facilities were added to the list. Third, state and local governments, as well as the public, have identified many additional hazardous waste dump sites and facilities. The CERCLA Information System (CERCLIS) data base incorporates information for approximately 30,000 sites. Additionally, the USEPA established specific reportable quantities (RQs) for the release of hazardous substances. Under the provisions of the Superfund, any releases of hazardous substances larger than these RQs must be reported to the National Response Center by the owners and operators of the facilities or vessels that release hazardous substances (except under federal permit). The Center then notifies other appropriate agencies and begins any necessary emergency response or cleanup actions. Once sites are identified, the USEPA is required to collect data and make a preliminary assessment report. The sites are then ranked using a risk assessment procedure for their likelihood of causing human health impacts or environmental damage. The sites with the highest ranking are placed on the national priority list (NPL) and are targeted for clean-up with monies coming from potentially responsible parties (PRPs) or from Superfund monies derived from taxes on chemical feedstocks and crude oil supplies. Therefore, a next step in the Superfund process is the identification of those parties responsible for site contamination. These "potentially responsible parties" and "responsible parties" identified by the USEPA and state agencies can be required to finance clean-up activities, either directly or through reimbursement of federal Superfund expenditures. The final steps involve the clean-up of contaminated sites. The program includes extensive provisions for site investigations, selection of clean-up sites, and establishing the levels of clean-up to be attained.[6]

Emergency Planning and Community Right-to-Know (SARA Title III)

The possibility that mishaps at facilities handling hazardous chemicals could harm neighboring communities was brought to worldwide attention with the December

1984 toxic gas disaster in Bhopal, India. In response to this disaster, U.S. federal, state, and local governments created a variety of laws to improve accident prevention and emergency response planning activities by chemical-handling facilities and local governments. "Right-to-Know" laws increased public access to information about the storage and use of hazardous chemicals.

Title III of SARA was the first congressional response to the Bhopal accident. A federal program subtitled the Emergency Planning and Community Right-to-Know Act of 1986 (EPCRA) was mandated by SARA Title III. Title III was one of those laws that was appended to the SARA legislation as way of getting it passed, and it really has very little to do with the Superfund program. SARA Title III is focused on worker safety regulations. The Occupational Safety and Health Administration (OSHA) took the lead in developing the regulations because these new requirements were extensions of OSHA's traditional work safety rules. Identical requirements have been adopted by the USEPA and are applied to state and local government employees (over whom OSHA has no jurisdiction). The purpose of Title III was to assure the public and emergency response agencies that information regarding hazardous chemicals would be made available to them, thereby lessening the likelihood of Bhopal-like disasters in the U.S. The USEPA and the individual states are responsible for these efforts. The USEPA enforces SARA's requirements for compliance. Any facility that stores or handles extremely hazardous substances (EHS) at, or in excess of, USEPA-prescribed threshold planning quantities (TPQs) must comply with SARA's release, planning, and release reporting requirements. This also covers unauthorized discharges (leaks or spills) of any EHS, or of any substance already subject to emergency notification requirements under section 103(a) of CERCLA. Any discharge of reportable quantity must be immediately reported to appropriate local emergency response agencies (police, fire department), to national and state emergency response agencies, and to state and district Title III agencies.

Title III reporting requirements (SARA sections 311-312) and the federal Occupational Safety and Health Act of 1970 requires all facility owners or operators to prepare or have available material safety data sheets (MSDSs) (Figure 13.2). Businesses are required to provide either an MSDS for each on-site chemical in excess of its threshold quantity, or a list of all such chemicals to state and district Title III agencies and to the local fire department.

The reporting requirements in Title III (Section 313) require selected businesses to file annual reports with the USEPA and the state on the USEPA "Form R." These reports must include the following information for the current and previous calendar years:

- Estimated quantities of both routine and accidental releases of listed "toxic chemicals"
- The maximum amount of listed on-site chemicals during the calendar year
- The amount of listed chemicals contained in wastes transferred offsite

These reports estimating releases during the previous calendar year are due each July 1.

Figure 13.2 Material safety data sheets (MSDSs) on hazardous chemicals in the workplace must be made available to employees in a convenient place.

In October of 1999 President Clinton declared that beginning January 1, 2000, the EPA would require industries to release fuller reports on 27 additional chemicals, including the endocrine disruptor, dioxin, for the first time. These new regulations require industries to report if they use 100 pounds per year. In the case of dioxin, the industry must report if one tenth of a gram is released.

Transportation of Hazardous Materials

The Department of Transportation (DOT) regulates the transportation of hazardous wastes and materials, and it is within this department that the Hazardous Materials Transportation Act (HMTA) is administered. States are allowed to adopt standards that are more stringent than federal regulations. State provisions are often similar to the federal standards. The regulations of the transportation of hazardous wastes can differ from state to state.[6]

What are hazardous materials under HMTA? As defined by HMTA, hazardous materials are those which are capable of creating an "unreasonable risk" to health, safety, or property while being transported. Any hazardous substance listed in DOT regulations (49 Code of Federal Regulations sections 172.101, 172.102) that is transported in a package containing a quantity that equals or exceeds the listed reportable quantities (RQs) is considered a regulated material. Each hazardous material is designated by DOT and, depending on its risk to health, safety or property, is assigned to classes and divisions. Each class has specific packaging, handling, labeling, and registration requirements.

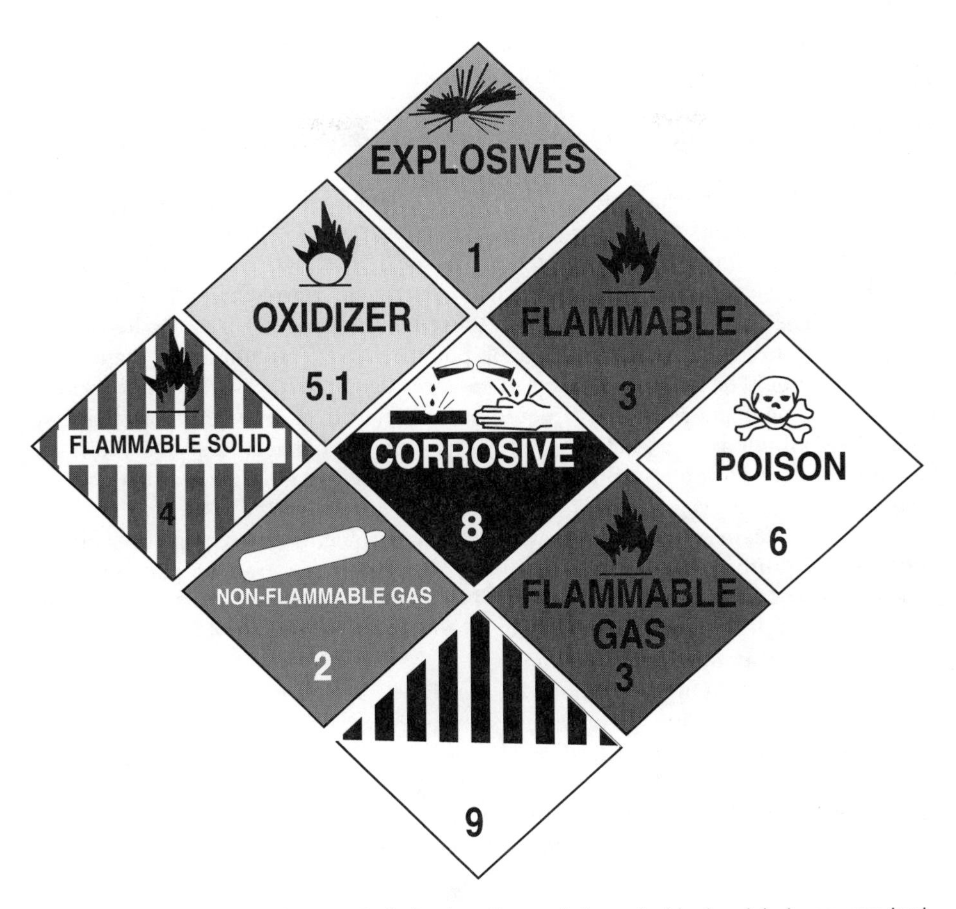

Figure 13.3 Examples of several distinctive diamond-shaped shipping labels as required under DOT.

Packaging, labeling, and construction requirements must now meet United Nations (UN) recommendations. There are three packaging groups under the UN-based regulations: Group I for the most hazardous to Group III for the least hazardous. Shipping papers must accompany every shipment. Basic information must be provided within the papers including name of shipper, shipping names of materials, hazard class, standard materials ID numbers under either the UN or NA (North America) system, packaging group (if there is one), quantity of each material being shipped, and the emergency response telephone number.

All containers being shipped must be appropriately marked and labeled. The labels must include the proper shipping name, ID number, and the consignee's name and address. The labels are distinctively diamond-shaped that must clearly identify the hazard class. Even the size, color, and text of each label is specifically laid out by DOT (Figure 13.3). Transport vehicles must also display placards in a specified pattern. Emergency response information must also accompany each hazardous material being transported. The transporting company must also provide a monitored 24-hour emergency response telephone number.[7]

Pollution Prevention and Improved Waste Management Programs

The Pollution Prevention Act of 1990 has become integrated into today's environmental regulations at both the state and federal levels. Pollution prevention, also referred to as "source reduction," is defined as the deliberate decrease in the amounts of hazardous substances which enter the environment via recycling, treatment, or disposal.[7] The emphasis of the Pollution Prevention Act focuses on remediation and treatment rather than the prevention of pollution. Any pollution that cannot be prevented should be recycled, or treated and disposed of in a way that is not harmful to the environment. Mandatory requirements as well as voluntary programs to minimize pollution were established. The Act provides financial assistance to states for their role in prevention. Businesses are encouraged to share and transfer source reduction technology.[8]

For example, in response to the environmental problems posed by methyl tertiary butyl ether, California has a new "biorefinery" to convert rice straw, orchard prunings, and other agricultural waste into ethanol. This will be used as an additive in gasoline in the place of methyl tertiary butyl ether. The plan is to phase out MTBE and replace it with the cleaner alternative, ethanol.

Source reduction must be reported by facilities generating over a certain amount of toxic emissions during the previous calendar year. The form, referred to as Form R, must be filled out for each toxic chemical to include information such as the amount of toxic chemical recycled and treated. Also included is the amount of toxic chemical that entered a waste stream, percent change from the previous year, and the amount anticipated for the following 2 years. Recycling efforts as well as source reductions must be reported for the previous calendar year.[4]

Toxic Substances Control Act (TSCA)[6]

There are over 64,000 chemicals listed in the TSCA inventory. TSCA mandates that chemical manufacturers develop safety and health data on chemicals and mixtures and requires the USEPA to regulate chemical substances and mixtures that present an "unreasonable risk of injury to health and the environment." The USEPA is also directed to act when chemical substances and mixtures pose "imminent hazards" and is authorized to restrict the production and use of such hazardous materials. The USEPA has done this by banning the use of polychlorinated biphenyls (PCBs) and asbestos-containing products. Congress intended that the USEPA not abuse this authority and should not unnecessarily impede or create economic barriers to technological innovation.

There are a number of reporting and record keeping activities and data collection procedures under TSCA. Additionally, there are mechanisms for public information and "citizen suits" to compel the USEPA to action when TSCA is violated. The basis of the reporting system is the TSCA inventory, originally compiled during

1977. This consists of two lists: a main inventory and a confidential inventory for all regulated substances that are protected as trade secrets.

Since 1977, each new toxic substance falling under TSCA regulations needs to be added to the list. Before new chemical substances may be introduced into commerce, TSCA requires premanufacture notification as well as notification of significant new uses of existing chemicals. TSCA regulations stipulate guidelines and standards for chemical testing and testing laboratories, rules for the reporting of chemical information, and the use of health and safety data.

TSCA gives the USEPA the power to regulate, restrict, and even prohibit the manufacture, importation, and use of certain toxic chemicals. Under this mandate, the USEPA has enacted strict regulations for:

- The phase-out, removal and disposal of PCBs
- Asbestos and asbestos-containing materials
- Fully halogenated chlorofluorocarbons (CFCs — once used as aerosol propellants, responsible for ozone depletion in the upper atmosphere)
- Dibenzodioxins and dibenzofurans (contaminants present at Times Beach, Missouri[2])

Regulation of Underground Storage Tanks

In the late 1970s and early 1980s, lawmakers became aware that thousands of tanks located underground and containing petroleum products such as gasoline were beginning to leak, or in danger of leaking, and contaminating groundwater. The lawmakers developed standards for storage tank construction, installation, and removal and defined a UST as any tank or underground piping connected to the tank that has a minimum of 10% of its volume located underground. The federal UST law is a component of the Hazardous and Solid Waste Amendments of 1984 (HSWA) under the Resource Conservation and Recovery Act (RCRA). This is known as Subtitle I, and this subtitle deals with USTs that store petroleum and other hazardous substances, but not hazardous wastes that are regulated by other sections of RCRA. The USEPA issued its final technical requirements for USTs on September 23, 1988. These regulations include detailed technical standards for new USTs installed after December 22, 1988, and retrofit requirements for tanks already in service on that date. These particular regulations only applied to the storage of petroleum or certain hazardous chemicals. The hazardous chemicals are the same as those listed under CERCLA, except for those chemicals listed as hazardous wastes. All of the UST regulations also applied fully to the underground piping associated with the tank(s) in question. These standards also included tank registration and monitoring. Regulations were initially enforced to tanks that were leaking, thereby causing contamination of groundwater. Aging storage tanks and steel associated with gas piping and oil become a hazard when these pipelines begin to leak and contaminate soil and water supplies.

There are several types of tanks and piping systems that are not included by UST regulations — septic tanks, tanks for heating oil for on-site use, and farm and residential fuel tanks under 1100 gallons.

Leak detection devices had to be installed in all USTs per federal mandates by December 1993. At the same time, federal and state regulations enforced the use of

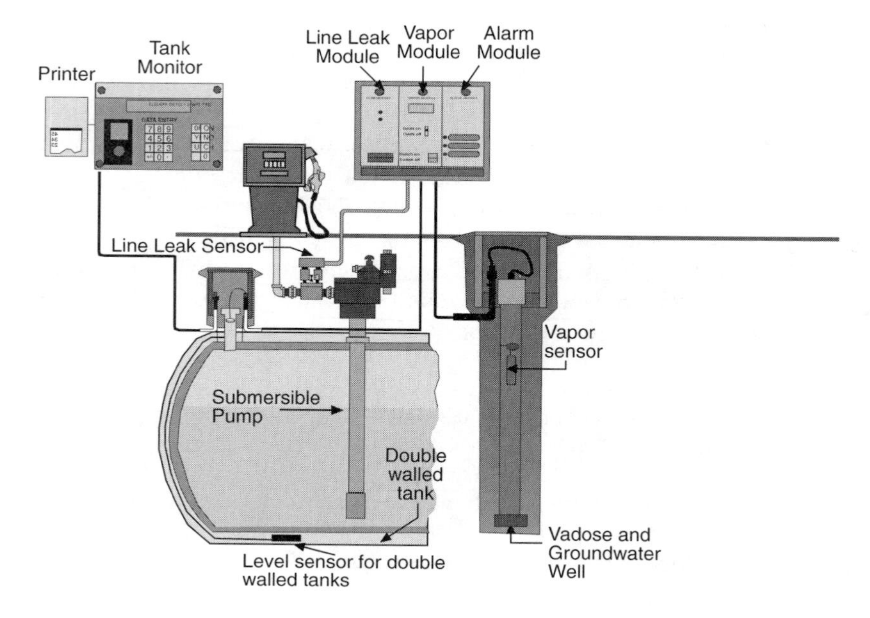

Figure 13.4 A typical underground storage tank installation meeting current regulatory requirements.

overfill protection or spill protection. Two categories of record keeping must be maintained including records generated at the point of old tank removal and new tank installation and records of proof concerning monitoring device maintenance, inventory reconciliation, registration, and permit renewal.

All tanks installed after December 22, 1988, are required to have the following construction standards: new underground piping must be corrosion resistant; and new USTs must have spill and overfill protection and leak detection provisions. Also included in the standards are regulations for new petroleum USTs. These USTs may continue to be single-lined, whereas all other USTs must now be double-lined for both primary and secondary containment (Figure 13.4).[4]

Pesticide Regulation

The basic national framework for pesticide control is provided by the Federal Insecticide, Fungicide, and Rodenticide Act (FIFRA), which is administered by the U.S. Environmental Protection Agency in cooperation with state and local agencies. Initially enacted in 1947, FIFRA has been amended several times. Pesticides are chemicals specifically formulated to be toxic to living things and are therefore subject to regulation. Pesticide regulations are intended to protect public health, livelihoods, and the environment by ensuring that workers are trained in proper application techniques; pesticides are properly handled and stored; the location or content of chemicals is made known; pesticides are known to emergency response units, medical personnel, and workers who may be exposed to the chemicals.[6]

Applicators of restricted-use pesticides must be certified based on demonstrated competency in the use of pesticides. Regulations also establish worker safety precautions, including requirements to warn field workers before pesticide applications and to prevent exposure to workers not actually involved in pest control activities. The USEPA uses the authority under FIFRA to collect information necessary to register and control the active ingredients in pesticides, while state and local agencies control the registration and actual use of the pesticides themselves. Congress has directed the USEPA to develop regulations for the design of pesticide containers to promote safe storage and disposal of pesticides, and standards for the removal of pesticides from containers prior to disposal. States that have been delegated primary FIFRA enforcement have the responsibility to incorporate and enforce these new requirements within 2 years.

Air Quality Control

A comprehensive national framework for protecting air quality was created by the Federal Clean Air Act (CAA). The CAA was first enacted in 1963, but its scope was significantly expanded in 1970 and 1977. In 1990, amendments were enacted to measure compliance. There are eleven sections, referred to as titles. The first seven titles are most pertinent to most facilities and are briefly described below. The 1990 CAA amendments include the following titles:[9]

Title I: Provisions for Attainment and Maintenance of the NAAQS

The USEPA monitors six indicator pollutants for the primary protection of human health: ozone, carbon monoxide (CO), particulate matter (PM_{10}), sulfur oxides (SOx), nitrogen oxides (NOx), and lead (Pb). Areas where pollutant concentrations exceed National Ambient Air Quality Standards (NAAQS) are referred to as nonattainment areas. These areas are categorized according to the severity of their air pollution problems. Deadlines by which NAAQS must be reached for these areas expired in 2001. Later deadlines have been assigned to those areas with the most serious pollution problems. Nonattainment areas that move too slowly toward attainment, or that fail to develop acceptable attainment plans, face the USEPA sanctions. Currently, there are roughly 100 nonattainment areas, home to 150 million people. Five classes of ozone violations (marginal, moderate, serious, severe, and extreme) and two categories (moderate and serious) for CO and particulates are established by these amendments. Compliance deadlines and control requirements exist for each category. Severely polluted areas are given more time to comply, but they must adopt more stringent controls for ozone, CO, and PM_{10}. The CAA directs the USEPA to set air quality and emissions standards at levels that protect public health and the environment.

Other Related Programs Established Under Previous Air Quality Regulations

The USEPA gives states with approved *state implementation plans (SIPs)*, as required under the 1970 CAA, the authority to run federal clean air programs. The air quality in designated regions, basic strategies to achieve or maintain NAAQS, and enforcement strategies are described in a SIP. Every state must submit a SIP to the USEPA. In various states, SIPs are the blueprints for achieving air quality goals. Standards more stringent than those required by CAA may be imposed by the states. Once a state submits an SEP or SIP amendment, the USEPA either approves it or orders the state to revise it. If a state fails (or refuses) to submit an acceptable SEP, the USEPA may either draft the SEP for that state or write (and implement) a federal implementation plan (FIP). SIPs are reviewed and revised continually. Each state must make sure it identifies its SIP emission controls and other measures that prevent significant deterioration (PSD) of air quality with respect to two criteria pollutants: SO_2 and PM_{10}. The *prevention of significant deterioration* of air quality in pristine areas is the focus of this CAA program. Special attention is directed to national parks, national wilderness areas, national monuments, national seashores, and other areas of special natural, recreational, scenic, or historic value by the PSD program. A region's classification (Class I, II or III) determines the allowable levels Of SO_2 and particulates for that region. The most severe restrictions are imposed on Class I regions; further construction of major stationary sources in these regions without a permit issued after PSD review is generally prohibited.

Emissions standards for new and modified sources of air contaminants are set by the USEPA. New and modified sources are forced to achieve the greatest emission reductions that are technologically and economically feasible by these *new source performance standards (NSPSs)*. States are required by CAA to develop emissions standards for existing sources. If a state fails to submit a satisfactory SIP, the USEPA may set these standards. New sources are typically subject to stricter emissions standards than existing sources because new sources are considered to have greater flexibility to incorporate innovative emission control technologies. The 1990 CAA amendments add important new controls on stationary sources.

The USEPA announced new NAAQS for ground-level ozone and particulates on July 17, 1997. Ground-level ozone is a major component of smog that is photochemically produced as a secondary pollutant of the stratosphere from the interaction of sunlight, nitrogen oxides, and hydrocarbons. The previous ozone standard was last revised in 1979 and set at 0.12 ppm for one hour. The previous standard did not adequately protect the public health, and the USEPA replaced it with an 8-hour standard set at 0.08 ppm. Similarly, the USEPA announced a new standard for particulate matter (PM) under the NAAQS based on the review of hundreds of scientific studies. The USEPA has added a new annual PM2.5 standard set at 15 micrograms per cubic meter ($\mu g/m^3$) and a new 24-hour PM2.5 standard set at 65 $\mu g/m^3$. The USEPA is retaining the current annual PM_{10} standard set at 50 $\mu g/m^3$. There have been several court cases aimed at preventing the EPA from implementing these new standards. The Supreme Court upheld the right of the EPA to promulgate these standards, but they placed restrictions on the implementation of those standards.

Title II: Provisions Relating to Mobile Sources

Pollution-reduction requirements for motor vehicles are defined by Title II. These requirements are of great significance to the facility and its large vehicle fleet. In areas with CO levels above 16 parts per million (ppm) or in areas designated as seriously, severely, or extremely above the limit for ozone, Title II requires centrally-fueled fleets to burn only clean fuels.

Vehicle Emissions Requirements — The 1990 CAA pollution standards are stricter than the 1970 Act for automobile and truck emissions. Tailpipe emissions of hydrocarbons, CO, and NO are expected to be reduced by these standards. New vehicles purchased by the facility (cars and light-duty trucks up to 6000 pounds) have stricter emission controls. The maintenance of appropriate emissions standards for non-methane hydrocarbons, CO, and NOx will affect fleet management.

Inspection/Maintenance Requirements — As mandated by the Clean Air Act, the Inspection and Maintenance (I&M) program requires periodic inspections of vehicles to ensure that emissions of specified pollutants are not exceeding established limitations. All ozone and CO nonattainment areas in the U.S. are required to have an I&M program. Annual inspection programs are required unless a state can demonstrate that a biennial program is just as effective.

Reformulated Gasoline Requirement — Since 1995, reformulated gasoline with a 2% minimum oxygen content is required during the winter months in non-attainment areas for carbon monoxide. Presently being sold as gasoline, this "oxygenated" fuel contains 10% ethanol or 15% methyl tertiary butyl ether (MTBE) as additives.

Title III: Hazardous Air Pollutants

Hazardous air pollutants (HAP) are defined as those substances that are hazardous to human health or the environment but are not specifically covered under other portions of the Clean Air Act. A list of 189 toxic air pollutants for which emissions must be reduced is included in this law. A more comprehensive program for regulating HAP emissions was introduced by the 1990 amendments. All new and existing stationary sources that emit HAPs must comply with air toxics emissions standards to be adopted by the USEPA.

Maximum achievable control technology (MACT) will have to be applied by stationary sources to control their HAP emissions. MACT is a standard that the USEPA uses to define various categories of sources. The USEPA is required to produce a list of all categories of stationary sources that emit HAPs. The USEPA was required to adopt MACT standards for at least 40 categories by November 2000.[6] The strategy is to identify at least 30 HAPs emitted by area sources presenting the greatest threat to public health in most areas, and identify source categories emitting 90% or more of each of the 30 HAPs. The strategy is to show how the USEPA intends to reduce by at least 75% the cancer risk posed by the release of

HAPs from stationary sources. Sources will have 9 years to comply with the strategy. Reports to Congress that identify high-risk metropolitan areas and describe actions being taken to reduce health risks posed by HAP emissions are due finally in 2002.[6]

Title IV: Acid Deposition Control

Title IV regulates the sources of acid deposition. Most of the acid deposition in the U.S. comes from the burning of fossil fuels. The operation of facility boilers is affected by these Title IV regulations. Emissions of SO_2 and NO from fossil fuel-fired electric utility plants are the leading cause of acid rain. An ambitious program to reduce and control these emissions is included in the 1990 amendments. However, the program does not affect existing units with a generating capacity of 25 megawatts or less, qualifying cogeneration units, or qualifying small-power producers.

The reduction of SO_2 and NOx emissions was to be done in two phases. Each SO_2 source was given tradable allowances equal to the number of tons of SO_2 it is permitted to emit. If emissions are reduced below this amount, a source can sell its extra allowances to another source. New SO_2 sources will have to purchase allowances from existing sources for their emissions. Certain boiler units in Phase I and others in Phase II must meet the USEPA emissions standards to achieve NOx reductions.

Plans for achieving the required emission reductions must be prepared and submitted to the USEPA by units covered in the SO_2 and NOx regulations. These same units must obtain air emissions permits (see above), install and operate continuous emission-monitoring systems, and submit detailed records of their SO_2 and NOx emissions. Clean-coal technology demonstration projects are exempt from new source performance standards and non-attainment requirements, therefore, SO_2 and NOx sources are encouraged to switch to clean-coal technologies.

Title V: Permits

Title V requires all major sources of air pollution to obtain permits and extend emissions controls to thousands of sources in many unregulated areas. For the first time, a nationwide program of air-emissions permits had been established by the 1990 amendments. Following the USEPA guidelines, the states were encouraged to set up their own permit programs which are subject to the USEPA oversight. Each state proposed a permit program and states faced sanctions if they failed to submit acceptable permit programs covering non-attainment areas.

Permit holders must certify that the facility complies with the terms of its permit on an annual basis. Under certain conditions, compliance with the permit terms will be considered compliance with CAA. Monitoring progress reports must be submitted at least once every six months by permit holders.[6]

Title VI: Stratospheric Ozone Protection

A complete production phase-out of ozone-depleting chemicals (especially CFCs and halons) is required by Title VI. A new national program designed to minimize human impact on the stratospheric ozone layer by phasing out the use of chlorofluorocarbons (CFCs) was introduced by Title VI of the 1990 CAA amendments. The Bush Administration accelerated the phase-out schedule for certain listed CFCs to 1995, based on new evidence concerning the deterioration of the ozone layer. The accelerated phase-out schedule keeps the U.S. in terms with the Montreal Protocol Ozone Treaty of 1987. Ozone-depleting chemicals are divided into Class I and Class II substances, and phase-out schedules for each class, are prescribed by Title VI. The greatest threat to the ozone layer is Class I chemicals, which were not phased out of use in the U.S. after December 31, 1995. Methyl chloroform is an exception, and its production may continue until January 1, 2002. In the interim period, specific annual percentage reductions are required. Beginning January 1, 2030, all production of Class II substances is banned by Title VI.

Title VII: Provisions Relating to Enforcement

The Amendments contain a broad array of provisions that bring the law up to date with the other major environmental statutes and thus makes the law more readily enforceable.

Water Quality Control

The basic framework of the Federal Water Pollution Control Act, passed by Congress over President Nixon's veto in 1972, remains today a part of the modern Clean Water Act. Effluent limitation guidelines, water quality requirements, and the permit program are continually enforced. The goals of the Clean Water Act are to eliminate the discharge of pollutants into surface waters and to achieve water quality that "provides for the protection and propagation of fish, shellfish and wildlife," and "for recreation in and on the water."[10] The Act also incorporates the prohibition of the "discharge of toxic pollutants in toxic amounts."

The Clean Water Act (CWA) only applies to surface waters and the groundwater, which is hydrostatically connected to surface water. The Safe Drinking Water Act provides federal protection of groundwater sources. One priority of the CWA is the discharge of mainly industrial pollutants into our nation's waters, referred to as "point sources." These controls are also known as technology-based effluent limitations. The CWA ensures that the water quality standards are met for the waters receiving the discharge, and that the discharger applies the required technology-based standards to the facility through application of a comprehensive permitting scheme known as the National Pollutant Discharge Elimination System (NPDES).[6]

The second major focus is water quality-based controls. The CWA describes three categories of pollutants including: (1) conventional pollutants (biochemical oxygen demand, total suspended solids, pH, fecal coliform, oil, and grease); (2) toxic pollutants, including the 65 pollutants identified by the USEPA; and (3) nonconventional pollutants (all pollutants not categorized as either toxic or conventional).[4]

The CWA allows states to apply to the USEPA for authorization to administer various aspects of the federal NPDES program. Generally, for a state to obtain the USEPA authorization to administer all or part of the permitting program, it must match the requirements mandated by CWA or apply more stringent ones. The owner/operator of a facility must obtain a permit prior to discharging any pollutant into the waters of the state from any point source under the NPDES program requirements. Both CWA and most state acts broadly define these terms to regulate almost all activities that result in the release of pollutants from a discrete point (e.g., a pipe).[6]

Stormwater[6,11]

Water quality may be degraded and pollutant concentrations may exceed water quality standards as a result of polluted stormwater. Pollutants may dissolve in the stormwater, become suspended, or float on the surface. Pollutants that could potentially occur in stormwater draining from sites include sediments, solids, nutrients, oxygen-demanding materials, bacteria, hydrocarbons, heavy metals, herbicides, pesticides, and other organic substances. The runoff and pollutants enter storm drains (Figure 13.5) and are then discharged into receiving waters, such as streams and lakes. During dry weather periods pollutants can also enter the storm drain system as a result of improper connections or illegal dumping.

In 1987 Congress amended the Clean Water Act requiring the USEPA to establish phased NPDES requirements for stormwater discharges. NPDES permits are required for the discharge of pollutants from any point source into waters of the U.S., which include almost any surface water or wetland. These permits have provisions for monitoring the discharge of pollutants to U.S. waters and for establishing source controls where necessary. On November 16, 1990, the USEPA published a final rule in the Federal Register (55 CFR 47990) that contains permit requirements for stormwater discharges. Facilities regulated under the new federal storm water program must comply with the state and/or federal storm water regulations. There

Figure 13.5 Stormwater runoff is a significant contributor to water pollution and is regulated under amendments to the CWA.

are three distinct types of storm water permits established by the federal regulations: individual, general, and group.

Oil and Hazardous Substance Spill and Reporting Requirements[6]

CWA contains specific provisions regulating the handling of oil and hazardous substances. Specific penalties and rules of liability for unauthorized releases of these materials are established by CWA. Similar to the requirements outlined under Superfund, these provisions focus on reporting unauthorized leaks, spills, and discharges to water (discharges without a permit). Under the CWA, any person in charge of a vessel or an on-shore or off-shore facility who has knowledge of an unauthorized release of oil or any hazardous substance must immediately report the release in a manner consistent with the National Oil and Hazardous Substances Pollution Contingency Plan (commonly called the National Contingency Plan).

The Oil Pollution Act of 1990 was created in response to the Exxon Valdez oil spill in 1989, which dumped 11 million gallons of oil into the waters off south central Alaska. Media coverage of black oily beaches and the resulting deaths of thousands of birds, sea otters, and whales among other wildlife, created public outrage. This, in addition to the spill's size and location, led to OPA 90. This act provides tougher penalties and liabilities for oil spills, allocates more funds and other resources for dealing with spill cleanup and places more responsibility on the executive branch to respond to these incidents.

COMPLIANCE STRATEGIES

Since the 1970s, employers have been given increased responsibilities. Managers are becoming more responsible for environmental health and employee safety. Various agencies have implemented regulations that are enforced to further protect the employee and the environment. Employers have developed worker safety programs, and the method in which these programs are enforced can directly impact worker health and safety, liability and insurance premiums, worker's compensation, remediation costs, and fines and costs for meeting regulatory requirements. However, it is not uncommon that such programs result in a negative company image due to the inadequacy of the implemented programs. Unfortunately, these programs are often not given enough attention and support. Additionally, supervisor and worker training programs are frequently insufficient.

There are criminal penalties for noncompliance with environmental health and safety programs. There has been a tremendous increase in liability for both the federal agencies and for individuals from the civil and criminal provisions of recently reauthorized environmental legislation. Even though the thousands of pages of regulations are difficult to interpret, they still apply. Some of the regulatory initiatives include pollution prevention and waste minimization in facilities. The regulations are enforceable throughout all facets of a facility's operation including air, water, and solid waste disposal. The Community-Right-to-Know Act plays a large role in this process, requiring facilities to publicly report their waste production for hazardous wastes, while the Pollution Prevention Act ensures that waste is reduced through raw material substitution, process modification, and recycling. The Resource Conservation and Recovery Act targets recycling programs.

After gaining a basic understanding of the systems of environmental laws, employers need to apply their knowledge in order to remain in compliance. The development of such strategies for compliance is based on the premise that the most effective protection against aggressive enforcement and other efforts to assess liability is aggressive compliance.

The Government Institutes has published a variety of books that discuss and detail the elements of appropriate compliance organization. There are several basic principles with which one should become familiar. The first principle is that environmental law compliance is the responsibility of everyone. The second principle describes the need to demonstrate a concern for compliance by providing appropriate education and training. The third principle focuses on the prevention of violations and minimization of liability through aggressive implementations of environmental objectives. The forth principle recommends periodic audits to verify compliance and identify areas that can be improved upon. These audits will be conducted to address environmental issues and controls, such as asbestos, indoor air quality, and UST regulations.[2]

Trends in Regulatory Compliance

The states are responsible for nearly 90% of regulatory inspections involving air, water, and hazardous wastes, with the USEPA conducting the remainder of the inspections. Routine inspections usually consist of sampling analysis and record keeping. Inspections may also include investigations of operation and maintenance programs. The USEPA has developed target strategies, based on the quantity and severity of toxic materials discharged, emitted, or disposed of by a facility as reported under the Community Right-to-Know Act. This data is compared to other self-reported data required by other acts, and with information stored in the USEPA files such as manifest reports and permit applications. The USEPA will likely focus on repeated violations and evidence of suspicious self-reporting.

Traditionally, regulators have focused largely on the use of standard enforcement tools, such as sanctions, to persuade industries found to be in violation of environmental laws to comply. However, another approach is being explored by some environmental experts. It is being suggested that this approach be combined with an effort to educate industry officials regarding environmental laws in hopes of inducing increases in the standard of current practice and promoting a healthier environment for the future.

A situation that has recently arisen is the question of whether U.S. companies may be sued for environmentally damaging acts committed in other countries. In San Carlos, Ecuador, carcinogenic by-products of extraction methods were poured into open pits and left to seep into the ground. They also contributed significantly to air pollution when the pits were set on fire to burn off their contents. As a result of this issue, 30,000 Ecuadorians are filing a class action suite against Texaco. It is being said that Texaco executives violated human rights and should be held accountable for their actions.

On the other side of the world, China has begun to deal with the issue of environmental destruction that is resulting from its own development. Over the past two decades China has implemented several pollution control programs to curb the amount of pollution produced by its growing industries. These include enacting emissions taxes and setting a general goal of stopping environmental deterioration before the year 2015.

REFERENCES

1. Burns, I.A., Peltason, J., Cronin, T.E. and Magleby, D.B., *Government by the People,* 15th ed., Prentice Hall Publishers, Englewood Cliffs, NJ, 1993, chap. 14.
2. Sullivan, T.E.P., *Fundamentals of Environmental Law,* Government Institutes, Inc., Rockville, MD, 1995, chap. 1.
3. United States Postal Service, A Proposal to Establish Environmental Coordinators in District Offices of the Northeast Area, Northeast Area Environmental Compliance Staff, Northeast Area Processing and Distribution, 6 Griffin Road North, Windsor, CT 06006-7030, Feb. 25, 1993.
4. Vance, B., Weinsoff, DJ., et al., *Toxics Program Commentary: Massachusetts,* prepared by Touchstone Environmental Inc., Specialty Technical Publishers, North Vancouver, B.C., Canada, V7M IA5, 1992.
5. United States Postal Service, Waste Reduction Guide Handbook AS-552, Northeast Area Environmental Compliance Staff, Northeast Area Processing and Distribution, 6 Griffin Road North, Windsor, CT, 06006-7030, Feb. 1992.
6. Vance, B., Weinsoff, D.J., Henderson, M.A. and Elliot, J.E., *Toxics Program Commentary: Massachusetts,* Prepared by Touchstone Environmental, Inc., and published by Specialty Technical Publishers, Inc,. North Vancouver, B.C., Canada V7M I A5, 1992.
7. United States Postal Service, Hazardous Waste Management Guide, Handbook AS-553, Northeast Area Environmental Compliance Staff, Northeast Area Processing and Distribution, 6 Griffin Road North, Windsor, CT, 06006-7030, May 1992.
8. Moeller, D.W., *Environmental Health,* Harvard University Press, Cambridge, MA, 1997, chap. 13.
9. United States Postal Service, Clean Air Act Compliance, Handbook AS-553, Northeast Area Environmental Compliance Staff, Northeast Area Processing and Distribution, 6 Griffin Road North, Windsor, CT, 06006-7030, April 1992.
10. Sullivan, T.F.P., *Environmental Law Handbook,* Government Institutes, Inc., Rockville, MD, 1997, chap. 4.
11. United States Postal Service, NPDESI Stormwater Guide, Handbook AS-554, Northeast Area Environmental Compliance Staff, Northeast Area Processing and Distribution, 6 Griffin Road North, Windsor, CT, 06006-7030, May 1992.

Acronyms and Abbreviations

AHERA (Federal)	Asbestos Hazard Emergency Response Act
ASHAA (Federal)	Asbestos School Hazard Abatement Act
AST	Aboveground storage tank
ATSDR	Agency for Toxic Substances and Disease Registry
BAT	Best available technology economically achievable
BOD	Biochemical oxygen demand
BPT	Best practicable control technology currently available
CAA	(Federal) Clean Air Act
CERCLA	(Federal) Comprehensive Environmental Response, Compensation, and Liability Act of 1980 (Superfund)
CERCLIS	CERCLA Information System
CFC	Chlorofluorocarbons
CFR	Code of Federal Regulations
CO	Carbon monoxide
CWA	(Federal) Clean Water Act
DEP	Department of Environmental Protection
DOT	(U.S.) Department of Transportation
DPH	Department of Public Health
DWPL	Drinking Water Priority List
EHS	Extremely Hazardous substances/Extraordinarily Hazardous substances
EIS	Environmental impact statement
EPA (USEPA)	(U.S.) Environmental Protection Agency
EPCRA	Emergency Planning and Community Response Act
FIFRA	Federal Insecticide, Fungicide, and Rodenticide Act
FWPCA	Federal Water Pollution Control Act
HAP	Hazardous air pollutant
HCFCs	Hydrochlorofluorocarbons
HIV	Human immunodeficiency virus
HMTA	(Federal) Hazardous Materials Transportation Act
HRS	Hazard ranking system
HSA	(Federal) Hazardous Substances Act

HSWA	(Federal) Hazardous and Solid Waste Amendments of 1984
HW	Hazardous waste
ICS	Individual Control Strategy
IPM	Integrated Pest Management
LAER	Lowest achievable emission rate
LEPC	Local emergency planning committee
LEPD	Local emergency planning district
LOAEL	Lowest observable adverse effect level
LQG	Large-quantity generators
LUFT	Leaking underground fuel tank
LUST	Leaking underground storage tank
MACT	Maximum achievable control technology
MCL	Maximum contaminant level
MSDS	Material safety data sheet
NA	North America
NAAQS	National Ambient Air Quality Standards
NESHAPs	National Emission Standards for Hazardous Air Pollutants
NIMBY	Not in my backyard
NO_2	Nitrogen dioxide
NOx	Nitrogen oxides
NOEL	No observable effect level
NOAEL	No observable adverse effect level
NPDES	National Pollution Discharge Elimination System
NPL	National Priorities List
NSPS	New source performance standard
NSR	New source review
OEM	Original equipment market
OPP	Office of Pollution Prevention
OSHA	(U.S.) Occupational Safety and Health Administration
PCBs	Polychlorinated biphenyls
PEL	Permissible exposure limit
$PM_{2.5}$	Particulates <2.5 microns in size
PM_{10}	Particulates <10 microns in size
POTW	Publicly owned treatment works
PPA	(Federal) Pollution Prevention Act
ppb	Parts per billion
ppm	Parts per million
PRP	Potentially responsible parties
PSD	Prevention of significant deterioration
RCRA	(Federal) Resource Conservation and Recovery Act
PWS	Public water system
R&D	Research and development
Reg	Regulations
ROD	Record of decision
RP	Responsible party
RQ	Reportable quantity

SDWA	(Federal) Safe Drinking Water Act
SIC	Standard Industrial Classification
SIP	State implementation plan
SITE	Superfund Innovative Technology Evaluation
SOP	Standard operating procedure
SPCC	Spill Prevention Control and Countermeasure (plan)
SO2	Sulfur dioxide
SQG	Small-quantity generator
Stats	Statutes
SWDA	(Federal) Solid Waste Disposal Act
TCLP	Toxicity characteristic leaching (leachate) procedure
TPQ	Threshold planning quantity
TPY	Tons per year
TSCA	(Federal) Toxic Substances Control Act
TSD	Treatment, storage and disposal
UIC	Underground injection control
UN	United Nations
USA	United States of America
UST	Underground storage tank
VOC	Volatile organic compound
VSQG	Very small-quantity generator

GLOSSARY

Abiotic. Components of the ecosystem that include such things as water, air, sunlight, minerals, and their interaction to produce climate, salinity, turbulence, and other conditions that influence the physical conditions under which organisms survive.

Abortion. Refers to the medical means of terminating a pregnancy.

Absorption. Passing through the skin.

Acceptable daily intake (ADI). An acceptable level of exposure or dose of a substance that is considered "safe," or poses little risk.

Acid. A material that has a pH of less than 7.0. Acids are corrosive to human tissue and some metals.

Acidic deposition. Characterized by emissions of nitrogen and sulfur oxides which are partially converted in the atmosphere to nitric and sulfuric acids that return to the earth in rain, snow, fog, and on dry particles.

Active transport. A process that involves ATP in conjunction with special carrier proteins to move molecules through a membrane against a concentration gradient (i.e., high concentration to low).

Acute effects. Immediate effects from a high-intensity chemical exposure that result in a change in health usually within 24 hours. Many diseases such as measles are *acute* and have a rapid onset, are usually self-limiting, and are of relatively short duration.

Acutely hazardous waste. Any hazardous waste with an EPA Waste Code beginning with the letter "P," or any of the following "F' codes: F020, F021, F022, F023, F026, and F027. These wastes are subject to stringent quantity standards for accumulation and generation.

Acute toxicity. The ability of a substance to cause poisonous effects resulting in severe biological harm or death soon after a single exposure or dose. Also, any severe poisonous effect resulting from a single, short-term exposure to a toxic substance.

Additive. Occurs when the effects of two substances taken together are equal to the sum of the two substances taken separately.

Adduct. Carcinogenic residues bound to DNA.

Adsorb. To collect the molecules of a chemical on the surface of another material.

Age distribution. Describes the percent of the population in various age categories from birth to death.

Age–sex composition. Sex ratio and age distribution of a population. The age–sex composition of the population has a profound effect on the birth and death rates of a country.

Air exchange rate. The rate at which air is replaced in the structure by external air.

Air pollutants. Refers to substances that enter the atmosphere from human activities and have the capacity to cause human disease and welfare effects resulting in disturbances to plant life and physical structures.

Albedo. The combined reflective ability of cloud cover and ground surfaces.

Alkali. A material that has a pH of greater than 7.0. Generally corrosive to human tissue.

Allowance. Defined under 1990 CAAA as the right to emit one ton of sulfur dioxide.

Anaerobically. In the absence of oxygen.

Anaphylactic. A Type I hypersensitivity that is associated with hay fever, asthma, and anaphylactic shock from insect bites, certain foods such as shellfish, and a variety of chemicals. The reactions are usually immediate, occurring within moments in persons previously sensitized by exposure to the offending antigen.

Angiogenesis. A process whereby malignant cells grow into a small mass that develops nutrient-bearing blood vessels.

Antagonistic. A combined effect of two or more substances that is lesser than the sum of those substances acting independently.

Anthropogenic. Generated or created by humans.

Anticodon. The recognition site for nucleotides on the sRNA molecule.

Apoenzyme. Protein component of an enzyme.

Apoptosis. A process of cellular self-destruction.

Ash. The residue that remains after burning.

Asphyxiant. A substance that causes suffocation.

Atoms. The basic units of elements that consist of a small dense center called a nucleus surrounded by a cloud of negatively charged electrons.

Autotrophic. Refers to organisms that synthesize their own food from inorganic substances and obtain energy from the sun. Most plants are autotrophs.

Best management practices. Schedules of activities, prohibitions of practices, maintenance procedures, and other management practices to prevent or reduce the pollution of U.S. waters, including treatment requirements, recycling, reduction, reuse, operating procedures, and practices to control facility/site runoff, spillage or leaks, sludge or waste disposal, or drainage from raw material storage.

Bias. A type of error introduced into epidemiological studies through differential treatment of the cases and controls in a study, either during the selection process or information gathering.

Bioaccumulation. Refers to the storage of substances in the body at levels above what would be normally expected.

Biodegradable material. Waste material that is capable of being broken down by microorganisms into simple stable compounds such as carbon dioxide and water.

Biogeochemical cycling. Refers to the recycling of nutrients.

Biological Oxygen Demand (BOD). A standard test to determine the amount of dissolved oxygen in water used by microorganisms that feed on organic matter in the water. The test is performed over a 5-day period under controlled conditions.

Biological transmission. A process in which the disease agent goes through some reproductive stage in the vector and is transmitted by a bite, fecal matter, or vomit of the vector in an active process.

Biological vector. Refers to vectors (usually arthropods) in which a microbial disease agent goes through development or multiplication prior to becoming infective.

Biomarkers. These are mutations occurring along a gene appearing at characteristic sites and with unique base sequences.

Biomass. The accumulation of organic material in an ecosystem.

Biomes. Based on the dominant types of vegetation that are strongly correlated with regional climate patterns.

Biosphere. That part of the Earth and its atmosphere that can support actively metabolizing life.

Biotic. Components of the ecosystem include living organisms and the products of these organisms including waste products and decay such as urine, feces, decaying leaves and twigs, bones, and flesh.

Biotic potential. The unrestricted growth of populations resulting in the maximum growth rate for a particular population.

Biotransformation. A process in which metabolic processes alter the structure and characteristics of a chemical.

Birth rate. Refers to the number of individuals added to a population through reproduction (live births) and is normally expressed as the number of live births per 1000 population (counting the population at the midpoint of the year).

Bogs. Usually a stagnant, acidic, nutrient-poor ecosystem filled with peat and sphagnum moss. Its abundance of decaying material consumes enormous amounts of carbon dioxide.

Boiling point. The temperature at which a liquid turns to a vapor at ambient pressure.

Bottom ash. Larger residue pieces that fall through the bottom of the boiler grate after burning is completed.

Bottomland hardwood forests. These are deciduous forested wetlands consisting of gum and oak trees adapted to areas of seasonal flooding or covered with water most of the year and found along streams and rivers in the Southeast and Southcentral flood plains of the U.S.

Building-related illnesses (BRI). A phenomenon characterized by well-defined illnesses that occur in a building that can be traced to specific building problems.

Buttress. A shallow root system formed by trees with bases that spread at the bottom of the tree. This growth provides support in thin soil and may enhance the ability of the tree to secure nutrients over a broader area.

Bypass waste. Waste that cannot be recycled, reused, or burned, but must be buried at the landfill.

Canopy. An almost unbroken sea of green produced by leaves of trees at 70 to 100 feet above the forest floor.

Carcinogen. Cancer-producing substance.

Carnivores. Animals that eat primarily animal flesh. These animals include lions, tigers, hyenas, frogs, or even ladybugs (they feed on aphids).

Carriers. Persons who exhibit no symptoms but harbor and transmit disease organisms.

Carrying capacity. The sum of those limiting factors that serve to control the numbers of a species that can survive in defined area over time.

Ceiling limit. The concentration of chemicals in air that should never be exceeded during any part of a working exposure.

Chaparrals. Areas of moderately dry climate and limited summer precipitation, further characterized by small (3 to 15 foot) shrubs with leathery leaves that contain aromatic and flammable substances.

Chlorofluorocarbons (CFCs). A family of inert, nontoxic, and easily liquefied chemicals used in refrigeration, air conditioning, packaging, or insulation, or as solvents and aerosol propellants. Because CFCs are not destroyed in the lower atmosphere, they drift into the upper atmosphere, where the chlorine is released and destroys ozone.

Characteristic waste. A waste classified as hazardous because it is ignitable, corrosive, reactive, or toxic as determined by the toxicity characteristic leaching procedure. It has an EPA waste code in the range D0001 to D0003. Each of these four characteristics is defined in 40 CFR 261.20 Subpart C.

Chemical formula. A listing of the atoms in a molecule.

Chlorinated solvent. An organic solvent containing chlorine atoms, for example, methylene chloride and 1,1,1-trichloromethane, that is used in aerosol spray containers and in certain paints.

Chronic effects. Effects of repeated, low-dosage chemical exposure that may not show up for many years or may last for many years. Diseases having a slow onset and lasting for extended periods are said to be chronic and may include cancer, emphysema, and some forms of heart disease.

Clean Water Act (CWA). Redesignated name for the Federal Water Pollution Control Act following the 1977 amendments.

Climate. Can be viewed as average weather within a geographical area over years or even centuries.

Climax community. This is identified as one that forms in an undisturbed environment and continues to grow and perpetuate itself in the absence of further disturbance.

Clonal evolution. A process whereby a single cell develops a genetic mutation that enables it to divide even when normal cells do not replicate.

Cloning vector. A fragment of a self-replicating DNA molecule.

Code of Federal Regulations (CFR). The detailed regulations, written by federal agencies, to implement the provisions of laws passed by Congress. Regulations in the CFR have the force of federal law.

Co-dominant. Refers to genes that are partly expressed if present as a single allele (an allele is a pair of genes situated at the same site on paired chromosomes) and fully expressed if present on both.

Codon. A group of three consecutive (triplet) mononucleotide units in the mRNA that specifies the position of one amino acid in a protein molecule.

Commensals. A relationship in which two kinds of organisms live in or near each other with one obtaining benefits, and the other is not damaged or benefited.

Complement. A system composed of several (at least 20) serum proteins that follow a sequence or cascade of actions beginning with the Ag-Ab binding. This sequence of action enhances phagocytosis, inflammation, and cell lysis.

Composting. A controlled process of degrading organic matter by microorganisms into a humus-like material.

Conditionally exempt generator. A hazardous waste generator that meets the following criteria: (a) in every single month of the year, the site generates no more than 100 kg (220 lb) of hazardous waste, no more than 1 kg (2.2 lb) of acute hazardous waste, and no more than 100 kg (220 lb) of material from the clean-up of spillage of acute hazardous waste.

Congenital defects. The defects occurring during the embryonic stages of development.

Contaminant. Any physical, chemical, biological, or radiological substance or matter that has an adverse affect on air, water, or soil.

Corrosives. Chemicals that are acidic or basic in pH and are destructive to human tissue and many metals.

Crossbreeding. This involves the selection of genetic and morphologic variants that possess one or more of the desired characteristics and then crossing the plants/animals in the hope that the most desired characteristics of the different species will appear in the new crossbred organisms.

Cytogenetic defects. Abnormalities in the number or structure of chromosomes.

Death rate. Refers to the number of individuals removed from a population through death and is normally expressed as the number of deaths per 1000 population (counting the population at the midpoint of the year).

Decomposers. Insects, bacteria, fungi, and protozoans that sequentially break down complex organic materials into low-energy mineral nutrients that once again may be reabsorbed and used by plants.

Definitive host. The host in which the parasite reaches sexual maturity.

Deforestation. The permanent decline in crown cover of trees to a level that is less than 10% of the original cover.

Demographic transition. The simultaneous decline of death rates and birth rates in a country, resulting in a diminishing difference between birth rates and death rates and a very low rate of natural increase, permitting a stable population with very long doubling times.

Denitrifying bacteria. Bacteria that live in the mud and sediment of lakes, ponds, streams, and estuaries and recycle organic nitrogen back to atmospheric nitrogen.

Deserts. Areas defined by their arid climates which average less than 10 inches of precipitation a year and where evaporation tends to exceed this precipitation, causing a negative annual water budget.

Developmental disease. A disease that occurs when faults or mistakes occur within the genes (or chromosomes) or stages in development of the fetus are disturbed.

Diffusion. A passive process that occurs when molecules move from areas of high concentration to ones of low concentration.

Direct contact transmission. Occurs when the disease is passed directly from a source to the host by physical contact without an intervening object.

Discharge. The discharge of a pollutant when used without qualification.

Disposal. Final placement or destruction of toxic, radioactive, or other wastes; surplus or banned pesticides or other chemicals; polluted soils; and drums containing hazardous materials from removal actions or accidental releases. Disposal may be accomplished through the use of approved secure landfills, surface impoundments, land farming, deep well injection, ocean dumping, or incineration.

DOT. The U.S. Department of Transportation.

Droplet infections. A process whereby the organisms are contained within mucous droplets that are expelled to the air in sneezing, coughing, or even talking, and travel short distances to be inhaled or ingested by persons very near by (usually less than one meter).

Doubling time. The length of time required for a population to double its size.

Dry adiabatic lapse rate. A theoretical condition in which a parcel of warm, dry air will cool at the rate of –10°C/km of altitude.

Dynamic equilibrium. Processes that tend to balance each other so that there is little net loss over long periods of time.

Ecosystems. Refers to identifiable areas within nature where the organisms interact among themselves and their physical environment and exchange nutrients largely within that system.

Effective demand. The "demand level" corresponding to purchasing power without regard to food requirements.

Effective dose. The amount of a chemical that reaches the general circulation and ultimately the location where a particular effect is seen or felt.

Effluent. The water and the quantities, rates, and concentrations of chemical, physical, biological, and other constituents that are discharged from a point source.

Emergent trees. Taller trees that penetrate the lower canopy to form an umbrella-like canopy over the trees below. These trees reach heights to 130 feet.

Emigration. This is a process by which some species disperse or migrate out of the area.

Emissions. Gases and particulates that reach the air from the combustion of industrial processes.

Endemic. Refers to diseases that already exist in a community where it is maintained in a low, but constant, incidence.

Endocrine disruptors. Synthetic chemicals that have been referred to as xenoestrogens, estrogenic, hormone mimicking, and endocrine-disrupting chemicals which have the ability to mimic or block hormones; cause unwelcome changes in hormone receptor sites; cause changes in the baseline production of hormones; or replace hormones on the carrier proteins. This makes some proportion of the population of native functional hormone unavailable and produces adverse health consequences that may be barely noticeable to very severe.

Endpoints. The final or targeted result of a health event such as death or recovery. Researchers will focus their interests on certain health responses or endpoints from exposure to biological, physical, or chemical agents.

Energy recovery. The generation of power in the form of steam or electricity from the burning of garbage.

Enforcement. Efforts made by official agencies to ensure agreement with environmental laws and regulations.

Environment. In its broadest definition, the term refers to personal and cultural behavior including smoking, diet, alcohol consumption, sexual and reproductive patterns, workplace, infections, food additives, and pollution along with the strictly physical environment.

Environmental disease. Refers to any pathologic process having a characteristic set of signs and symptoms that are detrimental to the well-being of the individual and are the consequence of external factors, including exposure to physical or chemical agents, poor nutrition, and social or cultural behaviors.

Environmental lapse rate. The actual change of temperature with altitude.

Environmental resistance. Refers to those pressures that limit population and may include such factors as disease, wars, predatory behavior, toxic waste accumulation, or species competition.

EPA (also USEPA). The U.S. Environmental Protection Agency. Responsible for making and enforcing the laws and regulations regarding protection of the environment.

EPA identification number. A 12-character number assigned by either EPA or the authorized state to each hazardous waste generator, transporter, and treatment, disposal, or storage facility.

Environmental tobacco smoke (ETS). The tobacco smoke consisting of side-stream smoke that originates from the burning end of the cigarette, and exhaled smoke.

Epidemic. A situation in which there is a marked increase in the incidence of disease within a limited area affecting ever-increasing numbers of people.

Epiphytes. Airplants that grow on branches high in the trees. These plants are supported by the tree limbs and extract moisture from the air, trapping leaf fall and windblown dust for nutrients.

Estuaries. Inland marine waterways found worldwide. They are dominated by grass or grass-like plants and lie mostly between barrier islands and beaches.

Exoskeleton. A hard, outer skin that protects internal organs of arthropods and serves as an attachment site for muscles to provide movement.

Exponential growth phase. That part of the growth curve where the population increases at a rate proportional to its size, or exponentially.

Extremely hazardous substances (EHS). Any of 406 chemicals identified by EPA on the basis of toxicity and listed under SARA Title III. The list is subject to revision.

Extrapolation. Refers to the use of animal data to predict human response to chemical exposure.

Facilitated diffusion. A process in which molecules such as amino acids and sugars require specialized carrier proteins to be transported across a membranes.

Fallowing. An agricultural process whereby fields remain unplanted every few years, and this resting permits moisture and nutrients to return.

Family planning. Programs that are directed at assisting couples in having the number of children they desire regardless of how many.

Flammable limits. The percent concentrations of gas or vapor in air that will support combustion.

Flashpoint. The temperature at which a flammable liquid will flash when provided with an ignition source.

Fly ash. Very small, solid particles of ash and soot generated by burning and carried in gases that fly up into the environmental control equipment.

Foodborne pathogens. Agents including parasites, viruses, fungi, and bacteria that have the potential to contaminate a food supply.

Food security. A situation that occurs when all people have physical and economic access to the basic food they need to work and function normally.

Food web. The complex and interrelated patterns of consumption among many trophic levels.

Form R. Toxic Chemical Release Inventory (TRI) Report Form. This form is required by Section 313 of the Emergency Planning and Community Right-to-Know Act (Title III of the Superfund Amendments and Reauthorization Act of 1986), Public Law 99~99.

Generation. Refers to the weight of materials and products as they enter the waste management system from residential, commercial, institutional, and industrial sources, and before materials are recovered or combusted.

Generator. A person, company, site, or mobile source that produces solid or liquid waste.

Gene transfer. The transfer of genetic information from one organism to another by artificial means in order to create new organisms with desired characteristics.

Greenhouse effect. A process in which heat energy (infrared) reflected from the Earth is absorbed by infrared-absorbing gases such as carbon dioxide, chlorofluorocarbons (CFCs), methane, and water vapor, thereby trapping the warmth.

Halon. Bromine-containing compounds with long atmospheric lifetimes whose breakdown in the stratosphere causes depletion of ozone. Halons are used in firefighting.

Hazardous waste. A by-product of society that, because of its quantity, concentration, or physical, chemical, or infectious characteristics may cause, or significantly contribute to, an increase in serious, irreversible illness. It may pose a substantial present or potential hazard to human health, safety, or welfare or to the environment when improperly treated, stored, transported, used, disposed of, or otherwise managed. Possesses at least one of four characteristics (ignitability, corrosivity, reactivity, or toxicity) or appears on special EPA lists.

Hazardous material. A substance or compound that is capable of producing adverse health and or safety effects.

Herbivores. Animals that eat only plants.

Holoenzyme. Refers to a complete and active enzyme consisting of an apoenzyme and its cofactor.

Humus. Organic litter that decays further to a rich organic mat.

Hypersensitivity. An exaggerated immune response to the presence of an antigen.

Immigrate. The process in which species settle in a new area.

Incineration. Refers to the burning of certain types of solid, liquid, or gaseous materials; or a treatment technology involving destruction of waste by controlled burning at high temperatures.

Incomplete demographic transition. A decrease in deaths, but without the time for adjustment in birth rates. The number of births remains high, and the combination of high birth rates and lowered death rates results in very high rates of natural increase.

Incomplete metamorphosis. Refers to insects such as roaches, body lice, and grasshoppers that go through three developmental stages including egg, nymph, and adult stages.

Indirect contact transmission. A process in which the disease agent is transmitted from a reservoir to a host by a nonliving object.

Infectious diseases. Result from the pathologic process occurring when a microbial agent invades the body.

Ingestion. Swallowing.

Integrated solid waste management. The practice of using several alternative waste management techniques to manage and dispose of specific components of the postal waste stream. Waste management alternatives include source reduction, recycling, composting, energy recovery, and landfilling.

Intermediate host. Host in which a parasite passes through its larval or asexual stage of development.

Inversion. A meteorological phenomenon in which a stable, slow-moving air mass results in the formation of a dense, cool layer of air near the Earth with warmer air above. The warming of air with increasing elevation is a reversal of the normal tropospheric cooling process.

Ionizing radiation. The energy resulting from radiant energy that interacts with matter to form charged particles, and may include electromagnetic radiation (gamma and x-radiation) or particle radiation (alpha, beta, and neutron).

Isotopes. Atoms of the same element that have have different mass numbers.

Kwashiorkor. A disease of dietary insufficiency usually seen in children one to three years old.

K-strategy. A reproductive strategy in which large organisms with relatively long life spans have only a few offspring, but devote their energies to protecting and nurturing the offspring to enhance their individual survival until they can reproduce.

Lag phase. The initial portion of the growth curve in which the organisms show no increase in growth rate but are undergoing changes that ready them for rapid proliferation.

Landfill. A huge basin ranging in size from 20 to 70 acres wide scooped out of the Earth. The sides and bottom are lined with layers of clay and geotextiles. Inside are also 12- to 14-inch layers of sand in which is buried a piping system that collects the leachate that runs through the garbage.

Large quantity generator (LQG). A generator of hazardous waste that meets any of the following criteria: (a) the site generates in one or more months during a year 1000 kg (2200 lb) or more of RCRA hazardous waste; (b) the site generates in one or more months during the year, or accumulates at any time, 1 kg (2.2 lb) of RCRA acute hazardous waste; or (c) the site generates or accumulates at any time more than 100 kg (220 lb) of spill clean-up material contaminated with RCRA acute hazardous waste.

LD_{50}. Defined as the dose of a chemical that will cause death in 50% of a defined animal population.

Lianas. Woody vines that grow quickly up tree trunks when a temporary gap opens in the canopy. Here they flower and fruit in the tree tops.

Leachate. The liquid formed as moisture and rainfall and passes through refuse in landfills.

Lichens. Mixtures of fungi and algae that grow together.

Listed wastes. Wastes specifically named in 40 CFR 261.3. These wastes are listed as hazardous under RCRA, but have not been subjected to the toxic characteristics listing process because the dangers they present are considered self-evident. They bear EPA Waste Codes beginning with the letters F, P, U, or K.

Loam. Soils best suited to agriculture that consist of sand, silt, and some clay in a homogeneous mixture.

Macronutrients. Nearly all life consists of large percentages of sulfur, phosphorous, carbon, oxygen, hydrogen, and nitrogen. These are chemical elements necessary for life in amounts usually larger than 1.0 ppm.

Marasmus. Nutritional marasmus may appear in children one year of age who lack sufficient food and show signs of wasting, failure to grow, and complications of infections with multiple vitamin deficiencies.

Material recovery facility (MRF). A type of solid waste recycling facility that separates out all usable products and usable parts from individual products.

Material Safety Data Sheet (MSDS). An information sheet, required under the OSHA Communication Standard, that provides information on specific chemical health and physical hazards, exposure limits, and precautions. Section 311 of SARA requires facilities to submit MSDSs under certain circumstances.

Maximum mixing depth (MMD). Refers to the height a column of air can rise. Generally, the higher the column of rising air, the better the mixing or dispersion of pollutants.

Mass number. The total number of protons and neutrons in an atomic nucleus.

Materials recycling facilities (MRFs). Facilities that perform the function of preparing recyclables for marketing.

Mechanical injury. Injury that results from the transfer of a damaging excess of kinetic injury to tissues resulting in abrasions, lacerations, punctures, contusions, broken bones, and projectile wounds.

Mechanical transmission. A process of passive transmission by which organisms such as arthropods transmit a disease by simply carrying the disease organism(s) on the feet or other body part to a person's food or an open sore on the person.

Metastasis. A process in which malignancies travel, producing tumors at distant sites.

Monoculture. The planting of single crops.

Monitoring. The measurement, sometimes continuous, of air or water quality.

Mucociliary streaming. A process whereby particles such as dust and pollen of 10 μm or larger are removed by the constant streaming of mucous propelled from the bronchial and tracheal passages by the cilia beating at over 1300 times per minute.

Mutagens. Agents in the environment including chemicals or radiation that promote mutations.

Mutation. A change in the nucleotide base sequence of DNA.

National Ambient Air Quality Standards (NAAQS). Refers to standards that have been set under the Federal Clean Air Act and its amendments for criteria pollutants, thereby establishing maximum concentrations allowed in the ambient or outdoor air environment.

National Pollutant Discharge Elimination System (NPDES). A provision of the Clean Water Act that prohibits discharge of pollutants into waters of the U.S. unless a special permit is issued by EPA, a state, or (where delegated) a tribal government on an Indian Reservation.

Necrosis. A process in which lesions are produced that are local and often result in killing areas of tissue.

Neoplasia. New and uncontrolled growth of abnormal tissue from the transformation of normal body cells (tumors).

NIOSH. National Institute for Occupational Safety and Health.

Nonattainment areas. Those Air Quality Control Regions (AQCRs) as designated by the USEPA that exceed the levels of one or more of the criteria pollutants during the year.

Nonpoint source. Refers to pollutants entering the environment from a broad area and may include scattered sources.

Normal lapse rate. A characteristic loss of temperature as air rises. Air will lose heat from such mechanisms as radiative cooling at an average of –6.5°C/km.

Omnivores. Animals that eat plants and animals, and include rats, bears, humans, hogs, foxes, and many others.

Oncogenes. Cancer-causing genes.

OSHA. Occupational Safety and Health Administration. Designed to monitor the working conditions of individuals and ensure that they are safe.

Off-site facility. A hazardous waste treatment, storage or disposal area that is located at a place away from the generating site.

On-site facility. A hazardous waste treatment storage or disposal area that is located on the generating site.

Oxidant. A substance that readily gives up an oxygen atom or removes hydrogen from a compound.

Packaging. A product container, product cushion or wrapping around the product or container. Packaging may also refer to preconsumer shipping and storage containers as well as packaging on the shelf.

Pandemic. An epidemic that spreads throughout the world, as in the case of various outbreaks of influenza.

PCBs. Polychlorinated biphenyls. A known carcinogen.

Permafrost. Lies as a permanent layer of ice or frozen soil beneath a thin layer of topsoil.

Permanent threshold shift (PTS). A permanent hearing loss resulting from damage and loss of hair cells in the hearing apparatus.

Permissible exposure limit (PEL). An exposure limit to a chemical, as determined by OSHA, below which most individuals can work day after day with no harmful effects. This limit is legally enforceable.

Pesticide. Substance or mixture of substances intended for preventing, destroying, repelling, or mitigating any pest. Also, any substance or mixture of substances intended for use as a plant regulator, defoliant, or desiccant. A pesticide can accumulate in the food chain or contaminate the environment.

pH. A unit for measuring hydrogen ion concentrations. A pH of 7 indicates a neutral water or solution. At a pH lower than 7, a solution is acidic; at a pH higher than 7, a solution is alkaline.

Phagocytosis. This process occurs when specialized cells such as phagocytes extend out streams of cytoplasm much like an amoeba and engulf solid particles to form a sac around the particle(s).

Photochemical. Refers to the initiation of chemical reactions by sunlight.

Picocurie. The equivalent of one trillionth of a curie representing slightly more than 2.0 radioactive disintegrations per minute.

Pinocytosis. A process in which cells absorb nutrients and liquids. A small drop adheres to the cell membrane surface where an indentation forms to surround the drop and form a vesicle which then detaches from the membrane.

Pioneer plants. These are early plants that establish the beginning stages of secondary succession, and may include wildflowers, followed by tall grasses and compact woody bushes.

Point mutations. Single-gene defects.

Point source. Refers to pollutants entering the environment from a specific point such as a pipe or a specific source such as a factory or treatment plant.

Pollution. Refers to the presence of a foreign substance — organic, inorganic, radiological, or biological — that tends to degrade the quality of the environment so as to create a health hazard.

Pollution prevention. The reduction of multimedia pollutants at the source and by the use of environmentally sound recycling. This prevention includes all regulated toxic and nontoxic substances.

Polymorphisms. These are genetic differences reflecting the existence of several variations of the same gene.

Population. Considered to be the breeding group for an organism. Different species not only have differing physical attributes, but they also differ in the population characteristics.

Population control. Government-directed programs that set a policy for establishing an optimum population size.

Population momentum. A situation in which a population has grown so much that even smaller growth rates lead to additions of larger numbers of people to the population.

Prairie potholes. Glacially created fresh water marshes found primarily in the upper mid-western U.S..

Pulmonary. Having to do with lungs and breathing.

Radioisotope. A radioactive isotope.

Rate of natural increase. Determined by subtracting the death rate from the birth rate.

Recovered materials. Waste material and by-products that have been recovered or diverted from solid waste, excluding materials and by-products generated from, and commonly used within, an original manufacturing process.

Recruitment failure. Refers to the failure of eggs, larvae, and juvenile fish to thrive at pH levels of water below 5.5. Fish populations decline with the most sensitive, juvenile fish first disappearing.

Recyclable materials. Materials that still have useful physical or chemical properties after serving their original purpose and that can be reused or remanufactured into new products.

Recycled materials. Materials, which otherwise would have been destined for disposal, that have been collected, reprocessed or remanufactured, and made available for reuse.

Refuse derived fuel (RDF). A pelletized fuel derived from MSW that can be mixed with coal and burned in a standard boiler.

Refuse reclamation. Conversion of solid waste into useful products, for example, composting organic wastes to make soil conditions or separating aluminum and other metals for melting and recycling.

Replacement level. Defined as a total fertility rate (TFR) that corresponds to a population exactly replacing itself.

Replication. A process whereby dividing cells provide each daughter cell with one complementary strand of the parent and then synthesize a complementary strand to produce double-stranded DNA that is a duplicate of the parent DNA.

Reservoir. Living organisms or inanimate objects that provide the conditions where organisms may survive, multiply, and also provide the conditions necessary for transmission.

Resource Conservation and Recovery Act (RCRA). The federal statute that regulates the generation, treatment, storage, disposal, or recycling of solid and hazardous waste.

Resource recovery. The act of removing materials from MSW for a productive use.

Resource recovery facility. Any place, equipment, or facility designed and/or operated to separate or process solid waste into usable materials or to incinerate solid waste to create heat and energy.

Respiration. A process in which the respiratory system supplies oxygen to the body's cells and expels carbon dioxide from the body.

Restriction enzymes. These are microbial enzymes that are used to cut DNA into several predictable and reportable pieces.

Retrogression. A process resulting in a new ecosystem state with fewer overall species.

Rotation. Agricultural practices that traditionally have included the planting of nitrogen-using (corn) and nitrogen-fixing (beans) crops in alternative years in the same space or planting them in alternating bands in the same year.

R-strategy. A reproductive strategy characterized by populations that are typically small, short-lived organisms, which produce large numbers of offspring that receive little or no parental care.

Saltwater intrusion. A condition in which saltwater can seep into an aquifer as fresh water and is pumped out of the aquifer, thereby polluting the fresh water.

Sanitary sewer. A channel or conduit that carries household, commercial, and industrial wastewater from the source to a treatment plant or receiving stream.

Saprophytic. Organisms that feed primarily on dead organic matter. As these organisms process components of the dead tissues, they leave behind undigested materials, that are in turn processed by the next group of decomposers until finally there is nothing left but energy-poor mineral nutrients.

Scavengers. Organisms that prefer to feed upon the dead remains of animals.

Scurvy. Develops from a lack of vitamin C (ascorbic acid) in the diet and consequently the synthesis of collagen and mucopolysaccharides is disturbed.

Sedimentary. Nutrients originating from soil or rocks.

Sedimentary cycle. A process by which substances are gradually leached from sedimentary rock by the actions of rain or erosion from water.

Sick building syndrome (SBS). A phenomenon in which occupants of a building display acute symptoms without a particular pattern and the varied symptoms cannot be associated with a particular source.

Sewer. A channel or conduit that carries wastewater and stormwater runoff from the source to a treatment plant or receiving stream. Sanitary sewers carry household, industrial, and commercial wastes. Storm sewers carry runoff from rain or snow. Combined sewers are used for both purposes.

Sex ratio. Refers to the number of males relative to the number of females in the population.

Sludge. A semi-solid residue from any number of air or water treatment processes. Sludge can be a hazardous waste.

Small-quantity generator (SQG). A hazardous waste generator that meets all the following criteria: (1) in one or more months of the year the site generates more than 100 kg (220 lb) of hazardous waste, but in no month generates 1000 kg (2200 lb) or more of acute hazardous waste; generates 1 kg (2.2 lb) or more of acute hazardous waste, or (3) generates 100 kg (220 lb) or more of material from the cleanup of a spillage of acute hazardous waste.

Solid waste. Non-liquid, non-soluble materials ranging from municipal garbage to industrial wastes that contain complex and sometimes hazardous substances. Solid wastes also include sewage sludge, agricultural refuse, construction and demolition wastes, and mining residues. Technically, solid waste also refers to liquids and gases in containers.

Solid waste management. Supervised handling of waste materials from their source through recovery processes to disposal.

Solvent. Substance (usually liquid) capable of dissolving or dispersing one or more other substances. Solvents include but are not limited to the nonspent materials listed by the USEPA.

Source reduction. Any action taken before waste is generated that reduces its volume and toxicity.

Spermicides. Chemicals that are designed to kill sperm.

Species. Normally considered to be a group of organisms that can breed together with the production of a viable and fertile offspring.

Stable ecosystems. Ones in which materials are constantly recycled within the system through growth, consumption, and decomposition.

Storage. Temporary holding of waste pending treatment or disposal. Storage methods include containers, tanks, waste piles, and surface impoundments.

Stratosphere. A region of the atmosphere extending from the tropopause to about 50 kilometers above the Earth. This region contains the ozone layer extending from 24 to 40 kilometers.

Succession. Refers to the predictable and gradual progressive changes of biotic communities toward the establishment of a climax community.

Succulents. Plants that store water in their tissues to be used during dry spells. The water may be stored in different parts of the plant, including stem, leaf, fruit, or roots. The familiar American cactus is a stem succulent. The aloe plant is an African leaf succulent.

Superfund. The program operated under the legislative authority of the Comprehensive Environmental Response, Compensation, and Liability Act (CERCLA) and Superfund Amendment Reauthorization Act (SARA) that funds and carries out the EPA solid waste emergency and long-term removal remedial activities. These activities include establishing the National Priorities List, investigating sites for inclusion on the list, determining their priority level on the list, and conducting and/or supervising the ultimately determined clean-up and other remedial actions.

Survivorship curve. A mathematical plot in which the fate of young individuals is followed to the point of death in order to describe mortality at different ages.

Swamps. Often transitional environments between large, open water and drier land. They are home to species from both extremes and are marked by flood-tolerant trees and shrubs.

Synergistic. The combined effect of two or more substances is much greater than the sum of the those substances acting alone.

Taiga. Boreal forest that is an almost continuous band of coniferous (cone-bearing) trees extending in a giant arc from Alaska, through Canada, Europe and Siberia.

Temperate broadleaf deciduous forests (TBDFs). Forests that are located in western and central Europe, eastern Asia and eastern North America. These are climax deciduous forests including oak, maple, hickory, chestnut, elm, linden, and walnut trees.

Temperate evergreen forests. Areas where soil is poor and droughts and fires are frequent. The predominant species tend to be evergreens including coniferous, needle-leafed trees with some broad-leafed evergreens. In such areas, spruce, fir, and ponderosa pine proliferate.

Temperate rainforests. In cool, coastal climates where there is considerable rainfall or frequent heavy fogs, one is likely to see forests of giant trees such as redwoods which characteristically reach heights of 198 feet (60 meters) to 297 feet (90 meters). Such areas exist along the Pacific Coast of North America.

Temporary threshold shift (TTS). A reversible hearing loss normally lasting from moments to hours, although longer periods of temporary hearing loss are possible.

Teratologic defects. Defects that arise during the embryonic period of development.

Terracing. An agricultural process where farmers must create level fields along the contour of the slope when they are driven to plant on hillsides.

Thermal injury. Injury that results when tissues are exposed to extremely low or extremely high temperatures.

Threshold limit value (TLV). The maximum concentration of chemical that a worker can be exposed to day after day as determined by ACGIH. This value is not legally enforceable.

Tipping fee. The cost for a truck to unload its waste at a landfill.

Total fertility rates (TFR). Represents the number of children a woman in a given population is likely to bear during her reproductive lifetime providing that birth rates remain constant for at least a generation.

Toxic. Harmful to living organisms.

Toxic substance. Those that: (1) can produce reversible or irreversible bodily injury; (2) have the capacity to cause tumors, neoplastic effects, or cancer; (3) can cause reproductive errors including mutations and teratogenic effects; (4) produce irritation or sensitization of mucous membranes; (5) cause a reduction in motivation, mental alertness, or capability; (6) alter behavior; or (7) cause death of the organism.

Toxic Characteristic Leachate Procedure (TCLP). This is a process described by the USEPA that requires acidified liquid to be percolated through waste in a laboratory-created landfill condition for 24 hours. Samples of the leachate are tested for the amounts of the target substance released.

Toxicity test. The means to determine the toxicity of a chemical or an effluent using living organisms. A toxicity test measures the degree of response of an exposed test organism to a specific chemical or effluent.

Trace elements. Elements required in tiny amounts for life such as zinc, manganese, chlorine, iron, and copper.

Tragedy of the commons. A behavior in which society will likely pass on the consequences of their destructive actions if they will benefit in the short term and receive little or no negative consequences from that action.

Transporter. A person engaged in the off-site transportation of hazardous waste by air, rail, road or water.

Transstadial. A form of transmission in which disease organisms are passed from infected adult ticks through the egg, larval, nymphal, and adult stages.

Treatment. Any method, technique, or process, including neutralization, designed to change the physical, chemical, or biological character of composition of any hazardous waste so as to neutralize such wastes; to recover energy or material resources from the waste; to render such waste nonhazardous or less hazardous, safer to transport, store, or dispose; or amenable to recovery, storage, or reduction in volume.

Treatment, storage, and disposal facility (TSD). Site where a hazardous substance is treated, stored, or disposed. TSD facilities are regulated by the USEPA and states under RCRA.

Trophic level. Refers to each stage of a food-web through which energy travels, consisting of producers, primary consumers (herbivores), and secondary consumers (carnivores). It is a system of categorizing organisms by the way they obtain food.

Tropical rainforest. The tropical rainforest is complex in terms of species diversity and structure. The combination of constant warmth, with average monthly temperatures above 17.8°C and precipitation greater than 100 inches per year, encourages rapid plant growth throughout the year.

Troposphere. A region of the atmosphere extending from the Earth's surface to approximately 10 kilometers. The troposphere is greater over the equator (12 km) and lesser over the poles (8 km).

TSDR. Treatment, storage, disposal, or recycling.

Tubal ligation. A surgical procedure requiring small incisions near the navel and just above the pubic area in which the fallopian tubes are tied off or the tubes cauterize. This procedure blocks the entry of eggs into the uterus.

Tundra. A biome limited to the upper latitudes of the Northern Hemisphere and forms a belt around the Arctic Ocean.

Understory. Area on the floor of the rainforest where there is sparse vegetation in the inner sections of the undisturbed rainforest, and these sections have to compete intensely in an area with little sunlight and poor soil.

Uniform Hazardous Waste Manifest. The shipping document (EPA Form 8700-22 or 8700-22a) that pertains to hazardous waste and is duly signed by the generator.

Urbanization. The mass migration of people to the cities.

Vapor density. The weight of a gas or vapor compared to an equal volume of air. Gases with a vapor density greater than one settle to the ground; gases that have a vapor density less than one rise in air.

Vapor pressure. The pressure of vapor above a liquid.

Vasectomy. Male sterilization accomplished by making an incision on either side of the scrotum and snipping out a piece of the *vas deferens*.

Vectors. Refers to animals that carry pathogens from one host to another, either from another human that is infected or from an infected animal.

Vehicle transmission. The transmission of disease agents by various media such as air, water, food, intravenous fluids or blood, and drugs.

Very small-quantity generator (VSQG). A conditionally exempt generator of hazardous waste.

Volume reduction. The processing of waste materials to decrease the amount of space the materials occupy, usually by compacting or shredding (mechanical), incinerating (thermal), or composting.

Waste. Any material discarded as worthless, defective, or of no further use that, when disposed of, may pose a threat to human health or the environment.

Waste management costs. The costs to procure, handle, treat, and dispose of wastes; a key factor in deciding to introduce source-reduction techniques.

Waste minimization. The reduction, to the extent feasible, of hazardous waste that is generated or subsequently treated, stored, or disposed. It includes any source-reduction or recycling activity undertaken by a generator that results in the reduction of total volume or quantity of hazardous waste; the reduction of toxicity of hazardous waste; or both, as long as the reduction is consistent with the goal of minimizing present and future threats to human health and the environment.

Waste reduction. Any change in a process, operation, or activity that results in the economically efficient reduction in waste material per unit of production without reducing the value output of the process, operation, or activity, taking into account the health and environmental consequences of such change.

Waste stream. A specific type of waste leaving a facility or operation.

Waste-to-energy (WTE). Combustion of refuse with the production of energy.

Water stress. A term defined as the ratio of water withdrawal to water availability.

Wetlands. Those areas of land where water saturation is the major factor influencing the nature of soil development and the communities of plants and animals that live in the soil and on the surface.

Windrow. A composting process in which rows of material are placed next to each other outdoors and periodically mixed to incorporate air into the mix.

Zoonoses. Diseases that are spread mostly from animals to humans.

Index

A

B

C

D

E

G

H

I

J

K

L

M

N

O

P

R

S

T

U

V

W

X

Y

Z